Running Out of Time—
Introducing Behaviorology to Help Solve Global Problems

Toward a better future

Burrhus Frederic Skinner

(1904–1990)

Conversing at a convention in 1982

The products of the contingencies of his life established behaviorology.

Running Out of Time—
Introducing Behaviorology to Help Solve Global Problems

Stephen F. Ledoux

Published by **BehaveTech Publishing**
Ottawa, Ontario, Canada

Running Out of Time—Introducing Behaviorology to Help Solve Global Problems

Stephen F. Ledoux, Ph.D.

Published by **BehaveTech Publishing** of Ottawa, Ontario, Canada

ISBN 978–1–927744–02–4

Cover design and book layout by Stephen F. Ledoux

Printed in the United States of America

Prospective cataloging–in–publication data (under "natural science" or "biology" but *not*—since behaviorology is not any kind of psychology—under psychology or "BF"; such cataloging could be like placing a biology book under creationism):

Ledoux, Stephen F., 1950–.
      Running Out of Time—Introducing Behaviorology to Help Solve Global Problems /
      by Stephen F. Ledoux
      xxx, 570 p. ill. 26 cm.
      Includes detailed contents, appendix, glossary, bibliographic references, index.
      1. Behaviorology. 2. Biology. 3. Behaviorism. 4. Consciousness. 5. Natural science.
          6. Human behavior. 7. Operant behavior. 8. Global Warming Solutions.
      I. Title.
      ISBN 978–1–927744–02–4 (acid–free paper)

This edition is printed on acid–free paper to comply with the permanent paper z39.48 standard of the American National Standards Institute.

Improved printing number: >10 9 8 7 6 5 4 3 2 1 0
 Nearest year of printing: >2050 2040 2030 2025 2020 2014

Summary Contents

Dedication

To the children of future generations:

The contingencies on these generations contain the products of our present behaviors, products born of the struggles between the pre–scientific and scientific contingencies that produce our present labors. May we leave a lasting legacy for these generations of children, a legacy worthy of their appreciation.

Acknowledgements

*M*any people deserve thanks and praise for early on catching many of the sins of omission and commission that crept into my massaging of a wide range of related materials into the manuscript that became this book. While any residual problems are of course my own, I want to express my deepest appreciation to a subset of all those deserving folks. This small group of friends and colleagues went above and beyond the usual levels of reviewer assistance by providing extensive, careful, detailed and exacting corrections and commentary that particularly improved the clarity, readability, and topical coverage of this work. In alphabetical order (and from the USA unless otherwise specified) these people are Karen Clemons, John Ferreira, Lawrence Fraley, Bruce Hamm (from Canada), Philip Johnson, Werner Matthijs (from Belgium), James O'Heare (from Canada), Jón Sigurjónsson (from Iceland), and William Trumble. They each and all have my sincerest thanks (and I hope we are still friends).

Having had the pleasure of editing Lawrence Fraley's 2008 *General Behaviorology* book for publication, I must admit to a certain awe at the extensive repertoire that Professor Fraley's contingencies built, and I owe to him (actually, to the published results that those contingencies produced) a substantial debt of intellectual expansion and emotional gratitude. Without accruing those debts, I doubt I could have written half as much, or half as well, in providing this short introduction to behaviorology, from which I hope all humanity may derive benefits. My thanks also go to the State University of New York at Canton (SUNY–Canton) for the adjustment in work assignments, in each of two semesters, that helped me make meaningful progress on the manuscript for this book. *SFL* ✌

Foreword

Change often happens at a rapid and, for some people, uncomfortable pace. People have probably felt this way since the time humans began using tools. Thought leaders in every generation have looked at the "Great Work" of their time, and attempted to understand how and why events manifest themselves in certain ways, usually because this improves their influence over the change occurring about them.

For Dr. Stephen Ledoux and his colleagues, the *Great Work* that they have undertaken is to understand how and why humans (and other animals) behave in preparation and response to given circumstances. Please note that a key word here is "understand." These scholars have named their discipline Behaviorology since, by design, it does not fall into the area of Psychology, not even as a subset. Behaviorology is a study of behavior, treated as a natural science, something I can appreciate as a biologist. Behaviorology, therefore, requires adherence to natural–science philosophy as well as the application of scientific methods, hypotheses, proofs, repeatability, and convincing demonstrations. This deliberately removes the elements of superstition, spirituality, and mysticism.

One might rightly ask if the elements listed as removed from the study of Behaviorology do not also influence behaviors. The answer is that, of course, they do; they are behaviors. Since a goal of Behaviorology is understanding (and prediction, and so on) however, superstition, spirituality, and mysticism, as concepts of Faith, are incongruous with a demonstrable understanding. By definition this must be so or we would not classify such beliefs as "Faith."

Behaviorology is also beginning to find a niche in mainstream education, 100 years after its conceptual introduction, with scholars and champions who have benefited from "standing on the shoulders of giants" as Newton remarked. Professor Ledoux's text on behaviorology now joins a collection of other books on the subject to suggest and complete a curriculum from undergraduate through graduate studies. This book presents examples and case studies that are illustrative and understandable. However, the integration of Behaviorology into the education curriculum of higher education has not been easy.

At the current time, many aspects of the higher education system at colleges and universities are being tested and may be found lacking and insufficient to the task of teaching a new generation of students in a new learning environment. This environment involves required remedial assistance, "flipped classrooms," and the emergence of Massive Open Online Courses (moocs) to name but a few of the newer components. What many institutions do not do well is to prepare students for jobs that currently do not exist but will in a few years, or to accept curricular changes designed to teach what students need as opposed to what faculty can and want to deliver. The discipline of

Behaviorology, ironically, not only offers new ideas to address behavior, but tools with which to better educate in many areas.

Clearly the *Great Work* of understanding human behavior underlies most of the other *Great Works* of our time. What importance may be placed on our behavior, as a people, as we deal with climate change, the availability of fresh water, appropriate health care, or wars? Behaviorology, as a natural science, offers us hope and new approaches to understand current and historical questions about behavior. Dr. Ledoux's clear prose, illustrative examples, and poignant storytelling make this book's message accessible to many levels of readers as well as useful to those of us seeking new understanding and the opportunity to make a difference.

Perhaps of even greater importance, this book will interest those listening to the natural scientists who recognize that our global problems, and any solutions, clearly involve human behavior. We all need the science that addresses this area. This book introduces that science, Behaviorology, which joins the natural–science team working to solve global problems. This book helps us all—natural scientists, engineers, and concerned citizens—both understand, and better deal with the connections between behavior and timely solutions to our *Great Works* problems.

Read this book with an open mind (which is not an entity inside you, as you will see). You will thus enjoy a journey into a new discipline of cutting edge knowledge about, and with extensive applicability to, some *Great Works* of our time.♣

William R. Trumble, Ph.D.   ***August 2013***
Biologist, Professor, University of New Hampshire, and
Provost/Vice President for Academic Affairs, Unity College, Maine (Retired)

Dr. Trumble currently resides in Canton, New York, and consults on higher education issues.☙

Preface

$\mathcal{D}$uring the last decades of the twentieth century, traditional natural scientists (e.g., physicists, chemists, and biologists) increasingly realized that solutions to the major (and minor) problems around the globe, to which they were turning their attention, needed changes in human behavior to work. Yet most were generally unaware that a natural science of behavior, now called behaviorology, was available to address its part in the solutions. The centenary year of behaviorism—2012—provided an occasion to review this discipline for them, if ever so briefly, in the form of an abridged article, "Behaviorism at 100," in the first issue (January 2012) of the centenary volume of the journal *American Scientist;* the unabridged, peer–reviewed version appeared in volume 15, number 1, of the journal *Behaviorology Today* two months later. To support the teamwork of all the natural sciences in solving local and global problems, this book elaborates on the content of the unabridged article.

Chapter 1 features an expanded version of that article as an overview of the first hundred years of the natural science of behavior and its emergence as the separate and independent discipline of behaviorology, which is not any kind of psychology. Much science and history happened in this first hundred years, which made Chapter 1 the longest chapter in this book. As a summary it touches briefly on most of the topics in the book thereby showing their interrelations. However, this brevity makes Chapter 1 seem a bit difficult.

Chapter 1 may also seem difficult for two more reasons. It begins the process of countering 50,000 years or more of pre–scientific cultural conditioning about human nature and human behavior. And it also begins some gradual shifts, throughout the book, in grammatical phrase forms and word usages—making more standard some otherwise non–standard forms and usages—toward a more efficient grammar (e.g., occasionally using nouns as adjectives) that better supports scientific realities by containing fewer inherent implications of ghosts (i.e., mystical—as in untestable, unmeasurable—behavior–directing agents) residing inside bodies, a point that also receives more complete treatment in various chapters. In the hope that reader's rank *science helping to save our planet* as more important than maintaining certain aspects of grammar that have been harming us for millennia by coincidently supporting superstition, I beg the reader's indulgence in support of these grammatical shifts.

The chapters, which stress the importance of a naturalistic philosophy of science in this and all natural sciences, divide into two parts. Part 1 (Chapters 1–12) pursues history, concepts, principles, methods, and practices. Part 11 (Chapters 13–24) pursues advanced developments and natural–science answers to some of humanity's long–standing questions.

Many options arose for the title of this book. The title needed to present not only the "running out of time" theme but also the introductory status of the book, and to what end. The book introduces the natural science of behavior, about which the editor of *American Scientist* said (at the start of that article that became Chapter 1) "Over its second 50 years, the study of behavior evolved to become a discipline, behaviorology, independent of psychology." This book aims to introduce behaviorology to traditional natural scientists and anyone who works to solve global problems, because the discoveries of this natural science contribute to the success of their team efforts to bring about solutions in a timely manner. I believe that each one would want to know more about behaviorology in the same way that each knows something about all the other natural sciences beyond her or his own specialization.

On that premise this book leaves out discussions of some basic science terms that professional scientists already know but with which students may as yet be unfamiliar. If someone uses this book to introduce the behaviorology discipline to first–year students, then a dictionary would likely be of more than the usual value for these students; of course, they would also have a professor to help them. Note, however, that this is *not* a textbook in or for psychology!

This book describes many of the discoveries and applications of the first 100 years of behaviorological science, particularly regarding *why* human behavior happens. The undercurrent throughout the book concerns the relevance of this natural science not only to helping build a sustainable society but also, in that process, to addressing the wide range of local and global issues confronting humanity, all of which have definitive behavior components the address of which benefits from a comprehensive natural science of behavior. These issues range from backyard burn barrels, through dozens of increasingly complicated and problematic concerns, to overpopulation, which is perhaps humanity's most fundamental practical problem. No book, including this one, solves such problems by itself, but the more humanity understands about the causes of behavior, which is the topic of this book, the more success humanity will have at solving such problems in the timely manner that prevents us from having to experience the worst effects of these problems.

While this book serves as a first book about behaviorology, some readers may find that this book does not go far enough, for them, into the systematic substance of the behaviorology discipline. To make going farther easier, this book by design introduces the topics that Lawrence E. Fraley's 1,600–page, three–course textbook, *General Behaviorology—The Natural Science of Human Behavior,* covers in comprehensive detail. Fraley's book, however, has too much depth and detail to serve easily as a first book on behaviorology. The present book addresses this need.❧

Stephen F. Ledoux   **Canton NY USA**   **March 2013** ☙

Detailed Contents

On Typography & Author Contact

This book is set in the Adobe Garamond, Adobe Garamond Expert, and Tekton collections of typefaces. In addition, a valuable basis for the typographic standards of this work deserves acknowledgment. As much as possible, this book follows the practices described in two highly recommended volumes by Ms. Robin Williams (both of which Peachpit Press, in Berkeley, CA, USA, publishes). One is the 1990 edition of *The Mac is Not a Typewriter.* The other is the 1996 edition of *Beyond the Mac is Not a Typewriter.* For example, on page 16 of the 1990 book, Williams specifies practices regarding the placement of punctuation used with quotation marks, an area in which some ambiguity has existed with respect to what is "proper."

You can address correspondence regarding this book to the author at SUNY–Canton, 34 Cornell Drive, Canton NY 13617 USA (phone: 315–386–7423; email: ledoux@canton.edu).

For more information, see the pages of *Behaviorology Today* (ISSN 1536–6669) which is the journal of TIBI (The International Behaviorology Institute). TIBI renamed it *Journal of Behaviorology* (ISSN 2331–0774) in 2013. Also visit www.behaviorology.org which is the TIBI web site. Both the journal and the web site contain additional material and works by this author.

Some related volumes may interest the reader. In addition to the three–course *General Behaviorology* textbook that Fraley authored (2008) and which the *Preface* mentioned, another volume is *Origins and Components of Behaviorology—Second Edition* (2002). Stephen F. Ledoux prepared this book, to which each of the four founders of TIBI contributed papers (and a *Third Edition* of this book, along with an updated book of study questions for it, should be available before 2015). A book of study questions, for *Running Out of Time...,* should also be available before 2015; contact the author for a progress report in late 2014. Books of study questions help interested readers master contents more thoroughly. Lawrence E. Fraley wrote two other volumes that also deserve mention. These are (a) *Dignified Dying—A Behaviorological Thanatology* (2012), and (b) *Behaviorological Rehabilitation and the Criminal Justice System* (2013). ABCs, of Canton NY, USA, is the publisher of all these books except *Running Out of Time...,* which BehaveTech Publishing (of Ottawa, Ontario, CANADA) publishes.

You can order additional copies of *Running Out of Time—Introducing Behaviorology to Help Solve Global Problems* from the distributor, **Direct Book Service, Inc.,** at 800–776–2665. They will likely answer the phone with "Dogwise," because one of their oldest and most popular specialities involves books about our canine friends; several of these books already specifically apply the laws of behavior that *Running Out of Time...* systematically introduces.

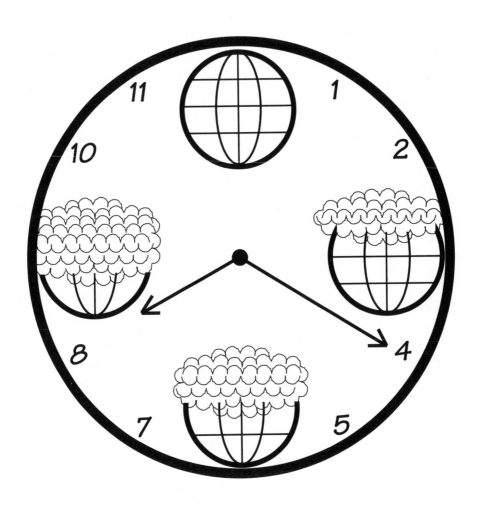

Running Out of Time—
Introducing Behaviorology to Help Solve Global Problems

Behaviorology:
The Natural Science of WHY
Human Behavior Happens

A Natural Science to Help
Build a Sustainable Society
in a Timely Manner

Behaviorology in its immediate scientific neighborhood

PART I
Some Philosophy of Science Plus Basic History, Concepts, Principles, Methods, & Practices

Chapters 1 – 12

Occasional blank pages provide extra space for:

Reader's Notes

Chapter 1
Reasons for Interest in Discoveries about Behavior

Where we are Going in this Book and Why

*W*elcome! Sit back. Relax. We are beginning a journey—sometimes uncomfortable, often fun, always revealing—to check out some fascinating and recent (in the last 100 years) scientific discoveries, and the benefits we derive therefrom, about the nature of nature, of human nature, of human (and other animal) behavior, even of consciousness and reality itself. Some other stops on our way include many topics, some of which may sound technical or never have obviously crossed your path before, but which have direct and significant effects on human existence and experience. These topics include shaping, fading, evocation, extinction, emotions, feelings, stimulus equivalence relations, schedules of reinforcement, personhood, attitudes, values, rights, ethics, morals, beliefs, robotics, the alternatives to coercion and punishment, the status of verbal behavior, and even life and death.

We will approach those and other topics as natural scientists. Traditional natural sciences cover the basic subject matters of *energy, matter, and life forms* as well as their related extensions. These thus include physics, chemistry, and biology, along with astronomy, geology, physiology, and so on. While we can see the beginnings of all these with Galileo's telescopic observations in the early 1600s, the natural science we examine in these pages concerns the subject matter of *life functions* that emerged over the last 100 years, and not only covers but clarifies and expands what we subsume under the ancient dictum to "know thyself." This natural science encompasses all behavior, especially human nature and human behavior, the behavior of you and me, of friends and trouble makers, of the ordinary and the extreme, of the subtle and the "in your face." Thus focused on behavior, we call this science "Behaviorology," and we will consider its experimental, conceptual, philosophical, analytical, and technological aspects along with its methodology and the basic and applied research that leads to its principles, laws, and applications, with a special interest in how it can contribute to solving local and global problems.

We begin this first chapter by providing an orientation to the historical and philosophical premises of the natural science that we will cover, along with a few choice examples of some of the discoveries we will meet along the way. This chapter seeks to inform readers, including—perhaps especially—readers from other natural science disciplines, about the many lines of progress occurring over the last 100 years in the natural science of human nature and human behavior, but it also informs them of some implications of that progress, implications probably crucial to humanity's future, crucial because the implications bear on a wide range of other natural–science subject matters that are concerned with the timely solution of some problems confronting the world, problems like global warming and overpopulation, problems whose solutions contain a substantial behavior component that currently receives inadequate attention. The rest of the book then elaborates that progress in detail appropriate to an introduction to the science, an introduction that enables far greater attention to the behavior components of local and global problems and solutions.

As part of this discovery quest, you may want to access additional resources that provide other details and verifications. In this book many such sources appear in the citation form of "author, year–of–publication" (usually in parentheses) and the full reference appears at the end of the chapter. To avoid the interference with reading that mentioning many citations can cause, the remaining chapters contain few citations and instead encourage the reader to follow up through the extensive resources listed in the bibliography.

Preview

Many human intellectual endeavors start out on the foundation of some basic assumptions. We use the term *philosophy* to encompass the combined variety of assumptions behind an endeavor, and we find that those assumptions actually exert a vital control over the behavior of those that the endeavor engages. This control occurs in both scientific and non–scientific activities, and even underlies many emotional activities. We describe this control by saying that the philosophy "informs" the endeavor. Some familiarity with philosophy and its effects in general, and with the specific philosophies related to particular activities, thus becomes a necessary part of any intellectual endeavor.

With scientific activities we use the term *philosophy of science* to denote the set of informing assumptions. We will cover philosophy of science just enough to clarify its significance both for the general endeavors we describe as the traditional natural sciences (e.g., physics, chemistry, biology) and for our more focused concern with behaviorology, the natural science of behavior, which is filling a chair at the roundtable of the basic natural sciences. We use the term *naturalism* for the general philosophy of science of all the natural sciences. Yet some of these sciences develop philosophical extensions that further inform work in those areas. The natural science of behavior is one of these sciences, because studying human nature and human behavior scientifically presents concerns that are specific to particular aspects of behavioral phenomena. We use the term *behaviorism* (actually, *"radical" behaviorism*) to denote the extensions of naturalism that inform the natural science of behavior.

To lay a solid foundation for covering our behavior subject matter, we devote this first chapter to expanding our understanding of some of the assumptions, and the rationale behind them, that inform the natural science approach to human nature and human behavior. Furthermore, to paint the big picture for the larger context encompassing our behavior–science study in this book, this chapter adds the historical perspective by summarizing the first 100 years of behaviorism along with some of the discoveries and developments in the natural science of behavior that behaviorism informs. Indeed, since the article that provided the basis for this chapter was written at the centenary of behaviorism, its original title was "Behaviorism at 100."

A quick abstract of this chapter will help: Summarizing 100 years of the philosophy of science that we know as behaviorism, and its impact, starts with a description of B. F. Skinner's 1963 article, "Behaviorism at fifty," which covered the first 50 years (1913–1962). Then, a review of the second 50 years presents interrelated developments in the philosophical, organizational, scientific, and interdisciplinary domains: (a) We consider Skinner's radical behaviorism, the philosophy that extends naturalism to inform the *natural* science of behavior and its emergence organizationally as an *independent* discipline that today we call behaviorology, after its separation from the non–natural, fundamentally mystical discipline of "behavior and the mind." (b) We evaluate some historical and organizational developments in this discipline. (c) We ponder samples of the experimental and applied advances of behaviorological science. And (d) we examine the interdisciplinary benefits and relations among all the natural sciences of energy, matter, life forms, and life functions (i.e., behavior) that accrue with the emergence of the natural science of behavior. These continuing developments improve the possibilities for reducing global superstition, extending global science, and solving global problems.

The Source of this Chapter

This first chapter has a background rather different from the rest. The 2012 centenary year of behaviorism coincided with several interrelated circumstances: Traditional natural scientists have been working to solve problems like overpopulation and global warming within the limited time frame available before we must experience their worst effects. In that process they have noted that the solutions require changes in human behavior. Yet they have lacked full access to the natural science of behavior. Thus natural scientists in general comprised one of the principal audiences for this material. Then, in 2012, *American Scientist,* the journal of one of natural scientists' major organizations (Sigma Xi, the Scientific Research Society) celebrated its centenary as well. This set of circumstances first put this chapter into the hands of the editor of *American Scientist,* David Schoonmaker, who introduced the *American Scientist* version to readers with the comment, "Over its second 50 years, the study of behavior evolved to become a discipline, behaviorology, independent of psychology." To emphasize the historical context of the article, he opted to have the paper (Ledoux, 2012a) begin bringing readers up to date, on the first 100 years of behaviorism, by appearing after some lengthy excerpts from B. F. Skinner's article, "The Experimental Analysis of Behavior," which had appeared in *American Scientist* in 1957. To get all the space he needed for these excerpts, he had to abridge the article by setting aside some early pages at the last minute. Yet these pages, mostly on the philosophical history of behaviorism, still comprised crucial content for the paper. After restoring this material, I submitted the longer version (Ledoux, 2012b) to *Behaviorology Today* which published it after formal peer review. What you are now reading is a modest expansion of "Behaviorism at 100 Unabridged" (with permission of both *Behaviorology Today* and *American Scientist*).

An Introduction to Behavior Science History

As a philosophy of science, behaviorism began with an article by John B. Watson in 1913, and its several varieties inform different behavior–related disciplines or sub–disciplines. During the last 100 years, disciplinary developments have led to a clarified version of behaviorism informing a basic, separate *natural* science of behavior. This recently emerged, independent discipline of life functions not only complements traditional natural sciences of energy, matter, and life forms, but also shares in solving local and global problems by showing how to discover and effectively control the variables that unlock solutions to the common behavior–related components of these concerns.

In 1963, B. F. Skinner published "Behaviorism at Fifty," reviewing the varieties of behaviorism and the directions of natural behavior science. By the 1960s common wisdom held that the experimentally discovered laws of behavior were largely irrelevant to normal humans, as researchers were then applying these laws mostly to treatments for "psychotic individuals" and to training for other animals. Skinner challenged that notion on scientific as well as philosophical grounds, and data accumulating over the next 50 years has validated his position that the natural laws governing behavior are relevant to *all* behavior of humans and other animals.

The 1960s were also a time when natural scientists of behavior were continuing their attempts to change the discipline under which many of them worked (i.e., psychology) into a natural science. Over the next 50 years, as recognition increased that resistance to those efforts was adamant, natural scientists of behavior gradually took their discipline outside psychology, establishing a separate independence for their natural science that some recognized formally in 1987 using the name *behaviorology*. That name is synonymous with "the natural science of behavior" and is conveniently shorter. While at present quite a number of possible natural scientists of behavior still use an older label for this discipline, only this name definitively indicates the discipline completely separated from disciplines accepting or espousing superstitious or mystical accounts for behavior.

Different Behavior Disciplines

Actually, several different disciplines claim behavior as a subject matter, including theology, psychology, and behaviorology. They are not equal. While the distinction between theology and science is already well established, some confusion lingers over the other two, so we should start by clearing that up: *Behaviorology and Skinner's behaviorism are not any kind of psychology!*

While the history that behaviorology and psychology shared some decades ago is not unlike the history physics shared with philosophy some centuries ago, the relation between behaviorology and psychology more closely approximates the relation between biology and creationism. Both psychology and creationism claim to be sciences because they use scientific methods, but neither qualifies as *natural* science because they both appeal to entities or events outside of nature (i.e., to mystical or supernatural or anatural or otherwise non–natural events). And natural scientists fail to see the *secular* mystical accounts of psychology as any sort of improvement over *theological* mystical accounts.

With behaviorism celebrating its one–hundredth year in 2012, a review of those developments, and their ramifications for other natural sciences and today's world, seems appropriate. If the implications of this review precipitate appropriate action, the results can elevate the status of naturalism and the natural sciences, lead to solving more human problems, reduce susceptibility to superstition and mysticism (both theological and secular), and improve human intellectuality, rationality, and emotionality.

Naturalism, the general philosophy of science in the natural sciences, has among its characteristics one that particularly helps achieve those outcomes, which is that natural scientists maintain a mutual respect for the continuous natural functional history of events. This enables their analyses to be more complete and to track well across their disciplinary lines. In contrast, ignoring the natural functional history of events often leads to unnecessary compromises between some natural sciences and non–scientific disciplines that make claims of mystical origination of events. An example is the early, ultimately unhelpful, and yet still extant, give–away to theology of human nature and human behavior considerations.

Related Natural Sciences

Conversely, here is an example of the cross–discipline tracking that respects the natural functional history of events: While behaviorology accounts for a stimulus evoking a behavior, such as a fast and close moving object evoking ducking the object, biology can provide details at the physiological level about how an energy change at receptor cells, such as how light from a close moving object striking the retina, sets off a cascade of changes through the nervous system that mediates—*not* originates—a behavior, like ducking. In addition, chemistry works on details at the cellular and sub–cellular level about those physiological events, while physics strives for atomic and sub–atomic details about those chemical events.

All four disciplines together provide a comprehensive account of the events. But if natural scientists instead allow claims that behavior in general— or ducking in this particular case—results from the spontaneous, willful act of some putative inner agent, then they lose the whole subject matter of human behavior—a subject matter whose application is perhaps vital for human survival—to purveyors of non–science. This loss occurs because an untraceable, untestable mystical account replaces the links in the natural functional history of the events, links that already trace more parsimonious, and more detailed, paths at those several different natural–science levels of analysis. When such compromises give undeserved status to mystical accounts, natural science loses ground, reducing its benefits. Maintaining respect for the natural functional history of events thus enables a more complete and consistent account of any phenomena including behavior phenomena.

To avoid unnecessary and harmful compromises, a greater availability of even introductory–level coverage of the natural–science study of behavior, in science degree curricula, would have been helpful to traditional natural scientists. Instead, many—perhaps most—have experienced exposure mainly to popular cultural and academic perspectives on human nature and human behavior of the kind that fundamentally mystical disciplines espouse, including some disciplines that use scientific methods to support claims of mystical origins of events. Some traditional natural scientists attempt to rescue the human nature and human behavior topics from theological and secular mysticisms by trying inappropriately to shoehorn them into strictly evolutionary, genetic, or physiological accounts. Exposure to popular mysticism can lure other natural scientists into accepting and repeating mystical accounts when they venture beyond the limits of their disciplinary training and specializations. Instead, as traditional natural scientists become more aware of the progress that other *natural* scientists have made on behavioral fronts, the accuracy of their work expands and the risk of resorting to mystical accounts shrinks. The point here is to provide some highlights of that progress.

Foundations in the First 50 Years

Skinner's 1963 treatment of behaviorism began by describing the primitive origins of mentalistic or psychic explanations of human behavior. His concern, however, was not the primitive origins of these explanations, but their continued use in a discipline, psychology, that began, in the 1800s, when its original philosopher members wanted to be "scientific." If one restricts the term *scientific* to the use of scientific methods, as they did, then much of the psychology discipline could be construed as scientific. But *science,* as we more broadly and typically construe it, also includes adhering to the fundamental philosophy of naturalism, a basic facet being that science deals solely with natural—real, measurable—events as independent and dependent variables, with no abiding place for non–natural—mystical, fictitious—events. Natural philosophy ties the natural sciences together, and distinguishes them from other disciplines.

However, as a discipline, psychology maintains a range of non–natural, even anti–natural, alternatives among its disciplinary schools; this alone is enough to maintain its exclusion from the natural sciences. In addition, psychological explanations, deriving from and so reflecting common, traditional cultural biases, ultimately trace to some type of mystical inner being or entity, sometimes little more than secular versions of the theological soul, like a behavior–originating mind or a behavior–initiating self agent, which spontaneously considers various factors, decides what to do, and tells the body to do it. With these processes, or sub parts like id, ego, motive, choice, or trait, we cannot trace the behaviorological, physiological, chemical, or physical links of a natural functional history chain; we can only trace the causal chain back to the putative spontaneous wilful act of the self agent. This breaks the chain of events in the natural functional history and thus, disrespecting naturalism, further excludes psychological analyses from natural science. However, the competition that these psychological analyses represent absorbs resources that could otherwise expand natural–science functional analyses and their related beneficial behaviorological–engineering applications. Such were some of the ongoing problems from the first 50 years.

Also in his 1963 paper, Skinner dealt explicitly with the question of "mind" (the quintessential self agent). His answer began with Charles Darwin's continuity of species. With humans and other animals qualitatively similar, some researchers looked for, and claimed to have found, human characteristics in other animals including consciousness and reasoning. Lloyd Morgan, however, pointed out that more parsimonious accounts than claiming an animal "mind" could explain such findings. And if that was so with other animals, then more parsimonious accounts could explain such characteristics for humans also. Trying to discover parsimonious accounts prompted the natural, experimental science of behavior that began with Skinner in the 1930s. The point was not merely to discover the naturalistic explanations of human behaviors, including complex behaviors such as consciousness, language, reasoning, imaging, and thinking (both verbal or visual) but to discover these naturalistic explanations initially as alternatives to, and ultimately as replacements for, superstitiously grounded explanations, and then to apply these explanations for humanity's benefit.

Early Natural Behavior Science
Behavior, a natural phenomenon, happens—and changes—because independent variables affect the particular body structures that mediate it. No

mysterious inner self agent *does* the behaving or instructs the body to behave. Instead the experimental literature describes two basic and continuously operating conditioning processes that we call respondent (i.e., Pavlovian) conditioning and operant (i.e., Skinnerian) conditioning. Both involve energy transfers between the environment (internal and external) and the body in ways that, as our physiology colleagues can show, trigger cascades of neural firings that variously induce both the greater energy expenditures involved in bodily movements, and the altered neural structures that constitute a different body; that different body mediates (not initiates) behavior differently on future occasions (in a process popularly called "learning," although no inner agent— no "learner"—is present to "do" the learning, so we seldom use these terms).

Early in the 1900s, Ivan Pavlov discovered respondent conditioning, which involves the "pairing" (i.e., the overlapping or successive occurrence) of neutral stimulus events with stimulus events that already (due to genetically determined neural structures) elicit responses. These pairings transfer energy to the body resulting in nervous system change that alters the way in which the neutral stimulus events function; they come to elicit responses also (because the body has changed, not the stimulus). Emotion and feelings and other reflexes and reactions involve respondent processes.

Some years after Pavlov's discovery, and acknowledging Edward Thorndike's work, B. F. Skinner discovered operant conditioning, which involves stimulus–evoked responses affecting—as in "operating on"—the environment in ways that produce, as consequences of those responses, environmental changes that transfer energy back to the nervous system thereby producing structural changes that establish the relative functionality of similar future evocative antecedent stimuli (e.g., see Seo & Lee, 2009). A wide range of such antecedent stimuli are usually present at any given time. Due to past conditioning, and either in a kind of competition, or sometimes singly but more often in combination, some among these stimuli evoke further behavior based on current neural structure derived from species and personal conditioning history. Additional stimuli may then consequate this behavior, altering the neural structures that mediate it and thereby changing how readily future similar situations will evoke it. Walking, talking, singing, dancing, loving, thinking, studying, working, fighting, planning, partying, publishing, and problem solving—including engineering and scientific research and writing—are all among the typical and ongoing results of operant processes.

We consider those two conditioning processes separately. The kinds of consequences occurring in operant processes appear to have little effect on the behaviors produced through respondent processes. Thus, respondent behaviors tend to appear more consistent or reliable.

Places for Physiology and Consciousness

The physiology that mediates behavior, and the genetic basis of that physiology are also important, but while both are real, access to them is generally unnecessary or unavailable when practical engineering interventions are altering environments at the behaviorological level of analysis thereby producing differences in behavior, such as when teachers change the instructional environment with the result that students' behavior repertoires expand. This is similar to chemists being able to account for changes in the properties of matter using the principles of physics while not needing that level of knowledge to engineer chemical reactions. Also, in the same way that nothing in the nature of chemicals requires involving "phlogiston" in accounting for a chemical reaction, nothing in the nature of behavior requires involving a homunculus, mind, or will in accounting for any behavior. Indeed, the spontaneous production of responses by self agents not only violates the conservation of energy principle but also mimics the spontaneous generation hypothesis in biology that Pasteur put to rest.

In expanding naturalistic explanations toward a more complete scientific account of behavior, Skinner discerned that the question of consciousness not only holds a central challenge but also attracts substantial attention and must be taken into account. This science begins its account of consciousness in terms of *neural* behaviors such as awareness, recognition, observation, thinking, and comprehension. While muscle behavior (actually, neuro–muscular behavior) is more familiar to us as it intertwines both neural–process components and the more observable innervated muscle–contraction components, the behaviors of consciousness happen as *pure neural processes* lacking muscle–contraction components (with possible exceptions like sub–vocal speech).

To help grasp consciousness as pure neural responses, consider describing consciousness behavior on several increasingly elaborate levels. We could first speak in traditional (i.e., agential, non–natural) terms, noting that people can observe and report to themselves or others the occurrence of some of their behaviors. To *restate* this point with simple behaviorological phrasing (i.e., naturalistically, without implying self agents) certain of peoples' behaviors *evoke* subsequent behaviors of observing and reporting the earlier behaviors. To *embellish* this point behaviorologically, the external and internal environments of peoples' daily existence feature the occurrence of energy exchanges that condition people such that some behaviors, including neural behaviors, function as real, independent variables evoking further behaviors, including more neural behaviors, as real, dependent variables that others describe, also due to past conditioning, as observing and reporting.

Actually many, perhaps most, moment to moment human behaviors play little or no part in evoking any consciousness behaviors that way, which may make the occurrence of consciousness behaviors even more discriminable (i.e., more evocative of further behavior). And if one's past conditioning history has made putative agential accounts reinforcing, then the occurrence of otherwise ordinary, natural consciousness behavior can seem particularly impressive as supposedly being, or showing, the activity of one or another inner agent. Through such processes the lore of traditional cultural contingencies builds this kind of inaccurate repertoire with large numbers of individuals, yielding the organized forces of theological or secular superstition and mysticism that oppose so much natural science today. This opposition occurs in spite of the opposers' thorough reliance on the vast array of products, from the opposed natural sciences, that enable living quality lives while rendering that opposition. For example the opposers take modern transportation to anti–science meetings held in beautifully designed and expertly engineered buildings with comfortable accommodations including flush toilets—and other personal–hygiene amenities—as well as healthy air quality and efficient food preparation areas, all of which are products of the opposed sciences which the opposers would not want to live without. Elaborating the natural–science analysis of behavior, including consciousness, helps counter these trends.

The elaboration of the consciousness behavior of observing/reporting some *other* behavior can also accommodate the physiological and genetic levels of analysis. While genes originally produce nervous–system structure, ongoing respondent and operant conditioning processes that occur throughout life continually change that structure as the internal and external environments exchange energies with the body on a moment by moment basis. Some of those changes to nervous–system structure are such that some structures now react in ways naturally mediating neural, consciousness behaviors that we describe as observing and reporting, and that differ from the observed and reported behaviors.

All entirely natural. When happening, all these intertwined environmental/neural/behavioral processes can move along at such a rapid pace that they may seem magical or undetermined, particularly when we try to encompass behavior in general across a time frame beyond a few moments, because events can quickly outpace our measurement technology. That, however, is a problem not with nature, which naturalism takes as lawful, but with our residual ignorance (see Fraley, 1994) which we then manage with a variety of techniques including probability and chaos theories. Meanwhile, consistent with Skinner's behaviorism, these processes are *still all entirely natural;* including self agents in accounts for behavior is not only redundant and misleading but also dangerous and irresponsible because the resulting reduced effectiveness in problem solving can cause people harm.

In his paper Skinner emphasized a range of concerns surrounding the question of consciousness, because this topic in particular seems to focus attention on the difference between science and non–science with respect to behavior. While Freud had assisted the behavioristic perspective by showing that consciousness was not required for other behavior to occur, the nature and occurrence of the consciousness itself presented a more difficult problem, namely the reduced access to real but private events (i.e., events that can function as stimuli only for the body in which they occur).

Inadequate behaviorisms. Some early behaviorists simply denied the existence of private events. For example, Skinner (1963) reports that Watson "tangled with introspective psychologists by denying the existence of images" (p. 952). Others accepted the public/private distinction but disallowed the inclusion of private events in scientific deliberations, because "science is public." Still others, while also accepting the public/private distinction, allowed such events in scientific discourse but only after defining out the private aspect. As examples, some might simply deny that hunger exists; others, accepting that hunger was private, disallowed its study due to its privateness; and still others, while also accepting that hunger was private, studied it only after defining away its privateness by defining it as some number of hours of food deprivation. Respectively, we allude to these three approaches to the privacy problem as Watson's *original* radical behaviorism, methodological behaviorism, and operational behaviorism. All are unsatisfactory because they sidestep the reality of private events, and thereby fail to deal with those events, particularly the events called consciousness.

Calling his 1963 article a "restatement of radical behaviorism" (p. 951), which contributed to calling his philosophy of science *radical behaviorism,* Skinner resolved the privacy problem by pointing out that, with the skin as a scientifically unimportant boundary, the physical dimensions of public and private events are the same; a natural science of behavior makes no assumptions that events inside the skin are of any special nature, or that we need to know them in any special way, different from the rest of nature. Instead, an adequate natural behavior science deals with private events as part of behavior itself. In the bulk of his article, Skinner went on to discuss the results of experimental science that had already accumulated in support of this privacy–problem solution, and many of its implications and ramifications.

That experimental science, however, was largely a natural science of the behavior of other animals; in the subsequent 50 years, it became both a natural science and an applied engineering technology emphasizing human behavior. During these years developments expanded in the organizational realm while also continuing in both the philosophical and experimental realms.

Developments in the Second 50 Years

The privacy–problem solution that capped the first 50 years prompted one of the major developments bearing on the question of consciousness in the years since 1963, namely a greater appreciation of the reciprocally valuable overlap between the separate yet complementary natural sciences of physiology and behaviorology. For example, to deal scientifically with emotion requires the different analytical levels of these two disciplines. *Emotion* refers to a release of chemicals into the bloodstream (an area of physiology) that external or internal stimuli elicit (an area of behaviorology). That changed body chemistry produces the reactions called *feelings.* Perhaps even more importantly, that changed body chemistry produces effects on other responses. When a bear startles you, you run faster than you run under more ordinary circumstances. Or, excising the fictitious inner agent that the word "you" can mistakenly imply, the sudden appearance of a big brown bear from behind a boulder only a meter away evokes faster running—due to the elicited emotional body chemistry change— than the running that more ordinary circumstances evoke, such as a clock showing the time that a jogging session begins.

The Value of Physiology

While behaviorology accounts for specific functional relations between real independent variables on both sides of the skin, and real dependent variables of behavior changes on both sides of the skin, brain physiology accounts for the structural changes that are occurring as those behaviorological–level independent and dependent variables interact. That is, brains mediate behavior that occurs as a function of other real variables; brains neither originate nor initiate behavior. Thus, the more brain physiologists work to account for *the mediation of behavior,* particularly *the mediation of neural behavior,* rather than for *mind* or *mystical mental operations,* the greater success they experience, and the more valid and valuable their work becomes.

Essentially then, behaviorology is not a natural science *of how* a body mediates a behavior (e.g., of how striated muscle contractions are a function of neural processes) which is a part of physiology; rather behaviorology is a natural science *of why* a body mediates a behavior (i.e., of the functional relations between independent variables, such as a boulder blocking a forest path, and the dependent variables of body–mediated *behavior,* such as the muscle contractions that the obstacle evokes thereby taking the body around the boulder). All the events at both levels of analysis are entirely natural; no mystical inner agent is considering options and then willing the body/ physiology to do the action that it has decided to take. "Considering" and "decision–making" are naturally occurring neural behaviors that happen as a function of real independent variables.

Also naturalistically, behaviorology has addressed some ancient and fundamental questions, leading to some exciting outcomes. This followed the enhanced accounting for complex human behavior, including consciousness, made possible by incorporating Skinner's analysis of verbal behavior (1957a) into the mix of more typical physiological and behaviorological variables: Since what scientists and philosophers and other knowers "do" *is* behavior— often verbal and stripped of residual agential implications—behaviorology, as the natural science of behavior, is providing scientific analyses of science, of philosophy, and of epistemology. By the 1990s such natural–science analyses also covered values, rights, ethics, and morals with important implications for a range of engineering concerns including robotics (see Fraley, 2008, Chapters 25 and 30). These kinds of scientific extensions of behaviorology led Lawrence Fraley, in Chapter 29 of his *General Behaviorology* (2008) to conclusions about reality that parallel those that Stephen Hawking and Leonard Mlodinow reached in their book, *The Grand Design* (2010) through the logic of naturalism in physics, that our neurally behaving reality is the sole source of knowledge (i.e., conscious/neural responding) about reality, because we can get no closer to reality than the behaviors that the firings of sensory neurons evoke.

Past and future. A related question arises, both on its own merits and due to its relevance to accounting for consciousness: How can events that seem to be in the past or future affect our behavior? The basic answer is that past or future stimulus events *cannot* directly evoke or consequate responses. Both responses and stimuli occur only in the present, an important implication being that all behavior is *new* behavior (with responses grouping into response classes for experimental analysis). Every behavior occurs under the functional control of *current* evocative—which includes eliciting—stimuli regardless of the complexity, multiplicity, or interactivity of those stimuli or responses. Even memories are not *stored responses.* They are new responses that current stimuli evoke and that current neural structures mediate, neural structures that have their current structure because conditioning processes changed them both at and since the time of the original instance.

Consciousness and community. With our now somewhat more informed perspective, we return to address consciousness more completely. Using the vision modality for convenience, Skinner had described consciousness as *"seeing* that we are seeing" (which we call "conscious *seeing*"). But he excised any mysterious implied inner agent who "does" the seeing by pointing out two general kinds of contingencies (i.e., functional relations between behavior and antecedent and postcedent variables). Our physical environment supplies the kinds of contingencies that condition seeing in the first place (called "unconscious seeing") while our verbal community supplies the kinds that condition both our conscious seeing and our reporting of what is seen. The thing seen evokes our initial unconscious seeing responses which in turn evoke the seeing/reporting conscious responses. Actually, the thing seen need not be present because other real variables, often as part of processes that some call imaging or picturing, can evoke the unconscious seeing response, which can then evoke conscious seeing/reporting responses. Equally pertinent, when current independent variables are insufficient to compel the conscious part to occur, it does not happen.

The verbal community conditions such seeing and reporting because benefits accrue to it when those events occur. In common terms (i.e., using the linguistic economy which the agential perspective sometimes provides) more effective social organization and discourse arise when the verbal responses (reports) of what we did, are doing, and are about to do, provide stimuli that share in evoking the responses of verbal community members. As members of that same verbal community, we also benefit when our own seeing and reporting share in evoking our own subsequent responding, for such reporting also evokes our own hearing responses which then naturally supplement the controls on subsequent responses. Often those reporting and hearing responses occur covertly as one type of the conscious neural behavior called thinking, a common and vital addition to the controls on subsequent behavior (since single stimuli seldom control responses). As with all neural behavior, this thinking behavior can be difficult to separate from the neural physiology that mediates it. Still, as with all behavior, independent variables must evoke the occurrence of neural behavior, including thinking. While all these events may be complex and occur so rapidly that they strain our measurement technology, they are nevertheless occurring entirely naturally. No inner agent of any sort is "doing" the seeing, observing, thinking, measuring, evaluating, reporting, or mediating.

While we sometimes benefit from its occasional economy (e.g., the "what we did, are doing, and are about to do" in the previous example) common language usually dampens or curtails scientific sensitivity to the natural status of human behavior. Having developed under primitive conditions that seemed to support the superstitions of personal agency, the common language *per se* unsurprisingly contains explicit and implicit references to inner agents (e.g., personal pronouns). Thus, avoiding common language in scientific discourse is often best, even though it seems comfortably familiar to most audiences. However, the technical language of natural behavior science, which works to exclude agential implications, can still sound overly complicated to new audiences even as these audiences experience an improving scientific sensitivity to the natural status of behavioral phenomena, including consciousness.

Some examples relating to the central concern about consciousness may help. While these use the seeing sense modality, other examples could use other sense modalities (e.g., hearing, taste, smell, touch). As an example of unconscious seeing, a hiker, engaged in a focused conversation with a companion, will step over an unconsciously seen football–size rock on the trail but later cannot describe that rock as it was not consciously seen. Conscious seeing examples are necessarily more complicated as they usually begin with unconscious seeing. For instance, under some current, relatively simple contingencies involving functional chains of external and internal (neural) stimuli and responses, one sees a favorite kind of car (i.e., physical stimulation in the form of light energy reflected onto the retina from a favorite car, perhaps on a dealer's lot, evokes an initially unconscious neural visual car response); it is a "favorite" kind of car, because it occurred (i.e., paired) with certain past variables. Later, unconsciously and consciously seeing the favorite car occurs again under other contingencies, often with that car absent, as when, unable to get to work, seeing our old, broken down, rusty wreck in the front yard evokes seeing the favorite car replacing the wreck. Still other variables can evoke such conscious seeing. When, at a grocery store, we see an acquaintance who sells cars, that person not only evokes consciously seeing both our wreck and our favorite kind of car (neither of which is present) but also evokes the responses of describing the favorite car, asking where to buy one, how much it will cost, and so on. Still other contingencies may evoke consciously seeing the favored car type, as when the clerk at an airport car–rental counter asks us, "What type of car would you like to rent?" The question may evoke only the verbal name of the favorite car, but often the question evokes consciously seeing the favored car which then evokes observing and reporting its type to the clerk.

Genes and conditioning and response chains. These responses—unconscious seeing then conscious seeing, and thinking, and sometimes reporting—are typical examples of the natural phenomena of responses chaining into response sequences (i.e., responses, as real events, evoking other responses; see Hayes & Brownstein, 1986) usually including neural responses, all in the present, all new, and not requiring the thing seen to be the current source of evocative stimulation. The same holds for other sense modalities. A physically present object transferring energy to neural receptors of any sense modality can be an evocative stimulus for either muscular responses or neural responses or both. And an evoked neural response can function evocatively for further responses either when *a directly genetically produced* neural structure mediates it or when a neural structure *that various continuously operating conditioning processes have changed* mediates it. If the necessary conditioning (i.e., neural restructuring) has occurred, then once some stimulation evokes a response, that response—as a real event—can evoke a further response, which can evoke yet another response, and so on, chaining according to the current set of operating functional relations.

Those cascading chains of sequential relations, though at times obscure, are not mystical. They merely involve two kinds of behaviors, covert and overt, each serving evocative stimulus functions for subsequent responses of either type. And, especially when contemplating interventions, behaviorologists quickly trace back the links in any causal chain to search in the *accessible* environment for the functional public antecedents of the covert events. By tracing the functional relations back to events in an accessible part of the environment, they locate potentially changeable independent variables. This affords control over the subsequent internal and otherwise less accessible parts of the sequence as well as control over the external parts. In this way the difficult problems of private events become quite manageable.

As those events unfold, no capricious inner agent makes these chained responses occur; any responses that occur are the only responses that can occur under the present functional independent variables. As natural scientists, we respect the natural functional history even of extremely complex and multiply–controlled response chains such as the text–composition responses that the physiology of this complex carbon unit (i.e., of the author) mediated under the then current conditions during the original moments (hours, weeks, years) of writing this chapter. While it would be quite economically wasteful to bother with the detailed analysis to identify and describe the range of variables compelling the present wording, B. F. Skinner (1957a) as well as Norman Peterson (1978) and Lawrence Fraley (2008, pp. 949–1094) have provided, in their textbooks, the foundations for making such an analysis, and it will occur under appropriate contingencies.

The Law of Cumulative Complexity. All that complexity in behavioral and physiological events can seem amazing, wondrous, awesome, even overwhelming, *but it is all entirely natural.* Perhaps summarizing the origin of such emergent complexity in the context of *all* the natural sciences as a theory or law will help. If it rates the status, let's call it the **Law of Cumulative Complexity:** *The natural physical/chemical interactions of matter and energy sometimes result in more complex structures and functions that endure and naturally interact further, resulting in an accumulating complexity.* Any origin of life is an outcome of the Law of Cumulative Complexity. On this planet other examples of this law include the vast range of life forms available for study, and the interrelations of physiology and behaviorology. *All these are cumulatively complex; all are entirely natural.*

All the developments that we have discussed involve extensions of the philosophy of science that Skinner called *radical behaviorism.* It is radical in the sense of comprehensive or fundamental, and it informs both the natural science that experimentally studies human nature and human behavior, and the derivative engineering technologies of that science for effectively addressing accessible independent variables in ways that bring about improvements in behavior at home and work, in education and diplomacy, in interpersonal relationships, and indeed in all applied behavior fields from advertising to zoo keeping. This philosophy, and the science and technology that it supports, first arose among a thoroughly naturalistic group of researchers and academics, working in early twentieth century psychology, that Skinner and his colleagues and their students best represent. However, this natural philosophy, science, and technology ultimately proved to be fully incommensurable with the more commonly available, non–natural, fundamentally mystical, agential perspectives of certain fields that popular culture currently supports, including psychology. As a result, a separation of disciplines had to occur.

Organizational Developments

Such full incommensurability with natural science begins when any discipline studying any phenomenon eschews the assumptions of naturalism that provide the foundation for all natural sciences. This is because *all* real aspects of the universe—including energy, matter, life forms, *and life functions* (e.g., human behavior)—are potentially within the reach of objective scientific accounts and applications. To qualify as *natural* science though, natural philosophy, rather than any sort of mystical assumptions, must inform the framing of research questions and the interpretation of experimental results.

That is basic to natural science, but difficult to maintain in a culture steeped in superstition, where some insist instead that accounts based on mystical assumptions are adequate. Of course, with much scientific activity being methodological, anyone using scientific procedures, even mystical people, can objectively collect data on any real phenomenon. But this is not enough; the inherently contradictory biases in any variety of mystical assumptions informing such data collection usually leads both to misdirected research questions and to misinterpreted experimental results either or both often aimed mainly at proving the mystical assumptions.

No one, however, can prove or disprove assumptions. The supporters of mystical assumptions only induce them off–the–cuff or, with allusions to tradition or common lore, from various kinds of faith beliefs that also result from researchable natural processes. In contrast, supporters of the assumptions of the philosophy of naturalism mainly induce them from the results and developments of several hundred years of accumulating experimental evidence and effective engineering practices with the wide range of phenomena that the various natural–science and engineering disciplines cover. Thus, these two types of assumptions, naturalistic/scientific and mystical/superstitious, are *not* equal in any meaningful way. They are incommensurable.

That incommensurability, and the growing pressure of expanding experimental and applied research, provided the principal driving forces behind reorganizing the natural science of behavior as a separate and independent discipline. The general result of this development is a foundation natural science related to all other natural sciences, not at the discredited level of body–directing self agents, but at the level of a body's physics–based interactions with the external and internal environments. Working in this natural–science tradition, Skinner's treatment of behaviorism in his 1963 article was well rounded but necessarily minimal. A decade later his book *About Behaviorism* (Skinner, 1974) provided details and helped pave the way for the sometimes controversial steps in this reorganization. In a long paper, Lawrence Fraley and I (Fraley & Ledoux, 2002) thoroughly discussed the issues and history of this transition, some highlights of which are relevant here.

Early steps toward independence. The movement to disengage from psychology began with several small independence steps. In *A Matter of Consequences* (1983, pp. 23–24, 44) Skinner reported that, around 1950, many of the Harvard psychology faculty considered his early course at Harvard, Psych 7, as too divergent, so he renamed it and transferred it to the general education area as "Natural Science 114: Human Behavior" where it proved very successful with a wider range of students. He subsequently used his notes for that course as the basis for his still highly regarded 1953 textbook, *Science and Human Behavior,* upon which Lawrence Fraley built when writing the first text (Fraley, 2008) explicitly delineating the natural science of behavior as the independent discipline of behaviorology.

Around the same time as his course transfer, Skinner and his behavioral colleagues were moving away from the label "operant psychology" which was their appellation in the 1940s and 1950s. The new label, "The Experimental Analysis of Behavior" (see Skinner, 1957b) implied their more independent direction. In the late 1950s, they founded a new *natural*–science journal, *Journal of the Experimental Analysis of Behavior (JEAB),* to publish the growing body of their experimental work. For decades papers in this journal have emphasized single–subject designs and eschewed inner–agent analyses.

By the 1960s, the growing number of researchers reporting experimental analyses in *JEAB* had begun to branch out into an increasingly wide range of application areas. Partially in recognition of the experimental *and applied* nature of their natural–science work, they began calling themselves "behavior analysts" and their work "behavior analysis" or, in therapeutic areas, "applied behavior analysis" (ABA). Given their ever growing amount of research, in the late 1960s they founded another journal, the *Journal of Applied Behavior Analysis (JABA).* In time, as research interests expanded further, they founded organizations as well as several additional journals.

With a core philosophy of science—Skinner's radical behaviorism—as a focus, the founding of several new organizations and their journals marked the expanding disciplinary independence of the natural science of behavior. During eight years of increasingly formal meetings started in 1966 (only three years after Skinner published "Behaviorism at Fifty") some natural scientists of behavior moved away from the Midwestern Psychological Association (MPA) which was not filling their professional disciplinary needs. In 1974 they formally established the Midwestern Association of Behavior Analysis (MABA) and began holding conventions separate from the MPA conventions. Margaret Peterson (1978) reported the importance of these events in a quote from an early MABA president, Nate Azrin: "What we are witnessing with MABA may be not only a distinctive type of regional convention organization, but also the birth of a new discipline... separate from Psychology, Psychiatry, Education, and other related areas" (p. 15). After four conventions, MABA was drawing members from the national, and even international, pool of behavior scientists, and it had also become concerned with professional certification issues. So in 1978 it began publishing *The Behavior Analyst* and changed its name first to the Association for Behavior Analysis, and later to the Association for Behavior Analysis International (ABAI).

A possible majority of those who remain natural scientists of behavior maintain membership in ABAI, whose activities have continued vibrantly. On its web site (www.abainternational.org) ABAI reported that in 2008 it had over 5,300 members, with 60 affiliated chapters worldwide reporting around 13,000 members. Over 4,000 members from about 40 countries attended the annual convention which featured over 1,500 presentations by more than 3,000 contributors and a convention bookstore of over 1,000 titles.

Independence–constraining problems. Organizationally, however, ABAI members had, from the beginning, increasingly emphasized political action on professional, social, and cultural fronts. This enthusiasm for political clout made a large membership seem more important than philosophical clarity and natural–science consistency. After decades of pursuing the political path, the ABAI membership now comprises a variety of philosophical and disciplinary perspectives. The commonality of its members currently inheres predominantly in efforts to produce good behavioral outcomes within their respective specializations, a circumstance that renders ABAI less of a basic–science organization and more of a behavior–engineering society.

Furthermore, as important as the political activities were, they distracted the organization from wholeheartedly pursuing its independence. As a result members lost touch with their independence origins, leaving the credibility problems—that inhere in gradually separating from another discipline, while still being seen as part of it—to remain and grow.

Exacerbating the controversy, behavior analysts took those and other independence–oriented steps while still closely associated with psychology. This allowed the psychology discipline to claim behavior analysis as part of itself. Decades ago some behavior analysts somewhat validated that claim when, apparently as part of a reasonable attempt to move psychology toward giving up mysticism in favor of natural science, they used the behavior–analysis label as the name for the journal of an official division of the American Psychological Association. As with all such attempts to change psychology, this one also failed, and later that same division took "behavior analysis" as its own name. With behavior analysts raising insufficient or no clamor of objections, the mystical discipline of "behavior and the mind" took over the "behavior analysis" label.

The valid psychology claim to behavior analysis and its label leaves others, including natural scientists in general, with the clear understanding that behavior analysis can no longer be trusted to be a natural science. The collegial relationships that some natural scientists of behavior have with some traditional natural scientists may buffer that mistrust a bit. But beyond such relationships, those natural scientists of behavior who remain "behavior analysts" invite avoidance, and even scorn, from traditional natural scientists. They see these behavior analysts as refusing to go independent and leave psychology—with its inherent secular mysticism—behind, as refusing to become behaviorologists and use the appropriate label to name their basic natural science.

On the other hand, the benefits of being part of psychology can include, among other things, an increase in job security and some safety from accountability. While problems from national economic swings can increase the value of such factors, a wide variety of contingencies can drive being part of psychology for different behavior analysts, including some—perhaps much— personal success in psychological work units. Even for those behavior analysts well trained in radical behaviorism and the natural science of behavior, these pro–psychology contingencies too often override the evocative and reinforcing potency of contingencies driving independence for natural behavior science. Could that include overriding the contingencies involving disciplinary integrity, and even the potentially greater benefits for humanity that could accrue from being able to work, with mutual respect, alongside traditional natural scientists to help solve global problems? (See Fraley & Ledoux, 2002, for details on the contingencies driving independence for natural behavior science.)

All that not only raises severe credibility issues for behavior analysts, issues that continue to haunt their efforts to collaborate with traditional natural–science colleagues, especially in efforts to solve global problems, but it also affects naming a separate natural science of behavior discipline. The emphasis on political power drove ABAI to maintain a very liberal policy on member qualifications, such that many members qualified for membership without the benefits that derive from full training or even interest in or agreement with the natural science of behavior (under *any* name), which is why multiple philosophical and disciplinary perspectives pervade ABAI, and why most members mainly emphasize the engineering of good behavioral outcomes without much concern for the basic science—or its disciplinary integrity or name—behind those engineering efforts. Some behavior analysts, however, *are* well–trained natural scientists of behavior, accepting naturalism over mysticism and agentialism. Yet so many of them, who could have objected, evidence so little concern over the status of the behavior analysis label, and of being under psychology's wing, that they have let the past status of behavior analysis as a natural science slip permanently away, thereby destroying any possibility of using the behavior analysis label as the name for a completely *independent* natural science of behavior. For such reasons formal separation of the natural science of behavior from psychology required adopting a new disciplinary name, one free of connections with non–natural disciplines.

In the years 1984–1987, an extensive debate filled the published behavioral literature regarding, pro and con, the question of fully and officially separating the natural science and philosophy of behavior from psychology (see Fraley & Ledoux, 2002, for a thorough review of this debate). Many discussants acknowledged numerous types of recognizable separation already present in varying degrees, including courses, journals, organizations, conventions, certification, accreditation, and even some academic departments and programs; but prior to 1987, none of these early types of separation occurred under full and formal declarations of independence, although some departments came close. For example, the "Department of Behavior Analysis"—named before the behaviorology label came into use—was fully separate from and independent of the Department of Psychology at the University of North Texas in Denton.

A more explicit independence move. The 1984–1987 debates culminated, in 1987, in a group of natural scientists of behavior meeting to reassess the situation and take action. They came to several conclusions. (a) If data from a half century of continuously attempting to change psychology into a natural science "from within"—by invoking standard, evidence–based methods that might take centuries and even then not work—showed failure to produce even slight movement in that direction, then changing psychology was not going to happen within a meaningful time span (e.g., before the opportunity passes in which to help humanity reduce global warming and so avoid its worst effects, a time frame of about 150 years as they rather optimistically understood it then). (b) Their natural science of behavior was not, and never actually had been, any kind of psychology as it had never accepted the basic psychological core of mystical agential origination of behavior. And (c) instead, their already well–established natural science would continue, at least in part, as a fully separate and independent discipline called behaviorology, a term first proposed in the late 1970s specifically to describe a natural science of behavior discipline completely separate from and independent of psychology, and the only term, from among all proposed names, to have survived and grown in use.

Based on those conclusions, these behaviorologists took several organizational steps establishing the independence of their natural–science discipline separate from all disciplines that espouse non–natural accounts for behavior. They founded The International Behaviorology Association (TIBA) and the journals *Behaviorology* and *Behaviorological Commentaries.* Ten years later, in 1997, they separated the research and convention functions from the education and training mission respectively by changing the name to the International Society for Behaviorology (ISB) and founding The International Behaviorology Institute (TIBI) and its journal *Behaviorology Today (BT)*, which became the *Journal of Behaviorology (JoB)* with volume 16 in 2013.

From the start, the behaviorologists also held annual conventions, with the first one in August 1988 at Clarkson University in Potsdam, NY. Figure 1–1 shows the attendees at that first convention. Most behaviorologists have also continued supporting the beneficial behavior–engineering efforts that ABAI disseminates, and in recent years, TIBI has expanded the convention offerings.

However, those disciplinary developments constrained other possibilities. For example, the behaviorology label usually elicits strong negative emotional reactions from some behavior analysts, especially those who forfeit claims to natural–science status—even if they adhere to natural science themselves— due to the formal connection that they support and maintain with non– natural disciplines like psychology. These or other behavior analysts have pointed out that the PsycInfo database yields only a handful of hits for the term behaviorology, while yielding thousands of hits for the term behavior analysis. Yet since that database is published by the same American Psychological Association that has a division named Behavior Analysis, while Behaviorology is instead a young natural science completely unconnected to psychology, such findings are quite appropriate. Similarly, some behavior analysts could claim that the lack of a Special Interest Group (SIG) in ABAI implies that behaviorology lacks any importance. But behaviorology lacks a SIG in ABAI because such SIGs are for various *parts of behavior analysis,* and a non–natural discipline fully owns behavior analysis and its label. On the other hand, behaviorology is the label, not for a part of something else but for the independently organized natural science of behavior. Participating as a SIG in that organization, or *any* organization that affiliates with psychology, is simply not appropriate for behaviorology.

Some benefits of independence. Those concerns may occur because behaviorology, now an independently organized natural science, arose

Photo by Stephen F. Ledoux

Douglas Greer	Ernest Vargas	Jack Michael		John Stone	Scott Beach	
		Lawrence Fraley		Laura Dorow		
Jerome Ulman	Al Kearney		Robert Spangler		Carl Cheney	
Sigrid Glenn	Guy Bruce				Julie Vargas	Juan Robinson
Stephen Ledoux		Jeffrey Kupfer		John Eshleman	Linda Armendariz	

Figure 1–1. Most participants at the first behaviorology (TIBA) convention, Clarkson University, Potsdam, NY, USA, August 1988.

historically from roots in the early behavior analysis movement, itself originally an expansion of *The Experimental Analysis of Behavior.* Ever since the 1980s, the behavior analytic literature has featured a broad spectrum of views not only on the emergence of behaviorology but also on other topics as behaviorologists view them (e.g., see Cheney, 1991; Eshleman & Vargas, 1988; Fraley, 1983, 1987, 1994a, 1994b). As some in organized behavior analysis turned increasingly *back* toward psychology, the behaviorologists instead moved to declare disciplinary independence. I suspect that some behavior analysts still adhere to naturalism and natural science, and reject any connection with psychology. They may even now be under contingencies to move away officially from psychology, with its mystical agentialism, and move slowly toward re–committing to Skinner's natural science of behavior. This would begin to qualify them as behaviorologists while they clearly and publicly re–declare their independence and adopt the behaviorology label as the name for their basic science. In this way they would regain recognized status as natural scientists of behavior; they might even join a behaviorology professional organization. Humanity might benefit more if they took that stand quickly, enabling more success in efforts to solve local and global problems, and faculty jobs for credible behaviorologists would become easier to fill.

One must wonder, for example, what path professional contingencies will induce with Board Certified Behavior Analysts (BCBAs). So long as BCBAs are natural scientists of behavior without independence, and use psychology's behavior analysis label, they cannot justify their separate credentials and so always face the threat of legal requirements for supervision by licensed traditional—and often anti–scientific—psychologists (a function that any local minister can just as inappropriately but perhaps more honestly serve). They also even face the threat of legal requirements to replace one half to two thirds of their natural–science and behaviorological engineering training with training in traditional psychology so that they can be both less effective and licensed as psychologists themselves. If enough qualified natural–science BCBAs re–declare their independence and adopt the behaviorology label as the name for their basic science, which need not otherwise change (beyond the self–correction inherent in science), then using the "Applied Behavior Analysis" (ABA) label for at least some of the engineering side of the independent discipline may succeed. "ABA" carries an engineering connotation, while behaviorology is a name for the basic science that informs this engineering. Based on the relative weakness of non–natural claims on the ABA label, less controversy has surrounded its use, contributing to its current wide recognition.

Meanwhile, behaviorologists are quite satisfied with the name of their discipline. That some who would not be strategically qualified to identify themselves as behaviorologists dislike that term is perhaps advantageous to all parties.

A related and more serious concern that some might express regards the current relative numbers of behavior analysts and behaviorologists. As this is written, behavior analysts are much more numerous than behaviorologists, and that may continue for some time. After all, engineers tend to outnumber basic scientists. However, and more importantly, the emergence of behaviorology has broken the monopolistic grip of the cultural forces of organized mysticism on behavioral phenomena. By its independence behaviorology definitively shifts that sphere of inquiry into the realm of the natural sciences. The widening disciplinary rift likely leaves many students of natural behavior science trapped on the non–natural side where mentors who wish to retain their allegiance may, unable to argue against natural science, resort to disingenuous tactics of distraction or career threats. To those students we can only shout back across the rift: "Examine the evidence; the contingencies therein may ensure that your maturing career investment remains inside the natural–science community."

Furthermore, the relative numbers carry little importance because, to natural scientists, numbers fail to trump scientific and disciplinary integrity. Behaviorologists need not be numerous; they need only be independent natural scientists addressing the relations of their discipline to the culture that it serves. Even if the number of behaviorologists were to fall to a mere dozen, those 12 would still represent the only independent natural science of behavior named using an established term uncompromised by any connections with fundamentally mystical disciplines. Such characteristics support the credible growth of this discipline.

Connections with non–natural disciplines, however, reduce the credibility of behavior analysts in the eyes of traditional natural scientists, making the behavior analysts' contributions to humanity's future more difficult to provide, contributions like helping with the behavior components of reducing global warming, within the limited time frame available before enduring its worst effects. And the clock is ticking. Behavior analysts who prefer both credibly helping their traditional natural–science colleagues make such contributions, and clearly increasing their distance from fundamentally mystical disciplines, can regain their credibility by adhering to naturalism and natural science while re–declaring their independence and, if qualified, using behaviorology as the name for their basic science. But if those steps are to help, they need to take them soon, as humanity is running out of time.

Scientific Developments

During the decades of gradual emergence as an independent discipline, natural scientists of behavior have reported a still growing range of experimental research and engineering applications. Consider, for example, this sample of behaviorological science articles from that period: William Baum (1995) further analyzed radical behaviorism and the concept of agency. Joseph Cautela (1994) elaborated on the importance of the concept of the General Level of Reinforcement (GLR). Carl Cheney (1991) summarized the controlling relations and sources of behavior. John Eshleman and Ernest Vargas (1988) promoted application technologies for verbal–behavior analysis. Philip Johnson (2011) and John Ferreira (2012) each analyzed aspects of clinical successes from Progressive

Neural Emotional Therapy (PNET). Lawrence Fraley discussed the cultural mission of behaviorology (1987), naturalistic consideration of uncertainties about determinism (1994a), behaviorological thanatology and medical ethics (2006), naturalistic analysis of the legal doctrine of *Mens Rea* (1983), and behaviorological penal corrections (1994b). Stephen Ledoux (2010) reported results on an experimental procedure to control simultaneously evocable and simultaneously occurring, complex human behaviors with multiple, even simultaneous, selectors, like reinforcers. And Jack Michael (1982) separated discriminative and motivational functions of stimuli.

Also, consider this sample of behaviorological science books. Charles Ferster and B. F. Skinner (1957) published the data from their extensive laboratory research on schedules of reinforcement. Robert Epstein (1996) and B. F. Skinner, through their *Columban Simulation Project,* reported the phenomena that they called *recombination of repertoires.* Murray Sidman provided works (a) on research methods (1960/1988), (b) on stimulus equivalence relations (1994), a work with considerable implications for the nature of research *as behavior,* and (c) on the detrimental effects of coercion and punishment at all levels of social interaction including families, classrooms, workplaces, and international relations (2001). Glenn Latham contributed works on positive practices for raising (1994, 1999) and educating (1998, 2002) children. Cathy Watkins (1997) clarified the value of *Project Follow Through* for improving regular education. Catherine Maurice, Gina Green, and Stephen Luce (1996) detailed best practices for work with autistic children. Aubrey Daniels (1989) addressed performance management in business and industry. Lawrence Fraley (2008) published his graduate level behaviorology course materials, from which he had been teaching for 25 years, thereby providing the first extensive and systematic book explicitly establishing the separate and independent natural–science discipline of human behavior. Fraley also provided a book–length work on behaviorological thanatology and dignified dying (2012) and another on behaviorological rehabilitation and the criminal justice system (2013). The philosophy and science discussed here even prompted two popular and educational works of fiction, one each by B. F. Skinner (1948) and W. Joseph Wyatt (1997).

Those sample lists could have included a far larger number of the thousands of available books and articles; the particular selections they include are attributable as much to the author's familiarity with them as to various opinions regarding their relative importance. The same applies to the selections, next, for detailed research and application examples.

Highlighting three of the many areas of experimental research can sample the range of important findings discovered in the last 50 years. These three areas are schedules of reinforcement (Ferster & Skinner, 1957), recombination of repertoires (Epstein, 1996), and equivalence relations (Sidman, 1994).

Reinforcement schedules. Basically, reinforcers are postcedent stimuli whose occurrence produces increases in the frequency of the behaviors that they follow, and schedules of reinforcement are the patterns of intermittently occurring reinforcers. These schedules are defined either in terms of the *number* of responses since the last reinforcer occurred (which we call *ratio* schedules) or in terms of the amount of *time*—plus a required response—since the last reinforcer occurred (which we call *interval* schedules). The values of either type can be fixed or variable, thereby defining the four fundamental intermittent schedules of reinforcement: fixed ratio (FR), variable ratio (VR), fixed interval (FI), and variable interval (VI). For example, on an "FI–30 second" schedule, a reinforcer would follow the first response to occur after each 30–second interval since the last reinforcer, while on a "VI–30 second" schedule, a reinforcer would follow the first response after each interval, with intervals averaging 30 seconds in duration. On a "VR–30" schedule, a reinforcer would occur, not after *every* thirtieth response—which would be an "FR–30" schedule—but after every set of responses, with each set *averaging* 30 responses; using one of several available methods to arrange this VR schedule, ten reinforcers would occur in 300 responses with between one and 60 responses occurring between reinforcers. Researchers often combine or otherwise rearrange the elements of these basic schedules to conduct studies with a range of more complex schedules (e.g., mixed, multiple, chained, tandem, and concurrent schedules).

Outside the laboratory VR schedules are common. They produce relatively rapid and steady response patterns, which we characterize as persistence. For centuries before science discovered and analyzed this schedule, these response patterns compelled purveyors of games of chance intuitively to arrange VR schedules for control of the behavior of their players. And still today VR schedule effects (not the "gambling habits" of fictitious inner agents) are responsible not only for much individual citizen wealth reduction but also for swelling government treasury coffers from lotteries and gambling taxes.

Overall, schedule research has repeatedly led to several general conclusions, including these three: Many features of behavior emerge as the effects of particular reinforcement schedules. Schedules with only subtle differences often produce distinctly different response patterns. And, the direct effects of schedules of reinforcement reduce a wide range of putative inner–agent emotional and motivational causes of behavior to misleading redundancies.

Recombination of repertoires. Next we consider two examples of the experimental research concerning recombination of repertoires, with important implications particularly for scientific, engineering, and educational problem–solving behavior. For over a decade starting in the late 1970s, Robert Epstein, with B. F. Skinner, coordinated some studies at Harvard called the *Columban Simulation Project* in which pigeon behaviors that were functionally related to explicit variables simulated complex human behaviors. Some of these complex behaviors concerned novel behavior, symbolic communication, and the use of memoranda and tools. Others were traditionally thought to arise from various mentalistic notions such as "insight" or "self concept." The result of this research was a more objective explication of complex human behaviors.

The same kinds of common contingencies that people observed producing the pigeon–simulated behaviors were at work with the human behaviors.

The pigeon simulations began with analysis of the complex human behavior of concern to surmise the minimum repertoire components likely needed for that complex behavior to occur when a challenge situation confronts the organism. Then, for each pigeon subject, after conditioning each required repertoire component (in isolation from other components, to avoid confounded results) the experimenters placed each pigeon in the challenge situation. The researchers found that, for different problematic tasks, if the conditioning of all necessary component behaviors had occurred, then (and only then) the challenge situations evoked successful responses appropriately combining the trained repertoire components.

The first of our two recombination of repertoire examples concerns the test for a supposed "self concept." In this test a young child faces a mirror with a rouge spot on his or her forehead, a location that makes the spot visible only in a mirror. If the image of the child–*cum*–spot in a mirror only evokes responses typical of the presence of another child, we are supposed to accept that the child lacks a concept of self. However, if the image of the child–*cum*–spot in a mirror evokes responses of touching the spot, then we are supposed to accept that the child's having a "self concept" caused those responses.

In experimenting to discover the actual variables involved in the mirror test, the researchers came upon two classes of responses that they needed to condition in their pigeon experimental subjects. They began by conditioning the birds *with no mirror present* to peck blue stick–on dots placed on virtually every part of the bird's body that it could reach. Separately, they also conditioned effective responding in a mirrored space, with no blue dots on the bird's body, by reinforcing each bird's pecking at each of the varying, correct locations on the unmirrored wall of the chamber upon which a blue light had briefly flashed while the bird faced a mirror and could only see the flash locations in the mirror image. Finally they placed a blue stick–on dot on the bird's breast along with a bib around the bird's neck that prevented the bird from seeing the dot directly, because any even slight lowering of the head moved the bib downward covering the dot. When in the chamber with the mirror covered, none of the birds tried to peck the dot, which was possible and likely if they could detect it in any way. With the mirror uncovered, however, every bird began bobbing its head as the dot, repeatedly visible in the mirror whenever standing erect, evoked repeated attempts to peck the dot, which each time disappeared from view due to downward bib movement. Does this mean that these birds had a self concept? Parsimony requires accepting that spot images in mirrors evoke spot–touching not as a function of self concepts, for pigeons or humans, but rather as a function of a relevant conditioning history and current evocative circumstances, a history and circumstances that, for these birds, is an explicit matter of record.

The other recombination of repertoire example concerns testing the "insight" account of some complex human behaviors. Consider that many proud parents have watched as their child, too short to get a cookie from a jar atop a table and having never faced this situation before, looks around and, spotting a chair, moves it over to the table, climbs on it, and retrieves a cookie from the jar, putatively due to something called *insight.* In experimenting to discover the variables involved in this situation, the researchers came upon three pigeon response classes using boxes and toy bananas. They conditioned the birds *with no banana present* to push a box around the chamber toward a target spot and, separately, to climb on a stationary box and, still separately, *with no box or target spot present,* to peck a toy banana within normal reach. These response classes (pushing boxes, climbing on boxes, and pecking toy bananas) approximated the components of the child's complex cookie retrieval behavior (pushing a chair, climbing on a chair, and getting a cookie). Finally, they placed each bird in a chamber with a box to one side and a toy banana *suspended from the ceiling,* a challenge situation that had never confronted the birds before. With some apparent confusion and sighting, like the child's behavior, the birds pushed the box under the banana, climbed on the box, and pecked the banana. Does this mean these birds showed "insight?" Was the child's behavior due to "insight," or was the child's behavior also an example of previously conditioned repertoires combining under novel circumstances? We seldom observe children closely enough to track the conditioning of various repertoire components. Still, parsimony requires accepting that the occurrence of the challenge–meeting responses is not a function of supposed higher mental processes like insight, for pigeons or humans, but rather is a function both of the organism's history having included the conditioning of all relevant repertoire parts, and of the current evocative control in the new pattern of related multiple stimuli in the challenge situation. (For organisms with the necessary brain structures, evoked neural responses of consciousness, like thinking, may also naturally be *supplementing* the principle and more obvious sources of control; more research may clarify this situation.)

The recombination of repertoires line of research benefits the analysis of problem solving as well as enhances the justifications for multi–disciplinary education in scientist/engineer training curricula. As the range of an individual's conditioned repertoire of behavior expands, so does the likelihood that needed parts will be available to combine successfully in new problem circumstances for which no previous conditioning has provided explicit solution responses. While welcome physiological research will show how nervous systems mediate the combining of behavioral elements, the recombination of repertoire contingency accounts replace the unnecessary and counterproductive traditional mentalistic accounts of these complex behaviors.

Equivalence. Apparently related to recombination of repertoires in ways still being explored, *stimulus equivalence* is the remaining experimental research area highlighted here. Under some circumstances, after explicitly conditioning some functional relations between environmental antecedent and postcedent stimuli and responses, the number of related behavior–controlling functional relations that we can successfully detect is greater than the number originally involved in the explicit conditioning. Agentially stated, subjects seem to "learn" more than they are "taught," although explaining these phenomena requires no inner agents. Researchers in this area have come to call these explicitly and implicitly (i.e., emergent) conditioned relations *equivalence relations.*

Stimulus equivalence can transpire in fairly simple circumstances. For example, to train a new cloakroom attendant, we might first reinforce (i.e., condition) his behavior such that when a regular customer, Ms. Minkowner, appears and puts her pink (i.e., polyester) mink coat among the coats on the counter, the Ms. Minkowner stimulus reliably evokes the trainee's response of picking up her mink coat. Then, we reinforce the trainee's behavior such that in the presence of the pink mink coat and several different coat–hanging cubicles, this mink coat reliably evokes the trainee's depositing it in a particular cubicle, say, number seven. *With no further training,* we find that Ms. Minkowner's reappearance at the counter reliably evokes the trainee's movement to cubicle number seven from which he retrieves her pink mink coat.

Beyond such simplistic examples (which actually pertain mostly to the part of equivalence relations that researchers call transitivity), researchers in this area have demonstrated the phenomena occurring in far more complex circumstances. Using, for example, six sets of three stimuli each, explicit conditioning of a particular 15 environment–behavior functional relations turns out implicitly to condition an additional 75 behavior–evoking functional relations. In this instance, conditioning 15 particular relations can produce a total of 90 testable relations!

The implications of equivalence phenomena for a science–based revolution in, say, education can be substantial. More careful arrangements, which research efforts uncover, of curricular components—what we would scientifically call educational conditioning programs—in, for example, history, language, math, and science, can economize by explicitly conditioning only certain evocative functional relations, relevant to the subject matter, in ways that virtually guarantee the implicit conditioning of many other possible, relevant and equivalent relations evocable by the same broad set of stimuli. Teachers already engage in approximations of these steps, but that occurs under haphazard and unanalyzed contingencies, which leaves them less effective. Imagine the progress that could occur if the vast numbers of teachers had this science in their training programs, enabling them to research and improve the abundant equivalence relations in their teaching activities, an outcome that harbors profound implications for changes in teacher training.

Although physiological research gradually continues to elucidate at the cellular and molecular levels *how* respondent and operant conditioning processes work, and contribute to equivalence relations, natural selection has produced the kind of bodies that these processes can change in varying degrees. For example, if their genes happened to include variations that produced neural structures enabling the mediation of even a small extension of equivalence relations through these processes, then proto–species members could benefit from any likely survival/reproductive advantages that these emergent equivalence–relation extensions confer. Over millions of years, the accumulation of such selected variations would result in genetically produced nervous–system structures of increasingly sophisticated potential. As a result, humans today genetically inherit neural structures that generally mediate a relatively extensive range of equivalence–relation phenomena.

With biological selection as the foundation of the physiology through which conditioning processes work, including equivalence relations, untestable mystical constructs—from autonomous behavior–initiating self agents to cognition and the secularized soul called the mind—all remain scientifically unparsimonious and redundant. Behaviorological science accounts for why these relations happen, and is bringing them under practical control. Still, an even more complete description of the nature of these phenomena must await the physiological research of neural scientists that will eventually show how, in stimulus equivalence, the conditioning of a subset of relations actually changes the nervous system such that the subset conditioning turns out to produce the remaining relations as well. While all these behavior–related processes have substantive societal implications, society has only begun to notice the potential benefits of applying all the processes and phenomena that basic behaviorological research has discovered in the last 100 years.

Nevertheless, beyond *experimental* research, the last 50 years have seen an explosion of studies applying natural philosophy and science to practical concerns. Touching on two applied research areas, *Project Follow Through* in regular education (Watkins, 1997) and the refining of best practices for work with autistic children (Maurice, Green, & Luce, 1996), minimally samples the extensive range of behaviorological engineering applications.

Best practices for regular education. *Project Follow Through* was the most extensive and expensive federally funded educational experiment in U.S. history. It looked at how the outcomes, on a variety of standard measures from children taught with a range of distinct instructional models that whole districts voluntarily sponsored, compared with the outcome measures from children whose school districts across the country had not adopted any particular model.

The results led to a major observation: While some models produced a range of poorer outcomes than those of the comparison group, other models produced consistently better outcomes, particularly the Direct Instruction and Behavior Analysis models. Importantly, these successful models explicitly derived from the application of the principles and concepts of the natural science of behavior. The Project Follow Through research had predictably revealed some science–based instructional approaches that work in education.

However, leaders in the education field little disseminate that revelation of some best practices for regular education to the very teachers who, along with their students, would benefit from implementing its findings, leading to the wide ignoring of those findings. When giving a workshop several years ago to about 100 teachers and staff at a public k–12 school, the author asked who had ever heard of *Project Follow Through;* only two answered affirmatively. Also, while the results of *Project Follow Through* focused mainly on student outcomes from the first several years of the project, the funding of various of its models continued for many years. Unfortunately, this funding was *not* limited to the models that produced improved student outcomes. Objecting to wasting funds on models that hurt children, Cathy Watkins (1997) concluded that suggestions to solve the problems of education include attempts to "change just about every structural and functional aspect of education except *how children are taught*" (p. 88). Sadly, ignoring *Project Follow Through* data not only indicates some blind respect for ineffective, agentially–focused methods that comport with popular mysticism but also indicates some persistence of the discredited notion that behaviorological laws are largely irrelevant to normal humans.

Best practices for autism. In contrast, the other applied research example, on best practices for work with autistic children, has achieved greater recognition than best practices for regular education. Much of the research initially applying core behaviorological principles and concepts to a wide range of practical concerns, including interventions for autistic children, occurred before the name *behaviorology* emerged to denote the separate and independent status of the natural science of behavior. Consequently, many people refer to such behaviorological practices with the older label *Applied Behavior Analysis* or ABA, even though today these terms may also cover some less well–grounded practices. Nevertheless, the extensive successes of the behaviorologically supported autism–related practices have made them the preferred intervention, especially for children diagnosed at a young age. For example, in 1999 the New York State Department of Health completed a multi–year project to evaluate the research literature on the numerous types of available autism treatments so as to make intervention recommendations based on scientific evidence of safety and efficacy. Its final report (NYS Department of Health, 1999) stated, for most evaluated interventions, either that the intervention was "not recommended as an intervention" or was "not to be used as an intervention" for young children with autism. The *only* fully recommended intervention was ABA: "It is recommended that principles of applied behavior analysis (ABA) and behavior intervention strategies be included as important elements in any intervention program of young children with autism" (*Quick Reference Guide,* pp. 33–51).

Interdisciplinary Developments

Contributions to green behavior. Based on its informing philosophy of radical behaviorism, and beyond experimental and practical contributions in general, behaviorology makes other important contributions to the capabilities of traditional natural scientists. One major current area involves *behaviorological green engineering* (e.g., working on overpopulation concerns as a foundation for achieving sustainable lifestyles). Behaviorological scientists and practitioners already work in this area, because so many of the seemingly intractable problems facing humanity today involve problems of human behavior as much as problems of physics or chemistry or biology. Examples include out–of–control population levels, increasing climate extremes from global warming, water and air pollution, potable water depletion and the rising risks of water wars, habitat destruction, resources depletion, and loss of species through higher extinction rates.

The solutions also involve human behavior. A special section in the fall 2010 issue of *The Behavior Analyst* begins to address this aspect. The section features ten articles devoted to "The Human Response to Climate Change" (see the Supplemental References). With introductory and closing remarks (Heward & Chance, 2010; Chance & Heward, 2010), the topics of these articles include recycling (Keller, 2010), buying green (Layng, 2010), procrastination management (Malott, 2010), increasing success by helping others (Neuringer & Oleson, 2010), driving green (Pritchard, 2010), cooperation (Nevin, 2010), and web–based children's environmental education (Twyman, 2010). Later, Grant (2011) extends these topics with good data on the negative effects of overpopulation and consumerism, and a range of positive and broadly scaled solution activities that go beyond individualistic interventions. (While worthy of scrutiny, the value of Grant's, and others' analyses suffers to the extent that they occur in unscientific, agential terms.)

After the introductory remarks, paleo–climatologist Lonnie Thompson (2010) sets the stage for that special section with his article entitled "Climate change: The evidence and our options." After reviewing the evidence and discussing the relative merits of mitigation, adaptation, and suffering, Thompson stresses the connection between human behavior and global problems, and their solutions, when he concludes that "There are currently no technological quick fixes for global warming. *Our only hope is to change our behavior* in ways that significantly slow the rate of global warming, thereby giving the engineers time to devise, develop, and deploy technological solutions *where possible*" (p. 168, emphases added). Similarly, Douglas Larson makes the behavior connection when discussing the problems and solutions regarding Devil's Lake in North Dakota (Larson, 2012).

Others have also made the crucial behavior–connection point, in some cases even earlier than Thompson. For example, in a 2007 speech, Frederick A. O. Schwarz Jr., the 17–year leader of the Natural Resources Defense Council, said, "Global warming is the greatest threat we face, but it is not the only threat... Too many wild places are disappearing, too many species are being snuffed out, and too many babies are being born with bodies and brains damaged by man–made chemicals and pollution... To win [these battles]... *we must change how people think—and how they act*" (p. 60, emphasis added). In acknowledging the importance of changing peoples' behavior as part of solving world problems, Schwarz was implicitly encouraging the traditional natural sciences to coordinate with an effective natural science of human behavior in green engineering efforts and the movement toward sustainable lifestyles.

Completing such tasks must be a team effort. The players are the natural sciences of energy, matter, life forms, and life functions (physics, chemistry, biology, and behaviorology) and all the natural science and engineering disciplines related to these, because the complex problems facing humanity, and hence the complex solutions, involve aspects of *all* these disciplines. Will we cooperate in time? In his paper Lonnie Thompson also pointed out, "… our future may not be a steady, gradual change in the world's climate, but an abrupt and devastating deterioration from which we cannot recover" (p. 165). As Thompson describes, we must mitigate the problems while that is still an option, or we will be stuck with adaptation and suffering. The message is clear; we are running out of time for efforts to solve world problems, including developing programs to train more people in *all* the relevant natural sciences, including behaviorology, so that they can work more effectively on solutions.

How much time remains before we are stuck with adaptation and suffering? Research continually shows estimates to be overly optimistic. The 150–year estimate, when behaviorologists were moving on formal independence in 1987, shrank to 100 years about a decade later, and more recently to 50 years *or less!* In a special article for Earth Day 2010—with the subtitle, "Want peace? Solve the energy crisis"—Walter Simpson (2010) makes this point: "Climatologist Jim Hansen said in 2006 that he believed we had just ten years to make substantial progress reversing current carbon dioxide emission trends or we would be unable to avoid the worst consequences of climate change" (p. G2). According to that math, we have until 2016. Hmmm…

However, the question can no longer merely be how much time is left to fix overpopulation and global warming before the worst effects overtake us; various media reports can leave audience members with the distinct impression that the worst effects are *already* beginning to overtake us. This could perhaps lead to various rather ultimate results including a deep population reduction stuck with surviving another 1,000–year dark age during which our best tool, scientific knowledge, may all but disappear and need reinventing while our unhelpful mystical assumptions endure with perhaps only some changes in flavor, a possibility that leaves quite a bad taste in my mouth, as the saying goes. In later chapters we will study knowledge that can help us avoid such scenarios. Right now, though, asking "How long do we have?" retains little value. Instead the reality that, in any case, *"We are running out of time!"* must prompt us continually to move ahead on solutions.

Those solutions require all natural scientists to work together. In part behaviorologists moved decisively for formal independence when they did, so that their science could contribute its share to the expertise and coordinated efforts needed to solve such problems within the necessary time frame; under these circumstances, they considered that their *not* declaring independence, and instead spending much energy over many more, likely fruitless years in further efforts to change psychology, would be essentially a mistake (and possibly irresponsible). In agreement, other natural scientists are welcoming behaviorologists in the coordinated efforts that solving major problems requires.

Further contributions to fellow natural scientists. The behaviorology discipline makes additional contributions to the capabilities of other natural scientists. After becoming basically familiar with behaviorology, scientists in many disciplines are more able to remain naturalistic in dealing with subject matters at the edge of, and beyond, their particular specializations rather than slipping into the compromising use of common, culturally conditioned, superstitious agential accounts. They may also add desirable details to accounts within their specializations. For example, when natural scientists (e.g., Sam Harris or Michael Shermer) say that science accounts for morals and values, mentioning the controlling relations that behaviorology describes for these topics strengthens their point. Also, behaviorology provides the students of natural scientists with a natural–science alternative to the non–natural disciplines that most of these students currently study when covering behavior–related subject matters.

For their part, other natural scientists can also help themselves by contributing to behaviorology not only through increasing their own familiarity with it—which may be particularly valuable in the efforts to solve world problems, as recombination of repertoires research indicates—but also, and especially, through support for the wider availability of academic behaviorology programs and departments. More of these are needed now to increase people's contact with behaviorology so as to reduce or avoid the increased difficulty in solving problems that stems from culturally conditioned susceptibilities to behavior–related superstition and mysticism. This need is difficult to meet because, as a result of the historical circumstances of the origins of their discipline, many academic behaviorological scientists and engineers remain scattered among academic departments of non–natural disciplines. Traditional natural scientists can quickly help solve this problem by promoting the addition of behaviorology courses and programs in their own larger academic units.

For most people a meaningful amount of contact with behaviorology will occur when behaviorology is a requirement in high school science curricula along with the other foundation natural sciences of physics, chemistry, and biology. To achieve that, science teachers must have behaviorology courses available in their college training programs. To make those courses available, faculty to teach them must be trained in this discipline. And for *that* to happen, programs and departments of behaviorology need to become more widely established at colleges and universities. This would also generate increased development of basic research and behaviorological engineering applications, including those contributing to solving personal, local, and global problems on which natural scientists in general are already working together.

While Ledoux (2009) reported the consensus among behaviorologists regarding some departmental curricula for various academic levels, one of the obvious places from which to grow behaviorology programs and departments is from within departments of biology, especially within strictly natural–science schools. Skinner recognized early in his "Behaviorism at Fifty" article that the *natural* science of behavior was an offshoot of biology. As he elaborated the connection in *The Shaping of a Behaviorist* (1979, pp. 16–76) even though he was earning his doctorate through the psychology department at Harvard University in the 1930s, much of Skinner's work occurred under W. J. Crozier who headed the physiology section of Harvard's biology department and who had been associated with biologist Jacques Loeb. Both Crozier and Loeb not only emphasized studying the whole organism, including its movement (behavior), but they also emphasized studying the causal mechanism of selection which Skinner subsequently adapted from biology and applied to behavior. While that process essentially started this natural–science discipline, modern behaviorology now features its own level of analysis and can stand alone on its own disciplinary merits. These disciplines complement each other, but are not logically dependent. Consequently, a biology department would be only a good temporary home for behaviorology.

Conclusion

In its second 50 years, the value and legacy of behaviorism broadened substantially. The natural science that Skinner's radical behaviorism supports and informs has emerged as an extensive, multi–faceted discipline, although its independence as behaviorology began only about a quarter–century ago. Its academic homes will continue to expand due to the repeatedly documented effectiveness of approaching human behavior naturalistically. Other disciplines also faced similarly difficult circumstances in their early histories and prevailed. The astronomical discoveries of Galileo 400 years ago helped move our human home, the Earth, beyond superstitious, mystical accounts. The biological discoveries of Darwin 150 years ago helped move the human body beyond superstitious, mystical accounts. And, based on the naturalism of Skinner's radical behaviorism, the current discoveries of behaviorological science help move human nature and human behavior beyond superstitious, mystical accounts. On this basis our continuing efforts not only improve effective scientific thinking about all subject matters including human nature and human behavior, but also reduce reliance on superstition across the worldwide culture, expand reliance on naturalism, science, and engineering, and increase success in solving personal, local, and world problems.❧

Endnotes

The original article in *American Scientist,* and the unabridged peer–reviewed version in *Behaviorology Today,* each featured a set of notes. Here are relevant portions of these notes:

The author (at ledoux@canton.edu) is Professor of Behaviorology at the State University of New York at Canton (www.canton.edu). He earned his ph.d. in *The Experimental Analysis of Behavior* in 1982 from Western Michigan University, and has taught behaviorology in Australia, China, and the usa.

Some of the papers and books we mentioned in this chapter, and links for many of the organizations and journals related to this chapter, are on www.behaviorology.org (the web site of tibi, The International Behaviorology Institute) and on www.americanscientist.org (the web site of *American Scientist,* the journal of Sigma Xi, the Scientific Research Society). Others are available through www.behavior.org (the web site of the Cambridge Center for Behavioral Studies) or www.abainternational.org (the web site of abai).

For helpful comments on earlier drafts, the author thanks many colleagues from over 15 institutions and agencies. These include Barry Berghaus (behaviorologist), Paul Chance (behavior analyst), Walter Conley (biologist), John Ferreira (behaviorologist), Lawrence Fraley (behaviorologist), Michael Hanley (behavior analyst), Feng Hong (physicist), Philip Johnson (*Journal of Behaviorology* editor), Joseph Kennedy (biologist), Marc Lanovaz (behaviorologist), Jerry Lin (mathematician), Werner Matthijs (behaviorologist), David Schoonmaker (*American Scientist* editor), Catherine Shrady (geologist), Donn Sottolano (behaviorologist), Jeffrey Taylor (biologist), Deborah Thomas (behaviorologist), William Trumble (biologist), and several anonymous reviewers.

The *Key Words* for indexing the article version of this chapter were behaviorology, behaviorism, human nature, B. F. Skinner, human behavior, evolution, neural behavior, consciousness, natural science, science education, and global warming.

References (with some annotations)

Baum, W. M. (1995). Radical behaviorism and the concept of agency. *Behaviorology, 3* (1), 93–106.

Cautela, J. R. (1994). General level of reinforcement II: Further elaborations. *Behaviorology, 2* (1), 1–16.

Cheney, C. D. (1991). The source and control of behavior. In W. Ishaq (Ed.). *Human Behavior in Today's World* (73–86). New York: Praeger.

Daniels, A. C. (1989). *Performance Management (Third Edition, revised).* Tucker, GA: Performance Management Publications. (This publisher, now in Atlanta, GA, released a fourth edition in 2006 with J. E. Daniels as coauthor.)

Epstein, R. (1996). *Cognition, Creativity, and Behavior.* Westport, CT: Praeger.

Eshleman, J. W. & Vargas, E. A. (1988). Promoting the behaviorological analysis of verbal behavior. *The Analysis of Verbal Behavior, 6,* 23–32.

Ferreira, J. B. (2012). Progressive neural emotional therapy (PNET): A behaviorological analysis. *Behaviorology Today, 15* (2), 3–9.

Ferster, C. B. & Skinner, B. F. (1957). *Schedules of Reinforcement.* Englewood Cliffs, NJ: Prentice–Hall.

Fraley, L. E. (1983). The behavioral analysis of *Mens Rea* (doctrine of culpable mental states). *Behaviorists for Social Action Journal, 4* (1), 2–7. (See Fraley, 2013.)

Fraley, L. E. (1987). The cultural mission of behaviorology. *The Behavior Analyst, 10,* 123–126. Expended versions appear in Fraley, 2012 and 2013.

Fraley, L. E. (1994a). Uncertainty about determinism: A critical review of challenges to the determinism of modern science. *Behavior and Philosophy, 22* (2), 71–83.

Fraley, L. E. (1994b). Behaviorological corrections: A new concept of prison from a natural–science perspective. *Behavior and Social Issues, 4,* 3–33. (See Fraley, 2013.)

Fraley, L. E. (2006). The ethics of medical practices during protracted dying: A natural–science perspective. *Behaviorology Today, 9* (1), 3–17. (See Fraley, 2012.)

Fraley, L. E. (2008). *General Behaviorology: The Natural Science of Human Behavior.* Canton, NY: ABCs. While this book details many topics, these may be of particular interest to readers, especially those who have mastered basic disciplinary principles, concepts, methods, and practices through *Running Out of Time:* explanatory fictions (Chapter 4); stimulus equivalence (Chapter 16); attitudes, values, rights, ethics, morals, and beliefs (Chapter 25); verbal behavior (Chapter 26); consciousness (Chapter 27); person, life, and culture (Chapter 28); reality (Chapter 29); and robotics (Chapter 30).

Fraley, L. E. (2012). *Dignified Dying—A Behaviorological Thanatology.* Canton, NY: ABCs.

Fraley, L. E. (2013). *Behaviorological Rehabilitation and the Criminal Justice System.* Canton, NY: ABCs.

Fraley, L. E. & Ledoux, S. F. (2002). Origins, status, and mission of behaviorology. In S. F. Ledoux. *Origins and Components of Behaviorology— Second Edition* (pp. 33–169). Canton, NY: ABCs. This multi–chapter paper also appeared across 2006–2008 in these five parts in *Behaviorology Today* (the precursor to *Journal of Behaviorology):* Chapters 1 & 2: *9* (2), 13–32. Chapter 3: *10* (1), 15–25. Chapter 4: *10* (2), 9–33. Chapter 5: *11* (1), 3–30. Chapters 6 & 7: *11* (2), 3–17.

Grant, L. K. (2011). Can we consume our way out of climate change? A call for analysis. *The Behavior Analyst, 34* (2), 245–266.

Hawking, S. & Mlodinow, L. (2010). *The Grand Design.* New York: Bantam. (Especially see Chapter 3.)

Hayes, S. C. & Brownstein, A. J. (1986). Mentalism, behavior–behavior relations, and a behavior analytic view of the purposes of science. *The Behavior Analyst, 9,* 175–190.

Johnson, P. R. (2012). A behaviorological approach to management of neuroleptic–induced tardive dyskinesia: Progressive neural emotional therapy (PNET). *Behaviorology Today, 15* (2), 11–25.

Larson, D. W. (2012). Runaway Devil's Lake. *American Scientist, 100* (1), 46–53.

Latham, G. I. (1994). *The Power of Positive Parenting.* Logan, UT: P & T ink.

Latham, G. I. (1998). *Keys to Classroom Management.* Logan, UT: P & T ink.

Latham, G. I. (1999). *Parenting with Love.* Salt Lake City, UT: Bookcraft.

Latham, G. I. (2002). *Behind the Schoolhouse Door: Managing Chaos with Science, Skills, and Strategies.* Logan, UT: P & T ink.

Ledoux, S. F. (2009). Behaviorology curricula in higher education. *Behaviorology Today, 12* (1), 16–25.

Ledoux, S. F. (2010). Multiple selectors in the control of simultaneously emittable responses. *Behaviorology Today, 13* (2), 3–27.

Ledoux, S. F. (2012a). Behaviorism at 100. *American Scientist, 100* (1), 60–65. *American Scientist* published this paper with introductory excerpts on pages 54–59, which the editor listed as an *"American Scientist* Centennial Classic 1957,"* from: Skinner, B. F. (1957). The experimental analysis of behavior. *American Scientist, 45* (4), 343–371. (Skinner's complete 1957 paper was also available online with this paper.)

Ledoux, S. F. (2012b). Behaviorism at 100 unabridged. *Behaviorology Today, 15* (1), 3–22. This peer–reviewed version of the paper included the material that the *American Scientist* editor had set aside at the last moment to make more room for the Skinner article excerpts that accompanied the original article in *American Scientist.* (*American Scientist* also posted this unabridged version of the paper online with the original abridged version.)

Maurice, C., Green, G., & Luce, S. (Eds.). (1996). *Behavioral Intervention for Young Children with Autism.* Austin, TX: Pro–Ed.

Michael, J. L. (1982). Distinguishing between discriminative and motivational functions of stimuli. *Journal of the Experimental Analysis of Behavior, 37,* 149–155.

New York State Department of Health—Early Intervention Program. (1999). *Clinical Practice Guideline: Autism / Pervasive Developmental Disorders, Assessment and Intervention for Young Children (Age 0–3 Years).* Albany, NY: New York State Department of Health. The complete report comes in three parts: (a) *Report of the Recommendations:* Publication No. 4215, (b) *Quick Reference Guide:* Publication No. 4216, and (c) *The Guideline Technical Report:* Publication No. 4217.

Peterson, M. E. (1978). The Midwestern Association of Behavior Analysis: Past, present, and future. *The Behavior Analyst, 1* (1), 3–15.

Peterson, N. (1978). *An Introduction to Verbal Behavior.* Grand Rapids, MI: Behavior Associates.

Schwarz Jr., F. A. O. (2008). *Onearth,* Spring.

Seo, H. & Lee, D. (2009). Behavioral and neural changes after gains and losses of conditioned reinforcers. *Neuroscience, 29,* 3627–3641.

Sidman, M. (1960/1988). *Tactics of Scientific Research.* New York: Basic Books. Authors Cooperative, in Boston, MA, republished this book in 1988.

Sidman, M. (1994). *Equivalence Relations and Behavior: A Research Story.* Boston, MA: Authors Cooperative.

Sidman, M. (2001). *Coercion and its Fallout—Revised Edition.* Boston, MA: Authors Cooperative.

Simpson, W. (2010 April 18). Earth Day 2010. *The Buffalo News,* G1–G2.

Skinner, B. F. (1948). *Walden Two.* New York: Macmillan. Reissued in 1976. (See the comments in the Bibliography.)

Skinner, B. F. (1953). *Science and Human Behavior.* New York: Macmillan. The Free Press, New York, published a paperback edition in 1965.

Skinner, B. F. (1957a). *Verbal Behavior.* New York: Appleton–Century–Crofts. The B. F. Skinner Foundation (www.bfskinner.org) in Cambridge, MA, republished this book in 1992.

Skinner, B. F. (1957b). The experimental analysis of behavior. *American Scientist, 45* (4), 343–371.

Skinner, B. F. (1963). Behaviorism at fifty. *Science, 140,* 951–958.

Skinner, B. F. (1974). *About Behaviorism.* New York: Knopf.

Skinner, B. F. (1979). *The Shaping of a Behaviorist.* New York: Knopf.

Skinner, B. F. (1983). *A Matter of Consequences.* New York: Knopf.

Thompson, L. (2010). Climate change: The evidence and our options. *The Behavior Analyst, 33* (2), 153–170.

Watkins, C. L. (1997). *Project Follow Through: A Case Study of Contingencies Influencing Instructional Practices of the Educational Establishment.* Cambridge, MA: Cambridge Center for Behavioral Studies.

Watson, J. B. (1913). Psychology as the behaviorist views it. *Psychological Review, 20,* 158–177.

Wyatt, W. J. (1997). *The Millennium Man.* Hurricane, WV: Third Millennium Press. (See the comments in the Bibliography.)

Supplemental References: Special Section on the Human Response to Climate Change (in order of appearance)

Heward, W. L. & Chance, P. (2010). Introduction: Dealing with what is. *The Behavior Analyst, 33* (2), 145–151.

Thompson, L. (2010). Climate change: The evidence and our options. *The Behavior Analyst, 33* (2), 153–170.

Keller, J. J. (2010). The recycling solution: How I increased recycling on Dilworth Road. *The Behavior Analyst, 33* (2), 171–173.

Layng, T. V. J. (2010). Buying green. *The Behavior Analyst, 33* (2), 175–177.

Malott, R. W. (2010). I'll save the world from global warming—tomorrow: Using procrastination management to combat global warming. *The Behavior Analyst, 33* (2), 179–180.

Neuringer, A. & Oleson, K. C. (2010). Helping for change. *The Behavior Analyst, 33* (2), 181–184.

Pritchard, J. (2010). Virtual rewards for driving green. *The Behavior Analyst, 33* (2), 185–187.

Nevin, J. A. (2010). The power of cooperation. *The Behavior Analyst, 33* (2), 189–191.

Twyman, J. S. (2010). TerraKids: An interactive web site where kids learn about saving the environment. *The Behavior Analyst, 33* (2), 193–196.

Chance, P. & Heward, W. L. (2010). Climate Change: Meeting the challenge. *The Behavior Analyst, 33* (2), 197–206.✑

Chapter 2
A Natural Science of Behavior

Resources and the Nature of Science

Living on the edge of a small town, I write this as a bright February sun illuminates nearby farm fields that are already devoid of snow. Deer pass through the unfenced yards between well–spaced homes in the twilight hours of morning and evening, on their way to or from an undeveloped area at the center of the next block, and one or another dog can always be seen pulling its "master" along the street. The resplendent view and clean air invite one to take a walk even without a pet, but a short distance in the stiff breeze, with the temperature below freezing, quickly compels a return to the warm side of the window with a cup of hot cocoa in hand. This is no idle scene; after most chapters, you will be able to return to this location and count the increasing number of our topics that make cameo appearances in this event snapshot just as they appeared more explicitly in the historical overview of Chapter 1.

Before we begin investigating our topics, however, we first need to consider a problem that we will face regularly regarding the examples that we will use, particularly our human–behavior examples. Since behaviorology concerns all behavior, in early chapters and with basic principles and processes, simple as well as non–human behavior occasionally provides the examples best illustrating a particular point. However, the realistic elucidation of human behavior remains our primary focus. The difficulty with human behavior examples (and many simple and other–animal behavior examples also) centers on their inevitable complexity. Virtually every realistic example contains numerous factors and effects, many often interacting with each other. While we report an example to illustrate one or another concept, these other factors and effects *continue to demand explanation as well.* To meet such demands, we would have to put the whole book into each paragraph that contains an ordinary (i.e., a complex) human behavior example, which is of course impossible.

Resource Options

Instead we will remain ever patient, going without answers until satisfactory ones take a turn in this telling. We thereby avoid oversimplified examples that easily mislead by falsely implying that the principles and concepts covered in early chapters apply only to limited areas of behavior such as human abnormal behavior or circus–animal training. Many examples, for instance, will include some aspect of verbal behavior that is not itself the point. We gradually build up to a chapter on verbal behavior in the second half of Part II, as each chapter adds to the foundation for later chapters. Proceeding then through the complete sequence of chapters builds your repertoire respecting as many explanatory concepts and principles as an introductory book can cover and, as we proceed, you will become more comfortable that we have indeed covered an increasing number of those factors and effects, enabling you to supply the additional accounts. Like our opening event snapshot, you can even return to early examples after finishing the book to provide the rest of the story for nearly every particular example of interest.

For examples with complexities beyond the scope of this introductory behaviorology book, you can turn to more elaborate treatments. While this book contains many helpful examples, it is not loaded with numerous intricately detailed examples, because these exist elsewhere, and at appropriate points I will provide explicit references. But the point of this book is not to tell the total story; instead the point is to survey the variety of basic topics in preparation for the more thorough treatments later that others supply.

The most extensive examples of very complex human behavior best serve the role of conditioning the kind of comprehensive repertoire that a reader would need to teach, or to apply, behaviorology. For that kind of elaborate treatment, for which this book prepares a reasonable foundation, I recommend turning, after this book, to the 1,600–page, 30–chapter, three–course textbook by Dr. Lawrence Fraley (Fraley, 2008). That book is entitled, *General Behaviorlogy— The Natural Science of Human Behavior.* The book you hold introduces the topics of this more comprehensive text, in a somewhat similar order, but with less depth of coverage and less extensive examples. Both these books, however, feature some repetitious examples. Such a pattern enables a gradual buildup of the necessary components of complex principles and concepts, rather typical of the overall pattern of most natural–science curricula.

However, responses among natural scientists—since they are behaving organisms whose behavior is a function of the same laws that govern the behavior on non–scientists—seldom show perfect agreement. Differences can even grow into disagreements, sometimes loud and even rancorous, which shows another very human side of science. Such situations fail to fall apart completely though, because natural science maintains an abiding characteristic, namely that it is *self correcting;* in time the adherence to naturalism coupled with the accumulation of additional data steers the general consensus of the often vast majority of natural scientists toward the best knowledge available at that time. This process works, because it continually cycles through resolution to more differences to better data to further resolution to more differences, and so on, all under the ongoing guidance of naturalism as the basic philosophy of natural science. This even applies to me and Professor Fraley as the respective authors of this book and the *General Behaviorology* book.

Out of respect for full disclosure, no one should be surprised if Fraley and I have in the past shared some few small differences on certain technical aspects of our natural science discipline; after all, we have known each other for several decades, and in that time we have collaborated on numerous projects and publications, large and small. We still share some such differences. What could be more naturally human? Scientific self–correction has resolved the earlier differences, and is in the process of resolving currents ones. Meanwhile, I will alert you when we cover topics about which differences currently exist, even though none of our differences has ever grown into any sort of volatile disagreement. Perhaps such examples will help readers manage differences that they experience, before the differences grow into problematic disagreements.

For readers who relish real challenges beyond these books, a number of other available books, building on these introductory and general disciplinary texts, handle the details of carrying this science into particular behavior areas. As one example, the topic of dignified dying—considered from the perspective of a behaviorological thanatology—affects everyone everywhere at some point, especially the terminally ill because our culture currently provides little respect and dignity for those in such circumstances (Fraley, 2012). Another example involves a behaviorological analysis of criminal justice and rehabilitation that may directly affect fewer folks; but at times, and in certain ways, this topic also proves to be of vital social significance to everyone (Fraley, 2013).

The **Behavior** *Nature of Science*

We begin our topics with the general notion of science. When most people hear the word "science," they immediately think of one or more of the traditional *natural* sciences (e.g., physics, chemistry, biology, astronomy, geology). As our coverage is more explicit than that general usage, most of the time we will use the complete term, "natural science." On those occasions when we simply say "science," we still mean natural science, unless something else is clearly specified. With that as a given, consider what is perhaps a most basic and yet profound characteristic of science, a characteristic that only began to demand attention in the last few decades, especially with the rise of the natural science of behavior. What is this characteristic?

Science provides the foundations for a broad range of both beneficial—and occasionally damaging—products for humanity. The news media keep scientific and engineering developments in the public eye, although sometimes they focus excess attention on occasional squabbles over who reached some milestone first, or the luckily even more rare occurrences of data faking. Why do such things happen? All of these—developments and squabbles and faking—happen because they all share that fundamental characteristic of science demanding our attention: *Science is behavior,* and the variables responsible for behavior are responsible for those things happening. As you become more familiar with behavior and the variables responsible for it, those things happening will become less surprising and more comprehensible.

Consider science as behavior more specifically. Science is the thoughts, emotions, feelings, and muscle movements—all of which are behaviors—of those whom we call scientists (including me and maybe you and also the applied scientists we call engineers), and all these behaviors occur under control of the same sorts of "causes" (technically, functional independent variables) that affect the behaviors of everyone else (including other animals) as well. Recognizing the nature of science *as behavior* helps us examine some aspects and constraints not only of scientific endeavors but also of any and all human endeavors, from the most mundane to the truly grand.

As with all behaving organisms, scientists are organisms whose behavior is entirely natural. That means it is neither magical nor spontaneous. Instead it occurs due to real, measurable variables. Nature contains a wide range of variables that naturally affect behavior in ways that scientific behaviors can discover and apply for humanity's benefit. These scientific behaviors include both respecting naturalism (the name of the general philosophy of science of the natural sciences) and experimentally analyzing phenomena, as well as both disseminating findings and developing them into practical products or procedures. While touching first on the question of philosophy, and returning to it as needed, we will soon begin with experimental analysis (while this book exemplifies the dissemination that prompts practical developments).

Constraining Assumptions Enhancing Science

How does experimentally analyzing any phenomenon happen? Before answering that, we should first consider some of the assumptions behind that activity. First we examine some assumptions behind the natural sciences in general; then we will look at some assumptions behind the natural science of behaviorology in particular.

Some Assumptions of General Natural Science

Taken together, the assumptions behind the natural sciences in general go by the name of *naturalism,* which arises not off–the–cuff or from appeals to tradition or common lore (as is the case with assumptions behind any variety of mysticism) but from several hundred years worth of continuously accumulating experimental evidence and successful engineering practices and products. While parts of our global culture are quite enamored of one or another flavor of mysticism or superstition (some theological and some secular) all parts of the culture depend heavily on natural science to deal with everything that is real. Some may quibble over whether the existence of the universe depends on the activity of some divine being or on the efforts of the flying spaghetti monster—although others see it as due to measurable physical processes—but we *all* turn to science as much as possible for food, clothing, shelter, medicine, sanitation, and so forth.

Measurability. A major starting point (i.e., a major assumption) in the effort to deal with whatever is real is that, at least in theory, things that are real are measurable along various dimensions such as distance, mass, size, and temperature, and things that are so measurable are real. If measurement of an event (or an item or a process) is not even theoretically possible, then the event (or item or process) is apparently not real. We say "apparently," because science *assumes* reality only for things that are measurable; a lack of measurability places the status of the event (or item or process) *outside* the realm of the natural, and thus outside the realm of science; there, no one is actually able to deal with it, although some make unsupportable contrary claims.

Natural functional history. Another assumption is that measurable events result not from magic or spontaneity but from natural histories that accumulate from the continuous chains of functionally related events behind the one of concern. And we can trace some of that recent history, which provides the basis for predicting subsequent events and, to the extent that we have access and can get our hands on the preceding events at some relevant level, we can exert control over the subsequent events. For example, say you walk into your kitchen and see a broken glass on the floor with milk splashed all over the mat in front of the sink. Any notion that the glass and milk magically got there spontaneously proves quite unhelpful if you are trying to figure out what happened as part of avoiding further occurrences. Doing that is easier if paw prints track across the counter above the spill; from such clues you can sort out the major parts of the event's likely natural functional history, and on that basis perhaps organize a predictably successful prevention strategy (while cleaning up the mess).

The direction of such natural histories not only defines time, which has existence only from the sequences of events that we say fill it (although there is no "it" to be filled) but the direction helps us to relate events to each other as well as to account for them on multiple levels of analysis as various disciplines coordinate their parts of the complete account of events. While you may have surmised from the evidence that the cat knocked the milk glass off the counter, and are now better able to predict and control whether or not that happens again, a chemist can tell you why the mat stinks later (after your cleanup left a milky residue in it) and a physicist can explain what ion exchanges are happening at the atomic level that leave you wanting to trash the mat.

Determinism. Sometimes the speed at which events transpire, or their complex interactions, outpaces our attempts to keep track of their measurements, making them seem chaotic or even somehow random. Another basic assumption comes into play here, namely that the universe and all the events in it are presumed to be orderly and lawful in their relations to each other. We presume that nature determines every event completely, that no event is arbitrary, spontaneous, or capricious. This does not mean that everything *is* orderly and lawful, which is something that we really cannot know for sure (the nature of knowledge being a very advanced topic for a much later chapter on reality). Instead we work under this assumption, because we have experienced far more success staying alive and dealing with the world when we have worked under it than we had ever experienced without it. The name for this assumption is *determinism*.

Furthermore, when speed or complexity makes events seem random or chaotic, we take the problem as *not* due to nature but as due to the limitations of our measurement repertoires and technologies, limitations which inevitably leave us with a certain amount of ignorance in that we cannot successfully account completely for the apparent randomness of the events. This difficulty endures because, even as we get better at measuring, our measurement technologies always seem to lag a bit behind the ever increasing complexity of the events up for measurement. Indeed this turns out to be such a common occurrence that we have developed several strategies to manage our residual ignorance, such as probability or chaos theories or statistical analysis. The competence of these strategies can itself be a source of errors, though, in that these strategies can leave us claiming that some merely fast or complex part of nature *really is* somehow random or undetermined and thus responsible for the ignorance that actually stems from our underdeveloped measurement techniques. This is something for which we must be continuously vigilant even though some events at the smallest levels of physical analysis may exhibit characteristics that could require certain allowances or adjustments in the basic determinism assumption (see Fraley, 1994, for additional details). Nevertheless, at levels beyond these, particularly at the levels of general experience, this assumption maintains a vital place in natural sciences.

While moving from the various assumptions behind the sciences in general to those behind behaviorology, consider this related connection. A couple of centuries ago, some cultural groups argued over whether or not we should obey the natural laws governing physical events, but gradually the reality settled in that we really have no option in the matter. More recently some similar groups argued about whether or not we should obey the natural laws governing behavior, but again the reality is settling in that we really have no option in this matter either. Behaviorology is the natural science that brings that reality to the forefront, enabling knowledge of behavior, and the variables of which it is a function, to help humanity stay alive and deal with the world.

All members of all species, including the human species—past, present, and future—are simply (actually, "complexly" would be more apt) wondrous sets of naturally evolved and elaborating physical and chemical processes and systems that unfold structures (i.e., a physiology) that share characteristics evoking biologically oriented verbal responses that summarize this complexity with exclamations including "life" and "it's alive!" Our interest in behaviorology resides in the way it clarifies through experimentation so many of the details, at the analytical level of environment/behavior interactions, regarding how natural laws pertain to the effects that energy changes have on some of those systems such that they mediate the phenomena called behavior. But what are the assumptions behind that clarification?

Some Assumptions of Behaviorology

While *naturalism* refers to those general assumption for all natural sciences, individual science disciplines often maintain further assumptions related to their particular subject matters. The assumptions of the natural sciences apply to the behavior subject–matter discipline as well, but it also has some subject–matter–specific assumptions. The systematized collection of additional assumptions behind behaviorology, as the natural science of behavior, go by the name of *radical behaviorism,* and several books elucidating its details are available for your perusal (e.g., Skinner, 1974; or perhaps Moore, 2008). Here we mention but a few of its salient characteristics.

Parsimony. One assumption of behaviorology, which is also shared with all other natural sciences, concerns the necessity to work with the simplest yet adequate explanation of an event. We call this the assumption of *parsimony,* and it helps prevent waste by dissuading researchers from falling for overly complex accounts too quickly, because complex accounts are usually more costly and more difficult—perhaps even impossible—to test or evaluate experimentally. On the other hand, when new data shows inadequacies in a current account, one should move to test only the next feasible and testable increment of complexity, and if it tests out as adequate, then the account with that increment becomes the current best explanation of an event. If it does not, then you test the next increment for adequacy, and so forth, until you can again account for the event adequately.

Parsimony then cuts one way or the other, because an account can be unparsimonious either by being an unnecessarily complicated (and possibly even untestable) account of an event, or by being, or becoming, inadequate as an account of an event. Of course, some accounts are by definition in violation of parsimony, such as mystical accounts, which offer events that are unreal as explanations thereby eliminating the possibility of testing. Behaviorology stands firm with all other natural sciences in refusing to allow metaphysical events to have any abiding presence in explanatory accounts. Be aware, however, that different disciplines have different needs for attention to parsimony.

Traditional natural sciences generally have little need for attention to parsimony; they take parsimony for granted, due both to their ingrained centuries–long history of practice abiding by it, and to the gradual rate at which they accumulate the range of phenomena that each one covers. For example, physicists in the 1700s barely had any notion of the vastness of the universe or the minuteness of the sub–atomic realm. These subject matter components accumulated gradually as research horizons ever expanded, usually due to improvements in measurement technologies; if you can measure something, then go for it, explore it, discover whatever you can about it. Of course, no inner agents are telling scientists to measure, explore, or discover. These activities are naturally occurring behaviors, and in later chapters we will consider the many variables of which such scientific behavior, along with all other behavior, is a function.

To anticipate the differences better, between traditional natural science disciplines and those certain disciplines needing greater attention to parsimony, consider that physicists in the 1800s operated on the level of Newtonian mechanics; their experience provided no need or option for anything like quantum mechanics or relativity. But their 1900s counterparts could not operate without quantum mechanics or relativity. The jump from Newton to quantum actually resulted from a precision inadequacy that developed for Newtonian mechanics due to measurement–technology refinements leaving the Newtonian formula predictions off by amounts greater than the error margins inherent in the newer measurement equipment. Quantum–mechanics equations produced more precise predictions than Newtonian–mechanics equations, and the newer equipment verified the quantum–mechanics predictions, leaving quantum mechanics more adequate, and hence parsimonious, than Newtonian mechanics. The process continues today (and tomorrow) with various string theories trying to manage the mathematical discrepancies between quantum mechanics and relativity. For an easy–to–read example of related phenomena, see Stephen Hawking and Leonard Mlodinow's *The Grand Design* (2010). The point is that all this happened without much reference to parsimony because physicists—typical of traditional natural scientists—would have trouble, due to their extensive history, dealing with such developments in any way other than parsimoniously, and because they never had to confront the *whole* range of their subject matter all at once, without tools to deal with it.

On the other hand, any discipline that deals with behavior (e.g., theology, psychology, or behaviorology) confronts the whole range of its subject matter right from the start, from eating to emoting to speaking to thinking, basically simple to impressively complex, all on comprehensive continua encompassing so–called normality to abnormality. Meanwhile society always and ever before and now demands complete accounts and effective applications regarding all these behaviors all at once, regardless of whether or not practitioners had workable tools to produce those "complete accounts and effective applications." Under contingencies like these, we should not be surprised when some disciplines studying behavior always fall incessantly for overly complex accounts (e.g., minds, psyches, selves, souls, or some other types of putative behavior–initiating self agents). These unparsimonious accounts can sound so impressive to those seeking answers that the seekers become satisfied, and even excessively defensive, about their new–found answers even though these answers actually provide little by way of working practical applications or interventions. Thus we can see why these disciplines need to give explicit attention to parsimony and adhere to it; but some of them either deny this outright, or perhaps they cannot afford the changes that adhering to parsimony would bring.

Behaviorology became an independent discipline in part because it opposed the commitments of non–natural behavior–oriented disciplines to unparsimonious accounts, and stood with the traditional natural sciences regarding parsimony as a component of their shared naturalism. In addition, though, behaviorology also has some further assumptions, specific to its subject matter, under its informing philosophy of science, radical behaviorism.

Four radical behaviorism characteristics. Of the various assumptions that characterize the radical behaviorist philosophy of science, four stand out for the role they played in the emergence of the behaviorology discipline. Due to the coverage they receive throughout the pages of this book, we will only touch on them briefly here for their stage–setting value. (Some of the papers in Ledoux, 2002a, also deal with these matters.)

(a) The first assumption regards behavior as a *natural* phenomenon. As part of respecting the continuity of events in space and time which, in natural sciences, accumulates as a researchable natural history, we see behavior not as originating in the machinations of one or another internal mystical self agent but as a naturally occurring phenomenon under discoverable laws of nature just like any other real phenomenon. This enables prediction and control of the behavior subject matter, which provides a firm foundation for developing beneficial behaviorological engineering applications and interventions.

(b) The second assumption takes *experimental* control of relevant variables as the most appropriate basis for analysis of environment–behavior relations leading to the application of that control in culturally valuable ways. Other kinds of control also contribute although in less extensive ways.

(c) The third assumption specifies that the same full range of variables affecting other people's behavior also affects the behavior—scientific or not—of scientists, because they also are behaving organisms, and that among those variables one of the most pertinent is scientists' philosophy of science, because it exerts a kind of quality control over other scientific activities. For example, preventing mystical assumptions from misinforming the selection of research questions, or the interpretation of research results, is a role that scientists' philosophy of science plays. This, of course, bears implications for other natural sciences as well as behaviorology. For example, centuries ago no fully developed naturalism informed the activities of the forerunners of physicists; instead the religious philosophy of the day informed their activities, such that they spent (Wasted?) plenty of time trying to figure out how many angels can dance on the head of a pin.

(d) The fourth assumption, perhaps the most important but also the most complicated, recognizes most private events, such as thinking, seeing, observing, or emoting, as most productively construed as covert, neural behaviors that are involved in the same lawful relationships that involve overt, muscle behaviors (and experimental data and interpretation, described in later chapters, have accrued in support of this assumption). Private events are lawful in the same way that one would regard public events. The skin presents no special boundary to the laws of the universe, and the nature of events on both sides of the skin is the same, with events measured in the same natural–science ways (although if we need access to the covert events, attaining that access usually involves welcome coordination with physiology colleagues).

However, behaviorologists cannot grant scientific status to private events that someone invents to be causes of behavior, such as internal hypothetical constructs conjured up conveniently with just the right characteristics to explain that behavior. Nor can we take real private events as the initiating causes of behavior, because we analytically pursue any causal chain to other, outside events for the sake of control in our subject matter. Instead, we see behavior, on the overt level, as neurologically based actions of the glands and muscles (both smooth and striped) while we see private events as covert behaviors observable and reportable only by the person, as a public–of–one, in whom they play an eliciting or evocative role. These covert behaviors involve only neurological–level events; the behavior of "seeing in the absence of the thing seen" is one example. (This coverage is brief, because you can find details appropriate for our current analysis level back in Chapter 1. In later chapters we will return to this topic for greater depth, a detail–building repetition pattern that we will follow for many topics, as is typical of education in the natural sciences.)

Single–subject designs. Other concerns inseparably intertwine with those characteristics of radical behaviorism, from sharing with other natural sciences the refusal to allow metaphysical events to enter explanatory accounts, and the necessity of parsimony in accounts of human behavior, to a preference for single–subject experimental designs rather than group statistical designs. While seeming more like a matter of experimental methodology than of philosophical assumptions, a vital philosophical connection supports the selection of research methodology. Single–subject designs and radical behaviorism have gone rather hand in hand throughout the history of natural behavior science, starting with the experimental preparations of Skinner's early laboratory at Harvard in the 1930s and continuing up through today's laboratory and applied research efforts. The most important reason for this interplay stems from the philosophical assumptions about the source of experimental variability that behaviorologists find inhering in different methodologies.

On the one hand, behaviorologists see some researchers, especially from non–natural disciplines (e.g., psychologists) assuming that observed variability arises from the vagaries and capriciousness of some putative inner agent posited as initiatively responsible for a behavior of concern. The mystical status of that kind of variability automatically curtails direct access to that variability. Thus, one cannot reduce this variability. The most commonly accepted method to deal with variability from this kind of source is a group statistical design of one sort or another. In these methods, the mathematical manipulation of the data supposedly spreads out the variability across a large number of randomly selected subjects so that functional relations can emerge in the results.

On the other hand, since behaviorologists, as natural scientists, cannot grant status to putative inner agents, those agents cannot be the source of observed variability, which implies that variability stems from something else. Whatever that is, can group statistical designs deal with it? Can a different methodology deal with it better, perhaps by reducing it?

Looking back as far as Skinner's early work, behaviorologists identify a different source for variability, one related to how thoroughly, or not, we exert experimental control over the functional variables relevant to a behavior under study. Starting out simply, we began working with handfuls of subjects, three to six "per experiment"; actually three subjects would really be three experiments, as we consider the behavior of each subject individually, because behavior is primarily a phenomenon of individual organisms. If our experimental arrangements actually controlled every variable relevant to the behavior of concern, the results would match predictions. But this is not the case, as we rarely control all those variables. So predictions are always off by some amount, and that amount indicates the variability, which we thus see as arising from the effects of the variables over which we did not exert experimental control.

This kind of variability is not a characteristic of nature; rather it derives from the incompleteness of our control over functional variables. The variability is larger when we control only a few relevant variables (leading to greater measured differences between predictions and outcomes) and smaller when we control more relevant variables (leading to smaller measured differences between predictions and outcomes) because the variability stems from the variables we are ignoring in the sense of not taking them into account.

Realistically, economics must enter this picture, because taking variables into account is expensive, not just in terms of energy and some other resources, but in terms of funding. Generally, the greater the number of variables taken into account (i.e., measured and either held steady or manipulated) the higher the associated monetary expenses. The more important the experimental question, the more of these costs we must bear to take more of the relevant variables into account to answer the question more thoroughly; this reduces variability and increases the success of prediction, control, interpretation, and application (and hence in some important ways justifies the increased funds society must authorize to cover the expenses). Conversely, for less important questions, since control is costly, we settle for controlling fewer variables and so must tolerate more variability along with the associated reductions in prediction, control, and so on. We must, of course, be careful not to let costs determine the importance of questions, but the point is that the problem of variability resides not in the magic of made up inner agents but in the amount of residual ignorance we tolerate stemming from the decreased amount of our experimental control over the variables in the natural history of some events.

Variability from that source then affects the selection of experimental methodology. We have no need to smooth out variability from discredited capricious inner agents across subjects in groups (which in any case tells us little about the individual behavior of those in the groups); rather we only need to adopt the methods that most effectively control, for a given amount of research funds, the most variables affecting a single subject (with three to six subjects studied, thereby providing a minimal level of replication, reliability, and generality for the results). Since the single–subject designs that we adopted early on fill this bill quite well, we continue to emphasize them in the natural science of behavior.

Anticipating some of our later chapters detailing methodology, the two most common and effective types of single–subject designs are the "ABAB" design and "Multiple–Baseline" designs. In both designs the subjects serve as their own control across different experimental phases containing different conditions, and we accept the data as documenting experimental control of the relevant variables when the data show that behavior changes when, and *only* when, the experimental conditions change.

In the ABAB design, the conditions of the experimental phases follow a repeating pattern, which we sometimes also call a "reversal" design. After condition A comes condition B; then conditions first revert to A and then revert back to B. Experimental control occurs when the data show the behavior changing when, and only when, the phases change. You will likely find the ABAB design emphasized in the repertoire of basic laboratory research scientists working, often with other animals as well as humans, to discover the various effects of all the variables related to behavior. Sometimes, however, the repetition of phases is unethical or otherwise not possible. The Multiple–Baseline design then becomes an even more valuable alternative.

In Multiple–Baseline designs, the experimental conditions need not ever reverse; instead, one variable changes in each phase, and the data track any effects on the behavior of concern across those phases. Again, as with the ABAB design, under Multiple–Baseline designs, the data document experimental control when they show that behavior changes when, and *only* when, the experimental conditions change in each phase. You will likely find Multiple–Baseline designs emphasized in the repertoire of applied research scientists working to develop and test procedures and interventions for cultural benefit.

Mostly through these kinds of methodology, behaviorologists work to discern and apply the variables of which behavior is a function, variables that generally reside in an organism's species history, personal history, current situation and, particularly for people, the cultural setting. But again, what are these variables and what relates them to experimental analysis?

Variables in General Science and in Behaviorology

Variables in Natural Science
Returning to the question of how experimentally analyzing phenomena happens—something many readers have likely covered in some detail and possibly more than once in their educational history—let's review some basics. Let's start by repeating a point too easily taken for granted: Successful experimental analysis begins with accepting the constraints on natural science contained in the philosophy of naturalism. One of the most fundamental of these constraints is that natural scientists deal *only* with natural events as

independent variables (IVs) and *dependent variables* (DVs). We exclude any untestable variables and accounts, including mystical, non–natural (i.e., anatural or supernatural or unnatural) variables and accounts, because they reside outside the range of real, at least theoretically testable IVs and DVs, a practice that has been vitally successful for at least the last 400 years in expanding our knowledge for dealing successfully with the world.

Let's look more closely at IVs and DVs and their relationships, because they are at the core of how experimentally analyzing phenomena happens. IVs are the ones researchers change (manipulate) to see if different values have any effect, or different effects, on what they are studying. DVs are the variables the changes in which *are* what they study. In other words, the effects of manipulating IVs are the changes that we measure in DVs, and those DV changes are what we study. When a particular DV reliably changes the same way, and in the same amount, every time we change a particular IV, in a particular way or amount, we say that this DV change is *a function of* that IV change.

For example, an electric stove usually has about ten buttons each tied to a different amount of electrical energy passing through and thereby heating a stove element. Those buttons therefore constitute the values of the heat IV, and we manipulate the heat when we push different buttons. (The word *manipulation* gets some bad press, sometimes deserved, but here in science it has a quite neutral meaning.) By manipulating the heat, we control the speed at which the water in our teapot boils. That speed (actually, the rate of change in the water temperature) can be our DV, and under conditions of guests getting impatient for their tea, we push the button that provides the most energy to the stove element, making the water heat quickly. Alternatively, if conditions are such that we cannot use the hot water for several minutes anyway (perhaps because we are preparing the crumpets or scones that accompany the tea) then we push a button that provides less energy to the stove element, making the water heat more slowly.

As usual that example is more complex in a couple of ways. The whole situation actually intertwines both physical variables (amount of energy and speed of water heating) and behavioral variables (conditions controlling pushing one button rather than a different button). And we excluded inner agents in the process, although at this point we still relied heavily on our agency–laden language to discuss the matter. All this anticipates our initial coverage of behavior IVs and DVs shortly. Also, the speed of water heating is a function of the amount of energy reaching the stove element, and the amount of energy is a function of which button we push. A reasonable summary of these relationships is that the fastest water–heating speed is *a function of* pushing the button that supplies the highest energy.

An additional observation regards the direction that each variable takes. In that example, both variables go in the same direction; when one goes up, the other goes up, and vice versa. We call this a *direct* functional relationship. In other functional relationships, the variables go in opposite directions, which we then call *inverse* functional relationships. For example (another with behavioral variables), a *decrease* in pupillary size (a studied, measured DV) is an inverse function of an *increase* in illumination (an IV), and vice versa.

We actually only observe the repeatable contiguity of these events in time, that is, one seems always to follow the other ("seems," because observing every possible repetition—past, present, and future—is not possible). Enough such repetitions breeds a certain confidence that the sequence is not occurring by coincidence; we may even begin using the overused (and possibly misused) term "cause" to describe the sequence, saying that a certain IV causes a certain DV. We may even reach the point where the relationship is so well established, with such a high confidence level, that we begin calling it a "law."

"Laws" are usually discovered by different disciplines, because different DVs and different scientific disciplines are intimately connected. DVs of a particular type make up the particular subject matter of a discipline. Generally, DVs involving energy constitute the basic subject matter of physics while DVs involving matter constitute the basic subject matter of chemistry, and DVs involving life forms constitute the basic subject matter of biology. As we will soon see, DVs involving life functions constitute the basic subject matter of behaviorology. What are some of the DVs, and IVs, of behaviorology?

Variables in Behaviorology

Behaviorology is interested in the direct and inverse functional relations and laws pertaining to behavior, especially human behavior. Since behavior in general, and changes in behavior, are our DVs and are the main topic of the next chapter, let's consider just a general conceptualization of the range of IVs that control behavior. The classification, evaluation, and application of these IVs constitute the main topics of most of our subsequent chapters.

An easy, and so possibly helpful in the brief term, initial view of behavior and its "causes" takes this very simplified form:

$$A \longrightarrow B \longrightarrow C$$

We call this the "A–B–Cs of behaviorology." Here the "—>" reads "functionally controls," while the "A" stands for antecedents, the "B" stands for behaviors, and the "C" stands for consequences; thus our equation, or formula, reads "antecedents functionally control behaviors (which) functionally control consequences." This indicates the most common kinds of variables surrounding behavior. A variety of stimulus variables comprise the antecedents of behavior; a variety of considerations surround behavior responses, and a variety of stimulus variables comprise the consequences, or "postcedents," of behavior. Stimuli are the vast range of real events in the internal and external behavior–controlling environments. More specifically they are various kinds, qualities, and intensities of energy changes at receptor cells. They occasion structural changes in the nervous system that result in the occurrence of behavior (or affect behavior in some other way) and we describe this interaction as the neural *mediation* of behavior under those environmental controls. All of this occurs entirely naturally. Internal and external environmental evocative or consequating energies affect the brain and nervous system, which thus neither originate nor initiate the occurrence of behavior. Instead, brains and nervous systems only mediate behavior as part of the physiology that is ever present and operating when behavior occurs. Essentially then, they are the locus at which environmental IVs and behavioral DVs interact. Similarly, the processes and outcomes, that the terms in our behavior formula imply, need no participation from mystical inner agents. Let's now take a brief look at some aspects of each term in our simplified behavior formula.

Antecedents. Several types of variables comprise the antecedents of behavior. Here we will mention the one that seems to be the most common, with the others coming up in later chapters. This most common antecedent stimulus is the "S^D." The pronunciation is "ess–dee" and it stands for "discriminative stimulus," which is an older term for "evocative stimulus." These stimuli function to evoke behavior. Any setting is filled with many of these stimuli, and which ones affect you next depends on several other variables, such as your history of past functional–relation conditioning. The result is that, according to the currently operating functional relations, at least one of these stimuli *will* be affecting you. This takes the form of its presence *evoking* the relevant behavior. That means the stimulus functionally controls the occurrence of the response; if the stimulus provides the appropriate energy change at receptor cells, the physiology of the carbon unit that is you mediates the response (i.e., in the presence of the stimulus, the response occurs). For example if, as you drive around a corner, you come into the presence of a flashing red light, you stop. Or rather, excising the inner agent implied by the word you, the presence of the flashing red light evokes the response of stopping.

However, in a sense, S^{D}s compete with each other. If in the presence of that flashing red light, an officer directs you to turn left, which means you are in the presence of two salient S^{D}s, the officer's directions evoke the behavior of turning left; at least, that is what happens if your conditioning history has been in keeping with the laws of the land where directions from a present officer take precedence (i.e., carry more severe consequences for non–compliance) over automatically operating red lights. (See all the levels of complexity we can get even in a simple example? And, this early in the book, I cannot even come close to mentioning all these levels. Behaviorology makes possible our managing many of those multiple levels simultaneously. Marvelous.)

Again, usually you are in the presence of a multitude of S^{D}s, and which ones affect you, and in what sequence, may or may not be important. As I craft these examples here at my desk, the S^{D}s that surround me include the keys on the computer keyboard, the computer mouse, the computer screen, the papers on the desk containing my notes for this chapter, the flash drive on which I backup my work, the USB hub into which the flash drive plugs, the clock, the telephone, the tissue box, the lights, the stereo, the CD player (and the remote control for each) and so on and on and on! Which one will evoke my next response? Survey your own desk or current setting. I am sure you can list a similar number of S^Ds and possibly still leave numerous ones unmentioned. Which one will evoke your next response? Right now the words in this book are again evoking your responses (called textual behavior, as covered in a later chapter). However, what if you looked, around and saw your clock showing that you have been enjoying reading for hours and are about to be late for an appointment? Now *that* would indeed induce behavior other than reading.

Let's turn now to the middle term of our behavior control formula. That term is ... behaviors.

Behaviors. In practice, we differentiate between behavior and responses. While *behavior* (B) is the general term, we use the term *response* (R) for particular instances of a behavior. Responses occur under the control of currently operating variables; this makes every response a new, and thus different, response. This could be a problem since science usually deals with repeatable phenomena. But responses naturally fall into "response classes" on the basis of sharing the same stimulus consequences. Thus our experimental analysis operates on the level of response classes. For example, in a well–managed classroom, students regularly raise their hands when the teacher asks a question. (What is the question? It is, in part, an S^D evoking the behavior of hands rising.) When the teacher acknowledges one of the students (and they all get acknowledged regularly but in turn) the acknowledged student provides an answer. (The acknowledgement is an S^D evoking the response of answering.) And when the answer is correct (which is quite often in an academic exercise well planned in light of the known laws of behavior and the reasons for classroom education) the teacher provides intellectually and emotionally honest and appropriate comments that function as consequences with effects on similar subsequent behavior. Those consequences occur for every student who provides a correct answer after raising his or her hand. Sharing those consequences thus makes all those instances of hand raising into members of a single, repeatable response class that we can study and that we might simply call "hand raising." We will address additional *behavior* considerations in later chapters.

Consequences. The third term in our formula receives much attention. It is not necessarily more important than the other two terms, but it is the part that helped differentiate operant behavior, to which this formula pertains, from the other kind of behavior, respondent behavior. (Both kinds have a summary back in Chapter I.) While the third term depicts events that follow behavior in time, the word *consequences* has a meaning far too specific to serve as the word for those events. Standing for the final "C," it helped ease the introduction to the formula. (The "A–B–Cs of behaviorology" is rather catchy.) But both it and the formula must develop further. Just as we call the events that precede behavior in time "antecedents," we call those that follow behavior in time *postcedents*. In both cases we are simply implying solely the relative time relation; at this level we could simply put the generic term "event" between them. Consequences, however, are but one kind of postcedent. Since in that status they make this part of our behavior formula receive the most attention, let's give them a bit of their due attention.

Consequences divide into various types according to three criteria (which we detail in a later chapter) that determine these three characteristics: (a) their status as reinforcing or punishing consequences, (b) their status as added or subtracted consequences, and (c) their status as unconditioned (i.e., primary) or conditioned (i.e., secondary) consequences. At this point we will touch only on the first of these characteristics as it pertains to the central role of consequences. That central role involves making some responses occur more often (in which case the consequence earns the title "reinforcer") and making other responses occur less often (in which case the consequence earns the title "punisher"). For example, at certain specially designed intersections in Europe, drivers are photographed. Yes, that is common, but what happens next is less so. Those drivers who stop for the red light receive a free lottery ticket while those who run the light receive a payable traffic ticket. The data, for all those who occasionally drive that route, and for the individual drivers who regularly drive that route, show that the number of drivers obeying the light increased and the number running the light decreased. Thus we can call the lottery tickets *reinforcers* and the traffic tickets *punishers.* However, we would prefer to see these stimuli occur immediately after the responses that produce them, and earn these titles at the level of the individual drivers rather than the crowd of drivers. We will see more individual examples in later chapters. We will also see that delays, such as from the mail delivery of these consequences, will compel an expansion of the details of our analysis. These real examples are always more involved than whatever particular principle we are using them to elucidate.

How does earning the reinforcer or punisher titles work? Recall that S^Ds work (usually due to a relevant conditioning history) by triggering changes in nervous–system structure that, on a moment by moment basis involving the more substantial energy resources of the body, mediate, as in evoke, the occurrence of the related behavior. Reinforcers work (again, usually due to a relevant conditioning history) also by triggering changes in the nervous system, but these changes are altered structures *that are more enduring* and thereby leave the body different, such that the occurrence of relevant evocative stimuli, like S^Ds, more easily—which we observe as more frequently—evoke the behavior again, with all this also involving the substantial energy resources of the body. And punishers work like reinforcers except that the enduring nervous–system structural changes leave the S^Ds *less* effective as evocative stimuli so that the behavior occurs less frequently.

Put another way, the consequences functionally feed energy back into the organism's nervous system, changing it so that the now different organism mediates behavior differently. When the type of antecedent stimulus that had evoked the response before again confronts the organism, the nervous system mediates the same kind of response more readily or quickly. Stimuli earn the title *reinforcer* when we observe their effect as enhancing the functional evocative control that the antecedent stimulus exerts on the behavior. We cannot call a stimulus a reinforcer until after we have observed this effect. And we call the process *reinforcement.* We see the behavior occurring more often and describe that by saying the consequence reinforced the behavior. Similarly, stimuli earn the title *punisher* when we observe their effect as diminishing the functional evocative control that the antecedent stimulus exerts on the behavior. We cannot call a stimulus a punisher until after we have observed this effect. And we call the process *punishment.* We see the behavior occurring less often and describe that by saying the consequence punished the behavior.

Of course much more happens as those events take place, but this early limited discussion still contributes to our growing understanding. The details of the physiological events for the operation of both S^Ds and consequences will come from cooperation with our colleagues in physiology.

All three terms together. We can elaborate these terms into a more clear statement of the fundamental formula for *initially* analyzing any behavior. The "A–B–Cs" formula readily becomes "S^D–R–S^R" ("S^R" = reinforcing stimulus). Using the more accurate S^{Ev} (i.e., evocative stimulus) instead of S^D, we have "S^{Ev}–R–S^R." We call this formula *the three–term contingency* for obvious reasons. We can (and will) write many similar formulas, with three or more terms, to describe a range of alternative contingency forms. However, this particular one summarizes our *starting point* for analyzing any bit of behavior that becomes of interest or concern for us. We start by investigating the stimulus variables in the setting; behavior never occurs in a vacuum, because numerous S^{Ev}s fill every setting (recall our "at our desk" example). Next—actually the order is not this strict—we examine the responses ("R" in the formula) for any characteristics that can affect outcomes. And we continue (as additional steps require further discussion later) by considering the many factors that affect the effective operation of the consequences. In due time we will find that our analysis of the contingencies in which behavior participates is deeper and far more extensive than our starting–point formula indicates (more material for later chapters).

By the way, you may feel uncomfortable with the notion of a "contingency." Perhaps I was just somewhat dense on the matter, but I recall that my first contact with that term in this scientific context (a rather long time ago) left me quite confused. I required about three weeks, many usage attempts, and someone finally making a particular point, before I felt comfortable with the term and could use it correctly. Why wait? Here is that point: A "contingency" may best be understood at this time as a *dependency:* The occurrence of the consequence *depends* on the occurrence of the response which itself *depends* on the occurrence of the evocative stimulus. More technically, the consequence occurrence *is contingent on* the response and the response occurrence *is contingent on* the evocative stimulus. For example, when out driving, as the car approaches some curves, those curves, as S^{Ev}s, evoke steering responses which staying safely on the roadway importantly consequates. That consequence of staying on the road depends (is contingent) on proper steering responses which depend (are contingent) on the S^{Ev} directions of the curves. I hope that helps.

Variables and Awareness

Here is something else that I find very important for thoroughly understanding human behavior. You should note (and possibly find fascinating and important as well) that *you need not be aware* of the curves, in that driving example, for them to exert their evocative stimulus function on your driving responses. Awareness is one of the neural behaviors of consciousness, and seems to operate as a single channel capability. If your single–channel neural–awareness responses are engaged, say, daydreaming about winning some lottery, the curves (as S^{Ev}s)—and continuing to stay on the road (as reinforcers)—will still effectively control your steering responses. (Up to a point, that is. "Don't try this at home" as they say—or alone, etc.). Ultimately you will observe (another neural behavior of consciousness) this not–needing–to–be–aware when, arriving at a red stop light further along the road, you find yourself wondering (yet another neural behavior of consciousness) how you got there, or where the last mile or two went. You neither crashed nor heard sirens wailing nor got pulled over, so your steering responses must have been adequate. However, those driving responses were occurring (under what we will later call *direct stimulus control)* right along with, that is, at the same time as, your daydreaming responses. Those daydreaming responses prevented your being aware of the steering responses because the needed channel, so to speak, was already occupied with those daydreaming responses. How often have you experienced something like that? It is not magic (although it seems wonderfully magical); it is just the behaviorological laws of nature at work, laws which operate without our needing to be aware of them or the variables involved.

The fascinating intricacies of complex and multiple motor behaviors occurring unimpeded alongside other complex neural behaviors, like consciousness, prepare us to consider some general aspects of behaviorology. Before going there, however, let's briefly consider a related topic, namely the difference between natural science and social science. Several commentators (e.g., McIntyre, 2006) have put forth calls for the development of a natural science of human behavior (like behaviorology) and the difference between natural science and social science is relevant to understanding the place of those calls. We will follow this with an example that can show the cultural value of a fully developed natural science of human behavior like behaviorology.

Natural Science and Social Science Defined

The main difference between natural science and social science pertains to the definitions of each, which need not be mutually exclusive. People define natural science and social science in various ways. Rather than review the range of definitions, here are workable definitions that avoid some all–too–common invidious comparisons, and instead highlight some positive aspects that help us sort out these two (see Ledoux, 2002b). We define *natural sciences* as disciplines that deal solely with natural events (i.e., with real IVs and DVs in nature) using scientific methods, while we define *social sciences* simply as disciplines that are interested in people. Social sciences also use scientific methods, but very few "deal solely with natural events"; those that do so thereby fill the definition of natural science as well, and so I think we should so designate them. For example, epidemiology (which is the scientific study of disease patterns, especially epidemics across a geographical area or society) is an offshoot of biology that deals solely with natural events using scientific methods; so it is a natural science. Yet it is clearly interested in people, and many universities teach it our of the Social–Science Department.

We can discern a similar pattern for behaviorology itself. Behaviorology (which is the scientific study of behavior, especially human behavior) is also an offshoot of biology and also deals solely with natural events using scientific methods; so we are correct to call it a natural science. Yet it is clearly interested in people, and sometimes it too is taught in a university Social–Science Department; so also calling it a social science is acceptable (although this is not my personal preference, even though SUNY–Canton, where I taught for over 30 years, administered the behaviorology courses—while acknowledging them as natural–science courses—from within the Social–Sciences Department).

What is curious is that calls for the development of a natural science of behavior have come as much from social scientists as from natural scientists. These other researchers, who made these calls, were apparently uninformed about the development, even existence, of behaviorology, perhaps because, prior to its declaration of independence in the late 1980s, behaviorology was hidden under a label (behavior analysis) that a non–natural discipline claims. Furthermore, the general lack of readily available access to any iteration of the natural science of behavior could have prevented researchers from contacting it. As one example of such a call, Lee McIntyre, a researcher at the Center for Philosophy and History of Science at Boston University, published a book–length call for a natural science of behavior (McIntyre, 2006) entitled *Dark Ages—The Case for a Science of Human Behavior.*

Are calls like that one justified? Clearly I think the natural science of behavior can make valuable differences. If I did not think so, why would I be telling you all this? Let's consider an example of potentially great cultural value, one about technologies and their use, like birth control technologies that can humanely aid the kinds of population reductions our world needs to improve the chances of human survival at a level above subsistence minimums. The still growing problem of the increase in greenhouse gases (regardless of whether or not these arise from human activity) leads to global warming, with its increasingly dangerous changes in climate and weather patterns that could effectively destroy our living standards. While we environmentally solve these problems with sustainable lifestyles, they are still among the indications that we have exceeded the carrying capacity of the planet. This includes the planet's capacity to provide not only enough food, clothing, shelter, and so forth, for all the people, but also enough resources for the range of other planetary plant and animal species that are so necessarily intertwined with human survival.

To return to levels of resource need and use that are compatible with the planet's carrying capacity, perhaps one of the most important steps, if not *the* most important step, is to reduce in humane ways the overall population level; the broad range of widely available birth–control technologies (e.g., various practices, compounds, and devices), and the various behavioral technologies that could support their use, would contribute much to our overall success. This is yet another example of the behavior components that suffuse our global problems and the apparently required part that our natural behavior science must play in the solutions.

Even though those birth–control technologies are available, who will use them? As a result of their past conditioning histories, many people throughout the world, perhaps at this point a majority, exhibit a hard–core, mysticism–based resistance not only to the birth–control technologies but also to their use and to the behavioral technologies that would enable their easier adoption. Indeed, that superstitious resistance even extends not only to our natural behavior science but also to natural sciences in general, even though the resistors cannot live without the products of those sciences, which may thus be safe from the destruction that those resistors sometimes talk of visiting on the culture of science. But is that the only threat to science and its products, many of which make the very development of sustainable lifestyles possible?

The risks are large and looming! The successful application of the technologies of birth control, and the sciences behind them, could enable the greatest single source of greenhouse–gas reduction, namely a smaller world population, one compatible with the planet's carrying capacity. However, if the resistance to these technologies and sciences succeeds in preventing their use, then that resistance may by default also succeed in preventing the overall and lasting solutions to global problems. If that outcome develops, we and our children's grandchildren for many generations to come—if our species does not succumb to an earlier than predicted extinction—may end up experiencing another thousand–year Dark Age in which science is again lost. Then people must gradually and painfully rediscover science, while old and new varieties of mysticism and superstition continuously abound.

Is this a kind of alarmism? Perhaps it is. But what if, too late, it turns out to be correct rather than alarmist. My own conditioning has left me leaning toward the side of solving problems faster than needed, emphasizing prevention and avoiding "too late." How about you? For instance, you may recall the importance we gave (and which many still give) to efforts to achieve "Zero Population Growth." The general global culture has made quite a bit of progress along those lines, quite a bit but not nearly enough. Indeed one can argue that mere zero population growth is today no longer a viable option.

Instead we really must somehow practice "Humane Population Reduction" on the broadest possible scale with every humane practice and technology at our disposal (i.e., no bombing or poisoning or other form of mass murder of whole societies or cultures). This is an especially important part of solving, as in mitigating, all interrelated global problems (e.g., over population, pollution, global warming, damaging climate change) in a manner that helps keep us within the still shrinking time frame that is available before we must experience the worst effects that these problems will throw at us, at which point we will be left to suffer while adapting in whatever ways may still be possible.

Humanity has already made some attempts at humane population reduction. For example, whatever the process, problems, or successes of China's one–child policy, we really must acknowledge not only the foresight but also the appropriateness of their attempt. And then we must coordinate even more successful practices, successful both at reducing the superstitious resistance to the birth–control practices of humane population reduction, and at implementing those practices, to bring about substantial reductions in the world population relatively quickly.

Remember: if we fail, other natural processes can and will inhumanely succeed at the population–reduction task, through famine, pestilence, war, and an increasing number and variety of natural disasters of increasing proportions (e.g., size, destructiveness). But events need not develop that way, especially if we increasingly take into account the behavior components that suffuse our global problems—of which overpopulation is but a particular example—by accepting the essentially required part that our natural behavior science must play in the solutions. Behaviorology is this science, and we examine some general aspects of it next.

Some General Aspects of Behaviorology

Definition of Behaviorology

We can state several definitions for behaviorology, some simple and others increasingly comprehensive. The simple ones may cause confusion as they are similar to the definitions that other disciplines use. The comprehensive ones may also cause confusion, but only due to their complexity, a problem we can easily solve by stressing the individual parts as we clarify the range of components that constitute the behaviorology discipline. Let's consider some of these definitions.

Simply put, behaviorology is the scientific study of behavior. Too simple? Let's try again. Behaviorology is the natural science—and its related engineering technology—of the interactions between behavior and variables in the internal and external controlling environment (i.e., the natural science of environment–behavior relations). Well, this definition seems pretty good. However, some readers demand full disclosure, so let's get really serious (i.e., complex):

Behaviorology, a comprehensive discipline with philosophical, experimental, technological, analytical, conceptual, and theoretical components, is the natural science among the life sciences, emphasizing the causal mechanism of selection, that discovers, interprets, and applies the single and multiple variables that are in functional relations with the simple and complex, overt and covert behaviors of individual organisms (especially people) during their lifetime (and beyond, with respect to cultural practices), and that takes into account the socio–cultural and physical variables from the internal and external environments as well as variables from the biological history of the species.

Now, perhaps that definition is all "perfectly clear." Wonderful! But on the off chance that it left some residual confusion, let's rephrase it to stress its individual parts:

⅋ *Behaviorology* is a comprehensive discipline with philosophical, experimental, technological, analytical, conceptual, and theoretical components;

⅋ *Behaviorology* is a natural science among the life sciences;

⅋ *Behaviorology* emphasizes the causal mechanism of selection;

⅋ *Behaviorology* discovers, interprets, and applies the single and multiple variables that are in functional relations with the simple and complex, overt and covert behaviors of individual organisms (especially people) during their lifetime (and beyond, with respect to cultural practices); and

⅋ *Behaviorology* takes into account the socio–cultural and physical variables from the internal and external environments as well as variables from the biological history of the species.

One can expand each of those parts, and even the parts of the parts, with extensive detail. But then, that is what we will cover in the rest of the book. Meanwhile, taken together, those parts clarify, at least in short form, exactly what our discipline is and does.

Emergence of Behaviorology

As we described in Chapter 1 (and as Lawrence Fraley and I elaborated far more thoroughly in our long "Origins, Status, and Mission" paper in 2002) behaviorology emerged as a separate and independent discipline, because the attempts to change the non–natural discipline of psychology into a natural science failed. Psychology had separated from philosophy in the second half of the 1800s when some philosophers, noticing the great strides of their physics, chemistry, and biology colleagues, and seeing the use of scientific methods by these colleagues, figured that even as philosophers they might also make more strides if they too adopted these methods.

In spite of many serious elaborations over the intervening two millennia since the time of our culture's favorite Greek philosophers (e.g., Aristotle) some have suggested that not a whole lot of actual advances had occurred in philosophy. A more charitable assessment would be to count as advances the emergence of some traditional natural sciences from philosophy in the last several hundred years.

In any case, their fellow philosophers essentially said, "No. Using scientific methods would not be philosophy; it would be a different discipline." So those favoring emulation of their scientific colleagues adopted scientific methods and called their new discipline psychology. However, they never also adopted what I would consider as one of the primary characteristics of natural science, namely the constraint to work only with real, natural events as IVs and DVs. Individually, particular psychologists may have adopted that constraint, but they evidenced little to no attempt to get their psychological colleagues to adopt this constraint generally, as a discipline, and instead agree to the general disciplinary rejection of that constraint, along with continuing to accept the extension of their philosophical outlook to the notion of fundamentally accounting for behavior by allusion to spontaneous internal agential machinations. As a result, while many aspects of this quick history are quite oversimplified, the psychology discipline nevertheless began as a non–natural discipline, and remains so today.

B. F. Skinner was an early exception to this operational pattern. In the 1930s, he brought the philosophy of naturalism from his work with W. J. Crozier (a leading professor in the biology department at Harvard University where Skinner was first studying and then working) to his study of behavior, which he acknowledged as a natural phenomenon. Adhering to the constraint to deal only with real, natural variables, he and his students and colleagues then built up this natural science of behavior over the next several decades. All the while they both worked in administrative units of traditional psychology and attempted to get psychology to change into a natural science. They tried to foster this change by successfully producing the standard kinds of formal evidence that one could normally expect to prompt that kind of change.

As these early natural scientists of behavior accumulated advances during this period when psychology and their natural science of behavior shared their history, the name they used to describe their work changed several times. After "operant psychology" came "TEAB" (The Experimental Analysis of Behavior) and "behavior analysis." In the process, they founded various organizations and journals in support of their natural science. Ultimately, however, they were unsuccessful in getting psychology to become a natural science, and some have acknowledged this.

In 1987, some of these natural scientists of behavior looked again at the data from the attempts to change psychology. Being under contingencies that compelled data–based actions, they saw that they had succeeded in setting up numerous arrangements separate from their psychology colleagues, including separate courses, programs, journals, organizations, certification, and accreditation. They also saw that the numerous, continual, and overlapping change attempts, over about five decades, had gotten nowhere, even as the global problems of our world accelerated, demanding a share of attention from a natural science of behavior. It was as if their traditional psychology colleagues had essentially said, "No. Accepting the real–variables constraint, and rejecting inner agent accounts, just to be a natural science, would not be psychology; it would be a different discipline." In appraisal of these circumstances, these natural scientists of behavior declared their independence from psychology, called their long established but newly independent discipline "behaviorology," and proceeded to found new professional organizations and journals.

So occurred the emergence of the natural science of behavior that we call behaviorology. While it started around 1913, about 100 years ago, with John Watson's description of a behaviorist perspective, and developed with the variously named natural science of Skinner and his colleagues and their students for the 75 years after that, it has been an *independently* organized natural science of behavior discipline since 1987 (i.e., for about the last 25 years). Let's consider now the kinds and areas of knowledge and skills that this recently emerged discipline expects its members to master.

Curricula of Behaviorology

By first clarifying the distinction between "discipline" and field," we will more easily see the patterns in various types of curricular content. A discipline consists of the accumulation, from the developed experimental and applied methods, of the discovered principles, concepts, theories, practices, and relationships that constitute the comprehensive knowledge and skill repertoires of disciplinary members. A field, however, is a thematically related set of concerns and problems in which practitioners apply whichever discipline is the most relevant to address these concerns and solve these problems. We would then say that the knowledge and skills of the discipline inform the problems and solutions of various fields. For example, behaviorology is the discipline that studies behavior and its functionally related variables, so its application is appropriate in fields wherein behavior comprises vital components. Thus behaviorology can and should inform such thematically integrated fields as diplomacy, education, entertainment, environmentalism, finance, law, management, manufacturing, nursing, politics, science, and writing (i.e., essentially *all* applied behavior fields from advertising to zoo keeping).

Furthermore, different disciplines may attempt, with varying degrees of success, to inform any particular field. For example, leaders in the field of education early on connected with psychology as the discipline to inform education. Given the alternatives available at the time, this may have made a lot of sense. Today, however, the alternatives include behaviorology, which could supplant psychology in informing education and so immediately provide substantive improvements for the children of our culture, and growing improvements for our whole culture in due time.

Recognizing behaviorology as a discipline that informs a range of applied behavior fields, we can sensibly return to questions of curricular content. What are the areas of knowledge and skills that comprise behaviorology curricula?

After my colleagues elected me the first president of The International Behaviorology Association at our first convention in August 1988, my topic for the presidential address for the second behaviorology convention focused on the answer to that question, that is, on the areas of knowledge and skills that could comprise the curricula of behaviorology. This constituted an early attempt to suggest a set of curricula for behaviorology programs at each level of higher education (see Ledoux, 2009, for the latest iteration of this paper). It was successful at least in the sense that the behaviorologists of the time supported it with a clear consensus about its contents. Here is a summary that takes into account further developments over the last 25 years.

Overall, our curricular designs reflect the usual repetition, at continually more sophisticated levels, of topic coverage as regularly found in the curricula of other natural sciences. However, the range of topics that a student should cover vary according to the level of the students' academic program (bachelor, master, or doctoral). Undergraduate programs would, of course, cover a solid and comprehensive foundation in the basic philosophy, principles, methods, concepts, research, and practices of the discipline, along with a wide ranging selection of the fields in which the students may later find themselves applying the discipline, including the basics of how the discipline applies to those areas. This assures that the curriculum complies with the results of recombination of repertoires research (introduced in Chapter 1) which shows that larger numbers of potential parts of problem–solving responses, conditioned now, more quickly lead to evocation of recombined–solution responses later.

Graduate programs, however, would not only reiterate the content of those foundation areas in more sophisticated terms, concepts, and relationships, but also would cover them in more elaborate depth, with greater detail, and with the latest developments and trends. More importantly, by the time they are in a graduate program, students are often more clear about their interests in particular application fields, so these fields receive the same kind of deeper and more detailed consideration as the foundation areas.

For example, the titles of typical undergraduate courses would range widely. They could include Introduction to Behaviorology, Introduction to Applied Behaviorology, History and Philosophy of Behaviorology, Experimental Behaviorology, Introduction to Verbal Behavior, Child Care Science and Skills, Behaviorology and Green Engineering, Classroom Management and Preventing School Violence, Performance Management and Preventing Workplace Violence, Companion Animal Behavior Training, Behavioral Medicine, Behaviorology and Education, Dignified Dying and Behaviorological Thanatology, Behaviorology and Criminal Justice and Rehabilitation, and Developmental Disabilities Intervention Methods.

Meanwhile, the titles of typical graduate courses would also range widely. They could include Advanced Behaviorology, Advanced Methodologies (for complex behavior and covert behavior), Advanced Experimental Behaviorology (with separate courses on such topics as Respondent Behavior, Stimulus

Control, Equivalence Relations, Reinforcement Schedules, Simultaneous Operants, Self Control, and so on), Advanced Verbal Behavior, Explanatory Fictions and Analytical Fallacies, Behaviorology and Physiology, Morals and Ethics, Consciousness and Reality, and Advanced Applied Behaviorology (with separate courses on such topics as Education and Teaching Behaviorology, Building Sustainable Lifestyles, Reducing Overpopulation, Enabling Energy Conservation, Behaviorology and Abnormality, Behaviorological Therapies, Autism Intervention Methods, and so on).

All those titles represent the kinds of topics and content that would go into constructing or, more accurately, conditioning the knowledge and skill repertoires of new behaviorologists (and to some extent the repertoires of interested others) at the various levels of higher education. Sound interesting?

Recognize, however, that no two institutions or programs operate under exactly the same local options and contingencies. So no two programs will be exactly alike. They will address the general disciplinary outlines as described here, but not likely with the same mix of courses either as described here or as compared with each other. That of course is quite normal and acceptable.

To maximize the students' repertoire expansion from each course, behaviorology faculty would design the teaching of their courses with practices applying the discipline's principles and laws of behavior to as many aspects of teaching the courses as possible. Likewise, graduate students, as likely future behaviorology professors, would take courses in the design and implementation of these teaching protocols and then practice them in courses that they teach.

Understanding and Managing Mysticism and Superstition

While mysticism and superstition are not among their teaching topics, behaviorologists will—for some time yet—have to teach what these are, as well as why and how to reduce their natural negative impact. Dealing with the world, day in and day out, compels plenty of interest in why events occur and how to make things better. You and I and everyone want answers, especially ones that seem to make us better at dealing with the world. The problem is, getting thorough and accurate answers seems to be directly proportional to the quality and amount of effort that we can expend in the process. In the remote past, staying alive under minimum subsistence contingencies occupied most if not all of peoples' energies. As the old saying goes, "Keeping body and soul together" was not easy. With current scientific knowledge, we would instead say, "Keeping body and behavior together" is easier today. In the millennia separating these two times, much has happened, even though too many folks are still today exposed to unhelpful accounts of behavior (e.g., astrology).

In those earlier times, people had little extra energy for answering the questions that bothered them. The social pressures to provide some account or explanation, indeed *any* account or explanation, far outweighed competing contingencies for accuracy. Accounts simply got made up, and if both plausible and not obviously contradictable, then these accounts gradually became intermixed, elaborated, and standardized as accounts for a culture. As these accounts accumulated such complexity over century upon century, some of them became institutionalized, often in the form of one or another religion.

Regardless of any contingencies compelling emotional support and intellectual rationalization for these theological accounts and explanations, they remained grounded in their mystical and superstitious beginnings. This is not to say that they were valueless; the rich and varied laws and codes of conduct that cultures derived from them were essentially the only game in town—better than bullying—for maintaining social organization during these periods. However, alternatives existed, but their discovery required the availability of energy for evaluation (i.e., testing, experimentation).

In the majority of major cases, those superstitious and mystical accounts went untested. Not only was testing energy–expensive, and thus unlikely to occur, but the accounts had grown layer upon layer of convoluted rationales and implications, many including a range of non–natural agents and components, such that they became thoroughly untestable. With their institutionalized supporters, they simply endured, becoming major sets of interrelated variables controlling peoples' behavior then and on down through today.

Then, a couple of thousand years ago, social contingencies got complex enough increasingly to compel the occurrence of simpler accounts that people could, and did, test. The result was the beginnings of science. But let's make this long story short. After all, readers have quite likely already encountered one or another of the many authors of every stripe who have covered the historical developments of both religion and science over the last several thousand years. Some of the most readable ones address modern reoccurrences of this struggle between superstition/mysticism and science. See, for instance, Carl Sagan's *The Demon–Haunted World* (1995) and James Randi's *Flim–Flam* (1982). Authors have also focused on particular episodes of this cultural war zone; see, for instance, James Randi's *The Faith Healers* (1989) and William Ryan's and Walter Pitman's *Noah's Flood—The New Scientific Discoveries About the Event That Changed History* (1998).

One of the two possibly biggest problems with superstition and mysticism involves their continued competition with fully, and today experimentally, tested accounts, which we thus can apply more confidently. This competition relies not on any demonstrable validity to superstition and mysticism but on the contingent potency of their thousands of years of history that induces the continuation of the cultural practice of conditioning (some call it indoctrinating) the young to respond early and often with emotional and (untestable) intellectual rationales in support of some particular version of superstition and mysticism (i.e., in support of some doctrine or dogma). For example, consider those whose religious conditioning leaves them refusing scientific medicine to relieve their children's suffering and disease.

Probably the other biggest problem with superstition and mysticism involves their direct and invidious opposition to science itself, which occurs in spite of some religious denominations maintaining official involvement in scientific endeavors. The current oppositional episode began most obviously with the arrest of Galileo some 400 years ago. Opposition has continued in the US even into the twenty–first century with our repeated "modern" courtroom trials over whether or not evolution can or should be taught in the schools. And religious opposition will probably continue against behaviorology next. Why? Here are just two reasons. This science can clarify the analysis of religion and religious behavior and religious faith and the explicit workings of religious contingencies throughout the culture. And this science can describe how to mitigate the many damaging effects of some religious contingencies as well as how to reorganize and improve some other contingencies that, while religions control them, still have beneficial effects.

However, religions are not the only mysticisms. In theology one of the main mystical categories is the soul, which theologians and their faithful followers accept as an active inner agent that supposedly exists and considers right and wrong and other relevant factors and subsequently directs feelings and actions and conduct. As such, many theologies attribute autonomous responsibility for human behavior to the soul, a prevalent philosophical position continuously available since before Aristotle, and an age–old basis justifying beating up a sinner whose soul directs transgressions against theologically couched social mores, in addition to condemning that soul to everlasting hell.

Some would argue that god, not people, condemns to hell, but god and hell have the same questionable status. (Just in which direction *does* the "spark of life" really travel in Michelangelo's Sistine Chapel ceiling painting? Who created whom?) Meanwhile, people have always told, and regularly tell, each other in no uncertain terms that they are going to hell, and often then move to make the life circumstances of the other a veritable living hell. I hope those who would condemn me to hell for saying such things are not reading this book. The contingencies under which they live are likely so stable that contingency changes like those that a mere book such as this could supply are unlikely to have much effect on changing them other than to make them uncomfortable and ticked off, neither of which is my point in writing.

Two Major Cultural Contributions of Behaviorology

Instead, my point right now concerns the perspective that psychologists take. When some philosophers began psychology, they maintained, from their philosophy background, the theologically grounded philosophical inner–agency category to which they attributed many phenomena including human behavior. In essence, as psychologists whom others see as members of an academic discipline unaffiliated with any theology, they secularized the theological soul, separating it from its theological history by renaming it, first calling it the psyche (thus, "psychology") and then calling it the "mind." Later they added "self" and "person" (and others) to the list of essentially equivalent appellations, supposed major or minor differences notwithstanding.

Those behavior–initiating, body–directing, inner self agents, and all their related and still mystically based sub parts and processes, continue formally in psychology today in spite of the nearly century–long accumulation of evidential material, from natural behavior science, undermining any support for their continued assumption. This is a discipline–level problem; some individual psychologists would just as soon see inner–agent accounts dropped in favor of natural–science accounts, but they do not openly or obviously seek, individually or collectively, this kind of change in their psychology discipline, and so share its non–natural status. The retention of non–theological inner agents as core explanatory concepts in psychology confirms its status as a secular mysticism, and its commitment to those agents makes it the leading light for a range of similar secular mysticisms in the form of other non–natural disciplines often administered in social–science departments solely for their interest in people, since they cannot qualify for natural–science status.

These putative inner self agents supposedly act in spontaneous ways similar to the ways their theological counterparts apparently act. Psychologists tell us that these secular agents consider right and wrong and other relevant factors and subsequently direct feelings and actions and conduct. Furthermore, these inner agents provide the new, secular basis justifying beating up criminals, because society can then treat criminals' inner person, psyche, mind, or self as agents autonomously responsible for violations of legal codes of conduct. Then society makes these inner agents bear the punishment through attacks on the bodies that the agents order around, punishment such as imprisonment or even death, punishment that their behavior directions earn in society's eyes.

As the natural science of human behavior, behaviorology is well positioned (a) to replace with science the superstitious problems of both religion's theological–mystical inner agents and psychology's secular–mystical inner agents, and (b) to clear the air about the mistaken compromises that can arise due to these two inner–agent sources. These compromises began several hundred years ago when religious authorities yielded the physical world to the emerging proto–scientists of the time; in return these proto–scientists yielded life, human nature, and human behavior to religious authorities. These proto–scientists assented to this compromise, perhaps observing the mistreatment of Galileo. (We say "proto–scientists," because the words "science" and "scientist" arose only after another hundred years or more.) Later, in light of scientific advances, religion grudgingly yielded the subject matter of life to science, but science still assented to the continuing compromise of religion hanging on tenaciously to human nature and human behavior.

Today, we can see those compromises as unjustified, because all real subject matters, including human nature and human behavior, are amenable to study by natural science. The historical jury is still out, though, regarding the outcome of the current struggle between the cultural forces of the natural sciences and the cultural forces of superstition and mysticism over whether or not to continue the unjustified compromises between them. We may also judge such compromises as counterproductive as they provoke debates that delay action, which is particularly problematic because we need action and we are running out of time. The outcome of this debate possibly puts human survival above subsistence minimums itself up for grabs due to the behavior components inhering in both global problems and their solutions, which raises the need for behaviorology. As natural behavior science, behaviorology can account for and deal with human nature and human behavior more accurately and effectively than religion can, which thus increases the chances that humanity will be able to solve these problems within the appropriate time frame before we must suffer their worst effects. Successful solutions could prevent another impending Dark Age, or worse.

Some circumstances can easily make those kinds of compromises look appropriate, which can compel our falling for them. For example, our culturally conditioned sense of fairness makes inner–agent accounts and natural–science accounts seem mere equals, which can make compromises seem appropriate. However, these two really are grossly unequal. The basis of both of these sets of accounts are assumptions, theological–style assumptions for psychology's inner–agent accounts, and natural–science assumptions for behaviorology's environmental–contingency accounts. However, recall that no one can prove or disprove assumptions. Theologians and psychologists induce mystical assumptions off–the–cuff or, alluding to tradition or common lore, from various kinds of beliefs. In contrast, scientists induce natural–philosophy assumptions from the consistency of the results and developments of several hundred years of accumulating experimental evidence and effective engineering practices that virtually everyone hopes would remain available. These two types of assumptions, naturalistic/scientific and mystical/superstitious, are indeed *not* equal in any meaningful way, and humanity can no longer afford the delays in problem solving that compromises between them can induce. We are running out of time.

Contributions summary. We have touched upon but two major contributions to the culture that behaviorology supplies. In the first we examined the problematic inner–agent accounts of religion and psychology so that we can replace them with natural–science accounts. In the second we recognized as unnecessary and counterproductive the tradition of polite but actually dangerous compromises between religion and natural science so that we can instead move forward, no longer burdened with unneeded compromises, to apply the more parsimonious, experimentally based findings of behaviorology to its share of contributions toward solutions of global problems.

The range of behaviorology's contributions, of course, extends well beyond those two. Earlier in this chapter, we touched briefly on some areas in this range by listing some of the human–nature and human–behavior fields to which behaviorology could, as an informing science, extend accuracy and effectiveness. And these build upon the end of Chapter 1 where we saw some of behaviorology's contributions to the work of other natural scientists.

Audience, Styles, and Topics of this Book

We will see those listed contributions again as they arise in relevant examples of the topics in this book. Let's now look not only at topics but also at the audience and some style concerns for this book.

The Audience for this Book

The two principle audiences for this book include global culture members either (a) who are already working natural scientists or engineers—especially those working on solutions to local and global problems including humane population reduction and developments supportive of sustainable lifestyles—or (b) who, while not holding official natural science or engineering degrees, nevertheless also see the value of the naturalism that makes possible such meaningful contributions from science and technology. These two groups share audience standing, because they might be among those most amenable to addressing a particular discrepancy that has arisen in the global culture, a discrepancy that Fraley (2008) describes, at the very start of his *General Behaviorology* textbook, as "a substantial imbalance in the objectivity exhibited by its people" (i.e., people around the globe). This imbalance, which behaviorology addresses, refers to the reliance on natural science that most of the culture exhibits regarding physical, chemical, and life subject matters, while large portions of the culture also continue to rely on mysticism and superstition regarding the subject matter of human nature and human behavior, an imbalance that we already took pains to begin addressing.

While I sincerely hope that someday soon those two audience groups will really encompass everyone, at present they at least comprise a vital and substantial portion of the global culture. That first group, of working natural scientists and engineers, is well defined, while the second group includes a wider range of people. Its range extends from those who regularly examine the contents of nutrition labels on food packages in their local grocery store, to those who need to hunt to put food on their table or who simply enjoy each other's company while target shooting at their local rifle or pistol range, to those who enjoy each other's company at a backyard telescope eyepiece observing the moon or the phases of Venus or other solar system objects or the vast objects beyond our local stellar neighborhood, to those who appreciate arts such as the arts of sight (e.g., European oil paintings or Oriental scroll paintings) or sound (e.g., popular or traditional music recordings or concerts) or combinations that include touch (e.g., Native American pottery or basketry or weavings), to those who read about or study or meet to discuss past, present, or even future scientific discoveries, developments, and applications.

My apologies if I left anyone off that list. My experience with many of those interesting activities biased their selection for inclusion, which carries the important implication that we on this planet are all far more alike than our conditioning may have left us believing or feeling. And that extends to our language usage, with implications for some concerns about writing style.

Style Considerations in this Book

In this book we avoid following a pattern of extra–detailed, merely reference–based, and overly reference–laced reporting that catalogs the small–scale results from the huge numbers of available and relevant research studies behind this discipline. Too often such a pattern slowly and incrementally compiles into a dry understanding and borderline appreciation of research findings. This book tries to rely instead on a conceptual treatment of topics, with numerous everyday examples, directly emphasizing meaningful understanding and appreciation of the foundation, and the application–scale findings, of the behaviorology discipline.

Nevertheless, all the discoveries, concepts, principles, processes, practices, and procedures that this book covers are available in the research literature of behaviorological science. This literature includes the vast number of articles and books that appeared during, as well as after, the administratively shared history between this natural–science discipline, as it was emerging, and the discipline of psychology. Those articles and books contain the details that relate the findings that they report to the efforts of their predecessor, as well as other circumstances leading to the findings and the methodologies behind them.

However, this book would become quite unpleasant to read if it included every relevant citation. Every sentence if not every line could contain citations, and then the book would necessarily contain dozens more pages filled only with the references for the citations. Books need not be—and this one is not—that way. Still, I include many of the most important references, and they are enough to give any intellectual sleuth a field day finding far more material in the reference list of each included reference.

To enable greater appreciation and retention of the information about the topics we cover, we will introduce the principles and practices related to each topic at their most basic level, and later reiterate these as appropriate at increasingly complex levels. This includes the overarching topic of naturalism itself as behaviorological science exemplifies it.

As we reach and cover the more complex, advanced levels, we will also employ our findings to explore and explicate some of the conceptually intricate and intriguing questions that have challenged human intellectuality and emotionality since ancient times. At such points I suspect you will agree that we can now begin scientifically to answer some of these ancient questions. To anticipate some of our topics, these questions relate to age–old interests including values, rights, ethics, morals, attitudes, belief, personhood, life, death, consciousness, language, reality, and even (of more recent interest) evolution, plus biological, mechanical, electrical and digital robotics.

One word of warning: If your past conditioning has left you likely to skip to the back of a book for a preview of the outcome, then you probably face some unusual risk with this book. Such ancient questions are truly compelling, but we cannot succeed yet in answering them in one of the next few chapters, because success requires the background of all of the intervening chapters between here and the back of the book. If you skip all, or even some, of these chapters and jump ahead, then a safe prediction is that you will be confused, and thus disappointed, due to missing the appropriate preparation. Instead, please be patient, and stay with us as we try engagingly to get through the intervening chapters to the really illuminating, ever more fascinating parts.

A particular style concern, however, compels us to preview a little of the ancient language question. Language is verbal behavior, and it is a function of the same kinds of variables that control all other behavior. One major class of variables controlling the kinds of phrasings we use in English stems from the reasonable agential viewpoint existing at the time of the origin of language; primitive animism was the most parsimonious view at that time and explained movements as the result of inner spirits animating both organic and inorganic objects. As languages evolved they often retained the certain economy of words that agentialism enables without reference to its shrinking accuracy. Thus today our language is laced with agential references, with personal pronouns as likely the most common. To say "I" or "you" or "he" or "she" is automatically to imply an inner agent of one or another variety (e.g., mind or psyche or self or soul or person). Thus, stylistically, we may try to engage phrasings that lack, or at least reduce, those pronouns; the result, however, while scientifically more accurate, may sound stilted, a result with which I hope you will be patient.

A similar problem confronts us over active and passive voice usage. Active voice, due to its direct subject–predicate–object structure, enhances clarity and readability, which accounts for its preference among authors, editors, and publishers. But look more closely; that very structure often implies agency. Whatever is in the subject slot comprises the agent of the action. Passive voice avoids this problem but at the high cost of reducing readability and clarity. In spite of those problems, some scientific disciplines expect their journal authors to rely on passive voice because the "who" is distracting. Stylistically, we will here continue to rely as much as possible on active voice but with regular reminders that the implied agencies actually contradict scientific realities.

In time, the increased quality of science and its products that may derive from recognizing and dealing with agential issues, including personal pronouns and active voice, will likely affect the way we regularly speak and write (i.e., the language itself, our verbal behavior). We may come to exhibit more and more verbal behavior that, consistent with scientific reality, lacks inner–agent connotations, and we will even gradually become more and more comfortable, individually and culturally, with this development. Meanwhile, when particular topics lead us to need more explicit scientific accuracy, some of the phrasings in this book may seem distorted to you due to the strength, only gradually reducing, of our lifelong agency–based verbal conditioning. Perhaps once again a new grammar is on the rise.

The Topics and How this Book Covers Them

Overall, the sequence of topics here in our introductory book focuses on increasingly complicated and testable functional relations between environmental and behavioral variables that extend into the technological applications of behaviorological engineering. We can appreciate the topics by considering what this book shares with, and how it differs from, the *General Behaviorology* book. That book (Fraley, 2008) extends our introductory emphasis in this book into a comprehensive educational mission.

Both books avoid the sometimes common, cookbook–style pattern of cataloging variables merely from simple to complex, which retards the appreciation of all the interactive complexity among the operative variables. Also, while both books share a regular set of disciplinary topics, substantial differences set them apart. *General Behaviorology* provides numerous and elaborate examples for each part of each topic as part of conditioning a comprehensive behaviorological repertoire. This book, however, attempts to introduce the science by using only the number of common, everyday examples that seem necessary to complete this task. By the end of this book, you will be in a good position to recognize how well it succeeded.

Here is the list of our topics, following the pattern of our chapter titles in each of the two Parts of our book:

PART I: Some Philosophy of Science Plus Basic History, Concepts, Principles, Methods, and Practices:

1 Reasons for interest in discoveries about behavior (an historical overview);
2 A natural science of behavior;
3 Kinds of behavior and functional relation analysis;
4 Fictions that fail to explain;
5 Basic contingencies in which behavior occurs;
6 Contingencies involving multiple *simultaneous* behaviors and stimuli;
7 Some complex analytical confusions and fallacies;
8 Basic laboratory equipment and methodology;
9 Some practical methodologies;
10 Postcedent processes that change behavior;
11 Varieties of postcedents plus superstitious behavior;
12 Context, stimulus control, and "Rules."

PART II: Advanced Developments and Answers to Long–Standing Questions:

13 Arranging consequences—differential reinforcement and shaping;
14 Arranging evocatives—backward chaining and fading;
15 Basic schedules of reinforcement;
16 Aversive control problems and alternatives;
17 Some applied behaviorological research considerations;

18 The stimulus equivalence relations horizon;
19 On attitudes, values, rights, ethics, morals, and beliefs;
20 Language is verbal behavior;
21 Accounting for consciousness;
22 Cultural concerns of life, personhood, and death;
23 The unexpected nature of reality and robotics;
24 Evolutions and types of selection.

You will find a detailed outline of each of those topics in the *Detailed* [table of] *Contents* at the front of the book. The range of explicit topics in each chapter is readily apparent there.

Chapter 1 and this chapter (2) both ran rather long in page count, because they both addressed broad, multiple, and interrelated foundational topics. Each of the remaining chapters addresses a more focused topic and so requires fewer pages, the better to maintain your reading pleasure and scientific interest.

However, let's end the chapter by recalling an ongoing activity. At the end of every chapter, review for yourself the chapter topics as part of keeping in view the pertinence of this discipline, behaviorology, the natural science of behavior, to helping solve global as well as local problems. Behaviorology shares the assumptions of the traditional natural sciences and, while no assumptions are provable or disprovable, the contingencies driving adherence to natural–science assumptions induced that adherence through at least the last 400 years of accumulating experimental evidence and effective engineering practices with the wide range of phenomena that *all* the natural–science and engineering disciplines cover, including behaviorology. These contingencies and assumptions form major foundations upon which all these disciplines team up to develop solutions to all global and local problems.♣

References (with some annotations)

Fraley, L. E. (1994). Uncertainty about determinism: A critical review of challenges to the determinism of modern science. *Behavior and Philosophy, 22* (2), 71–83.

Fraley, L. E. (2008). *General Behaviorology: The Natural Science of Human Behavior.* Canton, NY: ABCs. *Running Out of Time* introduces the topics that Fraley's 1,600–page book covers in far more depth and detail.

Fraley, L. E. (2012). *Dignified Dying—A Behaviorological Thanatology.* Canton, NY: ABCs.

Fraley, L. E. (2013). *Behaviorological Rehabilitation and the Criminal Justice System.* Canton, NY: ABCs.

Fraley, L. E. & Ledoux, S. F. (2002). Origins, status, and mission of behaviorology. In S. F. Ledoux. *Origins and Components of Behaviorology— Second Edition* (pp. 33–169). Canton, NY: ABCs.

Hawking, S. & Mlodinow, L. (2010). *The Grand Design.* New York: Bantam.

Ledoux, S. F. (2002a). *Origins and Components of Behaviorology—Second Edition.* Canton, NY: ABCs.

Ledoux, S. F. (2002b). Defining natural science. *Behaviorology Today,* 5 (1), 34–36.

Ledoux, S. F. (2009). Behaviorology curricula in higher education. *Behaviorology Today, 12* (1), 16–25. This article also appeared in S. F. Ledoux. (2002). *Origins and Components of Behaviorology—Second Edition* (pp. 173–186). Canton, NY: ABCs.

McIntyre, L. (2006). *Dark Ages—The Case for a Science of Human Behavior.* Cambridge, MA: MIT Press.

Moore, J. (2008). *Conceptual Foundations of Radical Behaviorism.* Cornwall–on–Hudson, NY: Sloan Publishing. As with Skinner's *About Behaviorism* (1974) but without the legitimate time–frame excuse of appearing before the 1987 adoption of the term behaviorology that applies to that book, this book continues to portray radical behaviorism as the philosophy of science of "behavior analysis" which people now see as a part of a non–natural discipline, psychology. Readers should not let that deter them from benefiting from the otherwise clear and careful coverage of this philosophy that this book presents. (The reference to Skinner, 1974, contains additional details along these lines.)

Randi, J. (1982). *Flim—Flam.* Buffalo, NY: Prometheus Books.

Randi, J. (1989). *The Faith Healers.* Buffalo, NY: Prometheus Books.

Ryan, W. & Pitman, W. (1998). *Noah's Flood—The New Scientific Discoveries About the Event That Changed History.* New York: Touchstone.

Sagan, C. (1995). *The Demon–Haunted World.* New York: Random House.

Skinner, B. F. (1974). *About Behaviorism.* New York: Knopf. With the legitimate time–frame excuse of appearing before the 1987 adoption of the term behaviorology, this book portrays radical behaviorism as the philosophy of "behavior analysis." Back then this could be accurate, because behavior analysts were then moving to become an independent natural science under the behavior analysis label. Sadly the efforts under *that* label were ultimately unsuccessful, because psychology usurped the behavior analysis label, without the kind of strenuous objections from behavior analysts that might have stopped it. Thus, psychology succeeded in making the term "behavior analysis" a part of the non–natural psychology discipline, something that cost behavior analysts much of the credibility they had—as independent natural scientists of behavior separate from psychology—with traditional natural scientists (which, unfortunately, makes their contributions to solving global problems more difficult to provide). Meanwhile, we use *behaviorology* to denote the independent natural science of behavior. Nevertheless, this book is a wonderfully thorough and readable treatment of radical behaviorism, the philosophy of science of behaviorology.ᥱᖯ

Reader's Notes

Chapter 3
Kinds of Behavior and Functional Relation Analysis

$\mathcal{L}$ast time, in Chapter 2, besides addressing some audience and style concerns at the end, we also considered several basic, background topics. After detailing the nature of science *as behavior,* and comparing and contrasting natural science and social science, we turned to some of the foundation assumptions of natural science in general (i.e., the philosophy of science called *naturalism)* and to some of the foundation assumptions of behaviorology in particular (i.e., the philosophy of science called *radical behaviorism,* a subset of naturalism). Then we peeked at the most basic level of organizing the variables that are functionally related to behavior, ending up with (a) the *three–term–contingency* (Recall what that is?) and, something that surprises some folks, (b) that awareness is not a requirement for these variables to affect us, but rather is just more behavior to be explained. Before finishing the chapter, we turned our attention to introducing the natural science of behavior as the *behaviorology discipline.* Thus, we covered its thorough definition, its emergence history, its training curricula, and some of its contributions.

Now, in this chapter, we get to examine two broad and related areas. We begin by considering some parameters about behavior, including its characteristics and classifications as well as some things that are not behavior (e.g., traits). Then we finish by considering some characteristics of the functional relations in which behavior is embedded, including the nature of control and the nature of the environment as well as some alternative accounts for behavior (e.g., genes). We will even touch on *love,* which is not a trait.

Characteristics of Behavior

One outcome of reading this book is that you may share the perspective of behaviorologists about the most parsimonious view regarding *behavior and its explanations.* We view both of these as both natural and lawful, even though they can be extremely complicated.

That status leaves alternative mystical or superstitious accounts unneeded. A completely natural account, involving only real, physical events regarding *behavior and its explanations* is both possible and sufficient, and so remains our focus. Since science is an ongoing work in progress, maintaining a focus does not mean that we can fully explain everything yet, but of one thing we can be sure: Those alternatives are not explanations at all.

The Definition of Behavior

We cannot separate behavior and the organism and the environment. Without an organism with the necessary and appropriate physiology, we have no behavior, and behavior always occurs in a setting or situation or context (i.e., behavior never occurs in a vacuum). This part of the environment exists outside the organism's skin while another part of that environment exists inside the organism's skin. Physical matter comprises all these three—organism, external environment, and internal environment—and they interactively exchange energies, with behavior arising from and contributing to the interactions.

Thus, a common starting definition is that behavior is innervated muscular movements. In previous chapters we have already touched on more sophisticated characteristics of behavior, and we will soon consider some of these in more detail, so let's try a more thorough definition. *Behavior is an overt or covert pattern of either innervated muscular movements or neural activity involving energy exchanges within the organism, and their effects, that other energy exchanges within the organism, or beyond it in the external environment, elicit or evoke, or consequate.* While this defines a general notion of behavior, we use the term *response* for an instance of behavior, and we analyze changes in behavior in terms of response classes... But we get ahead of ourselves. Having considered what behavior is, let's consider what is not behavior.

The "Dead Body" test. Given that we confront a vast range of possible examples, we would welcome wondrous tests that always clearly separate behavior from non–behavior. No such tests exist. However, the "Dead Body" test provides a handy test that well fits part of our needs. I say "part," because while it ignores what *is* behavior, it is still very good at indicating what is *not* behavior. The Dead Body test says this: *"If a dead body can exhibit it, then it is not behavior."* Sometimes we say that a live body exhibited a behavior, by which we mean that the physiology mediated the behavior; we certainly disallow "exhibit" implying that some self agent initiated the behavior. The Dead Body test works for folks of either persuasion. On the one hand, superstitious folks accept that a soul or mind or self or person no longer inhabits the dead body, so whatever that dead body exhibits cannot be behavior. On the other hand, natural–science folks accept that to mediate behavior, the physiology must be "alive," so whatever that dead body exhibits cannot be behavior.

We emphasize the latter, the natural–science, physiological behavior mediation. Behavior occurs as a function of energy changes in the internal and external environments that affect living physiological structures in ways that we see as mediating behavior. Yet we can have great difficulty separating this mediation either from the neural structural changes or from the behavior occurrence. In some important senses, the mediation and the behavior are the same thing, with the behavior part more easily accessible. Actually the neural structural changes and the behavior occurrence have already been separated for study along the lines of disciplinary fracture between physiology and behaviorology. Now we must put these two back together again (which would seem a little easier than restoring Humpty Dumpty) to study the mediation itself. We will consider it more in later chapters, but we must address many other points first. Let's start with a little more about those environments.

The External and Internal Environments of Behavior

While we presume the physiology of organisms, because another capable discipline covers that subject matter, *environment* provides the functional part of our behavior subject matter. Some disciplines have little need to take the environment as anything other than everything "around" us. Yet the environment has even more to it than this.

To grasp what more the environment has, recognize that you are "out there" as a part of my environment, and I am "out there" as a part of your environment. We call this part, for each of us, the *external* environment, the "totality of everything around" us, but it is not everything. If we were merely magical beings, we could stop here, as being "out there" changes nothing for magical beings. But we are not magical beings. ("Oh, no! Surely we are. If not, why, that also means Santa Claus is not real, right?") Whatever we are, we still are. What we are does not change merely because our past conditioning compels us to think that we are something else, something supposedly more interesting for cocktail party conversations.

Well, I think reality provides more interesting things to discuss. We are physical beings ("carbon–based units," according to a variety of sources) and science already makes clear that our skin constitutes no boundary of any sort to the laws of the universe. Part of that universe is inside our skin, which makes that inside part also a part of the overall environment, a part we call the *internal* environment. Now, I suspect that your insides make a more important part of *your* environment than they make of *my* environment, and vice versa; nevertheless, physical things related to behavior go on inside the skin of each of us. For that reason we usually speak separately of the external environment and the internal environment. The external environment certainly allows easier access than the internal environment to probes regarding functional relations. Still, the internal environment is not inaccessible, it is merely less accessible. The stimuli producing behavior may exist on either side of the skin, as can any environmental effects resulting from a behavior.

For example, in the external environment, the light switch on the wall at the top of the dark basement stairs provides the occasion for (i.e., evokes) flipping the switch. As a consequence of that switch flipping, the stairway becomes visible, enabling safer transit down it, and back up.

As an internal environment example, consider the behavior of seeing the flashing of a dashboard idiot light in the shape of a gas pump. Like all behaviors, that seeing is real. It is a real, internal event that can serve (according to your past conditioning) the stimulus function of evoking one or the other or both of two response chains, an external response chain of looking for a service station and speeding to it and filling your tank (if you don't have to rob a bank first—I mean, of course, your piggy bank) and an internal response chain of further waves of neurons firing that is the mediating of various behaviors such as seeing images of billowing air pollution bringing on more tornado filled storm clouds. As a consequence of those internal imaging responses (depending on your past conditioning) feelings of increased trepidation about, and worry over, global warming arise. Both the images and the feelings are real, but they occur internally, directly accessible only to a *public of one,* you, the body in which they occurred. Given the right past conditioning, these real events could also affect your subsequent overt behavior, including driving to the station without speeding, behavior that we otherwise say the idiot light evokes.

In general we presume that all real events that could potentially or actually control behavior constitute the environment. So when we behaviorologists say "environment," that term encompasses *all* of the environment including both the internal and external parts. However, sometimes we stress the actual control of responding by using the phrase *behavior–controlling environment* (external and internal). Now, all this responding to external or internal environmental stimuli requires energy. From where does that energy come?

The Energy for Behavior

Every interaction between any part of the total environment and the body or behavior involves energy. For any environmental stimulus to elicit a respondent behavior or evoke an operant behavior, that stimulus must transfer energy to the nervous system. Likewise, for any behavior to affect the environment, it must transfer energy to the environment. And, for any environmental stimulus, which an energy transfer from the body produces, to result in more or less behavior on future occasions, it must transfer energy back to the nervous system in a way that changes the body structurally, making the body different so that it is the different body that receives evocative environmental energy on future occasions and so mediates behavior differently on those occasions.

Most of these energy changes, however, especially those occurring only in the nervous system, are sufficiently potent only to *trigger* their effects; they are not potent enough to *fuel* those effects. The fuel for those effects comes from the reserves that the body builds and maintains as a part of the internal economy of the organism related to ongoing survival, to which behavior is an important contributing factor. For example, you attend an author's talk at your local bookstore, and the speaker asks a question to which you have a relevant answer. The sound energy of the question is sufficient to trigger the release of enough of the potential energy that your body stores *to fuel the extension of your arm* into the air above your head, which the question evokes. Also, the sound energy of the question is sufficient to trigger the release of enough stored potential energy *to fuel the chains of neural responses* that the question also evokes, and which we call your private verbal thinking responses—which may take less energy than hand–raising—regarding the answer and how it gets phrased, should your hand–raising result in the speaker calling on you.

The Continuity of Behavior

A difficult, and yet fun, part of following considerations about behavior, like those regarding energy, is managing the continuity of behavior. Behavior occurs on a continuous basis, and often has many parts. Some of these parts occur continuously at the same time, and other parts occur continuously in a sequence, and differentiating them may seem a bit arbitrary, especially when we can speak of the smaller parts making up a bigger whole. But the seeming arbitrariness causes no real difficulties so we simply acknowledge it, and live with it, while working to reduce it.

As an example of continuous responses occurring at the same time, you may be pushing your left foot down while simultaneously turning then holding a wheel about 70 degrees counterclockwise while also pulling back a little on the wheel—all with your left hand—while simultaneously turning your head back and forth as you visually scan your field of view while your right hand simultaneously pushes or pulls a small plunger in or out by a couple of inches. Now, no "you," as some internal agent, initiates or does any of these things. But the circumstances of piloting a small aircraft evoke all of these continuous, simultaneous, and overlapping responses in a coordinated fashion as part of the single behavior that we technically call a steep left turn. (Flying is fun!)

As an example of continuous responses in a sequence, when you dash off an e–mail (or compose a paragraph for a formal letter, or craft a chapter for a book) your fingers may fly over the keyboard buttons continuously, which we see occurring as part of the single behavior of writing, but we can also consider each press of any button as a separate response, many of which accumulate in the overall effort of writing.

Also, we consider some behaviors as "discrete" when they seem to show precise starting and stopping points, while we consider other behaviors as "continuous" when they flow on without obvious or precise starting and stopping points. For example, when a musician pushes down and releases the three buttons that operate the valves on her trumpet *one after another*, we see each of those pushes as discrete. Yet when another musician slowly draws some varying portion of the length of his bow over each viola string in turn, we see each of those draws as continuous. We also see in these examples the opportunity to be arbitrary; as the rigor and sophistication of our principles and examples accumulate, we should see those opportunities diminish.

Response classes. To account for behavior in that tangle of confusing possibilities, sometimes we defer to the concept of *response class,* which groups responses together, and which we can apply to all those possibilities in ways that mitigate the confusion. This works, because a response class is defined as a group or variety of responses that have in common the same eliciting stimuli (in respondent behavior) or the same effects on the environment (in operant behavior). This concept steers us clear of two additional concerns. One is that every response is a new response (see Chapter 1) while the other is that every response is of slightly different topography (i.e., every response differs at least slightly from all other responses in physical measurements, such as its extent, vector, or intensity).

Technically, each of those two concerns makes responses non–repeatable, which makes measurement of individual responses rather difficult as well as unhelpful for analysis. With response classes, we instead *measure the response class members* to enable analysis, such as counting the number of times members of a particular response class occur in a measured amount of time, which can give us the rate of the behavior. We continue measuring the rate while observing whether or not any changes in rate occur when one or another independent variable changes. From the results of such experiments, we can derive various principles and conceptual details of behavior along with a range of complex behavior interpretations and applied interventions.

As an example, consider a child's behavior of saying, "please" at the dinner table when requesting food or other table items. Each instance, each response, differs from the others. Some are sincere and respectful while others are bland and trite and still others snide and obnoxious. Any instance of these types is a member of the response class that we could call "saying please." Since our current interest is to get these responses going, we may ignore the differences; we just treat them all the same and let them all produce the same consequences of receiving the properly requested item. (Later we should intervene to reduce the inappropriate, impolite response class members, applying an intervention that we will call "differential reinforcement" in a later chapter.) Before turning to the classification of behavior, you can try out the response class concept. Ask (and answer) yourself what different name we could give for each (or some) of the response classes that we used in our examples for this section.

Classification of Behavior

People have traditionally divided behavior into a number of sometimes different and sometimes overlapping types, some of which relate directly to our natural–science interest in human behavior. We will consider several behavior types across two broad classifications: (a) descriptive classifications of behavior and (b) functional classifications of behavior. The former classification stresses thematic connections between behaviors related to particular characteristics that they share, while the latter stresses functional connections related to the types of independent variables that they share.

Descriptive Classifications

Here we cover motor behavior, emotional behavior, neural behavior, verbal behavior, and the overt/covert behavior distinction. In many cases the same behavior may appear under more than one of these descriptive classifications.

Motor behavior. When asked about behavior, most people respond in ways indicating that their past conditioning emphasized motor behavior as possibly the most common, if not the only, type of behavior. *Motor behavior* consists of behaviors involving mostly innervated skeletal muscle contractions visible as whole body or limb movements, although it also covers movements at small scales that reduce the ease of detection as, for example, would occur if someone were softly whistling into the wind while facing away from you; while you could not easily see them whistling, you might not hear them either. Since all such qualifying muscle movements require innervation, we will more accurately refer to this behavior type as neuro–muscular behavior. The examples are so numerous that particular ones may seem trite. Consider a knee bending or a hand reaching. Let's get extravagant; playing an original piano version of a Liszt Hungarian Rhapsody is, at least to some, a more interesting if rather vibrant and strenuous example of neuro–muscular behavior.

Note that to the performer of that Hungarian Rhapsody, however, playing it involves several other types of behavior as well. After all, while we can all rather easily observe each other's neuro–muscular behavior, we can individually, even more easily, and often more completely, observe our own neuro–muscular behavior, as is the case with the piano player, because we are privy to (i.e., affected by) a range of stimuli that others cannot so easily access. Examples include the proprioceptive stimuli related to balance responses and the kinesthetic stimuli related to limb movement and placement, as well as a variety of emotional behaviors and neural behaviors of consciousness during the playing process.

Emotional behavior. We speak of *emotional behavior* when the activity that the body mediates involves certain smooth muscles and glands in the body releasing various chemicals into the bloodstream. While we tend to say that the emotional behavior *is* the action of these smooth muscles and glands that *results* in the release of various chemicals into the bloodstream, these two are rather inseparable at our behaviorological level of analysis, and we graciously accept guidance on the matter from our physiology colleagues.

In any case, the chemical release occurs as a dependent variable in a functional relationship with certain independent variables. It occurs neither spontaneously nor because some mystical inner agent directed its occurrence. Furthermore it is not the expression of a putative inner agent's existence, nor is it the main cause of other behavior. Instead the chemical release occurs in reaction to the occurrence of various external and internal stimuli. Some of these stimuli produce the release, because the body is already genetically structured to produce the release when these stimuli occur (i.e., without needing any conditioning first). Other stimuli, however, produce the release only after some conditioning has also occurred (i.e., these stimuli must first be paired with stimuli that already produce the chemical release); this conditioning changes the structure of the body so that the energy from the occurrence of these other stimuli will also produce the chemical release.

The chemical release, which we call emotion, has some important effects that center on two functions. Most obviously, the released chemicals function to produce effects that, in conjunction with other real variables, are detectable as feelings of many different kinds, and we often give very specific names to these different feelings. Perhaps more importantly the released chemicals function to change the body in temporary ways, including calming but more often exaggerating or arousing, such that the body then mediates behavior differently, if temporarily, than it would have without the chemical release.

For example, let's say you stayed up much of Sunday night carefully reviewing previously well–studied material, because your schedule would not allow time later for such activity, and your professor scheduled an exam at the end of the week. Then, at Monday afternoon's class, you are dozing off from the lack of sleep when your professor announces a "pop" quiz. This announcement elicits an emotional chemical release. The first effect of this release, which typically dissipates after but a few moments (and so we often do not consider it as very important) is that, in this context, it gets detected as a confused mixture of surprise, anger, and pleasure—surprise because the professor has never before given a pop quiz, anger at being jerked around, and pleasure because last night's study has made you exceptionally well prepared. The second effect of this release, which can last for much longer due to its effects on external events (and so we see it as more important) is that it temporarily changes the body which then mediates behavior differently than it would have without the chemical release. In this case, it energizes the body, counteracting your fatigue from too little sleep for several valuable minutes, and so enabling your preparation to produce much higher grade results on the pop quiz (which could have who knows what further long term effects on the future).

As other body parts or processes remove the chemicals from the bloodstream, the body returns to its normal operating levels. The body then mediates behavior without the additional controls from an emotionally changed body. Note that once some stimulus elicits emotional behavior, the detection of feelings, and even some of the changes in behavior mediation, occur on the level of neural behavior. So let's review that kind of behavior a little.

Neural behavior. While motor behavior involves innervated muscle movement, which we more accurately call neuro–muscular behavior, when we speak of *neural behavior,* we refer to behavior comprised only of neurons firing. Recall that the skin represents no boundary to the laws of the universe, even though it may represent a boundary that reduces human access to what goes on inside the skin. Even then, some human activities, like physiology, have more access than others, like behaviorology. We manage these limits through interpretive analyses when we confront limited access to confirmable functional analyses. This helps dramatically in extending our understanding of human nature and human behavior, because we can, as we must for a complete analysis, include internal events like neural behaviors in our analysis.

What makes that so important? Perhaps it is not so important, but our past conditioning has convinced us that we cannot understand human nature and human behavior thoroughly enough unless we include the really difficult and limited–access topics like neural behavior, even if only at the level of interpretation. The well supported assumptions of our philosophy of science, radical behaviorism, enable us to reach some scientific grasp of phenomena like the status and function of consciousness, because we can begin by considering consciousness as real neural behavior. As behavior, we can then track as best as we can the extent to which the same variables that control more accessible behavior also control the behaviors of consciousness and the full range of other neural behaviors.

As with emotional behavior, neural behavior occurs as a dependent variable in functional relationships with various independent variables. Similarly, as a real event that our physiology colleagues are getting ever better at directly tracking, once some stimulation produces it, neural behavior can serve stimulus functions producing either other directly observable behaviors, or other neural behaviors, often in the kinds of chains of cascading neural firings that seem to make up the various behaviors of consciousness or awareness, and so forth. Neural behavior occurs neither spontaneously nor because some mystical inner agent directed its occurrence, nor is it the expression of an inner agent's existence. A large portion of verbal behavior makes up another kind of neural behavior, so let's look more closely at verbal behavior.

Verbal behavior. Language is perhaps the single most noticeably complex behavior repertoire that conditioning processes can leave a body's living physiological nervous system mediating. We use the term *verbal behavior,* however, to encompass more than just what we usually call language. Groups of people speaking the same language (i.e., a verbal community) initially condition—a far more complex process than this single word conveys—much of the basic verbal behavior of new community members during the several years of their early childhood, and then continue, for each other, to condition refinements and extensions throughout life, aiding its quality and continuation.

In the usual definition, verbal behavior is not involved in *directly* producing the reinforcing consequences that maintain it (the way turning and pushing or pulling a door knob directly produces an open door). Rather, verbal behavior *indirectly,* through the behavior of one or more other verbal community members, produces the reinforcing consequences that maintain it. For example, "Open!" or, in a more refined, later iteration, "Please open the door," in the presence of another verbal community member and a closed door, indirectly produces an open door through the other's behavior of door opening).

While interpreting verbal behavior in light of the basic laws of behavior, Skinner (1957) differentiated several types of verbal behavior based on increasingly complex functional differences among them. Since we will have occasion to mention one or another type of verbal operant at different points in these early chapters—before providing a more thorough account in the verbal behavior chapter later—we will mention a few of them here, along with a point or two about how they work. However, although the determining factor in how we define the different verbal operants resides with *how* certain variables, accessible to the verbal community, condition these operants, we will defer the full definition discussion until the later chapter. Note, though, that more than three decades of published experimental research on verbal behavior relations and applications has accumulated, and this research validates Skinner's original interpretive analysis. Most of this research appears in the journal, *The Analysis of Verbal Behavior* (TAVB).

Let's consider four of the most elementary verbal operants. We call these four mands, tacts, intraverbals, and textuals. *Mands* are verbal responses that occur due to, among other things, deprivations, and the occurrence of the deprived items reinforces the response. For example, under conditions of going without food, the mand "Let's eat!" occurs upon arrival home after a day's work. *Tacts* are verbal responses that occur due to the evocative functions of *non–verbal stimuli,* and the reactions of listeners reinforces the response. For example, if you are camping in the forest and wake up due to the noises of a big brown animal rummaging through your food stores, the whispered tact "Bear" occurs, and your companions provide reinforcers when they whisper their thanks to you as you all creep out the other side of the tent to safety. Note that your yelling "BEAR!" is also a tact, but perhaps less likely to produce such reinforcement. *Intraverbals* are verbal responses that occur due to the evocative functions of *verbal stimuli,* and the reactions of listeners reinforces the response. For example, under conditions of an American teacher saying, "Red, white and ____?" in the presence of a student, the student's intraverbal response "blue" occurs, whereupon the teacher's response, "That's right" (based, in this case, on a flag–based thematically related set of colors) provides reinforcement. And *textuals* are verbal responses that also occur due to the evocative functions of verbal stimuli, but in this case the stimuli are printed words on a page. Additional variables also apply, leading to the comprehension that we tact as *reading.* Your current responses that this page is evoking provide the example, with your increasing understanding providing the reinforcement.

Much verbal behavior is clearly neuro–muscular behavior. However, verbal behavior also comprises a substantial segment of the strictly neural behaviors of consciousness, including thinking, hearing, and comprehending.

Overt/Covert behavior. This dichotomy helps us manage those behavior types that have some neuro–muscular members and some strictly neural members. *Overt/covert* refers not to another type of behavior but rather to another way to talk about various behavior types. It combines and emphasizes both the general location in which behavior occurs and the amount of accessibility others have to the behavior. We consider some behaviors as public events, because they occur in the external environment and so others have ready access to them; hence we call them *overt* behaviors. Other behaviors, however, we consider as private events, because they occur in the internal environment and so others have perhaps severely limited access to them; these we call *covert* behaviors. While overt behaviors allow easy access for anyone in the relevant kind of range of the behavior, covert behavior is only accessible to a public of one. The same body that mediates covert behavior is the only public that has access to it (i.e., the covert behavior, as a real event, serves to affect the public of one by evoking chains of additional covert neural consciousness behaviors of observing and reporting the observed behavior). As an overt/covert example, we consider some verbal behaviors as overt and other verbal behaviors as covert; we consider neuro–muscular verbal behavior (e.g., speaking, writing) as overt and easily accessible to virtually anyone, while we consider strictly neural verbal behavior (e.g., much thinking and knowing) as covert and accessible only to the public of one in whom it occurs and whom it affects.

Our more extended concern with the overt/covert distinction involves avoiding private internal events in general in our analyses due to the accessibility issues. But simply calling behaviors public or private works poorly, because behavior is not the only kind of event that fits a public/private distinction; the overt/covert distinction separates behavior from those other events.

Regardless of whether we are dealing with stimuli or behaviors or something else as private events, we use the same methods to avoid all the private events in our analyses, including covert behaviors that start as real dependent variables and then become, as real events, real independent variables affecting subsequent covert or overt behavior. The main step in this method involves moving back the necessary steps in the functional chain of events, moving away from the interpretive analysis of private independent variables or covert behavior dependent variables, until we are dealing with *public* independent variables or *overt* behavior dependent variables. Since the functional chain usually travels from the external environment to the internal environment and then back to the external environment, we can keep our analysis in the external environment by jumping the internal environment. This takes the loose analytical form of "if A causes B, and B causes C, then A causes C," where A and C share the external environment. Furthermore this move often allows us to account adequately, if imperfectly, for both the internal and subsequent external events.

Let's consider an example, which also shows additional aspects of the analytical method. Your checking account statement arrives in the mail, and you face the challenge of checking your figures and balancing your checkbook. Speaking less agentially, the account statement evokes a little number crunching behavior (which can be overt or covert) the reinforcing outcome of which occurs when the statement's final figure and the checkbook record's final figure agree precisely. Presuming you wrote but a few checks this cycle, outsiders only see you sitting at your kitchen table staring back and forth between the statement and the records, both on the table before you, and occasionally placing a tick mark somewhere on the statement or on the records. After a final pause, the observers see you write "OK" on both the statement and the records, and leave the table, the reconciliation between the two satisfactorily complete.

Now, those observers could independently review the documents and math, and thereby confirm that assessment. What they cannot confirm is any claims you report, even as a legitimate public of one, about verbal behavior computations you thought through privately while at the table. We take your claims seriously, though, as we have no reason to disbelieve you, particularly when, as in this case, no contingencies exist that compel lying about your thinking math behaviors. While we as outside observers can accept that those covert behaviors occurred, we cannot confirm that they occurred, nor can we confirm that they did not occur. Accordingly, we instead let stand the external functional chain of events showing the account statement evoking document review behavior (of the statement and the records) leading to the reinforcing outcome of the statement's final figure and the checkbook record's final figure agreeing precisely. Nevertheless, we interpretively accept that the "document review behavior" legitimately involved scientifically acceptable covert behavior chains as real events that, at least theoretically, are of a type that may become more accessible in the future, perhaps through greater collaboration with our physiology colleagues. By the way, one of the reasons for accepting the reality of your covert math behavior, which you reported as a public of one, is that the outside reviewers, when they reviewed the documents and math for correctness, could have completed this task either overtly *or covertly.* In the latter case, they would accept their own covert behavior part of the reviewing chain, which they observed as individual publics of one. So why would they deny your public–of–one observed covert behavior? They would not. No good reason exists to accept the covert behavior in one case and deny it in another when all else is equal.

Functional Classifications

Every one of the behavior types that we considered under descriptive classifications involves one or both of the two kinds of behavior that we consider under *functional classifications.* While the descriptive classification relied essentially on either the type, or the location, of the behaving body part, the functional classification relies on the types of independent variables determining the behavior. Basically, eliciting stimuli control *respondent behaviors* while evocative and consequential stimuli control *operant behaviors.*

Respondent behavior. We credit Ivan Pavlov, a Russian physiologist researching digestive processes, with discovering what we now call *respondent behavior* early in the 1900s. Pavlov discovered that some functional stimuli, once they reach a certain threshold, completely compel the occurrence of a particular response. Other supplemental stimuli cannot prevent this from happening. We call the effect of these functional stimuli "elicitation" and say that the stimulus elicits the response. Since we are dealing with bodies that must mediate (not originate or initiate) behavior, what is really happening?

What is really happening is that the body is structured such that the energies from the stimulus automatically make the body mediate the behavior. If the body structure is the structure that the genes produced, then we call both the stimulus and the response that it elicits *unconditioned,* because no conditioning was needed for the stimulus to elicit the response. For example, if a companion surprises you by stuffing a small but quite spicy taco hors d'oeuvre into your mouth, you salivate. That little spicy taco is an *unconditioned stimulus* that elicits salivating as an *unconditioned response.*

Pavlov also discovered that other stimuli, ones that started out having no effect on the response under study, so we call them *neutral stimuli,* can come to have eliciting effects. Indeed, nearly any stimulus consistently paired with an unconditioned stimulus will start eliciting the response, but this requires other terms. Since we call pairing the stimuli *the conditioning process,* we call the no–longer–neutral stimulus the *conditioned stimulus,* and we call the response that the conditioned stimulus elicits the *conditioned response.* The energy transfers in the conditioning process change the body so that now the occurrence of the conditioned stimulus automatically makes the body mediate the behavior as a conditioned response. These terms, of course, are for our convenience; meanwhile, the body merely mediates. For example, if those tiny spicy tacos (an unconditioned stimulus) are only on the menu of a particular restaurant chain with a notable logo (originally a neutral stimulus) and you eat them regularly (inevitably pairing the tacos with the logo), then quite soon the logo itself (now a conditioned stimulus) will elicit salivating (now as a conditioned response). Whenever you see the logo, you salivate somewhat. Virtually everyone has had similar experiences.

The conditioned stimulus will continue to work this way so long as it and the unconditioned stimulus at least occasionally occur together. Happily (or perhaps not) if the restaurant drops those tacos from the menu so that they never again pair with the logo, then the logo will eventually stop eliciting salivating (unless, of course, the restaurant serves other mouth–watering delicacies that you keep eating). We call this process extinction; actually, we call it *respondent extinction,* because another type of extinction exists for which we traditionally reserve this term. In respondent extinction, the unconditioned stimulus and the conditioned stimulus no longer occur together with the result that the conditioned stimulus ceases to function as such and returns to being a neutral stimulus. Of course, the stimulus has not changed, rather the body has changed due to the changes in energy transfers.

Respondent behavior and respondent conditioning extend much farther than just tacos and slobbering. They are intimately bound up with the occurrence of all emotions and feelings, and with the internal economy of organism, and with phobias and so many other things.

Indeed, when discovered, researchers began to search for all the eliciting stimuli for every human behavior. However, they were unable to discover all those stimuli because, as they were—and we are—soon to discover, stimuli do not elicit all human behaviors. Instead, another kind of behavior exists, to which we now turn.

Operant behavior. We credit B. F. Skinner, one of the earliest natural scientists of behavior, with discovering this other kind of behavior in the 1930s. He called it *operant behavior* because, when stimuli evoke this kind of behavior, the behavior *operates* on the environment, changing that environment in ways that we call consequences. While Skinner understood that *stimuli evoke operant behavior,* he found that using the word "emit" increased the contrast between operant and respondent behaviors, which helped others to "understand" these behaviors (i.e., to respond correctly to the them). So he described operants as behaviors an *organism emits* while describing respondents as behaviors that *stimuli elicit.* However, *evoke* more accurately describes these behavioral events, while "emits" can imply a putative and unnecessary inner agent to "do" the emitting (or to tell the body to do it). For these reasons we eschew the term "emits" in favor of *stimuli evoking operant behavior.*

The evocation of operant behavior must happen, and always happens, but it is not the same as eliciting a behavior (a point to which we will return shortly). The consequences of the operant behavior transfer energy back to the body where this energy changes the body physically. The energy changes some neural structures so that the body now *more readily* mediates similar behavior *when similar evocative stimuli occur again.* Since we have less access to the body than our physiology colleagues, because that is not our level of analysis, we address that energy effect from consequences as *two* observable changes (the details of which those physiology colleagues will someday elaborate): After the consequences occur, (a) we see the evocative stimuli function more effectively in the future (meaning they more reliably evoke the behavior), and (b) we see the behavior increase in rate across future instances, which we sometimes describe as an increase in probability.

Rate, though, is just a potent and rigorous measurement method, and the stimuli are not what changes. What actually happens? The body changes. Let's repeat some details as a review: The energy from the consequences changes some neural structures so that the now different body *more readily* mediates similar behavior *when similar evocative stimuli occur again.* Note that this effect of consequences defines the consequential stimuli as reinforcers.

Consequences, however, can have another effect. Sometimes the energy from the consequences changes some neural structures so that the body now *less* readily mediates similar behavior. This effect of consequences defines the consequential stimuli as punishers, which also have respondent effects that generally are the opposite of the respondent effects of reinforcers (more later).

All these processes go on continuously, on a moment by moment basis, for our whole lives. (Many consider their cessation, and death, as the same thing.)

While we will cover all the technical details, principles, processes, and procedures in later chapters, these two types of behavior and conditioning, respondent and operant, provide the general structure for dealing behaviorologically with human nature and human behavior. Here are a couple more distinctions that will help us in this endeavor.

Elicit/Evoke distinction **and** ***respondent/operant distinction.*** The distinction between respondent and operant behavior and conditioning forms the foundation for effective understanding, prediction, and control of our behavior. Helping us make these distinctions is the *elicit/evoke* distinction.

While an elicitation is a kind of evocation, we restrict the term *elicit* to respondent behavior (which others call *reflex* behavior). We say "the stimulus elicits the response." That is, we use *elicit* when a particular stimulus transfers energy to the body and that energy transfer triggers neural structures—either that genes built or that earlier energy transfers have changed (i.e., whether of an unconditioned or conditioned status)—to mediate a fuel–supported response that *the occurrence of other stimulus energy transfers to the body cannot stop or otherwise change, and to which consequences are generally irrelevant.*

Let's repeat the core of that point with fewer qualifiers: We use *elicit* when a stimulus energy transfer triggers the neural–structure mediation of fuel–supported responses which other stimulus energy transfers cannot stop or change, and to which consequences are essentially irrelevant. More simply, some stimuli *elicit* responses, such as an increase in ambient light (a stimulus) compelling a decrease in pupil size (a response). For example, the sun's reappearance from behind a moving cloud *elicits* pupillary constriction; the occurrence of other stimulus energy transfers cannot stop the constriction, and any subsequent consequences make little difference in the constriction.

What differentiates respondent behavior and conditioning from operant behavior and conditioning is that with the former (i.e., respondent) *other stimuli cannot stop or otherwise change the mediation of the response, and consequences are essentially irrelevant,* while in the latter (i.e., operant) the occurrence of other stimulus energies *can interfere* with response mediation, resulting in a different response, and *consequences are vitally relevant.* Due to that differentiation, we use *evoke* (rather than elicit) with operant behavior.

We say "the stimulus *evokes* the response, which other stimuli then consequate." In our more comprehensive description, note the differences even though the phrasing is by design similar to how we just described "elicit," as this helps show the similarities between respondent and operant: We use *evoke* when a particular stimulus transfers energy to the body and that energy transfer triggers neural structures—whether from genes or already changed by earlier environmental energy transfers (i.e., whether of an unconditioned or conditioned status)—to mediate a fuel–supported response that *the occurrence of other stimulus energy transfers* **can** *stop or otherwise change* (interfering with response mediation and resulting in the occurrence of a different response) *and to which consequences are essential.*

Let's also repeat the core of that point without all the qualifiers: We use *evoke* when a stimulus energy transfer triggers the neural–structure mediation of fuel–supported responses that other stimulus energy transfers *can* stop or change, and to which consequences are essential. More simply, some stimuli *evoke* responses and, equally importantly, those responses produce consequences that alter the future occurrence of similar responses. That is, those responses produce consequences that change the neural structures such that they more readily (with reinforcers) or less readily (with punishers) mediate similar responses when similar evocative stimuli occur again.

As an example, particularly of how other evocative stimuli can interfere with the mediation of an evoked behavior, producing a different behavior, consider this. While enjoying a post–lunch siesta, your alarm sounds, and the clock reads 2:50. This gives you 10 minutes to get to your 3 PM afternoon class, and you start walking toward the classroom. The clock stimulus has evoked going to class (where more reinforcers are likely more contingent on your preparation than on your mere presence). But on the way to class you pass someone whose car is stuck in the mud and he gestures toward you for assistance. That gesture stimulus now evokes detouring to lend a hand. After helping, the sight of the classroom building might re–evoke going to class, although the amount of mud on your hands first evokes another detour, this time to the wash room. In any case, you can see how other evocative stimuli can interfere with the mediation of an evoked behavior, producing a different behavior. It also shows that the clock was not an eliciting stimulus as, in that case, none of the detours could occur. By the way, if your past conditioning has not left you ready to help "out of the goodness of your heart," then your past money–related conditioning likely leaves the $20 bill, which the car owner waves at you (as a last resort to get your help) more effective in evoking your aid, and we still see you detour to help move the car out of the mud. Also, "out of the goodness of your heart" is not an adequate scientific account; in due time you will be able to provide a proper scientific alternative.

Operant/Respondent distinction summary. Aside from operants and respondents that involve strictly neural firings, here are three of the differences that we have mentioned between operants and respondents. Operants involve innervated striated or striped or skeletal muscles, while respondents involve innervated smooth muscles and glands. With all operant behaviors, conditioning involves reinforcers following an evoked response with the effects that (a) the evocative stimulus function becomes more effective, and (b) the rate at which similar responses occur increases, while with all respondent behaviors, conditioning involves the pairing of two stimuli with the effect that the new stimulus now also elicits the response. And, evocative stimuli produce operant responses that affect the environment, producing consequences, while eliciting stimuli produce respondent responses, such as emotions and feelings and physiological changes involving the internal economy of the organism.

The description of the distinction between elicit and evoke will hopefully improve your distinguishing between respondent and operant behavior and conditioning, since these two form the foundation for effective understanding, prediction, and control of our behavior. For further clarification, however, let's consider what is *not* behavior.

What is *Not* Behavior

Some phenomena seem like behavior, usually because they involve some type of motion. However, they are not behavior, mostly because they are not innervated motion. Here are some examples pertaining to some structural changes, and to traits, all of which sometimes masquerade as behavior.

Structural Growth or Decay are not Behavior

Various chemical processes that accumulate deposits as a growth mechanism are not behavior. The apparent elongating movement of finger and toe nails, or claws, or hair or fur, are chemical–deposition growths, not behaviors. Also, changes in height and weight are not behaviors, even though some of the relations of which they are a function involve behavioral variables (e.g., exercise and eating). Changes in muscle bulk are not behaviors, even though they too involve behavioral variables. Increases in bulk can occur due either to exercising or to not exercising just as decreases in bulk can occur from exercising or not exercising. Similarly, inadequate dental–hygiene behaviors can lead to bad breath, but bad breath is not behavior.

Those are only some of many simple non–behavior possibilities that I hope never fool you. More complex ones, like traits, constitute a greater challenge.

Traits are not Behavior

We may generally describe traits not as behavior but as adjectives that others, who witnessed an event, turn into nouns when telling you about the event (which you never witnessed). If we take these nouns uncritically, they all too easily become a putative material, or an entity, present in the body that someone could claim account for the behaviors that occurred in the unwitnessed event.

In brief the adjectives may have originally described actual behaviors, but the nouns, at best, only label some characteristics of the behaviors. At worst the nouns become false causes of the behaviors. We say false, because the event reporter only invented the traits from the adjectives describing the original event, so the nouns have no independently verifiable existence. For instance a friend tells you that a mutual acquaintance revealed a lot of obnoxiousness and hostility after listening with great impatience to the remarks of an administrator at a meeting. Yet, except perhaps for "listening," your friend has not mentioned a single actual behavior. You are left to guess what the behaviors of the acquaintance actually were (as well as what the administrator said). Furthermore, whatever the acquaintance's behaviors were, they did not occur because the acquaintance supposedly harbored obnoxiousness, hostility, and impatience, as if these denoted characteristics or traits of some inner self agent (of whatever sort) in the acquaintance. Such mystical inner agents are, like traits themselves, irrelevant to a natural–science account of behavior, so having these characteristics, these traits, is of little import.

Love is not a Trait

To further clarify that traits are not behavior, let's consider *love*. When some people say, "He is such a loving father," or "She is such a loving wife," they think they are alluding to love as a trait, as a characteristic of some mystical agent inside him or her, a characteristic that someone might falsely take as a cause of his or her behavior. But love is not a trait. They are instead alluding to complicated aspects of the father's or wife's or anyone's behavior, in relation to others, by using that single word, *love,* as a sort of verbal shortcut that summarizes a deeply poetic, multi–faceted and scientifically explicable natural phenomenon. However, past cultural conditioning may disable appreciation of this interpretation in favor of one or another more traditional notion. Our treatment here begins to elaborate the details behind this verbal shortcut.

Love is the term we use to summarize some complicated effects. These effects occur, because the stimuli that elicit the feelings we call love, and the stimuli that reinforce the operant behaviors that we also cover when we speak of love, are the same stimuli (whether unconditioned or conditioned) and their interactions produce the reciprocating patterns involved in love. Before elaborating the science, let's consider some related points.

Traditional notions of love, coming in a wide variety nearly all of which are fascinating, charming, romantic, and so on, themselves derive from natural phenomena (e.g., the kind, amount, and specifics of one's past conditioning history) and so, along with love, are amenable to a scientific description. Since many of these traditional notions erupt from various conditioned mystical self–agent accounts in the common lore, they will conflict with any scientific account. But our focus will remain the scientific description of love.

In the past we have perhaps been under contingencies that left us fooled about what love is, and now we can begin to make elegant improvements. You see, love does not change character simply because we consider it scientifically; whatever love is, that is what love was, is, and remains after our scientific description. This exercise is worthwhile, due to the potential implications for life and living that become possible from an improved understanding of love when we consider it scientifically, even if our consideration here only interpretively expands our understanding on the basis of the few general principles of behavior that we have touched on so far.

Only ever so rarely will we find an instance—and be glad for it—when contingencies induce benefits merely through astute, intuitive observation rather than as an outcome of scientific activity. For instance, the author of the works of Shakespeare (poems and plays and all) had no science of behavior as a writing guide, and probably shared with contemporaries some mystical notions about love, and yet this writer was still an astute observer of human behavior and the contingencies in which it occurs, and well applied those observations in accurate, interpretive and poetic portrayals of human behavioral possibilities, including our averages, excesses, deficits, loves, faults, and heroics. Given the last several thousand years, we can gladly count a few other rare speakers, artists, and authors whose works—natural products of natural contingencies—have bestowed similarly astute and beneficial observations and applications on us, even though they were without the benefits that science can add. Let's elaborate a little about how science handles "love."

Consistent with the rest of natural behavior science, we may interpret the word *love* as referring to complex interactions of respondent behaviors and operant behaviors, along with their eliciting and evocative and reinforcing stimuli, all with particular characteristics. Perhaps as much from biological evolution as cultural selection, we value these interactions the most when they occur between two opposite–gender, or even same–gender, bodies, although we also value them when they involve more people, and we even often value them when they involve only one person plus some class of objects or events. In his 1953 textbook, B. F. Skinner introduced the topic of love with reference to its vernacular usage where love can refer to the predictability of some particular behaviors occurring more than other behaviors. For example, we can safely predict that someone otherwise describable as in love has a greater "…tendency to aid, favor, be with, and caress" rather than "injure in any way" their partner or other object of their love (p. 162). Elsewhere in his book, Skinner points out that "love might be analyzed as the mutual tendency of two individuals to reinforce each other, where the reinforcement may or may not be sexual" (p. 310). These points speak to changes in operant behavior that respondent emotional chemical releases support, something to which we will return shortly.

First, though, you should not take "in–love" as implying a bodily state, or a state of being, that is responsible for loving behavior. The temptation toward this implication stems not only from the general temptation toward mystical inner agents from our common, traditional cultural conditioning but also from the sometimes lengthy occasions when emotion–produced feelings linger, or when ongoing stimulus changes recharge such feelings. One might, as a verbal shortcut for the necessary details, call these occasions a "bodily state," but these states are otherwise superstitious and meaningless as causes.

Let's now examine love more closely by slowly detailing many of its interconnected parts. Generally, love involves certain respondent emotions (with their accompanying feelings) that alter the ways (e.g., arousal) in which certain neural structural changes mediate lots of related operant behaviors in a positive feedback loop that can continually increase these behaviors and their related emotions and feelings and reinforcing effects.

Say… What?! Perhaps combining some focused review will help. Reinforcing stimuli (both unconditioned and conditioned reinforcers) not only change neural structures such that the stimuli that evoked the behavior that produced the reinforcers become more functionally effective in evoking that behavior upon future occurrences, but (a) reinforcing stimuli also condition other stimuli to serve reinforcing functions (as additional conditioned reinforcers) because they reliably occur at the same time (a pairing process), and (b) reinforcing stimuli also often serve an eliciting function (as unconditioned and conditioned stimuli), an eliciting function that in the past we have treated too lightly as a mere side effect of reinforcement, an eliciting function that involves eliciting the release of chemicals into the blood stream, an elicited release that we label emotion and that, as real events, produces real detection responses that evoke the label "feelings," some of which are part of *love.*

Those feelings, as real events, also serve as consequences that alter the way neural structures mediate subsequent behaviors. Normally, any emotional arousal *only* temporarily changes the body, while the chemicals circulate, such that the body mediates behaviors (which other environmental events must still evoke) differently from the way it would mediate those behaviors if the emotional arousal had not occurred. However, when the feelings resulting from emotional responses consequate the behaviors that produced the stimuli eliciting those emotional responses, the resulting structural changes may endure beyond the moments when the chemicals course through the bloodstream. This may constitute an important method by which the occurrence of stimuli that we call reinforcers leave the body changed such that the body more easily mediates in the future the behavior that produced the stimuli. Future research will clarify to what extent this holds.

All the events in that series are *real* events, all are at least theoretically measurable, *and all occur together in love,* clearly giving love the status of a real event. When we speak of love, we speak at least of the feelings that elicited emotional chemical releases produce. But we really speak of so much more than that when we speak of love.

Consider two people in love, first on the level of operant behavior, and then on the level of respondent behavior. Injuring the other in any way is not a part of their mutual evoked behaviors. Instead, for each, the very presence of the other evokes responses describable as loving (i.e., they each aid, favor, stay with, and caress the other). And in love, *any sign that you reinforce the other is reinforcing to you.* (Technically, reinforcers *only* affect behaviors, not organisms; however, to simplify phrasing with this complex topic, we occasionally engage the verbal shortcut of "reinforcing people." For example we could have said that any sign *that your own behavior reinforces the behavior of the other* reinforcers your behavior.) A few such reinforcers hold unconditioned status, such as attention and copulation. (The notion of attention as an unconditioned reinforcer evokes considerable contention among some behaviorologists, contention which we consider in a later chapter.) However, most of the relevant reinforcers hold conditioned status—having occurred with the unconditioned reinforcers—including the sight and sound of the other, which both inevitably pair with the other during reinforcing activities. Indeed, such pairing is how the signs themselves, that one's behavior has reinforced the other, become reinforcers for the one! Furthermore, if the other's behavior *also* functions as a reinforcer for one's behavior, then we often speak of romantic love.

Many of these reinforcers end up being contingent on (i.e., delivered to the other on) occasions when the other's behavior supplies reinforcers to the one. In this way, both the other and the one reciprocally provide (often so smoothly and quickly as to be simultaneous) these reinforcers to each other. Taken together, these operant components of love involve an ongoing, mutually beneficial, reinforcing interaction that emphasizes evidence of each having successfully reinforced the other, with each finding that success itself reinforcing. All this supports continuing the mutually reinforcing interactions.

All those operant–level variables are not only operating at the same time that respondent–level variables occur, but some of those variables, particularly reinforcers, also function as eliciting stimuli (both unconditioned and conditioned) both for respondent behaviors and, when paired with other stimuli, for the making of (i.e., the conditioning of) additional conditioned eliciting stimuli. Apparently due to our genetically produced neural structure, unconditioned reinforcers already tend to elicit emotions as chemical releases into the blood stream that produce detection responses as agreeable feelings (whether these are elicited or evoked is not yet entirely clear). Similarly, when the course of events naturally pairs neutral stimuli (i.e., stimuli, many of which are present at any time, that currently have no effect on the behavior of concern) with other reinforcers, those stimuli loose their neutrality and begin functioning as conditioned reinforcers. In addition, however, since the unconditioned reinforcers also function as unconditioned eliciting stimuli, the new conditioned reinforcers also begin functioning as new conditioned eliciting stimuli. (Of course, the change is not actually in the stimuli, which remain as they were; rather the change is in the neural structures, which now mediate operant and respondent behaviors differently.) As a result the number and range of variables eliciting the emotions we detect as feelings of love multiply along with the operant reciprocal reinforcing events, all interactively present together, constituting what we call love.

Summary of love. Before taking up more general functional relations involving behavior, let me share with you yet another author's summary about the behaviorological analysis of love. This includes two familiar "what one does" agential verbal shortcuts that you can translate:

> Love is a term describing one's feelings that are elicited when one's own behavior is reinforced by evidence of its reinforcing effect on another organism—that is, when the reinforcer for what one does consists of any sign that what one does functions as a reinforcer of the other person's behavior. Romantic (i.e., mutual) love further requires that the other person's behavior be a reinforcer for your behavior. People "in love" with each other are reinforced when their own respective behaviors prove reinforcing to the other party. "I love you" means that my behavior toward you is reinforced by evidence that my behavior ranks high among *your* reinforcers. Also, we are especially likely to say "love" when the degree of respondent conditioning has been high, and the elicited emotional reactions are especially strong (Fraley, 2008, p. 929; emphasis in the original).

Functional Relations and Behavior

We already discussed not only independent variables and dependent variables and functional relations in some detail, but also our philosophy of science assumptions that behavior, as a natural phenomenon, is subject to natural laws. As behaviorologists we try to discover those laws. While they explain *why* behavior occurs, natural laws at the physiological level of analysis explain *how* those laws work.

We stress the connections between the physiological and behaviorological levels not because our accounts cannot stand alone, since they can stand alone and lead to extensions of knowledge, skills, principles, and practices regarding behavior. Rather we stress those connections as part of counterconditioning the traditional cultural lore that conditions people to accept various agential causes with which theological mysticisms (mostly external agents) or secular mysticisms (mostly internal agents) claim to cover the behavior subject matter.

Our ventures into the physiological level usually entail interpretive analyses that consistently overlap our functional analyses, helping us manage the limited access we have to the workings of the behaving bodies, especially the nervous systems that innervate and thus mediate the behaviors we study. For example, based on the observed reliability between the occurrence of some environmental stimulus and the occurrence of a particular response, we interpretively follow an energy trace from that environmental stimulus into the nervous system and out again as mediation of that particular response. Yet we usually lack direct access to that trace at our level of analysis. In this circumstance we use the term *evoke* to bear the weight of the functional relation. Let's consider some other aspects of functional relations and behavior.

Natural Laws Control All Behavior

Here is a very short but vital point, both philosophically and scientifically. It is short because we have covered it several different ways already. But is it so vital that it deserves its own section to stress its importance. It is so short that the side heading nearly says it all. To reiterate, *natural laws control all behavior,* which means that all behaviors, as dependent variables, occur as a result of the occurrence of other natural events as independent variables, all in the natural functional histories making up all events. For behavior, the controls reside in the functional relations that the terms respondent conditioning and operant conditioning summarize (with the term *contingencies of reinforcement* even better summarizing the functional relations of operant conditioning, even though that term is itself a summary term covering not just reinforcement but the full range of operant functional relations).

Environment–Behavior Functional Relations

As a short definition, some behaviorologists describe behaviorology as the science of environment–behavior relations. This covers the independent and dependent variables of most interest to behaviorologists, namely all the independent variables definitive of both the consequences and the settings in which a behaving organism exists, and all the dependent variables definitive of the innervated and so mediated activities of the organism. When we then observe a behavior dependent variable reliably changing each time a setting independent variable changes, we describe a functional relation between them. If our observation is accurate, then a change in a setting variable becomes the basis for predicting a change in the behavior variable. Such analyses provide the scientific alternative to mystical agential accounts.

Still, with so many setting variables existing in any local environment, we can have quite some difficulty tracking which one provided the energy trace that produced a particular behavior. Indeed, if our past conditioning has not made looking for those energy traces likely, then we may not even take their presence as a meaningful possibility, leaving us to take the behavior as appearing out of nothing, which leaves us susceptible to positing it as the product of some spontaneous internal activity. For example, if the energy that triggers a response enters the body through the eye, we are hard pressed to trace that energy from its source when the setting may contain thousands of potential sources, and that is only if we see the relevance of antecedent energy traces. With no antecedent energy traces to account for the behavior, nor even any conditioned tendency to look for them, we may accept or invent a mystical account with just the right characteristics to explain the observed behavior.

Compounding that kind of agential error, we may further attribute the behavior to some intrinsic variance in the operations of the inner agent. This is an all too common attributed source of behavior occurrence. Variance, however, stems not from the capriciousness of the inner agent but from the difficulty in searching for and tracing the energy transfers from the environment into the body. These energy transfers determine (i.e., control) the behavior. Any observed variance results from our incomplete accounting for *all* the relevant energy traces, which is not surprising given their great number *and* the great difficulty of following them; we manage that variance with different methods, such as describing it in terms of probability. Nevertheless, at our level of analysis, the behaviors, and the energy traces for that matter, are completely determined, and behaviorology strives to sort it all out as much as possible, and to apply the findings to benefit humanity. Let's peek a little at how that works.

General Analysis of Behavioral Events

The general approach to accounting for behavioral events proceeds by first analyzing the antecedent events (i.e., those occurring *before* the behavior of concern) for functional relations, and then, if appropriate, analyzing the postcedent events (i.e., those occurring *after* the behavior of concern) for functional relations. Sometimes the antecedent–event analysis is enough, as it documents *respondent* functional relations. When respondent relations are either not present or not a complete account, then we proceed further with the postcedent event analysis. That analysis usually documents *operant* functional relations. Either case lays the foundation for work on interventions.

An Analytical Example

Let's look at a "simple" example of analysis in some detail. We must consider the antecedent event, the behavior of concern, and the postcedent event.

Here is the stage. You have had in your cellar for some years now an uncommonly good bottle of port from your favorite South Australian winery. You hope to serve it someday to an uncommonly good Australian friend, although you have no clue when, if ever, she might arrive. You have not seen her for years past and you may not see her for years to come. And you worry that she might not like your special port and think less of you as a result (as she hails from Queensland and has definite wine preferences that differ from yours). Then one day quite unexpectedly she turns up on your doorstep...

Our first analytical job involves identifying ① the behavior of concern, ② the antecedent event, and ③ the postcedent (possibly consequent) event. Then we must, if possible, relate ② and ③ functionally to ①. If they are functionally related, we will use more accurate terms for them. We analyze, because we want to know what the variables are and how they are related. So we ask three questions. What is the behavior of concern? What evoked the behavior of concern? And what consequence did the behavior of concern produce?

Perhaps somewhat arbitrarily we take ① to be serving the special port. After all (to ask some of the questions differently) you have kept the port a long time, so why would you open it now? What difference might opening it make?

Using visual aids often assists analysis. Our first question is, "What is the behavior of concern?" Let's diagram the answer:

Antecedent Event	Behavior of Concern	Postcedent Event
?	① **Serving special port**	?
②		③

Our second question is, "What is the antecedent event ②?" Let's diagram this answer also, because it might show a functional antecedent event that could explain why you would, after so many years, break out the port. In this case the answer is your friend's surprise arrival:

Antecedent Event	Response* of Concern	Postcedent Event
Friend's surprise arrival	① Serve special port	?
②		③

*We change the general term *behavior* to the more specific term *response*, because this is a single instance of a behavior.

Our third question is, "What is the postcedent event ③?" Let's diagram this also, because it might show a functional postcedent event that could explain recurrences of the behavior of concern, port serving. As investigation reveals, your friend finds your special port one of the best she has ever tasted. So she compliments you profusely and, when she leaves, she says, ever so sweetly, "I'll be back." While each of these is a postcedent event, we predict that the one most relevant to future serving of special port is the profuse compliments:

Antecedent Event	Response of Concern	Postcedent Event
Friend's surprise arrival ②	① Serve special port	**Profuse compliments** ③

If we have access to the rate of your friend's return visits, we will discover if your receiving those profuse compliments increases the rate of your serving your special port. If we discover that pattern holding, then we can use more sophisticated terminology, which connotes the functional relations, to diagram the following three–term contingency containing two functional relationships (which the arrows indicate, pointing in the direction of the functional relation):

Evocative Stimulus	_Response of Concern_	**_Reinforcing Consequence_**
Friend's arrival →	Serve special port →	Profuse compliments

Reading that last diagram provides answers to all our analytical questions. Your friend's arrival evokes serving your special port which produces compliments for you and, as an outcome or result, the occurrence of those compliments increases the effectiveness of your friend's arrival as an evocative stimulus for port serving and thus you serve special port more often.

Our questions this time only focused on the events *immediately* preceding and following the behavior of concern. Clearly, earlier events led up to these events just as various later events may also affect similar future events. As our analytical knowledge and skills develop, we will deal with more complex events.

Meanwhile, questions from other readers about the brand of port, or how they can reach your friend, are not germane to this discussion. However, alternative accounts for behavior are germane to the general discussion.

Alternative Accounts for Behavior

We have touched already on some alternative accounts for behavior, mostly of the mystical agential variety. We will cover a range of others in the next chapter, but those are more of the fictional variety. Here we touch on a couple of non–

fictional accounts that nevertheless prove inadequate as accounts for behavior, namely, the passage of time and genetic endowments.

Some Caveats About Time

If you couple the ubiquity of clocks with the current relatively sparse cultural conditioning regarding the functional effectiveness of antecedent energy traces in producing behavior, then you are not surprised that many people rather casually attribute the occurrence of behavior merely to the passage of time. That is, their explanation–related conditioning has been deficient, and so they think that as time passes a behavior's turn arrives and so it occurs.

Time, however, has no such functional status, because time itself has no real, independent existence. It is an intellectual construct that, while useful in some contexts, we define as a non–spatial, non–reversible continuum consisting of a sequence of events. That is, a sequence of events defines time by filling it. Thus time itself, lacking physical reality, cannot transfer energy to the body and so cannot serve as a functional antecedent for behavior.

However, any of the physical antecedent events in a sequence that defines time *can* provide traceable energy transfers to the body and so serve as functional antecedents for behavior. These can be as complex as the covering of the usual series of sub–topics at a convention presentation about a research experiment (i.e., introduction, methods, results, discussion) or as simple as a progression of heartbeats or the sweep of the second hand on an analog clock face. Traceable energy transfers to the body from any such physical event in a time–defining sequence can produce behavior. But those events are producing the behavior; the mere passage of time is not producing behavior.

For example, your physician directs you to take a dose of medication everyday at 6 PM after you get home from work. Each day you get home sometime between 5:30 and 5:50, due to traffic. Once home, though, part of your behavior includes watching the clock. At 6 PM you take the medication. Time passing provided no energy trace to the body to evoke this behavior; rather, the configuration of the clock face provided the energy trace to the body and thereby evoked this behavior.

Note that your "clock watching" likely itself is more complicated than merely sitting and staring at the clock face until 6 PM arrives. You might more accurately describe it with the verbal shortcut, "keeping track of the time," as other stimuli evoke and consequate a series of other behaviors, such as setting the table for dinner. Meanwhile, as but one of many possibilities, the "keeping track of the time" behavior may involve regular, repetitive glances toward the clock interspersed among these other behaviors. However, regardless of the complexity, the passing of time still provides no antecedent energy trace evoking the medication–ingestion behavior. Instead the clock face finally reaching the 6 PM configuration provides that energy trace during the last of those glances, at which point you take your medicine.

Some Caveats About Genes

With time not an explanation for behavior, what about genes? When contingencies move people away from either obvious theological mystical accounts for behavior, or obvious secular (e.g., psychological) mystical accounts for behavior, many of these people derive some comfort in scientifically connected genetic accounts of behavior. Even gratuitously simplistic genetic explanations, such as "she does that because she was born that way" or "he does that because it's in his genes," carry some aura of scientific authority through a generalization process from real science. Sophisticated genetic accounts carry more such authority, even if unjustified. Consider examples like, "he works well with children because he inherited a very calm demeanor from his father," or "scientists say they have located genes for alcoholism and smoking."

One problem with such accounts is that, even if genes could cause behavior, interventions remain in the realm of science fiction. Although we cannot readily change our genes, even if we can turn some off or on, we still lack the technology to manipulate genes as independent variables on the time scales at which behaviors occur and change. However, the greater problem with all genetic accounts is that genes cannot exert any direct control over behavior, because *genes only produce physical body structures.* That is the biological role of genes. The body structures that result from genes then interact with antecedent environmental energy traces, and change in ways that produce (as in mediate) behavior. In that *indirect* way, we can say that genes affect behavior.

Some genetic body structures are already capable of mediating behaviors though always and only under the energy traces of unconditioned eliciting and evocative stimuli. For example, some neural structures connected to the iris are directly genetically produced, and these structures already mediate the behaviors of pupillary dilation and constriction, but *only* upon the energy trace of a change in the level of illumination (i.e., the pupillary reflex).

In other cases genetic body structures get changed by respondent and operant conditioning processes and then are capable of mediating behaviors, again though, always and only under the energy traces of eliciting and evocative stimuli. Recall our example of spicy tacos eliciting salivating. The likely genetically produced neural structures that already mediate that behavior got changed by the respondent conditioning pairing process such that they also mediated salivating in the presence merely of the restaurant logo.

Furthermore, the body structures of some people's genetic endowments may more easily change, relative to the body structures of other people's genetic endowments, due to environmental energy traces, and so they may mediate certain behaviors more easily. However, without environmental support, even a favorable genetic endowment cannot mediate inspiring behavior.

Consider this example. A couple (literally, a married couple) of colleagues, invites you over for dinner. They have three children, ages 15, 10, and 5. (They are mathematical physicists, OK?) Now, good music has a highly reinforcing effect on your behavior (i.e., when you hear some, you will go back for more, at least within the constraints of politeness). Well, over the course of the evening, you discover to your amazed pleasure that the two younger children play Chopin polonaises fabulously. Perhaps we could forgive you for jumping to the merely emotionally informed conclusion that these kids have inherited an amazing virtuosity. After all, both parents have played the piano a bit since childhood, and still do. So, what is going on?

What is responsible for the children's extensive and capable repertoire? What is the oldest son's excuse for not playing (an unfair but typical phrasing that translates as "why does the older son not play")? Are the younger son and the youngest, the daughter, just more musically inclined? What does that mean? What roles, if any, do environment and genes play in this playing?

Those are all typical questions for this type of situation. Let's answer those we can. We know that genes do not cause the great piano playing, because genes can only build body structure. However, these children's genes may have built structures that interact particularly smoothly with the musical stimuli in environmental contingencies, such that they mediate musical behaviors more easily than structures that the genetic endowments of others build. And what were the relevant contingencies? Investigation shows that, when the youngest turned three, the father got a good university professorship, and the parents bought a good piano and arranged for two or three lessons with a capable teacher, and several practice sessions, each week for both younger children for the last few years. They also arranged positive (i.e., non–coercive) contingencies supportive of effective lessons and practice–time usage, with lots of positive parental attention, all with the result that the children's playing expanded and improved in leaps and bounds, leading to the performances that you experienced. This shows the mutually beneficial interplay between gene–built structures that can facilitate some behaviors and the occurrence of those behaviors under the efficient control of relevant environmental stimuli. Also, note that the younger child's playing impresses us more than the older child's, even though they play equally well and have had similar lesson and practice experience. This bias stems simply from the younger child being younger, a bias we can note when attributing the playing not to age or genes or some ability trait, but rather to an opportune set of circumstances and contingencies.

Also, perhaps you think this example is a bit unrealistic, because the finger spread required to play the polonaises is rather wide. Well, the parents are quite tall and strongly built, and so are the children. Indeed their growth rates exceeded the norms for their ages, leaving them always larger, in a healthy sense, than their peers, such that even the youngest has just enough finger spread to meet the notation demands of the music. This shows another way in which gene structures can facilitate some behavior, even though the behavior actually occurs only under the control of relevant environmental stimuli.

By the way, why was the oldest child, also of capable body, not playing? The reason was probably not that he lacked the right genes; while he may have lacked them, we really have no way at present to know. The real reasons he was not playing were two, both pertaining to a lack of opportunity, which concerns the environment, not genes. For one, back when he was growing up, until he was 11 or 12, only the mother held a university professorship while the father was free–lancing, so lessons were not affordable. As for the second reason, due to that same lack of funds, the family lacked a piano for playing and lessons and practicing. So again, no practicing was possible even if lessons had been affordable. Ultimately, we can see a third reason as well. With piano lessons and practice impossible, contingencies compelled other behavioral investments. While his siblings excel at piano and scholastics (for similar reasons) he has excelled at football (i.e., soccer) and scholastics, again for similar reasons.

Here is another point to consider. Since genes have real, physical status, they *can* affect behavior. They can serve as evocative stimuli for behavior with respect to genes themselves, particularly verbal behavior. For example, when a police lab technician completes the necessary tests and concludes, "The genes from the crime scene match the genes from the suspect," genes as real environmental stimuli are evoking those verbal behaviors.

All these objections to genes as causes of behavior essentially repeat the objections that we must have to other explicitly physiological variables as *causes* of behavior (as opposed to mediating behavior). In particular, we cannot directly manipulate either genes or other physiological variables, as independent variables, to affect behavior, except that, particularly in the case of other physiological variables, we can sometimes change them through alterations of the environment (e.g., drugs) that change the physiology which then mediates behavior differently. However, as we will see in due course, drugs do not represent very elegant, and often not even very successful, solutions for behavior problems.

To finish this chapter, let's continue our ongoing chapter–conclusion activity of keeping in view the pertinence of behaviorology to helping solve global as well as local problems. Both of these kinds of problems and their solutions involve both of the kinds of behavior and conditioning, respondent and operant, that this chapter emphasizes.❖

References (with some annotations)

Fraley, L. E. (2008). *General Behaviorology: The Natural Science of Human Behavior.* Canton, NY: ABCs.

Skinner, B. F. (1953). *Science and Human Behavior.* New York: Macmillan. The Free Press, New York, published a paperback edition in 1965.

Skinner, B. F. (1957). *Verbal Behavior.* New York: Appleton–Century–Crofts. The B. F. Skinner Foundation (www.bfskinner.org) in Cambridge, MA, republished this book in 1992.ℰↄ

Reader's Notes

Chapter 4
Fictions that Fail to Explain

*I*n Chapter 3 we dealt with many considerations about behavior. These included continuity, triggering and fueling energy, the internal and external environment, and two types of classification. Common classifications covered motor (i.e., muscular) behavior, emotional behavior, neural behavior and, very briefly, verbal behavior, while functional classifications covered respondent (i.e., reflex) behavior and operant behavior. Besides some helpful distinctions (e.g., overt/covert and elicit/evoke) we also considered love, the "dead body test," and some things that are not behavior including traits, which cannot explain behavior and which we could have included in *this* chapter as a type of fictional account.

We finished Chapter 3 with two other factors that also cannot account for behavior, time and genes, but at least these are not fictitious. We consider several more fictitious explanations in this chapter. Note, however, that our most fundamental objection to fictitious accounts is *not* that they are fictional, that they do not exist, although that forcefully compels our attention. The real objection is that such accounts are irrelevant to the scientific understanding, prediction, control, and interpretation of behavior.

Setting the Fiction Problem Stage

First, though, let's start with a bigger picture. Recall our discussion of natural functional histories and levels of analysis. If we start well beyond the special considerations required at the sub–atomic level of quanta, then, as we traverse the continuum of analysis levels from physics to chemistry to biology, we approach our level, the behaviorology level, at which assemblages from atoms to molecules to compounds reach the expanding point where the accumulating complexity of natural chemical interactions, among all of these, builds structures that we call single–celled organisms. These continue naturally to accumulate complexity, building multi–celled organisms some of which contain sufficiently complex neural structures for natural responses to stimuli to occur. This is the level of behaviorological analysis where the interactions of these complex chemical systems (i.e., these organisms, including humans) with each other and with all the various aspects of their surroundings, produce the behavior phenomena. The "how" of these phenomena resides in the details of physiological structure and function occurring as (i.e., mediating) neural and muscular responses, while the "why" of these phenomena resides in the details of the continuously accumulating (throughout life) reactions of (i.e., changes in) those physiological structures to the energy exchanges—which we call stimulation—between parts of the physiological structures and parts of the internal and external environment. As implicit in our law of cumulative complexity, all of this raises a cacophony of reciprocal interactions that are nonetheless lawful, orderly, and, for this analysis level, predictable and controllable. Let's go to an example at this analysis level.

Consumer Reports regularly carries articles on the benefits of maintaining a healthy weight, even if that requires some special diet. All too commonly, people claim that to maintain a diet one both must *restrain oneself* from all the chocolates (Heaven forbid!) and must exert all the necessary *will power* to eat only the right things in the right amounts. These fictitious causes of dieting share common themes with the ones at the focus of this chapter. In an article in the February 2009 issue, analysts at *Consumer Reports* "were able to identify six key behaviors that correlated the most strongly with having a healthy body mass index (BMI), a measure of weight that takes height into account" (p. 27). The six behaviors were (a) watch portions, (b) limit fat, (c) eat fruits and vegetables, (d) eat whole grains rather than refined grains, (f) eat at home, and (g) exercise, exercise, exercise! Just reading the list of these behaviors helps bring the reader under some of the contingencies relating these behaviors to the reinforcing outcomes to which the phrase "successful dieting" alludes.

Dieting is predictable and controllable, and behaviorological principles apply to generating and maintaining this behavior (see Stuart & Davis, 1972). In a pattern typical of fictional accounts, the fictitious causes that we call self restraint and will power are not only unnecessary and redundant but also can cause the harm of continued ill–health, because they contribute nothing to dieting. Any dieting success actually accrues to a range of real independent variables; while behaviorology, like all natural sciences, lacks explanations for everything, *fictitious alternatives are not explanations at all.*

Let me repeat that in a different but related sense. Due to conditioning of the traditional cultural type, or to later, theological or secular non–scientific educational conditioning, many who swell the organized ranks of cultural superstition and mysticism unsurprisingly reject behaviorological explanations, finding them somehow degrading or blasphemous. The alternative accounts that they would support, however, are unnecessary, redundant, harmful, and not explanations at all. On that we must stay focused.

By now I hope you are becoming quite comfortable with the contingencies this book provides regarding naturalistic explanations for all behavior. You may find an increasingly compelling allegiance to natural philosophy and science characterizing both your intellectual and emotional reactions. This will make material about the many pre–scientific, unscientific, and pseudo–scientific accounts for behavior easier to manage.

For all of history until about 100 years ago, fictional explanations for behavior constituted the main, if not the only, game in town. In his courageous book, *The Demon–Haunted World,* Carl Sagan (1995) referred to science "as a candle in the dark." We must turn that candle into a floodlight exposing the whole variety of unhelpful accounts for behavior while illuminating the helpful accounts from natural behavior science, and thereby support the role of this science in helping solve local and global problems. Along with other authors, Sagan has exposed a multitude of false and otherwise invalid explanations for many natural phenomena. Here we focus on the false and otherwise invalid accounts, particularly of the fiction kind, that pertain to behavior.

Behavior Science Fiction

Of course, our concern differs from the concern of the literary fiction author. Many science fiction novels anticipate later scientific developments, or explore ramifications of current developments. In the 100 year history of natural behavior science, at least two authors have published such science fiction novels. B. F. Skinner wrote *Walden Two* (1948) and W. Joseph Wyatt wrote *The Millennium Man* (1997). These novels, based on accumulated philosophical and experimental results, interpret developments in ways that can make the behaviorology discipline more accessible for anyone, while also providing a good foundation for books with more comprehensive coverage. Both books also provide considerable coverage for the question of fictitious explanations of behavior. However, in the context of their contributions, each of these novels also has its own problematic concerns; discussing these concerns will reduce the trouble they might otherwise cause for readers today.

Walden Two. Skinner wrote this novel, which he considered an example of utopian literature, when someone challenged him to describe the kinds of benefits that a science of behavior might bring to society. Most utopias, however, differ in a major way from *Walden Two,* as they are set far away in either time or place—or both—while Skinner set his utopia in the here and now. However, as with other utopias, the author never gets around to telling us explicitly how to go from our current conditions (i.e., from our society) to the utopian conditions he describes (i.e., to *Walden Two* society) although his basing the novel on the experimental natural science of behavior gives a far bigger hint on that score than the authors of other utopian novels provide.

Yet not everyone agrees with characterizing Skinner's story as utopian. Some people evidence strong negative emotional reactions to *Walden Two,* and call it disutopian, often due to the scientifically grounded portrayal in the novel of freedom as *only a lack of coercive controls* on behavior, rather than the scientifically inaccurate but traditionally culturally conditioned appreciation of freedom as a *complete lack of controls* on behavior, a topic Skinner further explored some decades later in his book, *Beyond Freedom and Dignity* (1971).

Others, however, agree with Skinner that the story is utopian, because it describes a place mostly better than the one that readers inhabit, while also accounting, scientifically, for those occasions when we call what we have better. Perhaps the most important characteristic of *Walden Two* society resides not in all its curious, different, and often labor–saving practices, a few of which some have described as shocking—more so in the 1950s than the 2010s—but rather resides in the explicit appreciation of the value of experimentation in the design of cultural practices. The alternative of letting things develop without such experimentation has produced our current culture, replete with theological and secular unscientific accounts for our behavior and culture that interfere with successful problem solving.

The difficult problem for readers today, however, lies in a shift in terminology since the publication of the novel. When Skinner wrote *Walden Two* in the late 1940s, the possibility may still have existed that efforts to change traditional psychology into a natural science might succeed; this was still the period when traditional psychology and the natural science of behavior shared their history (see Ledoux, 2002a). Most natural scientists of behavior had only traditional–psychology units available as work places. So, in the novel, Skinner regularly used the term "psychology" to refer to the natural science of behavior that his own research efforts had established a mere decade earlier.

Forty years later, however, by the late 1980s, traditional psychology had made clear that it was not giving up its mystical agential core, precluding it from natural–science status, so natural scientists of behavior founded the independent behaviorology discipline. The point is that, without this information, readers of *Walden Two* today can be left thoroughly confused by Skinner's usage of the term "psychology" wherever he means the natural science of behavior rather than traditional psychology. Now, readers can reduce that confusion by thinking "behaviorology" whenever they come across these usages of "psychology" in the novel, a solution that readers can also apply as needed when reading Skinner's other works, all of which I highly recommend.

The Millennium Man. The twentieth century produced more scientific advances than any past century. As but one example, for thousands of years before 1900 CE, the horse set the speed and distance constraints for most transportation, yet only 70 years later, well before the century ended, people walked on the Moon. Joe Wyatt wrote *The Millennium Man* to help readers appreciate such vast changes in science and technology as well as the scientific world view, all while still entertaining the reader (Wyatt, 2002). In spite of all the advances in transportation, communication, construction, medicine, and so on—all wrought through the natural sciences—the author uses some clever plot devices to expose the lack of change in certain areas, such as the easy allegiance so many people still give to various pseudo explanations of their own behavior, comforting but essentially shallow and untestable and so essentially unhelpful explanations that are available in the occult, mentalistic, and astrology sections of bookstores, as well as much of the self–help section.

As counterpoint, the main character also quickly recognizes the development of a natural science of behavior, and therein rests the only problem for *The Millennium Man*, a problem similar to *Walden Two's* problem, and similarly resolved. The problem is that, in counterpoint with "psychology," Wyatt employs the term "behavior analysis" as the name for the natural science of behavior. This is an older name for that science, and one that behavior analysts let psychology claim, to the point that a Division of the APA (American Psychological Association) now goes by that name. This makes that term quite unsuitable as the name for the natural science of behavior, because many people now react to "behavior analysis" as part of traditional, non–natural–science psychology. Without this information, though, readers of *The Millennium Man* today might find Wyatt's usage of "behavior analysis," wherever he means the natural science of behavior, rather confusing. Again, readers can now reduce that confusion by thinking "behaviorology" whenever they come across these usages of "behavior analysis" in the novel, a solution that may also apply to some occurrences of "behavior analysis" elsewhere.

Before continuing with fictional explanations, let me conclude this literary fiction topic with a comment adapted from my earlier review of *The Millennium Man* (see Ledoux, 2002b). If you combine Wyatt's plot devices with (a) the twentieth century's scientific "miracles," (b) the continued clinging to pre–scientific notions about human behavior and its causes, and (c) the value of the discoveries and applications of behaviorology, then you have the mix that I think Carl Sagan was correctly addressing when he described this novel as "an excellent device to view our time." (His comment is inscribed on the front cover of *The Millennium Man*.)

Common Explanatory Fictions

Our treatment of fictitious explanations for behavior will not be exhaustive. More complicated classes of fictional accounts remain for others to cover, or for analysis in a later, more thorough chapter. Here we will consider the more common classes of fictitious accounts regarding behavior.

Some Origins of Fictitious Explanations

The origins of most explanatory fictions reside in the verbal conditioning most cultures put their members through early in life. That conditioning built up over many, many generations, from times when language was a new behavior for humans. The only accounts for nearly everything then, and for many thousands of years, were versions of primitive animism. Refinements occurred and expanded but the core superstition and mysticism remained. Science made several appearances over the last few thousand years, only to be mostly lost each time, until about 400 years ago. As science expanded, fictional accounts retreated. Today fictional accounts generally still thrive only with respect to human nature and human behavior, and they are gradually but surely giving way in these areas also.

Linguistic Fictional Explanations

Language actually plays multiple roles in the continuation of fictional explanations. It is the vehicle by which most cultural conditioning of new members occurs, conditioning that includes both operant components (e.g., language itself and traditional, usually mystical, cultural lore) and respondent components (e.g., emotional support and alteration of operants). Language also contains inherent allusions to mystical inner–agent fictional explanations of behavior. What led to these allusions?

The usual energy–conservation contingencies on the development of language led to verbal response–pattern economies that reflected and supported whatever mystical accounting for events was prevalent at the time of that development, including inner person agents. As a result, in English, the most obvious of the verbal response–pattern economies that are still with us, and that contain inner–agent allusions, comprise the set of personal pronouns (i.e., I, you, he, she, it, we, they, me, her, him, us, them, my, your, his, its, our, their, etc.). Each of these quite short words economically reduces a subject, object, or possessive to just a few letters, compared to the several syllables or words that might otherwise occur. However, these personal pronouns inherently imply, and even seem to refer to, the existence of internal mystical behavior–directing self agents. Such verbal shortcuts need not cause problems; conditioning could manage them, making them merely refer to bodies mediating evoked behaviors. We will, of course, continue using these energy–saving verbal shortcuts, but we must excise the inner–agent implications through education in behaviorological repertoires. Until so excised, these personal pronouns probably comprise the most ubiquitous form of inner–agent explanatory fiction, but not the only linguistic one.

Another agential verbal response pattern involves the active voice. Passive voice leaves the agential "doer" of the verb action either out completely or gives it only the status of an afterthought. For example, in a passive–voice construction, the "by" clause can often be skipped. (See!) This constitutes a kind of improvement over active voice in terms of fictional causal agents, because active voice puts the agent right up front, indeed rather in–your–face if conditioning that favors science has sensitized you to these issues. (Of course, no inner agent "you" resides in a reader's physiology...—See the problem?) But many people find passive voice at least a bit awkward and confusing.

So who uses passive voice? As one example, many scientific circles prefer passive voice for experimental reports, ostensibly because the passive voice helps emphasize aspects of the experiment instead of the researcher who is not the focus of these reports. Also, science maintains the struggle against unhelpful, untestable mystical accounts, a struggle the passive voice supports through de–emphasizing the agential "doer" (which, recall, exists not). While this also would at least intuitively tend to encourage the use of passive voice, especially among scientists, the passive voice construction generally makes for rough reading, and thus poor writing style. The conundrum for authors then, especially behavior–science authors, involves crafting sentences that are well written, easy to read, *and* non–agential. I doubt that my writing has succeeded very often with all three! Very likely, all facets of the culture could benefit from some linguistic changes in what passes for good grammar and writing style, changes supportive of the scientific realities benefitting the culture.

Managing linguistic fictitious explanations. Many scientific and other authors would welcome linguistic changes that allow them to write without automatically implying inner agents. They currently cope by trying to avoid personal pronouns and by carefully composing active but inner–agent–less sentences, all without that much success (as my writing here may show only too well). The results trend toward a cumbersome and impractical writing style. What we are going to need is a new, scientifically consistent set of linguistic practices (e.g., a new grammar). Linguistic change of this magnitude, however, generally accrues so gradually as to be unnoticeable across even several human generations. For example, mostly under the contingencies of a series of population and cultural conquests, the English our ancestors spoke a thousand years ago changed, along with other cultural practices, into the quite different English of today. Similarly, under the contingencies of scientific realities, the English we speak today faces a continuing gradual change, reducing its inherent agential supports, as it evolves more scientifically along with other cultural practices over the next thousand years, assuming that we solve some other problems in a timely fashion.

Meanwhile another kind of change helps us substantially, right now, to counter not only the implied linguistic types of fictional explanation for behavior but also the many other types that we will cover as well. This change arrives with the increased availability of behaviorology. Any educational experience necessarily involves conditioning respondent and operant responses related to the subject matter. The conditioning that is automatically involved in any behavior–science educational experience, like reading this book, necessarily and naturally leads to the generation and maintenance of responses that are consistent with, and appropriately reflect, the scientific realities of behavior causation. After such conditioning, the contingencies of these scientific realities effectively counter the prior traditional cultural conditioning that respected what we now see as fictitious explanatory accounts. Thus we manage personal pronouns and the active voice by seeing them for what they really only are, and using them only that way, namely, as energy–saving economic language tools (i.e., verbal shortcuts). As a residual problem, those throughout the rest of the culture, who as yet lack access to conditioning regarding behaviorological science, continue not only to see ghosts behind every personal pronoun and active voice usage, but also to respect other fictitious accounts for behavior.

Cause and effect review. Before we visit more of the common explanatory fictions, let's review our use of the terms "cause" and "effect" as this will facilitate our efforts. Throughout this book, we use these terms only as verbal shortcuts for the functional relation between the longer terms "independent variable" (cause) and "dependent variable" (effect). Of course, we only contact pairs of variables in which one changes systematically with changes in the other. Any functional relation we describe is not necessarily something we actually see; instead, the functional relation arises implicitly, from our contact with the consistent temporal contiguity of the pairs of variables, as an increase in our confidence in the reliability of the pattern of occurrences of the variable pairs. While arising implicitly, such functional relations, which occur throughout science, are not mystical, because they involve traceable energy transfers between the variables, transfers that may require a shift in analytical level to trace. For example, our behaviorological–analysis level may find a functional relation between an uncomfortably high indoor room temperature and turning on the air conditioner. But tracing the relevant energy flow requires a shift to the physics analytical level. We can make those shifts easily enough when we need them to better understand all the functional relations that actually account for behavior, but we have no requirement for those shifts at the practical level of the environmental control of behavior (although we may occasionally employ them). However, being *unable* to make such shifts implies a breach in the functional history of an event, and that usually leads to discovering a fictional rather than a real explanation, because real accounts make no such breaches while fictional accounts make them inherently or automatically. So, let's move on and deal with some fictional accounts for behavior.

Fictional Explanations and Some Constraints

Of the many types of fictional explanations for behavior, our focus here pertains to some of the more common types. These include reification, turning adjectives into nouns, nominal fallacy, circular reasoning, gratuitous physiologizing, and teleology. Even though many of these categories overlap each other in various ways, we will consider them separately. (Would you like a not–so–difficult challenge? Working through the various overlaps on your own, after you finish this section, might be a helpful exercise for you.) However, before describing these explanatory fictions, here is a brief review of parsimony, and a word about description versus explanation, both of which constrain fictional accounts and help us better understand them.

Parsimony. Attending to parsimony plays a big part in dealing with fictional accounts of behavior, because all fictional explanations violate parsimony in one way of another, and so, again, are not really explanations at all. Rather than covering these violations separately with each fictitious account in turn, we cover them all by recalling that *parsimony means working with only the simplest yet adequate explanation for an event.* Thus an account can violate parsimony either by being unnecessarily complex and untestable at this point (e.g., a mystical or superstitious account) or by being simply too scientifically *in*adequate to fill an explanatory role (e.g., time). Some accounts, especially some fictitious accounts, violate parsimony both ways.

Description versus explanation. Description differs from explanation, and so is not explanation. Description involves elaborating or detailing *one* of the variables in a relation, usually the dependent variable, the effect, the behavior. This variable cannot make an explanation. Worse, such descriptions often mean talking about the behavior in such a general way that its response status disappears, so you end up with zero variables, even though linguistically the description still sounds like a plausible explanation; closer scrutiny shows otherwise. Explanations involve *two* variables, each of which is, at least theoretically, both detectable and measurable (i.e., real) and together they are established or scientifically interpreted as in a functional relationship. The functional relationship then *is* the explanation. (It may or may not be valid; that is a different question.) With only one variable (or none), descriptions cannot be explanations, which require two real variables in a functional relationship.

For example, say that we see the responses of grabbing several cookies without hesitation whenever someone first offers them, and grabbing other children's toys whenever they show up with toys. Someone might describe these behaviors as impatient or showing impatience. That gives us some description. But have they explained the responses we saw? Indeed, where are the responses we saw? Impatient and showing impatience are so general that we have lost the responses (which can make "impatient," and "impatience," available for use as one type of fictional explanation involving adjectives and nouns, but that is a further reality we will get to shortly.) However, seeing just those responses provides insufficient information for an explanation; the responses are still only one variable, so even describing them leaves us short on explanation. Are there other variables we saw but failed to include in our analysis? Yes, indeed. Beyond the grabbing responses, we saw the cookies, and we saw the toys. Each of these provides a second variable for our analysis of the cause of the grabbing responses. (I bet you can predict where this is going.) While we would need to test them, one explanation is that eating the cookies causes future cookie grabbing, and keeping the toys causes future toy grabbing, both of which provide explanations without descriptions confounding us. We could make a further analysis by noting that the cookie grabbing only occurs in the presence of cookies, and the toy grabbing only occurs in the presence of toys; the cookies and toys could be evoking the grabbing responses. This would again provide us with explanations (still needing testing) without descriptions confounding us.

Then again, say that we also see adults making inappropriate but polite comments after each grab for cookies or toys, along the lines of, "That was quick," or "How cute." This provides plenty of adult attention consequent to the child's responses, which also provides yet another variable for our analysis of the cause (Causes?) of these grabbing responses. Both cookie grabbing and toy grabbing could be occurring at least partly as a function of adult attention. While we would also need to test this account, at lease it too is an explanation without descriptions confounding us.

Alternatively, someone might describe these behaviors as greedy or showing greed (instead of impatient or impatience). Can you correct and complete the necessary analysis? Give it a try. We must all watch out for descriptions masquerading as explanations, especially since some subject matters rely heavily on descriptions, often as explanations (e.g., some approaches to child development) and the sophistication of some descriptions–as–explanations can make spotting them no easy task.

Some contention over attention. Since our cookie–grabbing example mentioned attention consequences that involved verbal behavior, because such verbal behavior often has attention as well as verbal components, let's take the opportunity here to consider the contention over attention as an *unconditioned reinforcer.* Unconditioned reinforcers are stimuli that function as reinforcers without prior conditioning. On the other hand, *conditioned reinforcers* are stimuli that do not function as reinforcers until they have occurred along with other reinforcers, and that pairing (that conditioning) makes them function as reinforcers. I find that distinguishing between *attention* as an unconditioned reinforcer, and *praise* as a (verbal) conditioned reinforcer, helps people respond correctly to the distinction between unconditioned and conditioned reinforcers.

As a verbal behavior, praise can affect a child as a reinforcer only after a year or so of accumulating the effects of verbal–behavior conditioning. Attention, however, affects a child as a reinforcer much earlier, perhaps as an unconditioned reinforcer. That is, attention might affect a child as a reinforcer without any prior pairing with other reinforcers (which is the kind of conditioning that would otherwise make attention a conditioned reinforcer). Confusingly, verbal comments, as in our example, also have attention components.

Even though praise and verbal compliments and comments may be inseparable from some element of attention, they can easily be shown to be conditioned reinforcers (as they depend on prior verbal–behavior conditioning). However, while attention *can* be conditioned (i.e., attention often does occur with other reinforcers) trying to show that attention *must* be conditioned is much more difficult. Parents who attend in various ways to relative new–born infants when they cry are often left distraught when crying increases "for no reason" (as they describe it, although the actual reason involves the attention reinforcing the crying). Was such attention already paired with other reinforcers, making it a conditioned reinforcer? Perhaps. Perhaps not.

However, in my experience, describing "attention" as an unconditioned reinforcer and "praise" as a conditioned reinforcer definitely helps people distinguish between attention (implying *no* verbal component) and praise (with a definite verbal component). These descriptions also help people distinguish between unconditioned reinforcers and conditioned reinforcers.

That pedagogical consideration is presently valuable even if the self–correction of science might later show that "all attention is conditioned." For now, though, let's move on and consider some fictional accounts for behavior.

Fictional Explanations: Reification

The first fictional explanation we consider is *reification,* which comes from the verb "reify." Both words mean taking the kind of abstraction that involves something lacking physical quantities or qualities, and treating it as if it were real, with a concrete, material existence. While initially an abstraction may begin as a mere conceptual device to help explore a topic, often metaphorically, the mere repetition of the abstraction often leads to its use as something real, as if it indeed had physical status. But it lacks any such status, and so its use as an explanation is scientifically inadequate, since scientific explanations must involve events with physical status, detectable and measurable.

For example, consider the abstraction that we call the "mind." In the process of secularizing the soul (a mini deity powered by a maxi deity) people began by using the "mind" to explore the possibilities metaphorically. The body behaved *as if* the mind was telling it what to do. Soon, however, they were using the "mind" as the secular but still non–physical controller of the body. Losing the metaphor left us saying, "The body behaves the way that the mind tells it to behave." This standard kind of comment reifies the mind.

Such talk began thousands of years ago when no behavior science was yet available to countercontrol the contingencies driving explanation development that occurred regardless of physical reality concerns. Over this time frame, right up to our own, people—ordinary and professional—repeated the term so often that they began uncritically to consider it as, and use it to stand for, a real thing inside you. But, as a continuing abstraction, it still lacks any physical status, and so its use as an explanation is an example of reification, and as such it is scientifically inadequate because, again, scientific explanations must involve events with physical status, detectable and measurable.

For another example, consider personality, which also is an abstraction (i.e., something lacking physical status). People used this word originally as a verbal shortcut for a consistent set or pattern of behaviors characteristic of a particular individual. The main problem then was its descriptive status, which fails to convey explanatory status. However, people repeated it over and over, using it with other words, each denoting one or another type of personality, such as an introverted personality or an aggressive personality. Rather than being helpful, though, this compounds the problem. The adjectives, sharing the problems of descriptions, are now supposed to describe the kinds of behaviors that the personality compels from the individual (e.g., he slugged the other guy because of the aggression that his aggressive personality compels). Furthermore, the adjectives make the noun, personality, seem all the more real, which it is not! "Personality" still lacks any physical status, along with any parts or processes attributed to it, and its use as an explanation is an example of reification. As such, personality is a scientifically inadequate explanation.

Fictional Explanations: Converting Adjectives to Nouns

You may have noticed a common theme running through many explanatory fictions so far. They often involve *turning adjectives into nouns.* The adjectives may or may not be helpful, but the nouns simply become one or another explanatory fiction. Consider some typical adjective/noun pairs taken from our examples, noting that the pair members need not maintain the same form, although that is common. The adjective "impatient" becomes the noun "impatience." The adjective "greedy" becomes the noun "greed." And the adjective "aggressive" becomes the noun "aggression." Here are some other examples. The adjective "intelligent" becomes the noun "intelligence." The adjective "paranoid" becomes the noun "paranoia." The adjectives "hostile" and "antagonistic" become the nouns "hostility" and "antagonism."

In every one of those cases, turning the adjective into a noun leaves the impression that we now have a cause for some behavior, when actually we have added nothing new to our analysis. Instead, we have a bigger problem, as we are now likely to stop analyzing, because we think we have a cause and need no further analyzing. But all we have is yet another fictional explanation.

That problem is so common to all types of fictional explanations that it bears repeating. As soon as you think you have a cause, you tend to stop looking. But if all you have is one or another type of fictitious cause, then you still lack what you need if your job is to intervene regarding the behavior for which you are seeking a cause. You still need a real, accessible cause. Worse, lacking a real cause can lead to harm if, without it, no effective intervention is forthcoming; this makes all explanatory fictions at least potentially harmful.

Converting adjectives into nouns characterizes a large portion of fictional explanations, because our language so easily allows these conversions. And our traditional cultural conditioning, which takes the appropriate search for causal connections and perverts it into an acceptable search for mystical causes inside us, makes these conversions particularly likely. Yet all such causes are still explanatory fictions. We must be particularly careful to consider only causal variables that have the requisite physical status if our scientific attempts to understand, predict, control, and interpret behavior effectively is to have any beneficial effects. After all, *that* is the point of scientific behavior itself.

Fictional Explanations: Nominal Fallacy

We use the term *nominal fallacy* for another common fictional explanation for behavior. This is a particular version of changing adjectives into nouns, because "nominal" means naming things, which linguistically leaves us with nouns. After we observe a behavior, we often describe it. Then, our past linguistic and traditional cultural conditioning leaves us both first pinning a name (i.e., a noun) onto our observation, and then easily taking that name as the cause of the behavior we observed. Nominal fallacy, then, really just means giving a name to something and taking that name as the thing's cause.

Here is an example. You observe a skillful performance. Then you speak of the performer—as if you were referring not to the "body that performed" but to a mysterious (i.e., mystical) inner agent inside that body—as having talent. You might first describe the performance with the adjective "talented," before changing it into the noun "talent," but for many who engage fictitious explanations, contingencies often compel skipping this step. Finally, you explain the skillful performance in terms of that possessed talent saying, "She acted so well because she enjoys a wealth of talent." Yet you have only named your observation, and we still have only that one variable, so we cannot yet have a cause, which makes "talent" a nominal–fallacy type of fictional explanation.

Here is another example. You observe a young teenage boy snapping constantly at everyone around him. Then you speak of him as holding a lot of hostility. Finally, you explain the snapping in terms of the hostility saying, "He snaps at everyone because he suffers much hostility." Yet you have only named your observation, and we still have only that one variable, so we cannot yet have a cause, which makes "hostility" a nominal–fallacy type of fictional explanation. (Fear not the repetitious phrasing. It can speed your understanding.)

Here is a final example, although you could easily come up with plenty more. You observe a sixth grade girl achieving good grades in high school level science and math. Then you speak of her as endowed with great intelligence. Finally, you explain the achievement in terms of that intelligence saying, "She achieves such good grades well above her level because she commands so much intelligence." Yet you have only named your observation, and we still have only that one variable, so we cannot yet have a cause, which makes "intelligence" a nominal–fallacy type of fictional explanation. (For a challenge, consider the nominal fallacy of paranoia causing paranoid behavior.)

Fictional Explanations: Circular Reasoning

We use the term *circular reasoning* for yet another common fictional explanation for behavior. Upon observing a behavior, something from our traditional cultural conditioning history—often in conjunction with some characteristic of the behavior—evokes some causal statement about the behavior that sounds at least linguistically satisfying. But this supposed cause gives us no new information about the behavior, and has no separate status as a variable. This gives the supposed cause the status of a fictitious explanation. We call it circular reasoning, because the cause appears as an inference from the very behavior that it is supposed to explain, in this common format: "Why did he take that action?" "Because of this cause." "How do you know this cause is operating?" "Because he took that action." "But why did he take that action?" "Because of this cause." "But how do you know this cause is operating?" "Because he took that action." … And so on, around we go, getting nowhere!

Consider this example. A non–behaviorological practitioner or equally scientifically uninformed parent might be faced with a student's earning poor math grades. We have a fictional account if past cultural or educational conditioning evokes statements like "a 'mental block' within the student causes the poor math behavior." Some folks explicitly call it a *math* block. Either way, this statement gives us no new information about the poor math behavior, and has no separate status as a variable. Instead its circularity is clear to those whose conditioning has prepared them to watch out for fictitious accounts:

"Why does the child do poorly in math?"

"Because he has a mental block to math."

"How can you tell [or, How do you know] that he has a mental block?"

"Because he does poorly in math."

"But, why does he do poorly in math?"

"Because he has a mental block ."... [and so on...].

In his book, *Science and Human Behavior,* B. F. Skinner (1953) referred to a wide variety of fictional accounts for behavior as invalid and false "inner causes" of behavior, pointing out that the major (and often only) source of knowledge or information about them turns out to be the very behavior that they are supposed to explain. In a circular sequence, past conditioning not aligned with scientific realities leads to stimuli evoking behaving these accounts as *inferences* from the same behavior that the inferred causes are then supposed to explain, a pattern that some call *inferential circularities.*

We can often analyze fictitious explanations of other types as also fitting the pattern of circularity. To see this in operation, consider some of the examples we already used with other types of explanatory fictions, and ask the circularity–exposing questions for each:

"Impatience or greed causes grabbing cookies or toys..."

"Talent causes skillful acting performance..."

"Intelligence causes getting consistently good grades in school..."

"Paranoia causes paranoid behaviors..."

"Hostility causes the aggressive behavior of snapping at everyone..."

For that last one, the circular reasoning would be ridiculously obvious if someone said, "Hostility causes aggression." This is actually worse, however, because it removes the behavior, which leaves us without any real variables.

In each of those cases, first ask, "Why did that behavior occur?" Then ask, "How can you tell that *that* causes the behavior?" You can easily go through all of those examples. Here is the first one: "Why did cookie grabbing occur?" "Because the child is greedy" "How can you tell that the child is greedy?" "Because of all the cookie grabbing." Note how this phrasing avoids most of the additional fictional causes of the usual inner agents, especially those that personal pronouns imply. Making the phrasing of *your* questions and answers for each example also devoid of inner agents represents even greater progress. In anticipation, congratulations!

Another Inner Cause Commonality and Global Danger

Circular reasoning will also crop up in at least one of our remaining types of explanatory fictions. Something common to *all* fictional accounts, though, pertains to the resources lost when educational programs remain disrespectful of science. Societies that allow such programs waste laboratory methodologies to investigate what is not really there, while mis–training legions of students to replace the professionals expended on the crusade, all of which diverts resources that could otherwise support the development and dissemination of the findings and applications of natural behavior science, behaviorology.

That leaves resources in short supply to replace fictional inner accounts for human behavior with scientific accounts. The importance of providing those resources grows exponentially as efforts to mitigate environmental destruction languish for lack of widespread access to natural behavior science, a requirement the traditional natural sciences recognize as vital for the team effort necessary to solve these problems due to the extensive behavioral components in both the problems and their solutions.

Here is a little sample of scientists recognizing the behavior connection. Dai Qing is a Beijing–based water policy specialist. Concerned about the chronic water shortages for Beijing's 17 million people, she said, in a *USA Today* article (20 June 2008, p. 9A), that there "will never be enough unless the citizens of Beijing change their behavior and water usage." Also, articles in publications of the *Union of Concerned Scientists* regularly delineate the kinds of behaviors needed to achieve solutions to some problems. An article under the title "Climate Action in Your Hometown" (*Catalyst,* Spring 2008, p. 20) mentions *launching* a local shuttle service (and getting people *using* it), and *conserving* energy, and citizens *agreeing* on, and *participating* in, solution development and follow through. All of these involve behavior, and increasing them benefits from a behaviorological analysis. Perhaps most important are the calls that traditional natural scientists make for a natural science of behavior, as part of ending the grip that the forces of organized superstition have on the culture, especially with respect to human nature and human behavior. Authors like Richard Dawkins (who wrote *The God Delusion*) have argued persuasively that behavior must be approached naturalistically; but again, such authors often seem unaware that what they seek is already largely available through the behaviorology discipline. The continuing series of court cases over the teaching of evolution in biology classrooms also all too regularly reminds us of the difficulty of winning the cultural battle against superstition. Entrenching behaviorology thoroughly among the natural sciences provides a potent, pertinent, and practical partner in these struggles with superstition. (For a related and enlightening exposition, see Skinner's 1971 book, *Beyond Freedom and Dignity,* especially Chapter 1.)

In the mean time, we are running out of time. Way back in 1972, scientists at MIT released the book, *The Limits to Growth* (Meadows *et al,* 1972; also see Hayes, 2012, for an update) which examined many of today's worrisome environmental parameters while addressing how much time we had before the human population outgrew the carrying capacity of our world and its resources. They concluded about 100 years. That was over 40 years ago, and the parameters they examined have worsened faster than expected. In 2012 Megan Gambino quoted Dennis Meadows, one of the book's authors, as saying that already, "We're at 150% of the global carrying capacity." That cannot and will not last very long. We are quickly overspending our food, water, air, and other resources, and nature's accounting will soon cost us dearly as these resources run out and world population drops drastically from losses as people die of starvation, thirst, disease, war, and so on. Any chance we may still have to avoid or reduce such scenarios depends on our increasingly and comprehensively dealing *scientifically* with human behavior. In this chapter that means understanding and rejecting inner, and other fictional, causes of behavior. (We cover the natural science causes of behavior throughout the book.)

Fictional Explanations: Gratuitous Physiologizing

Here is a somewhat common fictional explanation for behavior of which my students are particularly fond, or at least they are after they become comfortable correctly pronouncing it, and spelling it. We call it *gratuitous physiologizing.* This involves inventing phony physiological accounts for behavior, in part because the physiological–analysis level makes the phony part sound more scientifically credible. But the physiology is merely made up; it is often circular; it also often fits under nominal fallacy, and it can involve changing adjectives into nouns. The mistaken physiological credibility makes these inner fictional causes look like variables separate from the behavior they are to explain, giving us two variables, but closer inspection reveals the lack of scientific status. They are not real. They are merely another type of fictional explanation for behavior.

For example, consider again a practitioner faced with a student earning poor math grades. Some types of educational conditioning might lead him or her to claim that this math repertoire deficit is due to a "minimal brain dysfunction" (MBD) within the student. (Some school systems have actually used this "diagnosis." I find that a bit scary.) And just why is the brain dysfunction "minimal?" Because the practitioner cannot find any real, physical dysfunction of the student's brain. But the poor math behavior *is* a problem and something must surely be causing it. Since nothing else seems responsible (not that much serious looking took place, apparently) the brain surely must be dysfunctional. If we cannot find the dysfunction, then it simply must be "minimal." As this argument goes, if it was not minimal, then we would be able to find it, and would have found it.

The problem is that the presumption of a brain dysfunction is gratuitous in the first place. It is unreal, invalid, and unwarranted. An initial assumption warranted by the situation, and worth exploring and fixing—as a behaviorologist would—is that something about the student's math–related environment is "dysfunctional." For some examples, consider these. Is the lighting adequate? Is the homework at the appropriate level? Is classroom assistance available? Is encouragement or help available at home?

That the brain dysfunction is not real but gratuitous also becomes obvious from further scrutiny. Ask the circular–reasoning exposure questions:

"Why is the child poor at math?"

"Because he has an MBD."

"How can you tell that he has an MBD?"

"Because he is poor at math."

"But, why is he poor at math?"

"Because he has an MBD ."… [and so on…]

Practitioners, especially those with the kind of education that would allow or encourage gratuitous physiologizing, have no intervention technologies, stemming directly from their mentalistic analyses, that are appropriate for dealing with MBDs (or, for that matter, with ineffective math–focused behaviors, or with dysfunctional math–related environments); they can only fall back on intuitive technologies or practices. That is, if these scientifically uninformed practitioners experience any success helping the math–poor student, that success must arise intuitively through practices that only coincidently are congruent with the natural laws governing behavior. Behaviorology enables such practices by explicit design rather than merely inadvertently.

Real physiological events are sometimes directly investigated (appropriately, if by physiologists). But this still has little bearing on the emergence of an environmental–change technology that affects behavior, math behavior in this ongoing example, as opposed to a more medical technology such as some form of drug therapy, which is quite inappropriate for this example, and quite often inappropriate for other examples as well, because less invasive behaviorological–level interventions would work if tried. Too often these putative physiological events are only supposed, hypothesized, invented, or theorized—a well–criticized pattern of non–explanation that we call gratuitous physiologizing (e.g., see Skinner, 1953, 1974).

Fictional Explanations: Teleology
A not–so–common, yet subtle as well as confusing, fictional explanation for behavior is one that we call *teleology,* which refers to the study of future causes. These are causes that supposedly occur in the future with respect to the effect that they are causing. But the future, by definition, has not yet happened, so no future event can cause a present event. This is not to say that the present has no effect on the future; this is only to say that an event in the future, because it has not yet happened and so cannot have the physical status required of scientific causes, cannot cause an event in the present. We refer to these future causes as teleological causes, and they constitute yet another type of fictional explanation for behavior, because they too lack the requisite physical status, at least while they are in the future.

You often find teleological causes imbedded with a phrase like, "in order to." For instance, "He cleaned up his apartment to [or, in order to] keep his parents happy when they arrive for a visit." Or, "She cooked a special meal in order to please her special friend." When you see or hear, "...to..." or "...in order to...," you very likely have a teleological fictitious account.

The example I use in my classes is one dear to my students' hearts. They work hard and, as a result, earn good grades (well, most of them anyway—after all, *by design* each of my behaviorology courses applies, in the teaching of the course, the same science that the course teaches to the students). Those good grades show a valid connection between present and future. Good grades are indeed contingent on working hard, and working hard produces good grades.

But just what causes students to work hard? When I ask them that, their initial answers inevitably include some variation of, "Why, to get good grades, of course!" But getting good grades later *cannot* be the cause of studying hard now; since the grades are in the future, they have not yet happened. (And will not happen, unless the students study hard. Confusing, yes?) A claim that "getting good grades at the end of the term causes careful and thorough study now" is teleological, as is claiming that, "I study hard in order to get good grades." Both statements comprise teleological fictitious explanations of behavior.

Before discussing with my students some of the actual causes of their studying efforts, I provide them a more complete account for why good final grades at the end of the term *cannot* cause their study behavior during the term. Faculty submit grades by taking their grade sheet to the administration building where the registrar's staff keys the grades into the most secure computer on campus. From there, the staff provides the grades to the students some days later. Using the fall semester for this example, consider that the campus entrance road has a fork where one way takes you to the campus maintenance center while the other way takes you past the administration building. (Both ways go around a loop road to various classroom buildings.) Now, the term ends in December, often with a few feet of snow on the ground and some ice on the road. A tanker truck arrives on campus to deliver a load of fuel to the maintenance center, but it takes the wrong fork and, slipping on the ice, crashes into the administration building while I am there turning in my grades. I am gone, the grades are gone, the computer is gone, even the records of the students' enrollment in the course are gone. They never get the grades! Yet all semester long, on a nearly daily basis, they worked at their studies earning the good grades that never got delivered. Why? The cause could not be getting the good grades, since the grades were in the future, and never occurred anyway. Such teleology is a fictional explanation.

After that story the students and I discuss some of the possible real causes of their studying efforts. These pertain mainly to past causal variables, because many present ones, concerning mostly evocative stimuli particular to each student, involve complexities that we cover later in their course (just as we cover these later in this book). The past variables are ones that affect essentially everyone who ever attended elementary school. By the time these students are in college, most of them have extensive conditioning histories, spread over 12 or more years, wherein academic work of varying qualities produced a variety of consequences, many of which made such academic work either increase in rate, or at least maintain the then current rate (i.e., these consequences demonstrated the appropriateness of calling them "reinforcing"). These ranged from stars and stickers for attempts and improvements in cursive letter forms and simple math facts and problems, in early grades, to compliments and letter–grade marks on completed homework assignments and tests—and marking period and final grades—at high school levels. That history leaves students' physically changed such that most assignments now successfully evoke—as real, testable causes—the appropriate study efforts that should produce—at the end of the term—good final grades.

The scientific inadequacy of teleological causes has no effect on the very real contingent connections between present and future events. Good grades are still contingent on good study. That is, because past good work produced good grades that occurred after the work, present good work produces good grades that may be delivered in the future. But the future good grades are not causing the present good work; saying so, as in "I study in order to get good grades," rates as a teleological fictitious explanation. (It also rates as mystically agential as no "I" inside the body "does" the studying, another example of the common overlaps among these fictional accounts.)

Accessible Control and Fictional Accounts

Again, we sometimes talk as though our biggest objection to explanatory fictions is that they do not exist. However, while certainly problematic—for them—this is not our chief objection. Rather, our most strenuous objection is that they are completely irrelevant to the prediction and control (and understanding and interpretation) of the behavior that they supposedly cause. They lack manipulable, independent–variable status. We have no access to change them in ways that bring about a change in the behavior they supposedly cause. One can claim that changing something else, such as an environmental variable, changes the fictitious variable, which then changes behavior in the sense of A causing B, and B causing C. But then the relation is similar to relations in mathematics: A causes C, and the middle term is at least unnecessary.

Another problem for fictitious variables, perhaps the worst problem for society, is that they leave the analyzers who invoke them comfortable; the analyzers seem to have found the sought–after cause when, in reality, they have added nothing to their analysis. But, thinking that they have found the cause, they stop searching for real causal variables. The result is that accessible causal variables that we might change to improve (control) the behavior in question remain unanalyzed. We seldom continue looking when a convenient answer is available unless separate, practical contingencies for such behaviors are in force. The search essentially stops, because the fictitious mentalistic or cognitive analysis provides no compelling reasons to continue. The search ends without actually attaining the capacity for control.

The dangers of inner causes may be less important when the behavior being explained presents no problems, such as excelling in math being explained by "intelligence" (inherited or not). But those dangers can be crippling in the opposite case, as we have seen. Yet one can still achieve a type of control. For example, the job specification of the practitioner with a math poor student may supply the *practical* contingencies that require effective environment–controlling, behavior–changing technologies. The specification may require him or her to document successful help for that student. If he or she finds that student parked in front of a television set for five hours each day to the exclusion of study on school assignments, he or she may intuitively change the student's environment by pulling the plug on the set. If the ultimate result of that action is that grades improve, then that *functional* control results from the environmental change. That functional control does not result from the practitioner's mentalistically or cognitively focused, verbal analytical repertoire (his or her theory) but in spite of it. Their analysis may inadvertently coincide with the successful intervention, but it is not functionally related to that intervention. And this fact remains unaltered even when the practitioner tries to tie his or her paradigm to the successful intervention by merely insisting that eliminating access to television must have diminished mental blocks or MBDs and thereby improved the math–grade outcomes.

Mentalists sometimes conveniently disavow control as their goal since their approach will not support analyses that backtrack through functionally related variables to the accessible environment. Mentalists' analysis precludes control so they argue against it. Their practical work is often confined to predicting behavior from other behavior—as when they predict behavior on a specific occasion from a putative trait that they infer from behavior on previous occasions. For example, the authors of educational psychology texts typically ignore direct measurements of the properties of behavior and instead focus on identifying and measuring the intensity of what they wrongly assume to be behavior–causing traits that they infer from previous occasions.

Alternatively, a behaviorological analysis of behavior cannot satisfactorily stop without pursuing the functional sequence of antecedent events further back. Behaviorologists pursue the sequence back until either (a) intervention becomes possible or (b) they can interpret the behavior in the sense of describing it in terms of relations of the same kind that on other occasions have permitted control under more technologically feasible circumstances.

Since control cannot stem from the mentalists' and cognitivists' analysis, they often regard the quest to demonstrate and apply experimental control of behavior as an allegiance to superficiality. However, failure to include control as at least a planned final step condemns a discipline to scientific immaturity (see Skinner, 1953, Chapter 2). Those who strive for control in their subject matter develop effective behavioral technologies based on the handling of real independent variables. They become behaviorological engineers discovering largely untapped markets for those skills in all facets of the culture, including helping to solve local and global problems. Those who would not strive for control in their subject matter can at best merely *interpret* life in various ways. But the point *is to change it,* a point that Karl Marx made long ago, and Stephen J. Gould (1987, p. 154) more recently echoed along with many major scientists.

Conclusion

Hopefully, our discussion in this chapter of some of the common types of scientifically *in*adequate explanations for behavior not only conditions what we might call a skeptical sensitivity to such accounts, evoking their rejection, but also prepares us well to proceed in later chapters with discussions of the major types of scientifically adequate explanations of behavior. However, the factual explanations and interpretations of later chapters are, and should be, just as subject to the effects of that conditioned skeptical sensitivity as are the fictitious explanations of this chapter. Let's make sure we get things as right as possible at this time, while remaining sensitive to additional data occasioning any necessary changes. This will prepare us best for the most pertinent applications of behaviorology related to helping solve global as well as local problems.❧

References (with some annotations)

Gambino, M. (2012 March 16). Is it too late for sustainable development? *Smithsonian Magazine* (at http://www.smithsonianmag.com/science-nature/Is-It-Too-Late-For-Sustainable-Development.html).

Gould, S. J. (1987). *An Urchin in the Storm.* New York: W. W. Norton.

Hayes, B. (2012). Computation and the human predicament. *American Scientist, 100* (3), 186–191. This article considers the impact of later developments in computer modeling on the conclusions in *The Limits to Growth* (see Meadows *et al*, 1972).

Ledoux, S. F. (2002a). An introduction to the origins, status, and mission of behaviorology: An established science with developed applications and a new name. In S.F. Ledoux. *Origins and Components of Behaviorology—Second Edition* (pp. 3–24). Canton, NY: ABCs. This paper also appeared (2004) in *Behaviorology Today, 7* (1), 27–41.

Ledoux, S. F. (2002b). Carl Sagan is right again: A review of *The Millennium Man. Behaviorology Today, 5* (2), 23–25.

Meadows, D. H., Meadows, D. L., Randers, J., & Behrens III, W. W. (1972). *The Limits to Growth—A Report for the Club of Rome's Project on the Predicament of Mankind.* New York: Universe Books.

Sagan, C. (1995). *The Demon Haunted World—Science as a Candle in the Dark.* New York: Random House.

Skinner, B. F. (1948). *Walden Two.* New York: Macmillan. Macmillan republished this novel in 1976 with a new introductory essay by the author.

Skinner, B. F. (1953). *Science and Human Behavior.* New York: Macmillan. The Free Press, New York, published a paperback edition in 1965.

Skinner, B. F. (1971). *Beyond Freedom and Dignity.* New York: Knopf. Bantam, New York, published a paperback edition in 1972.

Skinner, B. F. (1974). *About Behaviorism.* New York: Knopf.

Stuart, R. B. & Davis, B. (1972). *Slim Chance in a Fat World: Behavioral Control of Obesity.* Champaign, IL: Research Press.

Wyatt, W. J. (1997). *The Millennium Man.* Hurricane, WV: Third Millennium Press.

Wyatt, W. J. (2002). Critical thinking and a scientific world view: How students' thinking may be changed upon reading *The Millennium Man. Behaviorology Today, 5* (2), 25–27.

Chapter 5
Basic Contingencies in which Behavior Occurs

*H*aving considered the passage of time and genetic endowment as inadequate if *non*–fictional explanations of behavior at the end of Chapter 3, we moved on in Chapter 4 to consider a wide range of fictitious inadequate accounts for behavior. These went beyond the usual agential mentalisms, and included description versus explanation, nominal fallacy, circular reasoning, gratuitous physiologizing, teleology, and some problems inherent in our language. You, of course, recollect every detail of each one of these inadequate accounts, yes?

Here in Chapter 5, we begin to examine the range of natural variables that have experimentally demonstrated adequacy as natural–science explanations of behavior. However, since throughout this book we, as natural scientists, are going to deal exclusively with the natural variables functionally related to behavior, let's explicitly—rather than merely implicitly as has been the case thus far—revisit the word *natural* and some of its usages. One use of *natural* pertains to a general notion of nature, the "great outdoors," what you experience on vacation "away from it all," mountains and forests and rivers and skies and plants and wild animals. Another use extends this initial notion to include *everything*—asphalt and cities and airplanes and cell phones, all matter and energy—*except behavior*. As natural scientists we mean much more than any such usages; for starters *we include all things behavioral* as well, because no characteristic about behavior warrants excluding it from the rest of everything else in the universe. Furthermore, to us the word *natural* (e.g., as in natural variable, natural science, natural scientists) refers to the fundamental approach, to *all* real aspects of the universe, that involves searching for and applying objective and measurable principles, concepts, and laws—all of which are *of nature* as in affording no abiding place for mysticism or superstition—with respect to *everything* including behavior. And *that* means that mysticism and superstition are also researchable topics of natural behavior science (i.e., behaviorology). As well, saying, "all behavior is natural," is to say that all behavior is amenable at its level of analysis to the same laws of nature (i.e., natural laws)—of energy and matter and cumulative complexity—that affect everything else at all their levels of analysis.

So now, in this chapter, we begin our examination of the natural variables that affect behavior by first considering the concept of contingencies among environmental and behavioral variables in general, both antecedent and postcedent, all of which involve energy exchanges traceable between physical and physiological structures and processes. Then we focus on facets of the most common variables on the postcedent side of our behavior formula, the variables and processes that we call *reinforcers* and *punishers* as well as *extinction*. We will even touch on why these postcedent variables often get attention before the generally more crucial antecedent variables.

Environment–Behavior Contingencies

Recall that in Chapter 2 we introduced the concept of contingencies with what we called the "A–B–Cs" (antecedents–behaviors–consequences) of behaviorology. These contingencies pertain mostly to operant behavior and operant conditioning. After appreciating the convenience of the A–B–C initials, we introduced "postcedents" in place of consequences (to which we will allocate a specific meaning in this chapter) because, like antecedent, postcedent is also merely a term of placement in time; thus we have A–B–P.

Contingency Characteristics

Subject to the usual experimental verification, the most common antecedents in operant contingencies appear to be the stimuli that evoke responses, and the most common postcedents appear to be reinforcing stimuli. While we focus on postcedent stimuli in this chapter, a quick review of their connection with antecedents cannot hurt. Let's review these connections in terms of S^{Ev}s (the evocatives that we previously called discriminative stimuli) and reinforcers. We use the more general term S^{Ev}s, for "evocative stimuli," as part of our own, ongoing "ghost–reduction" process. You see, the term "discriminative" first came into use because in the laboratory these stimuli made the procedural difference between reinforcers occurring or not; these stimuli discriminated whether or not reinforcement occurred. But philosophical slippage led some people to talking about discrimination in ways implying that some ghostly inner agent in the organism was "doing" the discriminating. Speaking of "evocative" stimuli at this point helps sensitize us to that kind of mystical error so that, later, we can use either term appropriately. (We can consider ghosts and other spooky phenomena elsewhere.)

Let's return to the connections between S^{Ev}s (*evocative* stimuli) and reinforcers. With a relevant conditioning history, S^{Ev}s work by transferring, to neural receptors, energy that triggers changes in nervous–system structure (on a moment by moment basis involving the more substantial energy resources of the body). These changes include the neural firings that constitute the mediation of neural and motor behavior. Meanwhile reinforcers work also by transferring, to receptors, energy that triggers changes in the nervous system, but these changes are altered structures *that are more enduring* and thereby leave the body different, such that the reoccurrence of relevant S^{Ev}s more readily— which we observe as more frequently—evoke the behavior again, all this also involving the substantial energy resources of the body. Thus, the effects of the energy traces that we call reinforcers *provide the conditioning history* that makes

S^{Ev}s evoke behavior. That is, the reinforcing stimuli functionally feed energy back into the organism's nervous system, changing it so that the now different nervous system mediates behavior differently. When the S^{Ev} that had evoked the behavior before confronts the organism again, the nervous system mediates the same appropriate behavior more easily or more quickly. (Stimuli called punishers work like those called reinforcers except that the enduring nervous system structural changes leave the S^{Ev}s *less* effective as evocative stimuli so that the behavior occurs less frequently.)

As we increase the precision of the basic formula for initially analyzing any behavior, the "A–B–Cs" formula first improves to "A–B–Ps (…postcedents)" and then improves to an explicit contingency such as "S^{Ev}–>R–>S^{R}." We call this "the three–term contingency." It is actually made of two "two–term" contingencies, a point to which we will return.

Why postcedents before antecedents. But first, we will emphasize details about the postcedent portion of some simple contingencies. Are you wondering why we consider details about postcedent variables first, before we consider antecedent variables? Seems reversed, right? Well, while Pavlov discovered respondent stimuli, which are all antecedent, shortly before Skinner discovered operants, perhaps the main reason is that people discovered some postcedents first, so we cover them first. For example, people have employed punishers for millennia. Then Skinner's systematic discovery of postcedent reinforcers opened the era of operant conditioning research. That kind of history has left a tradition about what to cover first. For another reason, Skinner's operant research began in a context of contrast with Pavlov's respondent conditioning research; the presence of distinctive differences between these two made clarifying each one easier. Compared to the antecedent nature of operant evocative stimuli, the postcedent nature of operant reinforcing stimuli made for a greater contrast with the antecedent nature of respondent eliciting stimuli.

However, a better reason for covering postcedents before antecedents concerns accessibility, both for discovery and for application. We find observing the energy trace connecting postcedent stimuli to the nervous system generally more obvious and easy to follow—more accessible—than the energy trace connecting antecedent stimuli to the nervous system. For example, from the back of a classroom, we can easily see a student at the front facing the class and reciting various numbers each of which is followed by one or another student raising a hand in a reinforcing "thumbs up" gesture that produces similar gestures from the whole class. Tracing the energy from these postcedent gestures to the reciting student's eyes is relatively easy; everyone in the room sees the gestures. But what evokes the number reciting? Finding out that various simple multiplication problems—on printed cards (with answers on the back) that other members of the class hold toward the reciting student—are providing the stimuli evoking the recited numbers, takes considerably more effort (i.e., is less accessible) particularly since the light energy reflecting off the cards into the reciting student's eyes is not as directly observable as the gestures. The reciting student sees the problems on the cards, but no one else can see them. Similarly, in applying a therapeutic intervention, successfully arranging a postcedent, reinforcing stimulus occurrence after a target behavior is generally easier than successfully arranging an antecedent evocative stimulus for that behavior. For the evocative stimulus to be as successful in affecting the behavior as the reinforcer, separate behaviors such as head orienting toward, or eye contact with, the evocative stimulus may require conditioning. Adding such steps is more complicated than presenting a food reinforcer after the behavior. Given these considerations, we start with postcedents before covering antecedents.

Fact and theory differences. Since we focus on postcedent stimuli—reinforcers and punishers—in this chapter, we should review their definitions. "Reinforcer" is the term we use when we not only observe a stimulus occur right after a behavior but we also observe that behavior subsequently occurring *more* often, while "punisher" is the term we use when we not only observe a stimulus occur right after a behavior but we also observe that behavior subsequently occurring *less* often. These terms are based on experimental demonstrations; we cannot technically call any stimulus a reinforcer or punisher until after the required experimental demonstration of the definition–related outcome (more behavior, for "reinforcer," and less behavior, for "punisher"). This leaves the status of stimuli as reinforcers or punishers determined according to observational fact (which, by the way, prevents the involvement of reinforcers or punishers in circularities, a point to file away for future reference).

That restriction of reinforcers and punishers to observational fact differs from *theories* about how reinforcers or punishers work, which rely on the self–correcting nature of science for improvement. To be scientifically legitimate, any such theory must be grounded in the physiological mediation of energy exchanges between the environment (internal or external) and behavior, and thus any such theory must be falsifiable through experimentation. Experimental findings help us deal more effectively with the world around us, improving the quality of our lives and the chance that we (i.e., our species) will survive.

Science, Economics, and Contingencies

In science, however, no finding is ever the last possible word, because we always have additional details to examine, and we can always investigate back another link in the functional chain leading to any event—at any relevant level of analysis—and we might at any time also discover an entirely new phenomenon with implications for other phenomena that we might have thought we already understood. Past contingencies always provide plenty of reasons to suspect that some completely new and beneficial discoveries are likely always awaiting scientific attention. And all this is part of what we mean by the self–correcting nature of science.

Some effects of economics on discovering functional relations. Nevertheless, due to economics, scientific discovery and self–correction can be expensive processes. Since the scientific research that leads to discovery also involves at least testing, replication, and development, and all these require sources of monetary support, discovery can carry quite an economic price tag. Yet past reality shows that funding always has limits. We always reach some point at which uncovering and accounting for the remaining independent variables, responsible for variations in a behavior, become too costly, at least for that moment (or decade or century, depending on funding).

Generally, though, we have to settle for knowing just enough about behavior phenomena to deal with them at an acceptable level of effectiveness. For example, we may not yet have teased out every variable responsible for every percentage point of the observed variations in human–subject responding on certain schedules of reinforcement, but for now we know enough to enable managing those schedules effectively when an intervention requires it.

Similarly, throughout this book we describe many known independent variables of behavior that we understand functionally at that kind of acceptable level. Skinner and others reported the original discovery of these variables in books, and in journals like *Journal of the Experimental Analysis of Behavior* (which started in 1958). All that research continues. However, when the next step needs to go beyond the available funding, we must accept that being unable to account for the remaining variation rests on our economics–induced ignorance, rather than on any capriciousness about behavior, or on any of those inadequate explanations we surveyed in previous chapters. As research funding continues, so too the search for further functional details continues.

The generic meaning of "contingencies of reinforcement." In a different way, economics also affects our contingency terminology. We describe each of the wide range of functional relations between environments and behaviors as a contingency, but we lack a separate name for each contingency type. Many postcedent contingencies involve reinforcers while others involve punishers. Some antecedent contingencies involve evocative stimuli while others also involve function–altering stimuli. And the list goes on. Rather than name each type, we use a convenient, linguistically economic verbal shortcut for all these contingencies; retaining a usage that our predecessors (e.g., Skinner) favored, we generically refer to them together as *contingencies of reinforcement.* We manage the small but inherent confusion simply by pointing out, repeatedly if necessary, that this term covers *all* behavioral contingencies including the majority far beyond those that directly involve only reinforcers. For example, the term "contingencies of reinforcement" even covers extinction contingencies.

One reason we can use such a generic term is that all these contingencies share in common the fundamental participation in the exchange of energy traces between internal and external environments and behaviors via the natural nerve and muscle processes that we call *mediating* behavior. This physical reality underlies each and every variable in each and every contingency that we consider. Let's start by examining some basic characteristics of simple two–term and three–term contingencies, including how to indicate parts and relations.

Paying terminology dues. First, however, here is a little word of caution about this chapter (and perhaps some later chapters). Since this book gives an introductory level treatment of its subject matter, some readers may need to pay a certain amount of intellectual dues, through the cost of some possibly tedious reading, while accumulating a background repertoire in basic disciplinary terminology. Still, I am sure you will manage, and this dues–paying preparation provides major benefits down the road.

Two–Term and Three–Term Contingencies

As part of introducing two–term and three–term contingencies, a few words about the difference between a normal verbal–episode description (spoken or written) and a symbolically written string of contingent relations will be helpful, along with some of the basic conventions of our symbolic notation system. We describe the aspects of a behavioral episode as we would narrate a story; we use normal words and phrases to convey the parts, their relations to each other, and how the parts and relations change and develop and possibly even conclude. Analysis, however, benefits from the clarity that derives from the brevity and precision provided by symbolically representing those parts and relations and developments. For this we use a simple symbolic notation system. To show the basic conventions of this system, let's develop a simple example.

Beginning with narration, when Jack spots Jill on the other side of several fresh–produce aisles at the grocery store, he politely but somewhat loudly calls out her name, and she smiles and waves to him. To analyze this narrative episode (i.e., to try to increase our understanding of what is happening and how the parts relate to each other) we must first determine whether the behavior that concerns us is Jack's or Jill's. Since we are usually interested in both, we start with the first behavior as our initial behavior of concern, and we portray the relevant events regarding this behavior as briefly and abstractly as possible under symbolic headings typical of contingency descriptions. Thus, in this case, we would write:

$$\underline{S^{Ev}} \qquad \underline{Response} \qquad \underline{Consequence}$$

$$Jill \quad \rightarrow \quad \text{"Jill!"} \quad \rightarrow \quad Jill\ smiles\ and\ waves$$

[This is a three–term contingency in which
<u>Jill</u> = the person, and <u>"Jill!"</u> = the shout of her name.]

Jill's appearance across the aisles is the antecedent stimulus controlling, by evoking, Jack's shout, "Jill!" Jack's shout is the response. And this response produces its consequence, which requires a phrase due to its several parts, namely Jill's smile and wave. Actually, Jack's shout is not the first behavior in this episode; we might have been interested in the behavior of his *seeing* Jill. And we might also be interested in what happens after Jill smiles and waves, which all constitute possibilities for further examples.

Other conventions in this kind of symbolic notation include arrows, as well as other symbols, that this example lacks. Arrows indicate functional relations and should only be used when such relations are present. If none are present, then no arrow appears between the variables. In cases where a previously operative functional relation no longer holds, then some slashes or an "x" should appear on the arrow shaft (e.g., —//—>, —X—>).

If we have not yet identified one of the parts of a contingency, or will not identify a part due to its unimportance or irrelevance to our concerns, we can still include that part by indicating its status with "——?——"; this symbol simply serves a place–holder function for us, showing that some response or stimulus was an operative part of the depicted contingency. Keep track of these conventions, and additional but relatively rare ones that crop up, as we turn to two–term and three–term contingencies.

Two–Term Contingencies

Recall that in general we analyze contingencies to understand, predict, control, and interpret behavior. More specifically, we analyze contingencies to answer two vital questions: First, we ask the basic question: *Why does the behavior occur?* The answer to this first question leads to identifying the independent variables (i.e., causes) of which the behavior is a function. Second, we ask the practical question: *Is access to changing the causal variables available, or must we change analytical levels (e.g., from the internal to the external environment) to make access available?* The answer to this second question determines the possibility of therapeutic interventions, should these be or become necessary, since effective interventions *require* access to the causal variables (as a change in variables leads to different behaviors).

Note that our basic *"Why?"* question can have several different answers. The answer can be only an antecedent stimulus of the eliciting type (S^{El}) as occurs before respondent behavior. Or, the answer can be just an antecedent stimulus of the evocative type (S^{Ev}), or it can be both an antecedent stimulus and a consequential stimulus, as in operant behavior with, for example, its antecedent evocative stimuli and its postcedent reinforcing stimuli. The answer can even be more complex; the antecedents could also include a function–altering stimulus (S^{FA}) or even a neutral stimulus (S^N), all of which are things we will cover in due time.

A respondent contingency example. Since we will examine many operant contingency examples that have antecedent stimuli and consequential stimuli, here is a familiar respondent example with only an antecedent stimulus. If you go into your bathroom in the morning and are facing the mirror when you flip the light switch, you will see (if you are awake enough) your pupils shrink in size. This is the pupillary constriction reflex, and it is an *unconditioned* reflex, meaning that the response follows the stimulus without any previous pairing of stimuli—that is, without any conditioning—needing to occur first (i.e., the stimulus—the increase in illumination—elicits the response—the decrease in pupillary size—due to affecting some neural structures that derive directly from our genetic endowment). Here is how we would analytically diagram this respondent contingency:

<u>Eliciting Stimulus</u> (S^{El})   <u>Elicited Response</u>

Increase in light level $\rightarrow$ Decrease in pupil size

The presence of this reflex helps prevent blindness. Blindness from its absence could easily have led long ago to a body's becoming old sabertooth's lunch prior to puberty, showing us how contingencies of survival would, in our species history, select for the genes that build the neural structures that mediate this reflex. As a result you would be rather hard pressed to find any human being (read: human body) alive today lacking this reflex. This example shows that we will find some of the causes of human behavior in our species history. Other causes occur in each individual's own life history, often called one's "personal" history. These include *conditioned* respondent reflexes along with a variety of operant variables. Let's consider some of those operant variables.

Operant contingency examples. Ultimately we must answer both our basic *"Why?"* question and our practical *"Access?"* question by making repeated observations with, and changes in, *real* variables, although we may get along with only describing or discussing them when simply completing legitimate interpretative analyses. Of course, in a book like this one, such legitimate interpretative analyses are exactly the point; this is not a laboratory manual that merely describes a series of experimental setups and experiments, and then you (the reader) construct each setup and complete each experiment, which provides actual data that you can then report.

So let's begin some interpretative analyses by recycling our Jack and Jill example, first analyzing its parts as a pair of two–term contingencies. Recall that when Jack saw Jill from across several fresh fruit and vegetable aisles at the grocery store, he rather loudly called out her name; looking up, she saw Jack and smiled and waved to him. Let's analyze several things that are happening.

The first contingency involves the functional relationship between the stimulus, Jill, and Jack's response, "Jill!" This would lead us to write:

S^{Ev} <u>Response</u>

Jill $\rightarrow$ "Jill!"

By the way, what if Jack was *not* seeing Jill when he called out her name? What if he was gazing at the perfect watermelon for their patio party, an item for which they were both looking, and besides smiling and waving, Jill also said, "Over here." In this case we would write the first contingency this way:

$\underline{S^{Ev}}$ $\underline{Response}$

Best watermelon $\rightarrow$ "Jill!"

This shows that not all contingency parts of an episode need to change just because one or another part changes.

The second contingency in our original episode involves the functional relationship between the response, "Jill!" and the consequence, which is Jill's smiling and waving. This would lead us to write:

$\underline{Response}$ $\underline{Consequence}$

"Jill!" $\rightarrow$ Jill smiles and waves

In our alternate, watermelon episode, the second contingency would only differ slightly. We would write it this way:

$\underline{Response}$ $\underline{Consequence}$

"Jill!" $\rightarrow$ Jill smiles, and waves,
                                and says, "Over here."

Sticking with our original episode, if we put these two separate two–term contingencies together, we get our original three–term contingency. As before we would write this pair together this way:

$\underline{S^{Ev}}$ $\underline{Response}$ $\underline{Consequence}$

Jill $\rightarrow$ "Jill!" $\rightarrow$ Jill smiles and waves

Were you to write out the combined contingencies for the watermelon version of the episode, here is what you would write:

$\underline{S^{Ev}}$ $\underline{Response}$ $\underline{Consequence}$

Best $\rightarrow$ "Jill!" $\rightarrow$ Jill smiles, and waves,
watermelon   and says, "Over here."

Since these examples move us into three–term contingencies, let's make a more careful analysis with these diagrams to see better what is happening and how the variables relate to each other.

Three–term Contingencies

Contingency diagrams not only depict what is happening, but we can also expand them to indicate how the depicted functional relations operate. Recall that the variables we depict in these diagrams all participate in the exchange of energy traces between internal and external environments and behaviors via the natural nerve and muscle processes that we call mediating behavior.

Diagramming strengthening. The diagrams can convey some of how these processes work without getting into the physiological details. As a temporary aid to help us work through such diagrams, let's add numerical markers (i.e., numbers in circles) to some of them. In general, the numbers follow the sequence of antecedents, behaviors (or responses, if we are being specific), and consequences. Here is a generic three–term contingency:

Beyond the conventions introduced so far, this diagram adds the feedback loop that shows the occurrence of the consequence affecting the functional relation between the stimulus and the response. The diagram shows that not only is ② a function of ① while ③ is a function of ②, but it also shows the strengthening effect of ③ on the relation between the S^{Ev} (①) and the response (②). We can show this effect by enhancing the function arrow, giving us this diagram:

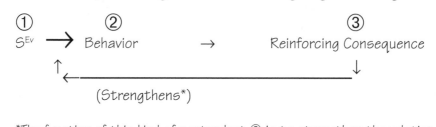

The function of this kind of postcedent ③ is to strengthen the relation between ① and ②. However, other than physiologically, we can detect such an effect only by observing an increase in the occurrences of the behavior ② across future occurrences of the S^{Ev} ①.

The best way to show that strengthening effect would be to repeat the diagram for each reinforcement, with the function arrow increasing in size with each cycle (i.e., getting stronger, as in changing neural structures such that they more readily evoke the response when the S^{Ev} occurs in the future). We should note that, while several others have made similar points, this clarifying analysis of the function of reinforcers, describable as strengthening the functional effect of the S^{Ev} rather than describable as directly strengthening the behavior, is a significant contribution that Lawrence Fraley made to behaviorology, based on his careful and detailed analysis of tracing physical energy changes between environments and behaviors and neural structures, which is the naturalistic basis relevant to behavior and its mediation (see Fraley, 2008; also see Fraley, 2013, Chapter 2, on the physics of behavior).

Before looking at one more diagramatic convention, we must also make this observation: A particular consequence on a particular occasion *cannot* affect the particular response that *produces* that consequence (i.e., a particular ③ cannot affect the particular ② that produces it) because the consequence comes *after* the response; to claim that this consequence affects that earlier response is to engage in teleology. (What's that? It's a fictional explanation that we had fun covering back in Chapter 4.)

Diagramming multiple antecedents. In anticipation of future complexities, one other diagramatic convention deserves our attention. This one pertains to increasing numbers of antecedent variables. Generally, each analytic diagram contains only one behavior or response; with more of these, you need multiple diagrams. (Dealing with multiple consequences occurs on a case by case basis, in one or more diagrams as appropriate.) Multiple, and functionally related, antecedents, however, are rather common, because some stimuli are effective only if other stimuli are functional first. For instance, a stimulus that occurs alone may be neutral (i.e., an S^N) with respect to the behavior of concern, but this stimulus may become evocative if it occurs in the presence of some other particular stimulus. We would designate this other stimulus as a "function–altering" stimulus (an S^{FA}) because its presence alters the function of another stimulus. (You might sometimes see the term "enabling stimulus" [S^{En}] which other researchers use instead of the term *function–altering stimulus.*) Using just our diagrams, here is what can happen.

We start with a lack of functions (i.e., no arrows) since the antecedent stimulus is neutral (so nothing gets evoked):

$\underline{S^N}$ $\underline{Behavior}$ $\underline{Consequence}$

[a neutral stimulus]   [no effect on the behavior]   [so no consequence]

The appearance of a function–altering stimulus, however, brings about changes that can establish a series of functions:

$\underline{Function-altering}$
   $\underline{Stimulus}$ $\underline{Changes\ S^N\ to\ S^{Ev}}$

      $\Downarrow$

   $\underline{S^{FA}}$ $\rightarrow$ $\underline{S^{Ev}}$ $\rightarrow$ $\underline{Behavior}$ $\rightarrow$ $\underline{Consequence}$

Diagramming combinations. Let's first combine the feedback loop with our Jack and Jill example, and then develop the function–altering diagrammatic conventions and add the feedback loop. Here is the first combination:

Jill $\longrightarrow$ "Jill!" $\rightarrow$ Jill smiles and waves

↑ ←—————————————————————— ↓

(Strengthens)

To show the function–altering effect, an elaboration of the watermelon version of our Jack and Jill episode, wherein Jack only looks up and sees Jill after spotting the best watermelon, would look like this:

First we have the neutral stimulus, which by definition has no functions:

$\underline{S}^N$	<u>Response</u>	<u>Consequence</u>
Jill	[no effect on behavior]	[so no consequence]

Then we add the function–altering stimulus and see the resultant changes:

<u>Function–altering</u>
<u>Stimulus</u>   <u>Changes S^N to S^{Ev}</u>

⇓   ↘↘

$\underline{S}^{FA}$ $\rightarrow$ $\underline{S}^{Ev}$ $\rightarrow$ <u>Response</u> $\rightarrow$ <u>Consequence</u>

See best $\rightarrow$ Jill $\rightarrow$ "Jill!" $\rightarrow$ Jill smiles, waves,
watermelon   and says, "Over here."

Now let's add the reinforcing consequence feedback loop:

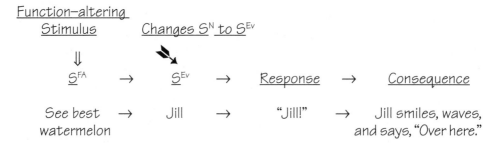

$\underline{S^N \text{ to }}$ ↓
$\underline{S}^{FA}$ $\rightarrow$ $\underline{S}^{Ev}$ $\rightarrow$ <u>Response</u> $\rightarrow$ <u>Consequence</u>*

See best $\longrightarrow$ Jill $\longrightarrow$ "Jill!" $\rightarrow$ Jill smiles, waves,
watermelon   and says, "Over here."

↑ ←——————— ↑ ←——————————————— ↓

(*Note that the reinforcing consequence strengthens all antecedent functional relations.)

Our exposure to all these basic contingency–diagram conventions provides a foundation that will make more complex contingencies easier to manage when we come upon them. Now though, let's move on to the terms that laboratory work has conditioned us to use regarding the range of parts participating in postcedent contingencies, because these terms help us deal more effectively with the world of behavioral phenomena.

Postcedent Contingencies

In this section we cover the terms that most commonly appear in the postcedent part of contingency diagrams. We take up general postcedent terms after considering two criteria pertinent to them. Then we turn to three other criteria related to the most common and specific postcedent terms. This approach should reveal for you not only the definitions of all these postcedent terms but also the logical connections among them.

Terminology Problems and Solutions

By the way, out of respect for full disclosure, this section of the chapter adapts and improves parts of a previous article (Ledoux, 2002) that introduced some new terms to avoid the confusion, discussed next, that some older but workable terms had with their common vernacular usages. Unfortunately, this paper appeared about the time that Lawrence Fraley was *finishing* his comprehensive *General Behaviorology* book (Fraley, 2008). Thus the new terms appeared too late for him to incorporate them into his work. Perhaps the new terms, were insufficiently reinforcing. He had worked on that book, each week revising the part covered that week in his graduate behaviorology course, week after week, semester after semester, for over 20 years. So, since they are still workable, the older terms appear throughout his book. Fortunately, by the time a reader finishes his book, that reader *will* be using even the older terms correctly and without confusion.

The problem with "positive" and "negative." Nevertheless, when behaviorologists need to describe research variables and experimental findings, and the resulting implications and technological applications, to persons not yet fully familiar with fundamental behaviorological laws (e.g., students) they need technical terms that minimize the possibility for confusion. Confusion occurring at this early point can lead to substantial misunderstandings later when more complex issues receive scrutiny. Some of the terms that we use, however, leave room for confusion merely because they have vernacular meanings that are at odds with the technical meanings. The terms *positive* and *negative* can lead to this kind of confusion.

Historically, we have used the terms *positive* and *negative* to describe some postcedent variables. Positive describes variables that the contingency *presents* after a response occurs, while negative describes variables that the contingency *withdraws* after a response occurs. In either case, presentation or withdrawal, if that kind of response subsequently occurs *more* often, we speak of reinforcers, and if that kind of response subsequently occurs *less* often, we speak of punishers. This means we must be clear about positive reinforcers and negative reinforcers and positive punishers and negative punishers.

That would be easy if positive and negative only had the meanings of present and withdraw. Sadly, however, this is not the case, because the terms positive and negative also have connotations in non–technical language that compete with their technical usage. In everyday usage *positive* connotes good or pleasant while *negative* connotes bad or unpleasant. And the everyday usage occurs far more often in one's experience. As a result people get confused and have some difficulty with the concept of a *negative (Bad?)* reinforcer strengthening behavior. They have even greater difficulty with the concept of a *positive (Good?)* punisher weakening behavior; they have trouble imagining much that is positive about punishment. The resulting confusion can take years to overcome, and even some professional authors still occasionally get the terms wrong due to the confusion. Since you would not want to get terms wrong that way, I am sure you are on the edge of your seat right now, waiting to find out how to get these terms right.

The solution of "added" and "subtracted." The solution to the problem of the terms positive and negative involves replacing them with terms that provide the same technical connotation but lack the competing vernacular meanings. For this purpose we now use the term *added* to replace positive, and the term *subtracted* to replace negative. These terms lack the complicating "good" and "bad" connotations of positive and negative. Yet at the same time they remain consistent with the signs (i.e., + and –) used in the symbols for the several types of reinforcing and punishing stimuli. Furthermore, by using the terms added and subtracted, the replaced terms of positive and negative are still available for their non–technical sense without confusion. For example, using the common, non–technical connotations of the terms, one could speak non–technically of rewards and punishments as positive and negative consequences respectively without fear of automatic confusion with technical terms.

Stimuli, events, and processes. Certain other terminological parameters are worthy of our attention. For one, we tend to speak of stimuli as if they were *things,* and in some ways this makes sense. But the reality is more complicated. To respect the interconnection between behaviorology and physiology, the term *stimulus* actually refers to an event, an energy change that affects a functioning receptor cell (or bundle of such cells) at a nervous system entry point. We can call the thing related to that energy transfer a stimulus, but this is a convenience, a verbal–shortcut usage.

The difference between things and events arises in another context as well. For example, we usually speak of the thing, money, as a reinforcer. However, money is not always a reinforcer; sometimes it is a punisher. To get right whether some instance of money is a reinforcer or punisher, technical accuracy—which we sometimes need—demands that we clarify what is *happening* to the money, and that means addressing money the *event*, rather than money the *thing*. For instance, if we *present* extra money to the high school senior who mows our lawn because she did a particularly good job (and find that she takes extra care on future occasions) then that *presentation of money* constitutes a reinforcer. (For practice, can you correctly say what kind of reinforcer, considering the presentation?) On the other hand, if a police officer *withdraws* some money from your wallet because you were speeding—

this is what happens in some jurisdictions (although later you can argue your case in court)—and we find that you speed less on future occasions, then that withdrawal of money constitutes a punisher. (Can you correctly say what kind of punisher, considering the withdrawal?) The example of electric shock works the opposite way. Normally we think of shock as a punisher, but only the presentation of shock is a punisher. (Can you say what kind of punisher, considering the presentation?) Thus we can also honestly say that shock is a reinforcer, but only in that the termination of shock is a reinforcer. (Can you say what kind of reinforcer, considering the termination?) The accurate extrapolation from all these examples is that any particular *thing* can end up as either a reinforcer or a punisher, depending on what happens to it, depending on the added or subtracted status of it as an event.

Let's address one more terminological parameter, this one pertaining to how we linguistically denote things and events and processes or procedures, differences that can apply to other categories of variables as well as to our immediate concern with postcedents. For postcedent *things* we use the "–er" suffix, hence reinforc*er* and punish*er* as things. Sometimes we treat events this same way, but the form "–ing event" is common also, as with reinforc*ing event* or punish*ing event*. On the other hand, for *processes* or *procedures,* we use the "–ment" suffix, hence reinforce*ment* and punish*ment* as processes or procedures.

Two General Criteria for General Postcedent Terminology

With those terminological parameters in hand, let's consider some general postcedent terms and the two criteria that inform their usage. Which postcedent term or terms we use for any particular stimulus (thing, event, or process) depends on whether or not it meets one or both criteria. As a result of applying the criteria, a stimulus might earn several partially overlapping designations. One difficulty is that, while we discuss the criteria one at a time, we get the terms by applying the criteria together.

Affects behavior? One criterion concerns whether or not a stimulus *affects future behavior.* We apply the criterion by asking a certain question, and the answer determines which term we use to describe that stimulus. The question is this: Does the occurrence of the stimulus affect *subsequent* responding?

If the answer is yes—that the occurrence of the stimulus affects later responding (in terms of increasing or decreasing the rate of the type of behavior that the stimulus followed, the way extra money increased the lawn mower's careful mowing)—then we call this stimulus a *selector.* (If the answer is no, then we call the stimulus a *non–selector,* but this is of little analytical importance.)

Products of behavior? The other criterion (which really applies only to various types of selectors, again because non–selectors are of little analytical interest) concerns whether or not the stimulus is *a product of the behavior it follows.* We apply this criterion by asking a different question. Again, the answer determines which term we use to describe the stimulus. The question is this: Does the response *produce* this stimulus?

If the answer is yes—that the response directly changes the environment in ways that produce the occurrence of the stimulus (e.g., the way flipping a switch, or asking someone else to flip it, produces a lighted room, which otherwise stays dark)—then we call this stimulus a *consequence.* If the answer is no, if the response did not produce the stimulus, which instead occurred coincidently, then we call the stimulus a *coincidental–selector* (and we use the term *superstitious behavior* to name the type of behavior that this type of selector affects—but more about this later).

"Coincidental" rather than "accidental." Historically, when a reinforcer occurs, and the response it followed did not produce it, we called it an "accidental" reinforcer. Here we will call such reinforcers *coincidental* instead of accidental. Let me explain why with an example. Having avoided the outdoors for several months due to the winter cold, you are now walking across campus, as no indoor route is available, to a Monday morning meeting you must attend. On your way, you notice a couple inches of green sticking out from underneath a snow pile and, upon pulling it out, you find a $50 bill in your hand. For the rest of the week, you walk outside to your classes and meetings around campus, even when buildings are connected, and even despite the persistent cold. (You may or may not be gazing strenuously at the bases of every snow pile you pass.)

Let's face it; your walking outside did not put that $50–bill reinforcer under the snow pile. (If only finding petty cash was that easy, right?) In that sense your walking did not produce that $50 bill. But calling that reinforcer an "accidental" reinforcer has a problem. To some folks "accidental" implies a certain capriciousness, some sort of chance, random, or spontaneous occurrence lacking a natural functional history. But all of these are wrong, because such reinforcers do have their own natural functional history—somehow that $50 bill got under the snow pile; it did not materialize magically out of nothing— even though we may be ignorant of whatever the actual functional history was. Nevertheless, if an available term carries fewer such implications, it would be worth using to replace "accidental." *Coincidental* is such a term. It merely means "happened at the same time as…," which puts the stress on the time relation, and so we instead use *coincidental.*

General Postcedent Terms

With the two relevant general criteria in hand, let's now define the general postcedent terms of "selectors," "consequences," and "coincidental selectors," along with some unimportant terms so that we can logically complete the set. Again, which term applies to any particular postcedent depends on (a) whether or not a stimulus affects future behavior, and (b) whether or not the stimulus is a product of the behavior it follows. For each term we will also provide some other considerations.

Selectors. Selectors are postcedents that affect subsequent responding regardless of whether or not the response they follow produces them. Thus both consequences and coincidental selectors fit under the more general category of selectors (just as *all of these* fit under the even more general category of postcedents).

Consequences. Consequences are selectors (affecting subsequent responding) *that the preceding response* (i.e., the response that they follow) *produces.* This constrains our use of the "consequences" term, which no longer covers all postcedents that affect responding. Now we only use it to tact postcedents that both affect later responding and that the preceding response produces. (By the way, *tact* is a behaviorologically more accurate word than "name," and we detail it in the verbal–behavior chapter.)

Coincidental selectors. Coincidental selectors are selectors (affecting subsequent responding) that the preceding response *does not produce.* Some other functional chain produces these selectors, and the behavior that these selectors produce is called superstitious behavior.

Completing this terminology set. Certain other terms are analytically unnecessary, but mentioning them enables us to complete the logic of our terminology set. This may make mastering our main terms easier, because it provides a relevant set of non–examples for comparison. Three terms are involved, and they are (a) non–selectors, (b) non–selecting consequences, and (c) coincidental non–selectors. Before we describe and define them, see if you can figure out to what each one refers.

Here are the details. (a) "Non–selectors" are the opposite of selectors; non–selectors are postcedents that *do not affect subsequent responding* regardless of whether or not they are produced by responding. (b) "Non–selecting consequences" are the opposite of consequences; non–selecting consequences are non–selectors (not affecting subsequent responding) that *are* produced by responding but, again, they have no further effect on the behavior that produced them. And (c) "coincidental non–selectors" are the opposite of coincidental selectors; coincidental non–selectors are non–selectors (not affecting subsequent responding) that *are not* produced by responding. Yes, a lot of "nons" and "nots" make up the descriptions and definitions of these terms that logically complete our set of postcedent terms but are otherwise unimportant. Some students in my introductory behaviorology course seem happy to find something that they need not know. How about you?

Outlining these terms provides another, more visual way to see their connections. Using simply the terms, an outline would look like this:

I.   Antecedents (the other category of variables beyond this chapter).
II. Postcedents.
    A. Selectors.
        1. Consequences.
        2. Coincidental selectors.
    B. Non–selectors (the opposite of selectors).
        1. Non–selecting consequences (the opposite of consequences).
        2. Coincidental non–selectors (the opposite of coincidental selectors).

Those last three ("B," "B1," and "B2") are the logically valid but otherwise unneeded sub–categories. Ignoring them, we could further list *all* the terms for the various types of reinforcers and punishers under *each* of the "A1" and "A2" sub–categories in this outline (consequences and coincidental selectors). The next subsection covers these various reinforcer and punisher types, and discusses why *all* of them would go under *each* of the consequence and coincidental–selector sub–categories.

Three Specific Criteria for Specific Postcedent Terms

Moving on, let's consider the more common, specific postcedent terms and the three *specific* criteria that inform their usage. Which postcedent term we use for any particular stimulus (thing, event, or process) depends on what combination of the three criteria that stimulus meets. Whereas the stimulus might earn a couple of partially overlapping general postcedent designations from the first two *general* criteria (of "Affects behavior?" and "Is a product of behavior?") we find that with specific postcedents, as a result of applying three *specific* criteria, the stimulus will earn an additional but three–part designation, each part coming from a different criterion. Again, the difficulty is that, while we discuss the three criteria one at a time, we get the complete, multi–part term by applying the criteria together. Generally these criteria are straightforward, but simultaneously combining them can at first present a (Fun?) challenge. Here are the three criteria, and the terms associated with each alternative, presented in the order of their relative importance for effective analysis.

First criterion: Increase or decrease? This criterion is vital for analysis. *Does the behavior increase or decrease?* Or, more completely, does the behavior that the selector follows subsequently show an increase, or does it show a decrease? When the behavior shows an increase, the selector has met the definition of a *reinforcer,* and when the behavior shows a decrease, the selector has met the definition of a *punisher.* (Explicit definitions appear shortly, along with our application of the criteria and some examples.)

Second criterion: Added or subtracted? This criterion is usually helpful for analysis. *Is the selector involved in presentation or termination?* Or, more completely, does the selector have its effect, of increasing or decreasing the rate of subsequent behavior, when the contingencies present it, or when the contingencies withdraw it (or terminate it)? If the subsequent effect follows presentation, then we describe the selector as *added,* but if the subsequent effect follows withdrawal, then we describe the selector as *subtracted.*

Let's combine added and subtracted, from the second criterion, with the *outcomes* from the increase and decrease of the first criterion. This could give us added or subtracted reinforcers, or it could give us added or subtracted punishers, all of which are postcedent selectors (of either the consequence or coincidental–selectors type).

Third criterion: Unconditioned or conditioned? This criterion is sometimes needed for analysis, and other times not so needed. *Does the* *selector work without conditioning having occurred, or does it work **only** if some conditioning has occurred?* Or, more completely, does the selector have its effect of increasing or decreasing the rate of subsequent behavior without any previous relevant conditioning having occurred, or does it have those effects *only* if certain relevant conditioning processes have already occurred? If the effect occurs *without* previous conditioning, then the selector is called *unconditioned,* but if the effect occurs *only after* certain previous conditioning processes have occurred, then the selector is called *conditioned.*

Combining *unconditioned* and *conditioned* with the outcomes of the first *and* second criteria could give us the greatest number of regular stimulus (S) possibilities. With the common symbols for each, these are: (1) added unconditioned reinforcers (S^R), (2) added conditioned reinforcers (S^r), (3) subtracted unconditioned reinforcers (S^{R-}), (4) subtracted conditioned reinforcers (S^{r-}), (5) added unconditioned punishers (S^P), (6) added conditioned punishers (S^P), (7) subtracted unconditioned punishers (S^{P-}), and (8) subtracted conditioned punishers (S^{P-}). Any of these could be consequences or coincidental selectors, and all of these are postcedents.

The alternate terms of "primary" and "secondary." Note that some authors use the terms *primary* where we use the term *unconditioned,* and they use the term *secondary* where we use the term *conditioned.* These terms, primary and secondary, can seem easier to use, and they are not entirely arbitrary in origin. After all, we get secondary (i.e., conditioned) reinforcers when we apply the conditioning process by pairing neutral stimuli (i.e., those having no effect on the behavior of concern) with primary (i.e., unconditioned) reinforcers, which thus must function first. On the other hand, unconditioned and conditioned, while accurate in usage, can be a little confusing due to other uses of these terms. In any case, both term pairs represent legitimate and interchangeable usages.

The Analysis of Types of Reinforcers and Punishers

Now let's turn to the application of those three particular criteria. (Recall our earlier comment about some of this chapter being rather tedious. This subsection is where the most dues–paying perseverance pays off; these dues will pay you potent dividends later in a more easily expanded repertoire. I encourage you to persevere.) Along with some examples (in which the operative word will be in *italics)* we will define reinforcers and punishers in all their varieties. For pedagogical reasons we present each of these definitions in rather comprehensive form (see the Glossary) *and we also put into italics those parts of the definitions that demand our greatest attention,* at this early stage of our coverage of behaviorology.

Behavior increase or decrease. Here we consider the two categories of postcedents that depend on whether they increase or decrease the kind of behavior that they follow. Again, this characteristic is vital for analysis. The two categories are reinforcer and punisher.

A **reinforcer** *is a stimulus* (i.e., a relative change in the environment) that provides an energy change at receptor cells *during or immediately after a response* (with reducing effects as the time between these events increases) *that results in an **increase** in the rate of the behavior across subsequent occasions.* For example, as a result of a teacher *placing a star* on a young pupil's correctly completed sheet of simple math problems, the pupil asks for and completes several more sheets right away. (Is this example an added, or a subtracted, reinforcer? Is it an unconditioned, or a conditioned, reinforcer?)

A **punisher** *is a stimulus* (a relative change in the environment) that provides an energy change at receptor cells *during or immediately after a response* (with reducing effects as the time between these events increases) *that results in a **decrease** in the rate of the behavior across subsequent occasions.* An example would be a dog on a chain trying to attack a letter carrier who crosses a side lawn instead of walking around on the sidewalk; as a result the letter carrier returns to following the sidewalk. (Is this example an added, or a subtracted, punisher? Is it an unconditioned, or a conditioned, punisher?)

Added and subtracted events. Here we consider the four postcedents that we describe either as added or as subtracted. Again, this characteristic is usually helpful for analysis. Four postcedents share this characteristic: added and subtracted reinforcers, and added and subtracted punishers.

An **added reinforcer** *is the presentation* (i.e., addition) *of a stimulus* in the environment *during or immediately after a response that results in an **increase** in the rate of the behavior across subsequent occasions.* For example, when Jack rather loudly called out Jill's name from across the fruit and vegetable aisles, she *smiled and waved* to him (two reinforcers) and subsequently he began calling her name rather loudly even when she was right next to him. (Is this example an unconditioned, or a conditioned, added reinforcer? Hint: It could be both, but which reinforcer is which?)

A **subtracted reinforcer** *is the withdrawal* or termination (i.e., subtraction) *of a stimulus* in the environment *during or immediately after a response that results in an **increase** in the rate of the behavior across subsequent occasions.* An example would be a *reduction of penalty points* (perhaps that you earned from disciplinary infractions in class, such as cell–phone texting during a lecture). These penalty points were to be reduced by completing extra pages of practice word problems, but you helped a student after class so your teacher reduced them. Subsequently you helped students after class regularly. (Is this example an unconditioned, or a conditioned, subtracted reinforcer?)

An **added punisher** *is the presentation* (i.e., addition) *of a stimulus* in the environment *during or immediately after a response that results in a **decrease** in the rate of the behavior across subsequent occasions.* For example, on another occasion when Jack quite loudly called out Jill's name from across the several fruit and vegetable aisles, she *scowled at him and shushed him* (two punishers) and subsequently he no longer called her name so loudly. (Is this example an unconditioned, or a conditioned, added punisher? Hint again: It could be both, but which punisher is which type?)

A **subtracted punisher** *is the withdrawal* or termination (i.e., subtraction) *of a stimulus* in the environment *during or immediately after a response that results in a **decrease** in the rate of the behavior across subsequent occasions.* An example would be a child's friends simply *leaving the room* when the child starts to cheat at the card game they were playing, and subsequently the child no longer tried to cheat. (Is this example an unconditioned, or a conditioned, subtracted punisher? Actually, this is not a fair question this time, or perhaps it is too fair because, depending on the circumstances, either answer could be correct. Instead, can you suggest what kinds of circumstances could lead to each correct answer?)

Unconditioned and conditioned events. Here we consider the four postcedents that we describe either as unconditioned (i.e., operative due to genetically produced neural structure) or as conditioned (i.e., operative due to past pairing with—that is, occurring at the same time as—other reinforcers or punishers). We get conditioned reinforcers and punishers when we apply the conditioning process by pairing neutral stimuli (i.e., those having no effect on the behavior) with, respectively, unconditioned reinforcers and punishers. Again, this characteristic is sometimes needed for analysis, and at other times is not so needed. Here are the four postcedents that share this characteristic:

An **unconditioned reinforcer** *is a stimulus the occurrence* (i.e., the energy transfer) *of which already **strengthens** prior behavior, without needing any conditioning* (because genes presumably produced the operative neural structure that these stimuli affect). An example is the *shock stopping* as soon as the movement of your hand breaks contact with both the new light bulb and socket after your hand contacted them as you were screwing in a new light, to replace a burned out bulb, without following your OSHA (Occupational Safety and Health Administration) safety regulations by unplugging the lamp first; subsequently your hand moves quickly away from metal lamp parts whenever it even brushes against them. (Is this example an added, or a subtracted, unconditioned reinforcer?)

Note that for people, unconditioned reinforcers are often roughly summarized by counting them on the fingers of one hand as food, water, sex, salt, and attention. (We count salt separately from food, because people who are otherwise food–satiated often continue through half a bag of salty snacks after a meal, suggesting that the reinforcing capacity of salt endures beyond the satiation of food intake. Similarly, we can observe other animals, also otherwise food–satiated, still walking miles to a salt–lick location.)

As we said previously, *attention* might not be an S^R but may actually be an S^r. Indeed, some people, perhaps many, consider it an S^r. Unfortunately, we may never be able to carry out the experiments that might demonstrate satisfactorily which it is, S^R or S^r. If attention is an S^r for people, then it starts out neutral and, quite early in life, occurs with other reinforcers, which conditions it as an

S^r. However, the difficulty, including ethical considerations, of manipulating those variables, and observing the outcome, leaves its status unclear. In my list of unconditioned reinforcers, attention contrasts well with praise, which is not the same as attention, although casually it might look the same. Praise is clearly an S^r as it is based on the conditioning of verbal behavior. For my part, I will treat *attention* as an S^R, at least for the pedagogical value of its contrast with *praise,* which helps my students see the difference between S^Rs and S^rs.

A **conditioned reinforcer** *is a stimulus the occurrence* (i.e., the energy transfer) *of which **strengthens** prior behavior **only** if certain conditioning has already occurred* (i.e., only if the pairing process has happened, in which this otherwise neutral stimulus occurs at about the same time as another established reinforcer, so that either one now produces the operative changes in neural structure). An example would be *receiving a verbal compliment* at a party about a wildly colored scarf (that had been hanging around your closet for some years) and subsequently wearing the scarf to every class for three weeks. (Is this example an added, or a subtracted, conditioned reinforcer?)

An **unconditioned punisher** *is a stimulus the occurrence* (i.e., the energy transfer) *of which already **weakens** prior behavior, without needing any conditioning* (because genes presumably produced the operative neural structure that these stimuli affect). For example, while you cut some roses for indoor appreciation, *the thorns prick your hands,* and subsequently you only admired the roses—on the bush—when you were outdoors. (Is this example an added, or a subtracted, unconditioned punisher?)

A **conditioned punisher** *is a stimulus the occurrence* (i.e., the energy transfer) *of which **weakens** prior behavior **only** if certain conditioning has already occurred* (i.e., only if the pairing process has happened in which this otherwise neutral stimulus occurs at about the same time as another established punisher, so that either one now produces the operative changes in neural structure). An example would be the *occurrence of penalty points* following disciplinary infractions such as cell–phone texting during class, and subsequently fewer such infractions occurred. (Is this example an added, or a subtracted, conditioned punisher?)

A postcedent outline with symbols. Here is an extended part of our earlier postcedent outline. In it you can see both the hierarchical relationship of the general postcedent terms (entries A , 1 and 2) based on our two *general* criteria, as well as the interrelationships of our specific postcedent terms (all the remaining entries) from the three *specific* interacting criteria.

This outline includes the common symbolic notation that we use with each of the four basic reinforcers and the four basic punishers. Let's specify the notation. "S" indicates "stimulus." Upper case "R" indicates "unconditioned reinforcing" while lower case "r" indicates "conditioned reinforcing." Upper case "P" indicates "unconditioned punishing" while lower case "p" indicates "conditioned punishing." Also "+" and "−" respectively indicate added and subtracted. For example, "S^{r-}" reads as "subtracted conditioned reinforcing stimulus." Try reading out each symbol fully when you get to it.

Also, note that, as in mathematics, we seldom write the "+" sign in the symbolic notation; it appears here (and only in parentheses) for pedagogical reasons. Furthermore, since the eight symbolized terms in this outline apply as much to *consequences* as to *coincidental selectors,* one can conclude that these eight actually represent a total of 16 postcedent–selector types:

I.   Antecedents (the other category of variables beyond this chapter).
II. Postcedents.
   A. Selectors.
     BOTH 1. Consequences. **AND** 2. Coincidental selectors.
      (a). Reinforcers.
        (1). Added reinforcers.
          [a]. Added unconditioned reinforcing stimulus [$S^{R(+)}$].
          [b]. Added conditioned reinforcing stimulus [$S^{r(+)}$].
        (2). Subtracted reinforcers.
          [a]. Subtracted unconditioned reinforcing stimulus [S^{R-}].
          [b]. Subtracted conditioned reinforcing stimulus [S^{r-}].
      (b). Punishers.
        (1). Added punishers.
          [a]. Added unconditioned punishing stimulus [$S^{P(+)}$].
          [b]. Added conditioned punishing stimulus [$S^{p(+)}$].
        (2). Subtracted punishers.
          [a]. Subtracted unconditioned punishing stimulus [S^{P-}].
          [b]. Subtracted conditioned punishing stimulus [S^{p-}].

A postcedent matrix and tree diagram. Instead of an outline, we can visualize all the specific reinforcing and punishing postcedent stimuli in a four–box matrix (see Figure 5–1). Then in the boxes we can put some examples of each postcedent type. In any blank space after the references, you can draw a similar matrix for more examples. These can be postcedent examples that you have experienced or observed or simply ones about which you have wondered.

We organize the matrix according to the three interacting criteria. Whether behavior increases or decreases appears on the left side of the matrix. (Note that "increase" includes maintaining the behavior, as the occasional reinforcement seems to counteract the natural ongoing degradation in the physiological neural structures mediating the behavior.) Next, we simply put whether the postcedents classify as added, or subtracted, across the top. However, designating whether the postcedents classify as unconditioned or conditioned presents a more difficult task, because a third criterion like this really requires a three–dimensional matrix. Since that obviously exceeds the two–dimensional limits of our page, we instead place a dashed horizontal line across the box holding added and subtracted *reinforcer* examples, and another one across the box holding added and subtracted *punisher* examples. In both cases we then put unconditioned postcedents in the space above the dashed line, and we put conditioned postcedents in the space below the dashed line. Here is one last question. Can you apply the process that turns neutral stimuli into conditioned reinforcers to tell how "Jazz" can rightfully be listed in both the added reinforcer column and the subtracted reinforcer column? The results of that process help us account for the myraid combinations of individual differences that help make us all interesting to each other. Think about that. Also, think about how a change in the type of occurrence (added or subtracted) changes the effect of a stimulus (as explicitly mentioned in the bottom half of the matrix).

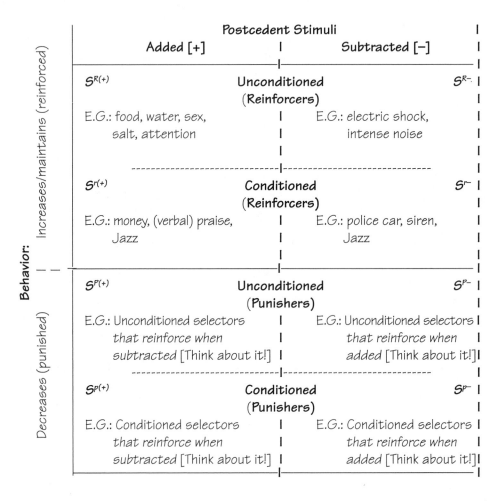

Figure 5–1. A Postcedent Matrix showing the effects on behavior of some common postcedent stimulus types (unconditioned or conditioned) and processes (added or subtracted) with examples.

You may also find Figure 5–2 helpful in reviewing the terms in Figure 5–1 and how they relate to some of the other terms that we covered in this chapter. Figure 5–2 provides a branching *Postcedent Tree Diagram* (one that improves on the version in Ledoux, 2002, p. 202).

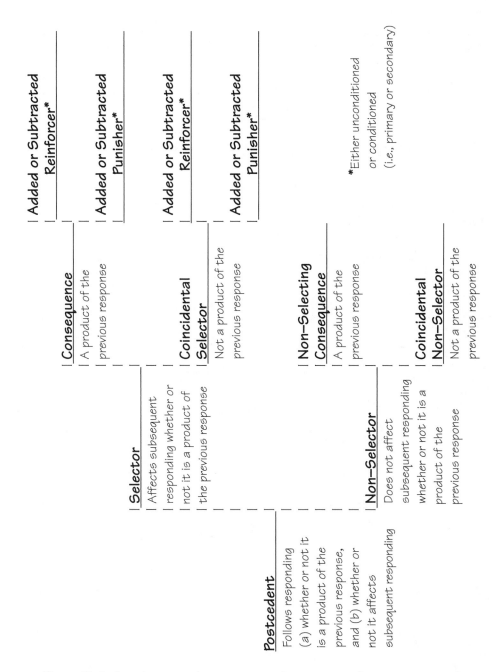

Figure 5–2. Tree diagram showing postcedent events of increasing specificity.

Remain focused on these terms as we move onto Chapter 6 where we consider some typical contingencies that involve multiple, simultaneous behaviors and multiple, simultaneous stimuli. Such contingencies are quite common with human behavior, and our coverage of them also introduces us to some real experimental research.

Take a moment, though, before going on to Chapter 6, to continue our ongoing chapter–conclusion activity. Keep in view the pertinence of behaviorology to problem solving, particularly our coverage of some fundamental contingencies in this chapter that constitute some basic components to our helping solve global as well as local problems.❧

References (with some annotations)

Fraley, L. E. (2008). *General Behaviorology: The Natural Science of Human Behavior.* Canton, NY: ABCs.

Fraley, L. E. (2013). *Behaviorological Rehabilitation and the Criminal Justice System.* Canton, NY: ABCs.

Ledoux, S. F. (2002). Increasing tact control and student comprehension through such new postcedent terms as added and subtracted reinforcers and punishers. In S. F. Ledoux. *Origins and Components of Behaviorology— Second Edition* (pp. 199–204). Canton, NY: ABCs. This article also appeared (2010) in *Behaviorology Today, 13* (1), 3–6.

Chapter 6
Contingencies of Multiple Simultaneous Behaviors and Stimuli

*T*he time has come to increase the complexity a bit more than usual. Chapter 5 described some basic two–term and three–term contingencies relating simple behaviors and stimuli. These covered a little of the range of contingencies that we know as "contingencies of reinforcement," which include not only operant behaviors under the control of evocative and selecting stimuli (e.g., reinforcers) but also respondent behaviors under the control of eliciting stimuli. We also introduced some conventions for diagramming the analysis of contingencies. Then we rounded out the chapter with a thorough list of the defined terms that we use in analyzing the postcedent side of behavior–controlling contingencies while also introducing some of the antecedent–side terms.

Now, in this chapter, we recover some of that ground while looking at more complex contingencies that include multiple behaviors and multiple stimuli. We will extend our analysis to contingencies containing not only simultaneously evocable multiple behaviors but also containing simultaneously occurring selectors like reinforcers. These kinds of circumstances begin to typify what we commonly find for general human behavior.

After looking at some of the parameters surrounding such concurrent contingencies, we will consider them in the context of actual experimental research, looking at methodology, data, conclusions, and extensions. This coverage provides a nice glimpse of behaviorological laboratory research. Furthermore, given the number of basic research questions possible, and with appropriate graduate–level training in behaviorology, some readers could carry on this research, extending it beyond whatever point it has reached.

Concurrent Contingencies

Given a whole, functioning physiology (i.e., an organism) behavior occurs, because eliciting or evocative stimuli occur. We call the functionally related combination of behaviors and stimuli *contingencies.* Occasionally a contingency occurs alone, but more often two or more contingencies operate at the same time, and we call these *concurrent contingencies.*

Concurrent contingencies come in a variety of sizes and types. The decades of ever expanding experimental behavior research since the 1930s have clearly established that behavior is an orderly process. With this demonstration in hand, as study and practice build our analytical and intervention skills, we gradually move from two and three term contingencies to contingencies with ever greater numbers of terms, which we call *n–term* contingencies.

Both the antecedent side of a contingency, and the postcedent side, can have multiple terms. But since our concern is with behavior, we deal with only one behavior per contingency. When multiple behaviors are involved, we analyze in terms of multiple, usually concurrent, contingencies. Alternatively, the same behavior can be under the control of more than one contingency, sometimes at different times, sometimes alternating, and sometimes at the same time. Adding to our analytical task, such contingencies can have one contingency that duplicates support for the effect of another contingency, or they can have one contingency that works against another contingency. Let's look at a small number of the many possibilities as part of the parameters surrounding concurrent contingencies.

Supportive Contingencies

While involving the same behavior, one contingency can *support* the effect of another contingency. This occurs when both contingencies feature the same kind of selector for the behavior, such as when both contingencies feature reinforcers. For example, when a fourth–grader completes an assignment, her teacher provides several reinforcers, which are parts of normal human interactions, and drive normal human outcomes (such as a smile, a compliment, a high mark, perhaps even a chance to participate in a drawing that selects which pupil gets to specify the flavor of ice cream for a class picnic). Around the same time, when the student completes an assignment at home, the student's parents also provide reinforcers, separate from the teacher's, which again are parts of normal human interactions, driving normal human outcomes (e.g., a smile, a compliment, perhaps even an after–dinner walk to the neighborhood ice cream parlor where the child gets her favorite ice cream flavor, a flavor which her past conditioning experience determines, not any putative agential inner self).

After both sets of selectors occur, contingent on the assignment completion once the teacher assigned it, we observe the student completing her work more often on later occasions. Both sets of reinforcers supported this effect. With any or all examples like this, you can benefit from drafting simple analytical diagrams of the involved contingencies using the conventions we covered previously. While sometimes you will need two (or even more) diagrams, here is one diagram, for this example, that applies both to the contingency at school and to the contingency at home:

S^Ev		Behavior		Consequence		[Result]
Work assigned	→	Complete work	→	Smile etc!	→	[Work completed more often]

Note that, to improve the clarity of what is happening, this and the next four diagrams show the outcome, the result, of the effect of the consequence, although we seldom include this in diagrams. These were supportive contingencies. What if the contingencies contradict each other? Then you would *require* two diagrams. Let's see how that can work.

Opposed Contingencies

In opposed contingencies, one contingency is incompatible with another contingency. The conflict can occur between behaviors, or it can occur between stimuli, and both situations often lead to an observed oscillation between or among behaviors. In any case one contingency *opposes* the effect of another contingency. Consider a case where both contingencies feature opposing kinds of selectors for the behavior, such as when one contingency features a punisher while the other features a reinforcer. As an example, let's visit a different fourth–grader under less favorable circumstances. When he completes an assignment, his teacher responds the same way as to his classmate, providing the kinds of reinforcers we already described. At home, however, when he completes an assignment, the parents ignore him (which involves a process we call extinction) or, worse, accuse him of trying to be a teacher's pet, which is a form of punishment or coercion.

After both sets of selectors occur, contingent on the assignment completion once the teacher assigned it, we observe whether the student completes his work more often, or less often, on later occasions. Only the outcome will show which contingency was more effective, the reinforcement contingency or the punishment contingency. Actually, since punishment elicits aversive emotional respondents, this situation can get pretty complicated, and a later chapter will consider such examples. In the mean time, we will patiently await future observations while we describe the outcome of such opposing contingencies as resulting from an *algebraic summation* of the separate effects, which may also induce some hesitancy about the work, or even some oscillation between working on the assignment or not.

Since this is the first time you would use two diagrams, here they are, one for the contingency at school and the other for the contingency at home:

❧ Contingency 1 of 2 (at school) *while both are operating:*

S^Ev		Behavior		Consequence		[Result]
Work assigned	→	Complete work	→	Smile etc!	—?—	[Work completed more often?]

✻ Contingency 2 of 2 (at home) *while both are operating:*

S^{Ev}	Behavior	Consequence	[Result]
Work assigned	→ Complete work	→ Called —?— teacher's pet	[Work completed less often?]

If we had more information, we might be ready to predict which outcome is more likely to result from the algebraic summation of the effects that these two contingencies produce. As it stands, though, we really have too little information and so need to refrain from trying too hard to second guess nature. This is a lesson best learned sooner than later.

With these topics and examples, we are beginning to see a little more deeply the importance and fascination of discovering why humans behave. So, let's charge on, putting some further considerations into the mix, like the difference in the analyses that we make from the different perspectives of a behaver (with no inner agent implied, but only referring to the functioning physiology mediating the behavior) and other observers (who are also only functioning physiologies mediating behaviors, without any inner agents).

Accounts from a **Public of One** and from a **Public of Others**

Contingencies produce people's responses, people's responses produce stimuli, and these stimuli affect us as members of the public at large, *a public of others.* These stimuli, from the overt responses of others, participate in contingencies driving "analysis responding," that is, our responses in which we analyze observed behavior. These analyses describe the most likely contingencies responsible for the observed overt behavior, and may provide access that enables helpful interventions when the overt behavior is in some way problematic.

Yet at the same time, each of us is also a *public of one,* a one–member public, privy to the occurrence of covert stimuli. Covert stimuli, which include the stimulus status of covert responses since these are real events, occur inside our body where others cannot be privy. Vargas (2013) calls this single–observer observation. With appropriate physiological measurement equipment, these others might gain some access.

We can make some valid analyses both regarding our own behavior, as a public of one, and regarding our own behavior and others' behavior as members of the pubic of others. But these two types of analyses differ, particularly in terms of accessibility. Being able to delineate clearly the *public–of–others* version provides a better foundation for interventions.

Let's look at an example that will show these differences and their implications. *The denial of access to a demanded snack for a child in a grocery cart at the store evokes the start of a tantrum and, in short order, the embarrassed mom or dad calmly wheels the cart out the door where gradually the tantrum fades.* Yes, I know, this is not what usually happens. The reinforcing value of relief from the screaming usually induces the behavior of giving in. Unfortunately this leads to the predictable result that tantrums increase across later occasions.

In that example, however, we are portraying scientifically (i.e., behaviorologically) informed parents whose education has conditioned them about how to avoid tantrums. Furthermore, once in a developing tantrum circumstance, this example shows one of the recommended steps for handling the situation: without delivering whatever the tantrum was about, calmly move from the current setting to a different one. Glenn Latham points out that a parent should never give in before the child gives out (see Latham, 1994, and especially 1999). After tantrums start, giving children what they want inadvertently reinforces tantrum behavior, something quite definitely opposite what would reinforce parental behavior. This clarifies a little about the child's behavior, but the parents' behavior deserves the attention of our comparative analysis of *public–of–one* and *public–of–others* perspectives.

So, consider that example again: The denial of access to a demanded snack, for a child in a grocery cart at the store, evokes the start of a tantrum and, in short order, the embarrassed parent calmly wheels the cart out the door where gradually the tantrum fades. Here the behavior of concern is leaving the store.

From the parent's public–of–one perspective, what happened? The parent might report, "The tantrum made me feel embarrassed, which prompted my leaving the store, which past experience taught me would make me feel better, which it did, and I will leave the store again the next time a tantrum starts." Here is how we might diagram this public–of–one reported contingency (noting that since this is a specific activity, we use the "Response" label rather than the more general "Behavior" label):

S^{Ev}		Response		Consequence		[Result]
Aversive feelings of embarrassment	$\rightarrow$	Leave store	$\rightarrow$	Aversive feelings diminish	$\rightarrow?$	[Leave store in future when tantrum starts]

We start with the aversive feelings of embarrassment, even though only the width of the page prevents us from including the tantrum as the stimulus that evoked the emotional response that the parent reported as those feelings of embarrassment. Also, notice that the reduction in the aversive feelings of embarrassment occurs upon leaving the store regardless of how long the tantrum persists before trailing off. So we cannot say that a reducing tantrum contributes to reduced embarrassment. But we could analyze in terms of some anxiety that the tantrum evokes; a reducing tantrum, whenever that occurs, leads to a decrease in the anxiety; consider this by writing such a diagram.

However, while this situation seems to lack any contingency that would induce a false parental report, being unable to be more confident about the accuracy of the report leaves a certain residual discomfort. In addition, we also lack ready access to observing emotions except one's own, and the same holds for access to changes in, or relief from, the reported emotion. Here, we lack ready access to the embarrassment (or anxiety) feelings, and the subsequent change in them, that the involved parent reported. We are stuck with this kind of difficulty with all public–of–one analyses, in spite of their inherent validity.

To improve this situation, we must move to a public–of–others analysis. So, from our public–of–others perspective, what happened? Other observers, upon witnessing the events, might report, "Shortly after the tantrum began, the parent pushed the cart holding the child out of the store door, after which the tantrum subsided; a prediction consistent with these events would be that the parent will leave the store again the next time a tantrum starts." Here is how we might diagram this public–of–others reported contingency:

S^{Ev}		Response		Consequence		[Result]
Tantrum begins	→	Leave store	→	Tantrum subsides	→?	[Leave store in future when tantrum starts]

While these overt events are related to the very real covert events that the participants experienced, far fewer accessibility problems inhere in this more public analysis. Moving to a public–of–others perspective renders our analysis practical. We can consider additional helpful intervention steps when the overt behavior turns out to be in some way problematic. But all this rests on a certain well–supported presumption about behavior, the presumption of *behavior passivity*, a natural science presumption to which we now turn.

Behavior Passivity

The benefit of public–of–others analyses will be evident in our continuing consideration of contingencies to cover more complex human behavior. Here, that behavior involves simultaneous stimuli and responses. However, before expanding into that area, we should more explicitly visit a principle that we have mostly taken for granted thus far. This is the principle of *behavior passivity.*

Behavior passivity refers to the nature of behavior, all behavior, including all human behavior. Like all real events in nature, on this planet, in this universe, behavior is natural, that is, it is a natural event. It is an inevitable reaction. It merely "happens"; it happens as a function of other also real variables. When these variables happen, as a result of their own traceable natural functional history, behavior follows. If a behavior happens, then it had to happen, but if a behavior does not happen, then it could not have happened. Various authors (e.g., Skinner, 1953, p. 112) have made this same point in similar ways.

As real events, behaviors participate in the accumulating natural functional history of yet other real events, some of which are also behavior. Neural behavior happens as the firing of neurons while neuro–muscular behavior happens as innervated muscle contractions. Since, in an appropriate time frame, we cannot really separate the observing of behavior and the firing of neurons along with the muscle contractions, we describe the situation in terms of the physiology *mediating* the behavior.

"Physiology mediating behavior" provides a direct description of those events that excludes traditional, pre–scientific (yet still all too current in some quarters) mystical notions about the origins of behavior. Some of these excluded notions involve physiology somehow originating or spontaneously initiating behavior, while other excluded notions involve a wide range of physiologically independent self agents that the cultural–level, even educational, conditioning of some people has led them to see as responsible for behavior essentially in scientifically untraceable or untestable ways. These mystical notions can be as specific as the notions that the agent does the behavior or tells the body to do the behavior, or decides which behavior will occur, or chooses the next behavior. Alternatively, these notions can be as general as the terms mind or psyche or self or soul. Yet no inner agent exists to do, tell, decide, or choose.

According to the evidence of our scientific experience with behavior over the last century (and our more general scientific experience over the last four centuries) all those mystical notions—specific or general—treat behavior as a magical event outside scientific reality. And allowing any part of those notions to surface in discussions about understanding, predicting, controlling, or interpreting behavior, which is of necessity our scientific agenda, greatly reduces the possibility of dealing with the behavior part of our world effectively, because those notions contradict the behavior passivity upon which our so far successful natural behavior science rests.

If we try to work outside behavior passivity, then we sacrifice our best chance to help each other or to improve our world and, especially regarding the latter, recall that we are running out of time. With this general look at some parameters surrounding concurrent contingencies in hand, along with this quick review of behavior passivity, let's now look at some concurrent contingencies in the context of actual research working toward discovering more of the properties of complex behavior.

Multiple Simultaneous Selectors in the Control of Multiple Simultaneously Evocable Responses

We now turn to examples of concurrent contingencies in the context of research with the kinds of multiple behaviors and multiple stimuli that are more typical of human behavior. We are aiming at extending our analysis to contingencies containing not only simultaneously evocable multiple behaviors but also containing simultaneously occurring selectors like reinforcers. These kinds of circumstances begin to typify what we find for general human behavior. (For a more thorough discussion of the issues and research that we describe in this section, see Ledoux, 2010; also, see Ledoux, 2013, which describes better equipment that behaviorological experimenters might use for the kind of research that we now consider.)

Over the last 75 years, as researchers gradually extended laboratory studies successfully to humans, most of the human–behavior studies followed the lead of the more common behavior studies with other animals, so they involved responses on a *single* manipulandum such as a single telegraph key, with a *single* reinforcer source such as a single point counter (with points exchangeable for money). These arrangements eliminated the confounding problem of a reinforcing consequence that follows one behavior reinforcing, coincidentally, another unrelated but concurrent behavior (a phenomenon we call concurrent superstition) because only one behavior was under study.

Concurrent Superstition and Response–Succession Practices

Gradually, some researchers began to look at multiple and concurrent behaviors. Moving with appropriate caution, they developed procedures to study these behaviors while still avoiding the confounding problem of concurrent superstition. These procedures work by forcing responses to occur in succession, which prevents responses from occurring simultaneously. To appreciate the historical context, let's list four common methods to arrange successive responses before going beyond them to allow and study response and consequence simultaneity. (You will find references for each of these response–succession procedures in the original study.) These are the four kinds of successive responses from the response–succession procedures:

❧ *"Biologically successive" responses:* These are responses that cannot occur at the same time due to physiological constraints, such as pigeons facing multiple pecking disks, but having only one head with which to peck.

❧ *"Structurally successive" responses:* These are responses on multiple manipulanda where the manipulana (e.g., levers protuding from a wall) are set far enough apart that the subject can reach only one manipulandum at a time.

❧ *"Contingently successive" responses:* These are responses on multiple manipulanda where the experiment includes a change–over delay (COD) contingency in which every first response on a different manipulandum starts a timer that delays the occurrence of the reinforcer for that response.

❧ And *"Instructionally successive" responses:* These are responses on multiple manipulanda where instructions specify operating only one manipulandum at a time.

Multiple Selector–Source Procedure

Again, the downside of such methods is that they exclude studying multiple, simultaneous behaviors that have multiple, simultaneous consequences. Yet we quite commonly observe the simultaneous occurrence of multiple responses, and the multiple and often simultaneous consequences which they produce, in the behavior of animals, including—perhaps especially—humans. But in life beyond the laboratory, each of the consequences usually *has its own source.* We now more fully appreciate the experimental importance of this reality.

Here are a few examples of these complex behaviors. You might be writing a note with your right hand about something I have written here while at the same time you are scratching the itch of a mosquito bite on your ankle with your left hand. Different stimuli are simultaneously evoking both responses, and both responses are not only occurring simultaneously but they are also producing consequences that are occurring simultaneously (in this case the appearance of a helpful memorandum and the reduction of a bothersome itch). The same holds for more complex examples, such as these: (a) smiling at an author's lame joke while thinking about whom you will invite to join you for dinner, (b) playing the pipe organ where sometimes each hand and each foot may need to deal with a different rhythm, as well as different notes, and (c) day–dreaming about winning some lottery while successfully negotiating a variety of curves on the highway (until noting that you are stopping for a red light, which interrupts you, at which point you wonder how the last several miles went by). While you and I could both easily fill books with such examples, no mystical body–directing self–agents are part of the account for any of these or other behaviors (and we consider the consciousness concerns of some of them in a later chapter).

Yet in all those examples, we would have trouble spotting if any one of the reinforcers, produced by the multiple–responses, is affecting the rate of any of the other, simultaneously occurring responses that did not produce it. We need to find a way to examine and gain experimental control over complex relations like these, without changing their character (e.g., without making them happen successively). One way comes from an experiment I conducted that devised and tested a procedure to enable studying simultaneous responses with simultaneous consequences. We call it the *multiple selector–source procedure;* in it each manipulandum connects with its own reinforcer source.

Before considering the procedure, here are some photos of the experimental situation, equipment, and data collectors (while the computer that managed the sequence of independent–variable changes was on another rack). The Figure 6–1a photo shows a typical subject seated at the console in the human operant chamber that we used in the test of the multiple selector–source procedure. The Figure 6–1b photo shows the layout of the stimuli and manipulanda console. The Figure 6–2 photo shows the two cumulative recorders that tracked the data, with some actual data showing on the extruded sheets.

Oh, wait! What is a cumulative recorder? Briefly, it is a roll of paper unrolling at a constant rate under a couple of pens. The lower pen, which we call the *event pen,* records the onset and offset of experiment–defined events by going up one step after which it can only go back down one step. As the paper unrolls, this pen marks a line at two slightly different levels according to whether it has moved up or moved down its one allowed step. The upper pen, which we call the *response pen,* marks one tiny, equal step up for every response

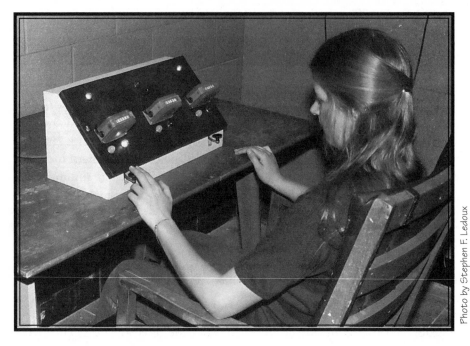

Figure 6–1a. Typical arrangement of subject, chair, table, and panel.

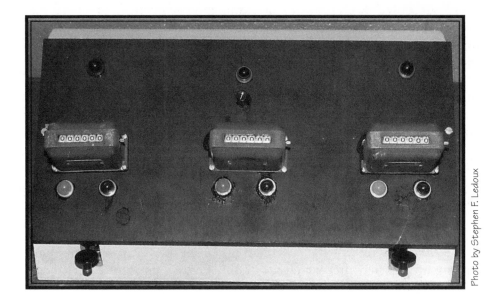

Figure 6–1b. Arrangement of the equipment panel.

that occurs. The width of the typical paper roll usually allows the response pen to step up a maximum of 400 responses before the pen resets to the bottom of the paper and starts up again. However, since the paper is always unrolling, this pen is always also drawing a horizontal line. These two motions (i.e., always *horizontal,* as the paper unrolls under the pen as time passes, and sometimes *vertical,* up a tiny step for each and every response that occurs) result in an angled line that we call the *response slope,* which directly provides the rate of occurrence for the behavior in responses per time unit (e.g., seconds). A steeper slope shows a higher rate, a shallower slope shows a lower rate, and a changing slope shows a changing rate.

When a response produces a reinforcer, the response pen also makes a tick, or hatch mark, off the response slope line at the same step that records the response occurrence. Since the ink line is thicker than the space between response steps, a run of reinforcers, when every response is earning reinforcement, makes all the reinforcer hatch marks run together as a thick and solid black line. We will see many of these features in our data figures.

At this point, however, my prediction is that adding more methodological details here would interfere with our focus. If you find that prediction wrong, you can preview the last section of Chapter 8, which contains more complete coverage of general laboratory methodological details, before proceeding.

Photo by Stephen F. Ledoux

Figure 6–2. Arrangement of the two cumulative recorders.

Here is a summary of how we evaluated the multiple selector–source procedure. College–aged human subjects pressed continuously available telegraph keys, one for each hand, that could be pressed simultaneously. On either key some presses produced point reinforcers on a single, center–of–console counter, while other presses produced point reinforcers on one of two side–of–console counters that only incremented points that the subject earned by pressing on the key that was below that counter. Simultaneous responses produced simultaneously delivered reinforcers.

Overall, the process involved seven normal human subjects (for two of whom we include data in this research summary) responding under separate contingencies, for each hand, of either (a) *every response* reinforced (denoted as CRF, for "continuous reinforcement") or (b) *no responses* reinforced (denoted as EXT, for extinction). What's extinction? Briefly, when reinforcers no longer follow a behavior, we speak of extinction, and we see a decrease in the rate of that kind of behavior. We say that the decreasing behavior is "on extinction," or that the behavior is "extinguishing." (Note that we *only* extinguish behavior; extinguishing subjects is unethical, immoral and illegal!)

For these two subjects, six sets of CRF/EXT contingency pairs occurred in each of three sessions (with each session lasting around one half hour). One session was a within–session comparison, called the "W" session, and the other two sessions comprised a between–sessions comparison, called the "B1" and "B2" sessions. The figures include the sequences of these CRF/EXT contingency pairs. Examining the data figures (i.e., Figures 6–3, 6–4, and 6–5) while covering these methodological points may help keep them straight.

In the W session, three contingency pairs occurred while point reinforcers, from presses on either key, accumulated on the center counter, which you see recorded in the figures as the thick black parts of the bottom horizontal line that the recorder event pen made as it produced a little hatch mark for each earned reinforcer; we made these reinforcers get recorded here so that we could instantly see that they had occurred on the center counter regardless of which hand produced them. Then the same contingency pairs repeated, except that this time the reinforcers that key presses produced accumulated on the side counter above the pressed key, which you see recorded in the figures as the thick black parts of the sloped lines that the recorder response pen made as it added the little reinforcer–indicating hatch mark to each vertical response increment that produced a reinforcer.

The B1 session involved a sequence of six contingency pairs with reinforcers, from presses on either key, accumulating *only* on the center counter. On the other hand, the B2 session featured the same six contingency pairs; however, this time the reinforcers, from presses on either key, accumulated *only* on the side counter above the pressed key. While I have described the rest of the methods and all of the results in excruciating detail in the original study (Ledoux, 2010) let's look at a sample of the results.

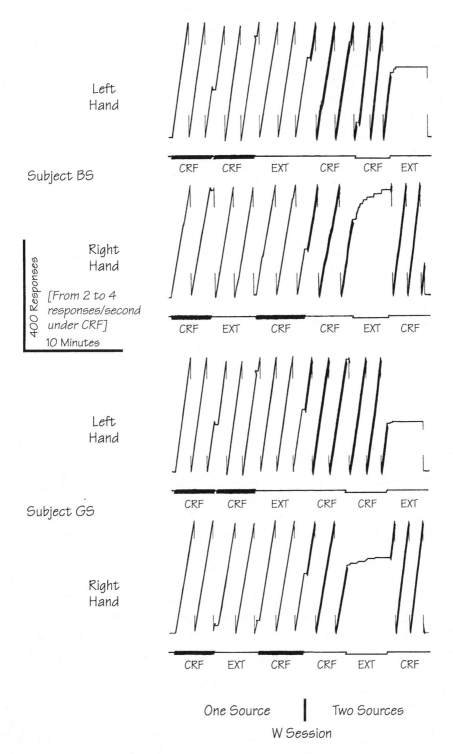

Figure 6–3. Cumulative records of left and right hand responding for subjects BS and GS during the W session.

Figure 6–3 shows cumulative records for the left and right hand responding of Subjects BS and GS in the W session. Note that when the single, center–counter reinforcer source is in use, for the first three CRF/EXT pairs (in which reinforcement hatch marks make the thick sections on the horizontal event pen line) the extinction contingency always proves ineffective at reducing responding. However, when the two side–counter reinforcer sources are in use, for the second three CRF/EXT pairs (in which reinforcement hatch marks make the thick sections on the sloping response line) the extinction contingency always proves effective at reducing responding. This Figure shows extinction occurring in several ways, only one of which looks like a curve. I hope future researchers are interested in discovering what variables lead to such differences.

Turning to Figure 6–4, we see the cumulative records for both hands for Subject GS across the two between–session comparison sessions, B1 and B2. And in Figure 6–5, we see the cumulative records for both hands for Subject BS across these same sessions, B1 and B2. Again, for both subjects, when the single, center–counter reinforcer source is in use, for the B1 session, the extinction contingency always proves ineffective at reducing responding. However, in both B2 sessions, when the two side–counter reinforcer sources are in use, the extinction contingency always proves effective at reducing responding.

The B2 session in Figure 6–4 also shows another phenomenon that I hope future researchers are interested in investigating. Look closely at the data for the four extinction contingencies. What you are seeing is that under these four contingencies, this subject pressed the non–productive key once, each time the other key had produced 100 reinforcers. It looks metaphorically like reality testing. At the time, someone suggested that this student subject might make a good graduate student!

In the interest of full disclosure, I am happy to report that, considering the data from all the subjects in the study (not just these two), in 9 of 36 possible cases (25%), the extinction contingency actually was effective in ending responding when only the single reinforcer source was operating. This is the kind of result that really grabs an experimenter's interest. Just what variables made that happen? Nevertheless, in terms of our interest here, in 36 of 36 possible cases (100%), the extinction contingency was effective in ending responding—as predicted—when the two reinforcer sources were operating.

The overall point is to note the lawfulness that the data show for the contingencies and behaviors studied. That lawfulness lets us conclude that we have a potentially valuable candidate procedure, this *multiple selector–source procedure*, available to study more thoroughly the multiple, simultaneous responses with multiple, simultaneous reinforcers of typical human subjects.

This study conveys some of the rigor and vigor of this laboratory science, which makes more obvious its relevance to understanding and solving global problems. Future studies will replicate this *multiple selector–source procedure* study, and apply the procedure to the full range of basic contingency principles and processes (e.g., conditioning, extinction, reinforcement,

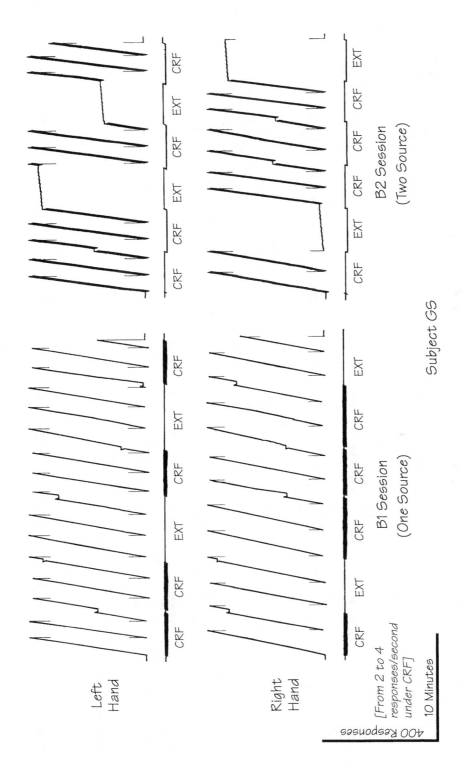

Figure 6–4. Cumulative records of left and right hand responding for subject GS during the B1 and B2 sessions.

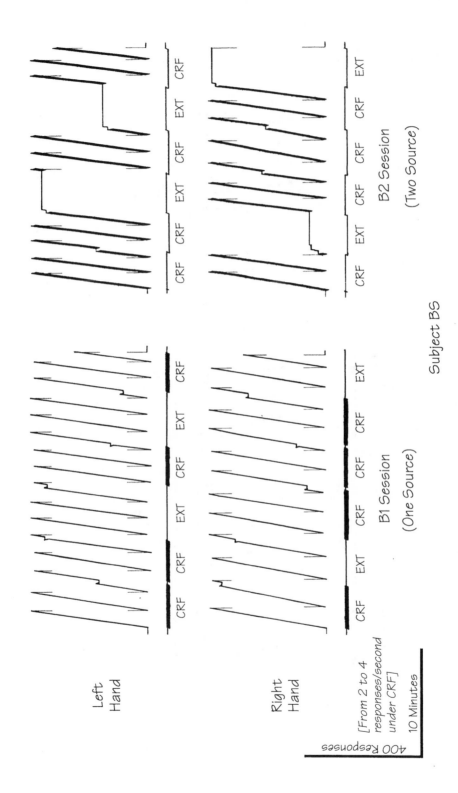

Figure 6–5. Cumulative records of left and right hand responding for subject BS during the B1 and B2 sessions.

evocation, generalization, chaining, fading, conditioned selectors, schedules of reinforcement, selector parameters, or punishment). We must wait until then to find out about either practical applications or if any further basic discoveries will turn up from this line of research.

You may feel that this study is far too complex to assimilate in a single read, and you may be right. But I think the multiple lessons it provides will serve you well as we move into related topics. Besides, with some careful or repeated review, I think you can get a good handle on it. To provide the opportunity for you to engage in such review, I kept this chapter on the short side.

Conclusion

Our examination of some actual research, data, methodology, and conclusions for discovering some properties of concurrent contingencies has also shown us a glimpse of behaviorological laboratory research. While in due course we will consider some basic and applied research methods in more detail, many additional research questions await experimentation on a wide range of disciplinary topics that may have implications for controlling our behavior in ways that protect our environment and otherwise benefit humanity. With appropriate graduate training in behaviorology, you could be one of those continuing this research.

However, even without such graduate training, you can contribute to the role behaviorology plays in solving global as well as local problems. Perhaps your professional interests already involve working on these problems. As a natural–science team member, keeping in view the constraints and enhancements that behaviorology provides helps that teamwork succeed.✵

References (with some annotations)

Latham, G. I. (1994). *The Power of Positive Parenting*. Logan, UT: P & T ink. A workbook is available for this textbook: Ledoux, S. F. (2001). *Study Questions for Glenn Latham's "The Power of Positive Parenting."* Canton, NY: ABCs.

Latham, G. I. (1999). *Parenting with Love*. Salt Lake City, UT: Bookcraft.

Ledoux, S. F. (2010). Multiple selectors in the control of simultaneously emittable responses. *Behaviorology Today, 13* (2), 3–27. This was a peer–reviewed reprinting of the paper with this name in S. F. Ledoux. (2002). *Origins and Components of Behaviorology—Second Edition* (pp. 205–241). Canton, NY: ABCs. This paper describes an operant research program, studying an analog of complex behavior with normal humans, that is appropriate for graduate behaviorology student participation, and perhaps even the participation of upper level undergraduate students.

Ledoux, S. F. (2013). Human multiple operant research equipment. *Journal of Behaviorology,* 16 (2), 3–9. This paper describes newer equipment that behaviorological experimenters could use in continuing the research program described in Ledoux, 2010.

Skinner, B. F. (1953). *Science and Human Behavior.* New York: Macmillan. The Free Press, New York, published a paperback edition in 1965.

Vargas, E. A. (2013). The importance of form in Skinner's analysis of verbal behavior and a further step. *The Analysis of Verbal Behavior, 29,* 167–183. This paper helps us appreciate the value of traditional linguists' work.℘

Chapter 7
Some Complex Analytical Confusions and Fallacies

*I*n Chapter 6, we addressed some ordinary human–behavior complexity that compelled considering some experimental research. This research involved contingencies on separate behaviors that stimuli often evoked at the same time, and that specific selectors often followed at the same time. The results supported a *Multiple Selector Source Procedure* that enables the laboratory study of typically simultaneously evoked and simultaneously consequated behaviors of humans and other animals.

As complexity in our subject matter increases, however, we begin to confront various confusions and fallacies that can arise in contingency analysis. *Time* is an example, one which, in a past chapter, we already noted lacks independent variable status with respect to behavior, because it cannot be manipulated on the level at which we analyze behavior. Of course, the events that define time are capable of independent–variable status, and we must consider the ones related to behavior and take them into account.

In this chapter, we go beyond time, and examine some other more complex analytical confusions and fallacies. We begin with the several different, and somewhat confusing—though all correct—usages of the term *conditioning*. Then we discuss (a) the problem of analyzing the non–occurrence of behaviors, (b) boredom, as an example of fictional constructs irrelevant to the variables controlling behavior, (c) bodily states, which are stimuli, not behavior, and (d) the difficulties with analyses involving *remote* evocatives and selectors. Lastly, we expand the complexity envelope a little more with a discussion of *macrocontingencies*. But first let's consider some inappropriate sources that encourage complex analytical confusions and fallacies.

Some Sources to Set Aside

Besides behaviorology, other disciplines claim behavior as a subject matter, including theology and psychology. The latter essentially features secularized versions of the former. These disciplines continue as major sources of many fictitious explanations for behavior, as we discussed in earlier chapters, due to their role in pre–scientific cultural conditioning. For this reason we should set these disciplines aside. Other related reasons justify that action as well.

Some of those reasons, including agential assumptions and legal restrictions, stem from the long history of ultimately indefensible harm that accrues to departures from the analyzed objectivity of natural science. As sources of fictitious explanations and analytical fallacies, we can now consider those disciplines as in danger of (if not already guilty of) fomenting actual harm to individuals and the culture, although I know of no *individual* practitioner from either of those disciplines for whom this is an intellectually or emotionally satisfying outcome; on the contrary, they adamantly insist that they and their disciplines are only trying to help. However, they are contingency–controlled behaving organisms like everyone else (including behaviorologists and all other natural scientists) and the contingencies that they are under continue compelling behavior in support both of their disciplines and of their practices, such as those agential assumptions and legal restrictions, that are proving harmful. Let's look a little more closely.

"Set Aside," Because of Agential and Legal Problems

The pre–scientific, misleading status of the fundamental agential principles and concepts of psychology and theology, which pervade the culture, allows practitioners from these disciplines only coincidental successes from mostly guessed–then–tried practices (a situation that might have been acceptable when they were the only games in town). Meanwhile, however, support for those agential principles and concepts interferes with the swift dissemination of the natural behavior–science principles and concepts the application of which could benefit humanity far beyond the effects of those intuitive practices, an outcome that pervades the contingencies evoking the writing of this book. That is, the pre–scientific, agential theories of theologians and psychologists only *intuitively* produce practices for behavior change, which some simply describe as *conjuring*. These agential theories cannot *directly* produce the kind of best practices that a natural science can produce, practices as direct implications and applications of the principles and concepts we derive from the scientific study of the natural laws governing behavior. The strength that some of the agent–based practices gain accrues only as an outcome that their occasional coincidental successes condition. Meanwhile contingencies involving variables like the cultural status of these disciplines—gained perhaps somewhat deservedly while they *were* the only games in town—now both compel interference with the successes of natural scientists of behavior, and make giving up their mysticism, to fully adopt natural science and philosophy, a virtual impossibility. For this reason also, should we not set these disciplines aside?

Furthermore, the often legally codified restrictions supporting psychologists, originally designed to protect them and their clients by preventing the operation of charlatans, now often prevent otherwise legitimate, more thoroughly scientifically informed behaviorological practitioners from broadly applying, to the solutions for a wide range of individual and cultural problems, the progress that natural scientists and engineers have achieved regarding behavior (e.g., see Maurice, 1993). Meanwhile, the successes of these scientific, behaviorological practitioners ironically turn psychology toward facing charges of charlatanism. For example, consider the negative recommendations regarding so many psychology–grounded therapies for treating autism. These appeared in the report of the New York State Department of Health multi–year project, completed in 1999, to evaluate the scientific research literature on the numerous types of available autism treatments so as to make recommendations based on scientific evidence of safety and efficacy. The final report (NYS Department of Health, 1999) either did not much recommend most of these interventions, or actually recommended *against* them (because they were harmful?) saying that they were "not to be used as an intervention" for young children with autism. Yet some psychologists and other therapists retain these so–called treatments and therapies, even though any continued offering of treatments and therapies with poor recommendations, or with recommendations against them, justifies charging psychology and these practitioners with charlatanism. For comparison, and to complete the record, the only fully recommended practices, in the NYS Department of Health report, were behaviorological practices, about which the report said, "It is recommended that principles of applied behavior analysis (ABA) and behavior intervention strategies be included as important elements in any intervention program of young children with autism" (*Quick Reference Guide*, pp. 33–51).

"Set Aside," Because of Harm

For all the reasons we consider here, and perhaps more, most psychological treatments and therapies may, in the long run, be more harmful than helpful. And the harm done to clients and the culture may be direct or indirect.

Direct harm. Direct harm occurs when a client's condition worsens as a result of implementing a practice. For example, such a worsening can occur when a practitioner only engages in talk sessions lacking a scientific therapy plan while having a depressed client merely otherwise ingest anti–depressant drugs (which may also constitute a misuse of the benefits that pharmaceuticals can provide). By not addressing the independent variables responsible for "depression," including the usual and observable reduction in, or loss of, reinforcers that often results from subtle environmental changes—a reduction or loss that can leave persons feeling depressed—these practices provide little help and so, over a period of time, the poor client either fails to get better or gets worse, both of which are forms of harm. Also, clutching at the truism that "these things take time" makes an inadequate excuse for being unprepared because one's training was in a discipline that emphasizes untouchable agential accounts while down–playing the importance of addressing the independent variables of problems as part of solving them. Meanwhile, behaviorological practices, such as those that raise the *General Level of Reinforcement* (GLR; see Cautela, 1994) could provide some success at addressing these independent variables. Being ***un***knowledgeable about the relevance, for reducing depression, of increasing the client's GLR is no excuse for any harm that results, even if the psychologist knows little or nothing about the GLR because it was reported in a different discipline (e.g., behaviorology) which brings us to indirect harm.

Indirect harm. Indirect harm accrues to clients when they cannot get help because cultural institutions, such as the law, bar the professionals who can help them from doing so. These same legalities often allow clients to see only those practitioners who have been culturally/legally designated (e.g., licensed) to help them, even if the training of these professionals derives from a discipline committed, not to practices grounded in a natural science of behavior, but to agential accounts of behavior, which remain irrelevant.

As an example for years New York State's laws stood in the way of clients receiving the interventions that the state's own Health Department recommends for autism; these laws limited the scope of practice for behaviorological professionals so that they could not easily provide these interventions. These behaviorological professionals have the education and experience in the relevant field—one that behaviorological science informs—that enables them to earn an appropriate credential for implementing effective interventions, but the laws recognized only "psychology" credentials. Similar laws in some other states—supported over the decades by psychological guilds and lobbies— also prevent behaviorological professionals from providing interventions that work for needy clients, which leads to harm for these clients. Some laws even require these behaviorological professionals to receive supervision from licensed psychologists the vast majority of whom have little or no training in the relevant natural science of behavior. Ignoring the natural science of human behavior provides little help for people. Purportedly, another New York law makes *changing behavior*, without holding a licence to practice psychology, a felony. If you think about that, you must ask the question: Who in the state, other than licensed psychologists, is *not* guilty? Certainly everyone, from the state's governor to its homeless, is guilty, including all parents, teachers, professors, ministers, firefighters, lawyers, judges, employers, politicians, police officers, business owners, medical doctors, and so on, all of whom change behavior on a daily basis. Is it sensible to make criminals of them all? Given such reasons, why should we not set problematic disciplines aside?

Revisiting a point we made several chapters ago, the theology/psychology disciplines, and the science of behaviorology, are neither equal nor culturally parallel. Over the last several hundred years, the distinction between theology and science has become quite well established. Similarly, over the last hundred years, the comprehensive research literature that thousands of natural scientists of behavior have produced, around the world, has equally established the distinction between the psychology and behaviorology disciplines. Specifically, *behaviorology is neither a part of, nor related in any meaningful way to, psychology of any kind!* And any contrary claim that psychologists or their discipline might make about containing behaviorology in any way would stand as a lie equal to some of those infamous whoppers certain politicians told during the last century (and this lie should share the same fate as those other lies).

Actually, the relation between behaviorology and psychology shares essential characteristics with the relation between biology and creation "science." Both psychology and creation science claim to be sciences because they use scientific methods, but neither qualifies as natural science because they both appeal to entities or events outside of nature as causes (i.e., to mystical or supernatural or otherwise non–natural events). And traditional natural scientists remain quite unimpressed by psychological accounts being *secular*–mystical rather than *theological*–mystical. On the other hand, those who study why behavior happens using scientific principles and practices that involve *only* natural events as independent and dependent variables—and completely forego mystical intrusions—have "behaviorology" as the name for their discipline.

For all those reasons, and possibly at this time even more importantly to help enable behaviorological contributions to solving the behavior–related components of major global problems, we need to set those disciplines aside. Consider a more general view (also, see Lucas–Clark, 2010). No one can objectively demonstrate the "mystical," whether theological or secular, even though it claims to offer certain explanations for everything. Science, on the other hand, remains inherently uncertain, and accepts that it can never explain all things, even as it strives to discover, understand, predict, control, interpret— in a word, explain—more and more. And natural scientists prefer to go without an explanation, while looking for a scientifically acceptable one, rather than accepting or conjuring some mystical account. The "mystical" insists that it accounts for everything, usually through a range of agential entities or processes, and that it is always right. Science, on the other hand, admits only to *natural* explanations of natural events, explanations that successfully derive from the best available experimental methods, explanations based on the informing philosophy of naturalism, explanations to which science admits only provisionally in the sense that science always remains open to new or better data that can alter or improve the previous best explanations.

In addition, the secular mysticism of psychology merely imitates, for its agential entities, a scaled–down version of the agential–entity power of theological mysticisms. Supposedly, for example, heavenly maxi–deities can move worlds but agential mini–deities (e.g., souls or minds or psyches or selves) can only move body parts. Such mystical power differentials may strike us as curious. Natural scientists, however, require experimental results that enable beneficial engineering applications, while they watch for, and avoid, analytical confusions and fallacies, to some of which we now turn.

"Conditioning" as an Analytical Confusion

In our analyses of behavior, we use the term *conditioning* in the names of two processes, respondent conditioning and operant conditioning. However, beyond this general labeling of these processes, we also use the term *conditioning* in several other possibly confusing ways. Let's sort these out through quick reviews of the two ways that we use *conditioning* in respondent conditioning and the two ways that we use *conditioning* in operant conditioning. As in past chapters, our analysis connects with physics and physiology through energy streams and mediation. (For more details on these connections, see Chapter 2 of Fraley, 2013, on the physics of behavior.)

Respondent "Conditioning"

In respondent conditioning the verbal response *"conditioning"* refers to (i.e., tacts) both the *pairing* of stimulus events and the *effects* of the process of pairing stimulus events. Any of these events could actually be behaviors serving stimulus functions as real events. This usage applies to any and all pairings of such stimulus events regardless of whether or not the natural history of, or the occurrence of, the pairing involves the intervention behaviors of humans.

More specifically, in the respondent conditioning context, *conditioning* tacts what happens as a result of two stimulus energy streams stimulating the nervous system at about the same time, which is the temporal contiguity we describe as a "pairing." The most obvious case starts with a stimulus energy stream (from the stimulus that we call the *unconditioned stimulus)* that already elicits a nervous–system change that mediates, due to genetically produced structures, a response (that we call the *unconditioned response)*. Think about the startle response that the puff of air, into one of your eyes, elicits during one of the tests that your eye doctor administers during a comprehensive eye exam.

That unconditioned energy stream occurs at about the same time (a pairing) as another stimulus energy stream from a stimulus that is originally neutral (and which at this point we call the neutral stimulus) in that it fails to elicit that nervous–system change (i.e., it fails to elicit the change that the unconditioned stimulus elicits, the change that mediates the response). One or more pairings result in an altered nervous–system structure (i.e., the *conditioning*, the pairing, alters nervous–system structure) such that subsequent occurrences of the previously neutral stimulus will also elicit a nervous–system change that also mediates a response (very much like the unconditioned response). This new nervous–system change occurs due to *conditioning*–altered structures (i.e., due to the occurrence of the two energy streams, the pairing, having the effect of altering nervous–system structure). When this happens, we call the no–longer–neutral stimulus a *conditioned stimulus.*

The conditioned stimulus continues to elicit a nervous–system change that mediates a response, one that we now call the *conditioned response* (which is essentially the same response that the relevant unconditioned stimulus elicits) as long as the conditioning process (i.e., the pairing with the unconditioned stimulus) reoccurs at least occasionally. Think about the startle response that the cold chin rest elicits when you contact it *again* when the doctor asks you to settle down for the test of your other eye. The chin rest was originally a neutral stimulus with respect to startle responses, but then it was present in a pairing with the puff of air into your first tested eye, so now when your chin contacts the chin rest, it elicits a startle response.

Respondent conditioning summary. In respondent conditioning we use the term *conditioning* to refer both to the pairing process and to the outcome of that process, which is the production of a new function for a previously neutral stimulus. That is, respondent *conditioning* is the production of a new function for a previously neutral stimulus, a function that produces a behavior, whereas before the stimulus pairing—the *conditioning*—the energy stream from the neutral stimulus produced no behavior like the unconditioned response.

At our analysis level, we simply say that, through the pairing/*conditioning* process, the originally neutral stimulus gains the function of the unconditioned stimulus. Respondent *conditioning* produces responses to previously neutral stimuli. That is how we use the term *conditioning* in respondent conditioning. Now let's look at some usages that occur in operant conditioning.

Operant "Conditioning"

With operant conditioning I will not repeat for you so many of the physiological details the way I did in our discussion of respondent conditioning. Instead, I want you to recognize that the stimulus changes we discuss here in operant conditioning also actually occur via energy streams affecting the physiological level rather than through changes in the stimuli themselves. After all, as in respondent conditioning, the stimuli in operant conditioning do not change. If you give your 18–month–old daughter a crisp $100 bill (an action which I cannot recommend, for several very good reasons) that bill will likely evoke responses like her putting it in her mouth. But if you give it to her when she is 18 *years* old, it will evoke very different responses. Yet it is the same $100 bill; *it has not changed.* But the physiology that is your daughter, especially the nervous system, has changed, due to 18 years of continuous, moment–by–moment conditioning (which continues...). Furthermore, you should carefully and explicitly try to review those physiological changes, at the same level we used in our respondent conditioning discussion, through your own thinking behavior before moving on to the next chapter section. (Completing such exercises builds your understanding—your verbal, intellectual repertoire—far faster than reading repetitive writing, which save several pages.)

In operant conditioning the verbal response *conditioning* can tact either of two different processes. The main usage of *conditioning* refers to the process of certain environmental independent variables changing behavior, regardless of whether or not the natural history of the environment–behavior relations involves behaviors of humans. The other, simpler usage refers to the process of pairing different stimuli (again, regardless of whether or not the natural history of the pairing involves behaviors of humans) that produces conditioned (i.e., secondary) reinforcers and punishers out of stimuli previously neutral with respect to the behavior of concern. Let's consider the simpler usage first.

"Conditioning" conditioned (secondary) reinforcers and punishers. One process in operant conditioning parallels the respondent conditioning pairing process. In this case *conditioning* tacts the process of pairing two stimuli, a stimulus that *already* has a reinforcing (or punishing) effect on behavior (and that we therefore call the *unconditioned,* or *primary,* reinforcer or punisher) and a neutral stimulus that has no reinforcing or punishing effect on behavior. After enough pairings, after enough occasions when the unconditioned reinforcer (or punisher) occurs close in time with the neutral stimulus (and sometimes only one pairing is enough) the cumulative result of the pairings is a change in the function of the neutral stimulus, a change in how its energy stream affects the nervous system. The previously neutral stimulus gains the functions of the unconditioned reinforcer (or punisher) with which it shared the pairing process. It now functions as a reinforcer (or punisher) and it continues to function this way so long as the *conditioning* process (i.e., the pairing with the unconditioned reinforcer or punisher) reoccurs at least occasionally.

Now, recall that the terms *reinforcer* and *punisher* can only be used to tact stimuli the energy streams from which produce nervous–system changes that mediate *more* responses (occasioning the tact *reinforcer*) or *fewer* responses (occasioning the tact *punisher*) to the energy streams of relevant evocative stimuli. After we verify that this function change occurs with the previously neutral stimulus, due to the pairing—*conditioning*—process, we can call the stimulus a *conditioned,* or *secondary,* reinforcer (or punisher). Of course, any of these stimuli can be the real event of a response, overt or covert, subsequently serving stimulus functions.

"Conditioning" changes in behavior. In operant conditioning the main usage of *conditioning* refers to a process that leads to behavior changing. In this usage *conditioning* tacts the functional effect of certain environmental independent variables happening in ways that bring about changes in behavior, changes that often involve both the generation of new behavior and the maintenance of ongoing behavior, changes that can include any of the functional outcomes of all the independent variable occurrences that we collectively call *contingencies of reinforcement* (which involve many more variables than just reinforcers). *Conditioning* then, in the operant sense, tacts the functional effect of stimuli, especially postcedent stimuli, in changing the nervous system such that it inevitably mediates either more of a previous behavior (leading us to speak of the postcedent stimuli as reinforcers) or less of a previous behavior (leading us to speak of the postcedent stimuli as punishers).

Recall that *postcedent* stimuli refer to stimuli that occur after a behavior (many of which the behavior physically produces). The energy streams from such stimuli induce changes in the nervous–system structure that mediates the kind of behavior that those stimuli have followed. These nervous–system structural changes are such that when energy streams from the stimuli that *evoke* the responses reach these changed structures on future occasions, the structures mediate the responses either more, or less, readily (leading us to speak of the related postcedent stimuli as reinforcers or punishers respectively). Remember, stimuli evoke *all* behavior, even though when they evoke respondent behavior, we instead speak of them as *eliciting* that respondent behavior.

Operant conditioning summary. In operant conditioning, beyond using the term *conditioning* in this name, we use *conditioning* in two ways. We use *conditioning* to describe the pairing of two stimuli, an unconditioned reinforcer (or punisher) and a neutral stimulus, with the result that the previously neutral stimulus now functions the same way that the stimulus that occurred with it functions. And we use *conditioning* to tact the change in evoked behavior, the increase or decrease in responding, that occurs when certain stimuli follow responding. The term reinforcer applies when behavior occurs more often, and the term punisher applies when behavior occurs less often; in both cases we speak of *conditioning* having occurred.

So, in operant conditioning, we use the term *conditioning* two ways. We use it not only in the sense of pairing (in respondent conditioning as well as in operant conditioning) but we also use *conditioning* in operant conditioning in the sense of postcedent stimulus energy streams changing the physiological structures through which antecedent evocative stimulus energy streams induce response mediation, producing changes of either more or less behavior.

If describing these various usages of *conditioning* still leaves you confused, try elaborating the summaries, in your own handwriting, along with whatever amount of physiological–level detail you feel comfortable. Assuming that leaves you less confused, let's move on to some analytical fallacies.

Other Analytical Fallacies

We already covered some traditional cultural perspectives on behavior. These derive from, and continue to support, pre–scientific analyses of behavior. We inevitably grow up under the contingencies of traditional culture, including its superstitious and mystical components, and these contingencies thus shape our early behavior, including repertoires of superstition and mysticism.

For some people those contingencies maintain such repertoires throughout life. For other people additional contingencies, particularly those of education, mitigate some—but seldom all—of the negative effects of early cultural shaping. Thus, that early–imposed traditional cultural legacy continuously compels some theoretical and practical fictions and fallacies that interfere with contingencies that condition more accurate and helpful intellectual and emotional responses regarding behavior. As we try, scientifically, to analyze behavior and solve behavior–related problems, we must recognize, and counter, those fictions and fallacies. We dealt with many of the fictions in an earlier chapter; *here* we deal with some of the remaining fallacies, including analyzing the non–occurrence of behavior, additional fictional constructs such as boredom, bodily states that are stimuli not behavior, and the problems with *remote* evocatives and selectors.

Avoid Analyzing the Non–Occurrence of Behavior

Have you ever heard, or said, "Is there no end to the bothersome behaviors of others?" Since examples abound, we will let a short list stand for them all. For instance, does your mate fail to treat the toothpaste tube the way you prefer? Does a neighbor's lawn left unattended for weeks or months disturb you? Does a dog owner ignoring a dog's leaving dung on your nice lawn leave you dealing with a severe negative emotional reaction? (More fun ways to say that are available, but they are not printable.)

While we will work with one of those relatively simple examples here, far more important, and complex, examples certainly exist. At some point these also warrant attention, such as the problems that stem from (a) someone not wearing a seat belt, or (b) someone not recycling (but merely throwing recyclable items away) or (c) someone not attending to driving (but phoning or texting instead) or (d) a group of professionals not changing a no longer appropriate name and, as a result, forfeiting some of the scientific credibility needed to make a full contribution to solving global as well as local problems (a concern to which we gave some attention in Chapter 1). In addition to the non–occurring behaviors that cause analysis problems, some of these examples contain the occurring behaviors upon which we should focus.

Those kinds of concerns often evoke the early steps of an intervention to end or change the offending behavior. If we bring behaviorology to bear at this point, these early steps will likely include trying to clarify the current contingencies, and plan the intervention contingencies. In analyzing this kind of activity, one thinks in general terms, so we refer to behaviors rather than to more specific individual responses. Also, sometimes we list emotional responses in these contingencies, while at other times we list feelings (effects of emotional responses) and at still other times we list the more public events available to observer. In all cases, however, we need to avoid writing contingencies that include *non–occurring* behaviors, because that causes analysis problems.

Let's take some of those early steps with the example of an owner's dog leaving dung on your lawn. The odor/smell of the dung, and the sight of someone leaving it on your lawn, are generally aversive stimuli that elicit various negative emotional reactions. The odor/smell is an unconditioned aversive stimulus while the sight is a conditioned aversive stimulus. *(Odor* refers to energy streams from particular molecular gas pressures to which nasal membranes are sensitive, while *smell* refers to the response that neural structures mediate when such energy streams contact nasal membranes.) The reduction of negative (i.e., aversive) emotional reactions functions as a reinforcer (of the subtracted type, often called *relief)* making the behavior that produces the reduction occur more quickly or more often whenever the evocative aversive stimuli occur. Let's consider two possible and interrelated contingencies for the lawn owner. The first is a three–term contingency with two behaviors resulting from the antecedent stimulus. The second is a three–term contingency that indicates a result as well, although *results* are *outcomes* of contingencies (i.e., descriptions of what likely happens in the future), not parts of contingencies. Here are the two possible and interrelated contingencies for the lawn owner:

S^{El} (Eliciting) & S^{Ev} (Evocative)	Emotion elicitation	&	Behavior evocation	Consequence
#1				
Dung on lawn →	Elicits an angry emotion	&	Evokes a behavior that removes dung from lawn	→ Clean lawn

S^{Ev}	Behavior	Consequence	[Result]
#2			
Owner of dung–dropping dog (aversive) →	Some activity that prevents dung deposit on lawn →	Reduced aversive emotions (relief) from less dung on lawn →	[Successful dung–reduction behavior likely to recur]

Note that the antecedent stimulus in that first contingency functions both as an eliciting stimulus and as an evocative stimulus in very complex ways. At a minimum the first function results in an emotional reaction that produces (a) feelings, (b) other neural responses of consciousness, and (c) exaggerated forms of the responses that the second function produces (e.g., *stomping* out, shoveling *more deeply than necessary* to collect the dung, and *strongly* flinging it into the trash bin). Beyond this minimum, the neural response of feeling the emotion—as a real event—can also evoke various neural responses of consciousness that also chain to other responses. Collaboration between behaviorologists and physiologists continues to sort out such possibilities.

The second contingency involves behavior that keeps dung off the lawn. But before considering it, we must also analyze the contingencies on the dog owner. First, remember that an observer of the lawn owner's reaction cannot directly see the emotions (for instance, the chemical dumps that the lawn owner feels as anger); only the lawn owner is privy to the emotions and some of the responses that they evoke. Still, since emotional arousal exaggerates the mediation of later responses for a short time, the observer reliably responds to (i.e., infers) the emotions from the wide range of events in the setting (i.e., the context) including, and perhaps especially, from hearing the string of hearty obscenities that seeing the dung on the lawn evokes from the lawn owner.

Also, recall that in many jurisdictions, leaving your dog's dung on a neighbor's lawn is illegal. The general history that we all observe of the punishments that the legal system provides for illegal acts can help condition aversive feelings that we tact as *guilt,* which any subsequently evoked engaging in illegal activities elicits. In some cases parents or peers, including neighbors, provide the punishment for activities, which leads to the conditioning of similar aversive feelings that we tact as *shame*. (Aversive feelings that we tact as a *sense of sin* occur when we engage in activities that religious authorities punish.) Even the stimuli evoking thinking behavior regarding punishable responses come to elicit the emotions that lead to feelings of shame, guilt, and sin. Such aversive feelings are often a part of currently operating contingencies. Here is one such contingency that affects the dog owner:

S^{Ev}		Behavior		Consequence		[Result]
Dung on lawn (aversive)	→	Walk away (with dog following on leash)	→	Diminishing feelings of guilt as distance increases (relief)	→	[Walk away more rapidly in the future after the dog leaves dung on a lawn]

Now, let's try planning some intervention contingencies that improve the dog owner's behavior and get the dung off the lawn. In an effort to describe a contemplated intervention, we might write this contingency for a dog owner:

S^{Ev}		Behavior		Consequence		[Result?]
Sight of dung on a lawn	→	**Not** attend to the dung	→	Feelings of guilt or shame remain	→	[Bags the dung for proper disposal?]

But wait! That middle term (the behavior) in this contingency is a behavior *that does not occur*. It is a *non–behavior*. But stimuli only evoke behavior; stimuli only stimulate nervous–system structures the operation of which is the mediating *of a behavior,* **not** of a non–behavior. Furthermore, non–behaviors cannot produce consequences; with a non–behavior, no behavior actually occurs, and so none produces energy streams that then affect the environment in ways that produce other stimuli (other energy streams) that further affect the nervous system as consequences. Thus, *this contingency shows function arrows that do not function;* since no stimuli can evoke the non–behavior, no behavior produces a consequence, so no result can occur. *All these arrows should appear crossed out* (i.e., —x—>). So, as a potential intervention, this contingency is useless, although it does show some of the kinds of reasons that should keep us from writing contingencies for non–behaviors.

Instead, we must write contingencies for behaviors that occur, and then implement the contingencies to establish those behaviors. In this dog–dung–dirtied–lawn case, one such intervention involves a couple of contingencies with stimuli evoking different dog–owner and the lawn–owner behaviors. Here is what we might design, beginning with one of two contingencies on the lawn owner (who might be oneself):

S^{Ev}		Behavior		Consequence		[Result]
Approach of dog and dog owner	→	Hand good grocery bags to dog owner and request their use	→	Dog owner puts dung into a bag, leaving lawn clear of dung	→	[Lawn owner hands out bags when they are needed]

Here is another contingency on the lawn owner:

S^{Ev}		Behavior		Consequence		[Result]
Approach of dog and dog owner	→	Hand good grocery bags to dog owner and request their use	→	Dog owner puts dung into a bag, leaving lawn clear of dung	→	[Lawn owner feels relief from aversive stimuli of dung on lawn]

A further contingency on the lawn owner might even involve the dog owner also saying, as a consequence, "Thank you" for the bags, which could produce the same results that these two contingencies list. (Test yourself by writing it out.) Meanwhile, here is one of two contingencies on the dog owner:

S^{Ev}		Behavior		Consequence		[Result]
Dung on lawn with bag in hand	→	Put dung in bag (leaving lawn clear of dung)	→	Dog owner feels relief from guilt or shame	→	[Puts dung in bags regularly (and even carries bags for disposal of dung)]

And here is another contingency on the dog owner:

S^{Ev}		Behavior		Consequence		[Result]
Dung on lawn with bag in hand	→	Put dung in bag (leaving lawn clear of dung)	→	Lawn owner says, "Thank you"	→	[Dog owner puts dung in bags regularly (and even carries bags…)]

Other interventions are possible. Try writing out the contingencies for one. Just remain wary of *non–behaviors,* and make sure your contingencies avoid them. While we know that many concerns about avoiding contingencies on non–behaviors remain for other books to discuss, let's now turn our attention to fictional constructs *in* contingencies, with "boredom" as an example.

Fictional Constructs Like Boredom

In a previous chapter, we touched on fictional constructs in our discussion of a range of explanatory fictions. Here we want to consider a problem that develops when fictional constructs appear in our analysis of contingencies. We avoid this problem by more thorough conditioning of appropriately constrained contingency–analysis behavior. The basic problem is that when people include a fictional construct in an environment–behavior contingency, they inevitably produce a fictitious account of the behavior, an account that is both false and quite unhelpful regarding interventions.

Consider the common behavior explanation that some behavior change occurred out of "boredom." We often hear this in the complaint form, "I'm bored" (although no inner agent, "I," exists to "be" bored). The complaint could easily substitute for a report that reinforcement density has become too lean to maintain the current behavior. Similarly *boredom* could make a reasonable verbal shortcut for the effect of a thin or thinning level of reinforcement. However, when an alternative, higher–density reinforcement contingency compels a behavior that replaces the behavior from an ongoing lower–density reinforcement contingency, people often say that the intruding behavior occurred not due to the improved reinforcement density but due to their boredom. Inevitably, though, when we try to analyze the contingencies that such boredom statements ostensibly describe, we find the result in error, we find the contingency that we write invalid, we find the account a fiction, and we find the explanation out of touch with real events.

While that also happens with the full range of explanatory fictions we covered earlier, let's elaborate boredom a bit. The feeling of boredom can be quite real. Reinforcement–density reductions commonly elicit aversive emotional reactions producing feelings that we tact as "boredom." But we need not take the emotion/feeling as inducing a change from one behavior to another; instead both the emotion/feeling and the behavior change are a function of other variables (e.g., the reinforcement–density change).

Let's watch the problem unfold as we analyze a situation and write contingencies about what is happening. Let's suppose that you are a data entry clerk at a big business where you and your supervisor are both rabid basketball fans. You even cheer for the same team (which may be how you keep your job). If we accept your being bored with the repetitive nature of your work, then we might write contingencies treating the fictional construct, *boredom*, as if it were real. Then we can use boredom for the cause when we want to explain why, when your boss checks up on you, she finds you looking out of your cubicle at the television she has set up to watch the final game of the playoffs at which your team finally landed a spot. Our first contingency, quite invalid and thus lacking functional–relation arrows, features boredom in its more common position, which is as the evocative stimulus:

X INVALID CONTINGENCY:

S^{Ev}		Behavior		Consequence		[Result]
Boredom	–X–>	Look at TV	–X–>	Exciting events happening at a fast pace on the TV screen	–X–>	[Continue looking at TV]

Alternatively, we might be tempted to put the boredom as the consequence, as in this also invalid contingency:

X INVALID CONTINGENCY:

S^{Ev}		Behavior		**Consequence**		[Result]
Full in–basket of un–entered data	→	Type in some data	–X–>	Get bored	–X–>	[Stop typing and switch from data entry to TV watching]

Both contingencies, however, are wrong, agential (Why is that?) and so on. At best, "boredom" simply implies that the evocative controls on your behavior are undergoing a transfer from one type of stimulus to another. A common source for such transfers resides in the different consequences for the two different behaviors. In this case the two behaviors are data typing and watching the television; these involve two different, actually competing contingencies. Here is the one on data typing, which conveys what your job entails:

S^{Ev}		Behavior		Consequence		[Result]
Full in–basket of un–entered data	→	Type in data	→	Rows of data appear appropriately on the screen	→	[Continue to type in data]

But here is the competing contingency, on television viewing:

S^{Ev}	Behavior	Consequence	[Result]
Final playoff → game on television	Look at television	→ Exciting events happening at a fast pace on the television screen	→ [Continue looking at television]

Note, however, that the reinforcement schedule for the two types of consequences differs substantially. The narrow columns and rows of repetitive data slowly scrolling on the screen, from the appearance of the characters as you type on the keys, provide essentially a CRF (continuous reinforcement) schedule with a constant stream of similarly unspectacular reinforcing stimuli. However, the constantly changing scenes of the game in progress on the television, occasionally and irregularly punctuated with surprising and exciting developments, provide more of a VR (variable ratio) schedule with fast moving, always different, and sometimes spectacular events. The difference in these reinforcement schedules lowers the evocative function of un–entered data and raises the evocative function of the playoff game. As a result, the boss observes you typing less and watching the game more. The reinforcement schedule difference determines this outcome, not boredom, so boredom should not appear in the contingencies. The occasional increase in the intensity of the television reinforcers, like the roars of the arena crowd, further supports this schedule effect as those often abrupt intensity increases easily win any contest with the un–entered data for control of your orienting and attending behaviors.

Traditional agential cultural conditioning can also lead to a notion that boredom, as in "being bored," is a sort of bodily state. But that would only mean that it has the problems of bodily states as well as the problems of fictional constructs. The time has come to consider bodily states.

Bodily States Function as Stimuli, not Behaviors

Like fictional constructs, states of the body also cause problems when they sneak into our analysis of contingent relations. A bodily state is not a behavior, but rather describes a stimulus function of a body, and so we exclude bodily states from the behavior position in contingencies of reinforcement. However, bodily states are real events that function as antecedent stimuli, and some might function as postcedent stimuli as well. Meanwhile any feelings or bodily sensations associated with a bodily state are behaviors, usually of the neural kind that, also as real events, might also serve stimulus functions. For simplicity, we usually leave out the "bodily sensations of" or "feelings of" phrase. Drug–induced physiological changes (including "bodily sensations of" drug–induced physiological changes) constitute a typical bodily–state example, wherein the changes usually last long enough to constitute a bodily state to which responding occurs differently from responses to the state of the body without the drug present. Similarly, the difference between other bodily states and their counterparts also evokes different responses to (e.g., different tacts of) these different bodily states, including (a) emotional arousal and relative calm, and (b) awake and asleep. Let's consider each of these, starting with sleep.

Sleep. The term *sleep* refers to a bodily state on a kind of continuum, and it comes after the bodily state of tiredness. Sleep pertains to a body that has temporarily lost the capacity to mediate many, perhaps most, kinds of behavior. Tiredness pertains to a body that is losing capacity to mediate behavior, perhaps due to excessive energy expenditures (e.g., a full day of hard work). Neither sleep nor tiredness are behaviors, and tiredness can function as an aversive stimulus that evokes the behavior of lying down, producing a consequence of relaxing relief from the aversive tiredness as a type of subtracted reinforcement.

We should not confuse sleep as a bodily state, which thus cannot be a behavior, with the various behaviors that can occur *during* sleep, such as turning over, breathing, and dreaming. However, when a body is sleeping, that sleep can function as a stimulus, perhaps evoking an observer's behavior such as the command (i.e., mand) for others to "Let him sleep." In that contingency sleep occupies the position of evocative stimulus, the first term in the contingency.

We face difficulties, however, when placing sleep in the consequence position in a contingency. Some may see sleep as a reinforcer for not working long into the morning. But even before considering the consequence status of sleep, I expect that you said, "Whoa! That is a non–behavior; lacking an energy stream, it cannot function in contingencies. Start again." (I hope you said that, because you would be right!) Instead, some may see sleep as the reinforcer for being tired, even exhausted. But again, before considering sleep as a consequence, I expect that you said, "Hang on. That's not right. Those sound like more bodily states, which also cannot be behaviors. So again, start again!" (Right, again!)

A behavior that actually precedes the state of sleep could be something like getting into bed, a behavior that a clock showing 11:45 P.M. could evoke. But is sleep really a consequential reinforcing stimulus? Can we trace an energy stream flowing from the state of sleep to nervous–system structures responsible for mediating climbing into bed, and changing those structures such that getting into bed more readily occurs in the presence of the 11:45 P.M. evocative stimulus? Stimulus–evoked changes in neural structures mediate the behaviors of lying down or getting into bed, but does sleep change these structures in more permanent ways? Until our physiology colleagues confirm otherwise (which is possible, because contingency claims are falsifiable) we must answer "No" to these questions. Furthermore the *time* involved in sleeping, and "going to sleep," makes manipulating them as independent variables rather difficult. So far this prevents us from testing sleep and observing sleep produce an increase in a behavior that it follows. Without such observations, calling sleep a reinforcer seems at present premature. We may regard one's sleep as able to function as an evocative stimulus (for others), but it cannot function (so far as we presently know) as a consequential stimulus.

Emotional arousal. Some environmental stimulus events, inside or outside the body, evoke—actually we usually use the term *elicit* in this context, so—elicit chemical dumps into the bloodstream. That is, the energy streams from these stimulus events compel changes in nervous–system structures, and these change mediate the glandular release of certain chemicals. We call that chemical *release* emotional behavior. The subsequent *presence* of those chemicals in the bloodstream is not behavior. However, this presence produces both the responses that we call feelings, and a general change in physiology that we call a bodily state of emotional arousal. When *other* stimuli occur during the aroused state (i.e., this general change in physiology) the behaviors that these stimuli evoke or consequate occur in exaggerated or more intense ways compared to the behaviors they evoke or consequate when the body is not emotionally aroused. The aroused body is not behavior but it mediates behavior in exaggerated ways. (Recall our example of aroused running after a bear elicits startle responses, compared with the normal running that a jogging schedule evokes.)

Being real, however, the emotionally aroused body can function as an evocative stimulus in the first term of a three term contingency. For example as a stimulus it can evoke changes in the behaviors of others that produce changes in distance from the aroused body. If the exaggerated responses that an aroused body mediates are aversive (e.g., dangerous) then they evoke increases in the distance between the aroused body and others, but if the exaggerated responses that the aroused body mediates are attractive (e.g., sexual) then they evoke decreases in the distance between the aroused body and others.

On the other hand, emotional arousal fits poorly as the third term in a contingency, for reasons similar to those that applied to sleep. Nervous–system structures mediate emotional behaviors but the subsequent emotional arousal is not behavior. These structures mediate the responses of releasing chemicals into the blood stream, an ongoing result of which is the chemically altered body that we tact as emotionally aroused. That arousal, as a general physiological change that we call a bodily state, alters—often exaggerates—the capacity of the body to mediate other behaviors, but that arousal seems unable to change the mediating structures in ways justifying calling arousal a reinforcer. However, as a real event, the glandular response of releasing chemicals, for example a release that leads to sexual arousal, might serve as a consequence that changes the behavior–mediating structures such that a behavior that the release follows more readily occurs in the presence of its evocative stimuli. Thus we must regard emotional arousal as able to function as an evocative stimulus, but it functions poorly (so far as we presently know) as a consequential stimulus.

Furthermore we should not confuse the *state* of the body that we describe as emotionally aroused, which is not a behavior, with various behaviors that can happen *to* an aroused body, such as other responses that stimuli elicit and which we can distinguish from the aroused bodily state. Stimuli elicit the glandular release of body–altering chemicals. With those chemicals in the blood stream, the now changed body mediates subsequent behaviors differently from the behavior mediation of the pre–release state. We note such bodily changes explicitly when writing contingencies, because the presence or absence of such emotional states readily accounts for the differences in behavioral effects during comparisons of two or more contingencies. Some of this applies to drug–induced physiological changes as well, to which we now turn.

Drug–induced physiological change. A wide variety of medical, recreational and illegal drugs induce a variety of physiological changes in the body. Some of these change are similar to the effects of some elicited glandular chemical releases except that elicitation does not introduce the chemicals. Instead swallowing, or injecting, or absorbing (e.g., from chemical patches) and so on, introduces the body–altering chemicals to the body. When any resulting state of the body with the drug present evokes a different tact (i.e., a different label) compared with the state of the body without the drug present, we are dealing with different bodily states. Of these two, the one that grabs the most attention is the drug–induced bodily state.

A contingency–related discussion of drug–induced bodily states shares similarities with our discussion of emotionally aroused bodily states. A drug–induced bodily state, being real, can function as an evocative stimulus, in the first term of a three term contingency. For example, as a stimulus it can evoke the verbal behavior of others, such as, "He is too drunk to drive!"

However, unlike sleep, a drug–induced bodily state might serve as a consequence, in the third term of a three term contingency. To the extent that the drug state functions to alter the nervous–system structures that mediate the evoked ingestion of the drugs (by whatever method) the drug state could serve as a reinforcer or punisher. Actually, the *induction* of the drug state, the *immediate* physiological effect of taking the drug rather than an ongoing residual state, might serve these functions better than the state itself.

As with emotionally aroused bodily states, if a drug state changes nervous–system structures so that these structures *more* readily mediate drug ingestion in the presence of the stimuli that evoke drug ingestion (including the effect of variables like deprivation) then we could call this drug state a reinforcing stimulus. Similarly, if the drug state changes nervous–system structures so that these structures *less* readily mediate drug ingestion in the presence of the stimuli that evoke drug ingestion (e.g., cause a "bad trip") then the drug state can be called a punishing stimulus. Recall, of course, that at our level of analysis, we lack ready access to what changes are occurring in the physiology of nervous–system mediating structures; we infer those changes most often from observed changes in the rate of the behavior of concern, in this case the ingestion of the drug, and our biology colleagues will explain in the necessary detail how these processes work physiologically.

The difficulties with bodily states, like the problems with fictional constructs, concerns whether or not we can include them in explicit contingencies. Some antecedent and postcedent stimuli, however, extend away in time from the occurrence of the behavior of concern, and this causes other difficulties for contingency analysis, difficulties to which we now turn.

Avoid Remote Antecedents and Postcedents

In analyzing a behavior of concern, especially when designing an intervention to help a person in need, we often must take into account not only some antecedent stimuli that occur back in the historical chain of functional events that led to the behavior, but also some postcedent events that occur long after the behavior of concern. However, while we are talking here about *real* events, both antecedent and postcedent, their remoteness in time from the current behavior necessitates their exclusion from the contingencies we write in our analysis of the *currently* functioning controls on the behavior of concern. We emphasize currently functioning controls, because this emphasis enables discovering the currently accessible independent variables that an intervention might change in ways that improve the behavior.

Put another way, we analyze contingencies for the antecedent and postcedent variables functioning *now.* Thus, we look for the antecedent stimuli that are *at present* functioning to evoke the behavior, while we also look for the postcedent stimuli that are *at present* functioning as immediate reinforcers or punishers for the behavior, especially postcedent stimuli that the behavior of concern is producing. If we write contingencies that include stimuli that are more remote in time (i.e., not current, present, or immediate) then this inclusion will share in controlling our intervention efforts, usually corrupting them—as in leading to reduced effectiveness and poor outcomes—by masking the more immediate independent variables.

For contingencies that have components removed in time away from the behavior of concern, a term that has seen some use is the term *defective contingencies.* Defective contingencies may contain real events and describe actual relations, such as the relation regarding effective study behavior now producing a good grade later, but they do not describe the present variables related to a behavior of concern. Thus, as *defective* contingencies, we exclude these contingencies when we speak of *contingencies of reinforcement.* For example, we exclude the relation between studying now and the later, delayed occurrence of a good grade not only due to the danger of teleology but also because the good–grade consequence occurs too far removed in time after the behavior to function in the manner of a stimulus that earns the title *reinforcer.* That is, every energy stream from the delayed appearance of a good grade happens long after the earlier study behavior, and so the appearance of that grade cannot affect the nervous–system structure as it mediated the study behavior; the grade appearance cannot make these structures mediate study behavior more readily in the presence of the relevant evocative stimuli; the grade appearance cannot *reinforce* the long ago behavior that produced it (although it may reinforce any behavior that *immediately* precedes its appearance).

By the way, does the term "teleology" evoke relevant responses? (Our discussion of teleology is in Chapter 4.) In any case, when analyzing the contingencies currently controlling a behavior of concern, we focus mostly on the immediately present antecedent and postcedent variables, and thus avoid the distractions and other problems of remote antecedents and postcedents.

Macrocontingencies Differ from Contingencies of Reinforcement

The last topic for this chapter to introduce concerns the concept of *macrocontingencies* (Ulman, 1998) which may prove vital to our applying behaviorology to help solve global problems. However, we risk some analytical difficulties if we confound macrocontingencies with contingencies of reinforcement, difficulties that could reduce our effectiveness.

When we analyze behavior in terms of contingencies of reinforcement, we stick either with the particular behavior in general or with a specific response instance of that behavior. We also stick with *currently* operating variables, antecedent and postcedent, controlling the behavior or response of concern. However, when we have occasion to consider a broad set of behaviors and controlling variables occurring together and ranging across a period of time, then an analysis in terms of macrocontingencies becomes more appropriate. After addressing concerns about some alternative terms and conceptualizations, Ulman (1998) says, "*Macrocontingency* may be defined as a set of differing actions (topographies) of different individuals under common postcedent control" (p. 209). He continues:

> Because ... the macrocontingency may involve any number of individual or collective actions—verbal as well as nonverbal and covert as well as overt—under the same postcedent control, the complexity of the contingency relations constituting a macrocontingency is unlimited. That is, the flexibility of the definition of the macrocontingency allows for descriptions of real correlated actions among any number of individuals and with any degree of complexity. Hence, as defined here, the concept of a macrocontingency is elastic—expanding or contracting with increasing or decreasing complexity so as to describe the actual material conditions as can best be determined from an empirical analysis of the nexus of contingency relations (consequences and other events) constituting the real sociocultural phenomenon under investigation. Due to practical constraints, of course, an analysis of a sociocultural phenomenon may need to be more conceptually that empirically based, in which case the positing of the macrocontingency as the controlling variable replaces notions of control by some ethereal inner or transcending (meta–) agency (p. 209).

For some macrocontingencies involving many individuals, or a range of evocative stimuli, or long–term as well as short–term outcomes, we also sometimes use the term *cultural contingencies.* This takes us into the realm of cultural practices and "culturology" (see Chapter 6 of Fraley & Ledoux, 2002) such as the cultural practice that we call *education.* An important factor is that such practices, which we often see as societal institutions, continue beyond the life spans of the individuals (such as persons either receiving or providing education) involved in the macrocontingencies controlling the practices.

Neither of our two analysis levels—contingencies and macrocontingencies, each complex enough on its own—benefits much from simplistically combining them in a single analysis. Doing so can be an analytical error leading to confusion, which is something we try to avoid.

However, let's end the chapter by continuing our ongoing activity. Before moving on to Chapter 8, consider how our coverage of all these complex analytical confusions and fallacies can help reduce errors and improve effectiveness when we participate with other natural–science team members in efforts to solve global as well as local problems.❧

References (with some annotations)

Cautela, J. R. (1994). General level of reinforcement II: Further elaborations. *Behaviorology, 2* (1), 1–16.

Fraley, L.E. (2013). *Behaviorological Rehabilitation and the Criminal Justice System.* Canton, NY: ABCs. (Chapter 2, on *the physics of behavior,* is of particular relevance to this chapter.)

Fraley, L. E. & Ledoux, S. F. (2002). Origins, status, and mission of behaviorology. In S. F. Ledoux. *Origins and Components of Behaviorology—Second Edition* (pp. 33–169). Canton, NY: ABCs.

Lucus–Clark, J. (2010). Framing the discussion: What to tell students about science. *Thought and Action, 26* (Fall), 123–125.

Maurice, C. (1993). *Let Me Hear Your Voice—A Family's Triumph Over Autism.* New York: Ballantine Books.

NYS Department of Health—Early Intervention Program. (1999). *Clinical Practice Guideline: Autism / Pervasive Developmental Disorders, Assessment and Intervention for Young Children (Age 0–3 Years) Quick Reference Guide.* Albany, NY: NYS Department of Health (Publication No. 4216).

Ulman, J. (1998). Toward a more complete science of human behavior: Behaviorology plus institutional economics. *Behavior and Social Issues, 8,* 195–217.☙

Chapter 8
Basic Laboratory Methodology

$\mathcal{F}$rom some valid but basic contingencies in Chapter 5, we moved on to consider some complex contingencies in Chapter 6 and a range of contingency examples in Chapter 7, some of which were valid while others were invalid and still others were valid but defective. Both the invalid and the defective contingencies stemmed from several confusions and fallacies, including non–behaviors, fictional constructs, bodily states, and remote evocative and consequential stimuli, all of which interfere with effective contingency analysis. We also considered some confusing uses of the term *conditioning,* and we ended Chapter 7 by introducing the concept of macrocontingencies.

The place and value of contingency analysis gradually arose from the outcomes of decades of successful investigations in laboratories and applied settings around the world. These investigations involved many independent variables that affect behavior, our dependent variable. In this chapter, we introduce some of the basic methodology of laboratory studies on why behavior happens. These studies contain implications for interventions not only that help people whose contingencies are compelling behavior excesses or deficits that are interfering with their daily living, perhaps even outside the norms of society or health, but also that help build and improve the ordinary and necessary repertoires that enable effective, competent, compassionate, skilled, just and happy living (e.g., education). Those kinds of interventions, and the experimental methodologies that we use in their generally non–laboratory investigation, come under our scrutiny in later chapters.

Typical of the processes that other natural sciences study, the processes that behaviorology studies operate on scales difficult—though in principle not impossible—to track continuously in human experience. In the past this has coincidently compelled the mystical notion that we are somehow above, or at least apart from, nature. However, according to our experimental research and naturalistic philosophy of science, we are not apart from, or above, nature; rather *we are inescapably a part of nature.*

As an example of one such broad–scale process, consider the overall effects of each of us inevitably accumulating a personal history. This history involves contingencies continuously accumulating—in the past, now, and into the future—the conditioning changes that produce the present body that mediates the current behavior that present stimuli evoke and consequate. Such histories carry some clear scientific implications about biological robotics (which we will explore in more detail elsewhere). For some folks the effect of those scientifically analyzable personal histories and implications leads to substantially accounting for individual yet scientifically orderly differences among people, including the extent, content, and competence of each body's behavior repertoire. For others, based on their also scientifically analyzable histories, the effect of those histories and implications leads to aversive emotional reactions against the very notion of personal contingency histories, their biological robotics implications, and science itself. These reactions raise spectres of suspicion, even fear, and evoke escape responses, including attacks on scientists and rejection of science (although usually without rejection of the life–enhancing products of science).

Yet our being *part of nature,* rather than being apart from it (or above it) is itself a healthy reality, and contingencies that generate and maintain complex response patterns consistent with this reality enhance the possibility of humanity surviving, including sustainably thriving, along with the rest of the planet. Research methodologies, including laboratory equipment and procedures, are both an outcome of, and a continuing part of, those contingencies. Since this includes the equipment and procedures by which any behaviorologist studies behavior, we should examine some of them. Before that, however, let's consider the history and benefits of methodology a little more closely.

Some Methodology History and Benefits

Over the last several thousand years, the contingencies of existence (i.e., the general contingencies under which a body continues to function, such as the contingencies compelling eating while not becoming a meal for another body's functioning) induced gradual changes in the technologies supporting cultural development. Note that all of these—changes, technologies, and cultural developments—are matters of human behaviors happening not due to the initiative of mystical inner agents but as lawful outcomes of these contingencies of existence. For some discussion of how this works, you may find the books of Marvin Harris insightful (see Harris, 1974, 1977, 1979; also, see Vargas, 1985).

Some of those changes proved beneficial for enough people, enough of the time, that they became cultural practices involving applied technologies in, for example, food procurement or production, metallurgy, and construction. These developments occurred rather haphazardly, that is, with little in the way of any systematic approach, and usually more superstition than fact informed them. However, additional contingency–related natural processes, such as generalization, also continued to operate, and these gradually brought into focus the contingent value of some systematic approaches, approaches that we see today as precursors of science. And over the last several centuries such changes have gradually accumulated—another example of our Law of Cumulative Complexity (from Chapter 1)—into the wide range of scientific methodologies from which we benefit today. That is, the contingencies driving investigations of various subject matters have given rise to the range of fairly standardized scientific methods.

Those standardized scientific methods often face subject–matter constraints such that a methodology that works well, under the constraints of one subject matter, may be quite inappropriate for other subject matters. While all natural sciences acknowledge experimental research (i.e., the direct manipulation of independent variables) as producing the greatest methodological benefits, the subject matters of only some natural sciences are amenable to experimental research. For example, while biologists and behaviorologists usually have direct access to the independent variables of their subject matter, astronomers seldom have direct access; while behaviorologists can directly move evocative and consequential stimuli around, observing any resulting changes in their behavior dependent variables, astronomers cannot directly move stars and planets around, which constrains them to use other methodologies for observing any resulting changes in their dependent variables. Nevertheless, to greater or lesser extents, all scientific methods provide certain benefits to researchers and their supporting publics. What are some of those benefits? Let's look beyond the common concrete benefits of better food, shelter, clothing, and so on.

Methodology and Confidence

One such benefit concerns scientific confidence. Many criteria contribute to determining the details of what methodology we use to study any subject matter. Also, as with methodology itself, while some criteria apply to all subject matters, other criteria may apply to only one or another subject matter. We, of course, are mostly concerned with the behavior subject matter.

Since scientists are, first of all, behaving organisms—just like everyone else, with the same natural laws governing all the behavior of *all of us*—some of these criteria may, but really should not, surprise you. These criteria can include rather less important ones such as mere convenience; some types of experimental subjects or techniques or equipment that are readily available may control our research behavior even though a different type of subject, technique, or piece of equipment would meet other criteria better but cost additional personal energy to arrange. Whether or not a particular type of subject or technique or equipment item is the most *appropriate* for answering our research questions always remains an important criterion. However, whether or not we can *afford* a particular type of subject, technique, or equipment item constantly looms as a bigger, though ultimately less important, criterion than appropriateness. The type of research question also enters the picture. Questions relevant to an outdoor setting demand some different methodological details compared to questions relevant to either an indoor or a laboratory setting. One of the most important criteria, in the contingencies determining the details of what methodology we use, concerns the extent to which our experimental methods can leave us not only more convinced of, comfortable with, and confident in the accuracy and adequacy of the conclusions arising from our studies of behavior, but also more confident that applying these conclusions will provide benefits in the short and long term.

Those kinds of increased confidence constitute one major kind of benefit accruing from the use of appropriate methodology. In addition, that benefit accrues when, even because, appropriate methodology enables precise replication of experiments. To the extent that such replication succeeds in leading to the same, similar, or even—preferably—extended conclusions, our confidence increases in the adequacy, accuracy, and applicability of our findings, well beyond the level that mere casual observation or untested guesswork might provide. Even when replication produces different, even contrary conclusions, we still benefit, because that is how science corrects itself, thereby avoiding ever growing errors. And sometimes the unexpected outcomes of experiments or their replications occur due to a large, and possibly too costly to deal with, amount of variability in the research topic. Let's examine more fully how methodology affects variability.

Methodology and Variability

In another major benefit, good (i.e., appropriate) methodology reduces *variability*. Variability refers to the spread of measurement values that occur when scientists measure the values of their independent variables and dependent variables (IVs and DVs). The existence of that spread indicates that the researcher's methodology is not fully controlling all the sources or conditions (i.e., all the IVs) responsible for variation in the subject matter (i.e., all the DVs). Those contingencies of existence that we already mentioned compel us to discover not only these sources and conditions, but also ways to constrain their effects. Recall that all these interactions, contingency generated and maintained, including the people parts, are occurring entirely naturally. Thus, we recognize that in all non–natural science disciplines, and in all natural sciences including behaviorology where we focus our attention, this connection between methodology and variability is as much a question of contingency–generated philosophical assumptions as it is of contingency–generated methodology, as discussed in an earlier chapter.

Contingencies, both cultural and educational, compel the assumption on the part of some researchers, especially those from non–natural disciplines (e.g., psychologists) that observed variability arises from the vagaries and capriciousness of some untestable (i.e., mystical) inner agent that contingencies have also compelled those researchers to posit as initiatively responsible for a behavior of concern. The mystical status of that kind of variable automatically curtails direct access to variability. Thus, one cannot directly reduce this variability. The most commonly accepted method to deal with variability from this kind of source is a group statistical design of one sort or another. In these methods, the mathematical manipulation of the data supposedly spreads out the variability somewhat evenly across a large number of randomly selected subjects so that the researchers can ignore it while functional relations can still emerge in the results (i.e., beyond the typical concern over the usual and arbitrary "significance" level of 0.05, and "type 1" and "type 2" errors).

Alternatively, since behaviorologists, as natural scientists, cannot philosophically grant status to putative inner agents, those agents cannot be the source of observed variability, which implies that variability stems from something else. Whatever that is, can a methodology different from group designs deal with variability better, perhaps by reducing it? Starting as far back as the early work of behaviorology precursors like Skinner, behaviorological scientists then and behaviorologists now continue to identify a different source for variability, one related to how thoroughly, or not, we exert experimental control over the functional variables relevant to a behavior under study. That is, we regard the variability that we observe in our data as the outcome of the extent of the thoroughness of our controls on all the relevant variables affecting the behavior under study. The more thoroughly we place more variables under control, the less variability we observe; conversely, if we exert less stringent control, or if we control fewer variables, then we observe more variability.

A variability–control example. For example, let's say we are studying how many equal–sized classrooms an employee mops when we assign him or her to spend two hours of her or his shift mopping. After two weeks of taking what we call "baseline" data, in this case counting the number of mopped classrooms at the end of each shift, we find that our janitor mopped from three to nine rooms each night. Of course, thus far we have not controlled any variables that might have an impact on this number; we cannot even guarantee that our employee spent the full two assigned hours mopping, or that his or her mopping did not consume six hours of the shift. Also, did the janitor's immediate supervisor make any comments, and if so were they complimentary or derisive comments? Did the mopping occur at the beginning, or the end, of the shift? Obviously we need more controls on some relevant variables.

Here are some of the controls we might institute. Not only might we instruct the supervisor in the relatively greater effectiveness, for increasing mopping and employee job satisfaction, of complimentary comments, but we might also instruct the supervisor to limit the mopping task to the first two hours of the shift when our janitor is likely to be less fatigued. And we should also have the supervisor record when in the shift, and for how much time, the mopping occurred, which better informs our data set and so better constrains the conclusions we draw from the data set. We may also set the same contingencies for the other four janitors working the classroom building, which would provide a small, inherent amount of replication in our study.

Let's say that data over the next two weeks showed that every day our janitor mopped from five to eight rooms each night, and only mopped for the first two hours of the shift. Perhaps, to this cooperative employee, we should give a small bonus, or just a well deserved compliment, to maintain such cooperation, rather than have to issue a pink slip after argumentative behavior occurs due to our ignoring (i.e., extinguishing) the cooperative behavior; that would also waste the money we spent on the state–mandated safety training about the chemicals that the job requires using. (Even these simple but real human examples can get so complicated.) And we still have no data on the supervisor's interactions with our employee. However, the variability did decrease. Did it decrease enough? If not, then we need to control additional relevant variables. Think about what might happen to the amount of variability if we worked on these variables, and how we might work on them: How many desks sit in each classroom? Do some classrooms contain easier–to–move integrated–seat desks while other classrooms contain two–piece desks with separate "tables" and chairs? And these are still not all the relevant variables. Can you think of some of the remaining ones, and how controlling them might affect variability? In any case, as the example indicates, the more thoroughly we place more variables under control, the less variability we observe; conversely, if we exert less stringent controls, or control fewer variables, then we observe more variability.

Preview of single–subject designs. Such controls inhere in methodology, which renders variability manageable, and we have developed methodologies that manage these controls and variability quite acceptably. To deal with variability, behaviorologists emphasize a methodology that we call *single–subject designs,* one of the many helpful terms that we gladly inherit from our precursor natural scientists of behavior. In these designs we work with handfuls of subjects, three to six per "experiment"; actually, three subjects would really be three single–subject experiments, which automatically engages a small, inherent amount of necessary replication (and reliability and generality) as we consider the behavior of each subject individually, because our main concern with behavior is primarily as a phenomenon of individual organisms. If our experimental arrangements indeed fully controlled every variable relevant to the behavior of concern, the results would match predictions precisely. But this is not the case, as we rarely if ever control *all* these variables. So predictions are always off by some greater or lesser amount, and that amount indicates the remaining variability. We see variability as arising from the operation of those variables over which we exert inadequate experimental control.

Remember, we cannot simply blame nature for this kind of variability as it derives from the incompleteness of our control of functional variables. The variability grows when we control few of the relevant variables (giving us larger measured differences between predictions and outcomes); variability shrinks when we control more relevant variables (leaving us with smaller measured differences between predictions and outcomes). Both cases hold, because the variability stems from the variables we are ignoring in the sense of not taking them into account, and controlling them, or controlling for them.

Variability and economics. Since taking variables into account is costly, economics enters this picture, not just in terms of energy or other resources, but in terms of funding. Generally, the higher the number of variables taken into account (i.e., measured and either held steady or manipulated) the greater the associated monetary expenses. Also, as the experimental question increases in importance, the more costs we must bear to take more of the relevant variables into account to answer the question more thoroughly, which reduces variability and increases the success of prediction, control, interpretation, and application (and thus in important ways justifies the increased funds society needs to authorize to cover the expenses of research). Conversely, for less important experimental questions, since control is costly, we settle for controlling fewer variables and so tolerate increased variability along with the associated reductions in prediction, control, and so on. Of course, we must resist letting costs determine the importance of questions. But the point is that the problem of variability resides, not in the magic of made up inner agents, but in the amount of residual ignorance we tolerate as a result of the decreased amount of our experimental control over the variables in the natural history of the events under study.

Seeing variability, then, as inhering in the incompleteness of our experimental control thus affects the selection of experimental methodology. We need not smooth out variability from discredited capricious inner agents across subjects in groups, which also tells us little about the individual behavior of those in the groups; rather we need only adopt the methods that most effectively control, for a given amount of research funds, the most variables affecting a single subject (with three to six subjects studied, thereby providing that helpful if minimal level of replication, reliability, and generality for the results). Since the single–subject designs that we adopted early on continue to fill this bill quite well, we continue to emphasize them in the natural science of behavior. With that background on methodology and its relations to variability, confidence, and economics, let's consider some basic experimental methodology of behaviorological science.

Basic Single–Subject Design Methodology

Beyond sharing with other natural sciences (a) the refusal to allow metaphysical events to enter explanatory accounts, (b) the necessity of parsimony, especially in accounts of behavior, including human behavior, and (c) the related philosophical assumptions and practical actions with respect to controlling the sources of experimental variability, behaviorological scientists have for sound reasons relied on single–subject experimental designs. Throughout the history of natural behavior science, starting with the experimental preparations of Skinner's early laboratory at Harvard in the 1930s and continuing up through today's laboratory and applied research efforts, disciplinary contingencies compelled development of single–subject designs. Here we consider some of the most basic types, and the associated laboratory equipment. (To satisfy curiosity, you can find the often fascinating historical details of the scientific invention process behind many of these developments in parts of Skinner, 1972, and Skinner, 1979; the references note the pertinent parts.)

The two most common and effective types of single–subject design are the "Reversal" design, for which we also use the label "ABAB" design, and the "Multiple–Baseline" design. In both of these designs, we usually present the data in one or another type of graph, although behaviorological data from twentieth–century laboratories most commonly appeared in the form of *cumulative records* (after Skinner invented the cumulative recorder in the 1930s); some studies today still make effective use of this graphic format. Also, in both of these designs, the behavior of the subjects during one or more periods serves as the control for the subject's behavior during other periods, across different experimental phases containing different conditions. The data best document experimental control of the relevant variables when they show that behavior changes when, and *only* when, the experimental conditions change.

Two Principal Single–Subject Designs

Let's look briefly at both basic single–subject designs. Then we will look at the Reversal design again in its common laboratory iteration (before considering the related laboratory equipment and data records and graphs).

In the *Reversal,* or ABAB, design, the conditions of the experimental phases follow a repeating pattern. We call the first phase the *Baseline* phase, or A phase. In it we measure and record the behavior as it happens prior to any experimental changes, and we always take as many steps as possible to prevent any effects on the behavior from the measuring and recording. After the A phase comes the B phase, which we sometimes call the Treatment or *Intervention* phase. For this phase we change the value of one independent variable and continue measuring and recording the behavior. Then, in the third phase—the second A–phase—the conditions revert back to the conditions that were present during the original data collection. That is, the conditions reverse back to A–phase (i.e., Baseline) conditions, which is why we call this the *Reversal* phase (leading us to call the whole design a *Reversal* design). Finally, the experimental conditions revert back to the B (intervention) conditions (hence, "ABAB"). When we use this design in applied settings, the first three phases document the effectiveness of the intervention independent variable, while the last phase reinstates the intervention for its beneficial effects for the subject. The use of the A and B letters to denote the different phases led to labeling Reversal designs as ABAB designs. Regardless of the label, we witness proper experimental control occurring when the data show the behavior changing when, and only when, the phases (i.e., the conditions, the contingencies) change.

Sometimes, however—especially in applied settings—the repetition of Baseline or Intervention phases would stand out as unethical or otherwise not possible. Under these circumstances the *Multiple–Baseline* single–subject design then becomes an even more valuable alternative. In the Multiple–Baseline design, the experimental conditions need not ever reverse; instead, one variable changes in each phase, and the data track any effects on the behavior of concern across those phases. Again, as with the ABAB design, under the Multiple–Baseline design, the data document experimental control when they show that behavior changes when, and only when, the experimental conditions change in each phase. You will find the Multiple–Baseline design emphasized in the research of applied behaviorologists working to develop and test procedures and interventions for cultural benefit in non–laboratory settings. We will elaborate on the Multiple–Baseline design under those circumstances in a later chapter.

Mostly through these kinds of methodology, behaviorologists work to discern and apply the variables of which behavior is a function, variables that generally reside partly in the organism's species history and personal history, but more importantly in the organism's current situation and, particularly for people, in the cultural setting. For now, though, let's turn to some details on how we use the ABAB design for experimental research in the laboratory, which is where this design originated.

The ABAB design in the laboratory. You will find basic laboratory research scientists regularly using versions of the ABAB design, often with other animals as well as with humans, in their efforts focused on discovering the various effects of all the variables, or particular variables, related to behavior. Usually in this context, however, the repetition of Baseline or Intervention phases is unnecessary, because a further sequence of phases, each with different conditions, may follow the A and B phases. Under this circumstance we seldom use the A and B letters to name the experimental phases; instead we name the phases by labeling them with the name of the independent variable operating during each phase, such as a "VR (variable ratio) reinforcement schedule" phase following a "VI (variable interval) reinforcement schedule" phase. Nevertheless, our standard control question still applies: Do the data show the behavior changing when, and only when, the contingencies change?

Perhaps thinking about the laboratory experimental design from another angle can clarify it even more. Most experimental spaces contain one or more *manipulanda,* such as a telegraph–type key or some sort of lever (with quite a number of manipulanda types available in laboratory equipment catalogs). Each *manipulandum* contains or connects with a switch such that, when a subject and a manipulandum come into contact with enough force, the switch closes causing recording devices to record a response. When we first introduce a subject, human or other animal, into an experimental space, various stimuli in the space—sometimes only due to their novelty—elicit and evoke a range of immediate responses, including aversive or otherwise unhelpful emotional responses, most of which are irrelevant to the experimental question. Some researchers begin recording responses right from the start, while waiting for the extinction of these irrelevant responses, a phenomenon to which we loosely refer as the subject "acclimating" to, or "adapting" to, the experimental space. Often the original baseline involves merely recording the manipulandum–related responses until the data show at least some semblance of relative stability. After this the series of experimental phases begins, with each phase generally also continuing until some stability occurs in the data.

For example, in reinforcement–schedule research, an experimental question might concern what variables contribute to the kind of behavior that we might describe as "compulsive" or that others might describe as "workaholic" (i.e., behaviors occurring at such relatively high and steady rates, apparently with few or no reinforcers, that they could have detrimental effects on health and other aspects of daily living). In this case one of those variables could involve how many responses occur, and across what time span, under an extinction contingency (i.e., the reinforcer no longer occurs) after various reinforcement schedules such as a VR–10, VR–100, and VR–1,000. For each of these schedules, while a researcher would record the data during each increment in stretching the ratio to the required level (such as to VR–100) the first experimental phase of interest is the "Stable VR–100" phase in which the data show stability at the VR–100 level. The next experimental phase of interest is the "Extinction of VR–100" phase, which starts when the experimenter turns off the reinforcer–delivery equipment. Along with the extinction contingency, data recording continues during this phase until the relevant stimuli stop evoking responses. The data then not only show how many responses occurred and across what time span, under the extinction contingency, but also the gradual shifts in the response rate from higher to lower, that is, the curve that the extinction contingency produced, information relevant to designing interventions to manage some types of compulsive behavior (behavior that "compulsions" do not cause).

As you can see from that example, when we use the "ABAB," or "Reversal," design in the laboratory, we can better characterize it many different ways, including "AB," "ABA," "ABCB," "ABAC," "ABCAD," and so on, according to the actual phases in our experiments. Now, however, let's turn to some details on the equipment and data records that we use in laboratory research.

Equipment and Records for Laboratory Methodology

For better or worse, experimental outcomes owe as much to the experimental equipment as to the experimental methodology. Witness, for example, the difficulties—particularly with concurrent superstitions—that researchers face in trying to study ordinary, multiple–operant human behavior in the laboratory, when their equipment limits them to using only a single reinforcer source both for when the operants under study can occur simultaneously and for when the consequences of those operants can also occur simultaneously. Effectively studying this kind of everyday human behavior only became possible after the development of equipment and procedures featuring multiple selector (e.g., reinforcer) sources (as we described back in Chapter 6).

The inclusion of multiple selector sources in experimental equipment extends the range of equipment component options available for the typical experimental chambers that behaviorological laboratories keep on hand for studying operant behavior. However, beyond certain vital features, rather little is typical of such chambers. They can range from less than a cubic foot of space to the size of a small bedroom or larger. They can feature bare surfaces rather devoid of stimuli beyond those we explicitly need for an experiment or, if the point is to study typical human behaviors in a recognizable human space, they might contain all the stimuli that are typical of a bedroom or office as well as those we explicitly need for an experiment. The complexity of experimental controls, data recording, and analysis may increase, but our assumptions about the thoroughness of how the laws of nature govern behavior continue. In spite of all the fun that accompanies such complexities, our experimental chamber description here will remain on the level of simpler equipment for simpler experiments. In due course I hope you graduate to the pleasures and benefits of greater complexities.

The Operant Experimental Chamber

B. F. Skinner invented the *operant experimental chamber* early in his career. For many decades, people have referred to this chamber as "the Skinner box." However, Skinner disliked that name, and never used it. What led to this name? Past professors provided a fun account, although Skinner's autobiographies (1976, 1979, 1983) and other writings may shed more accurate light on the question. Still, as the story goes, some students, who were studying under other professors, were running rats in mazes for their research, a rather complicated and time–consuming procedure both on a day–to–day basis and for how long (perhaps weeks or more) before you could analyze a publishable data set. These students, however, observed Skinner (and his students) putting experimental animals into an operant chamber, turning the equipment on, going to lunch and, upon their return an hour later, analyzing a possibly publishable data set. To no one's surprise, these maze–using students exclaimed to their professors, "We want to use 'Skinner's boxes'!" In any case, this name stuck.

By the way, do not confuse the operant chamber with the "Aircrib." This specially constructed and normally used crib, which we also know as the Baby–Tender or, jokingly, as the Heir Conditioner, offers particularly beneficial features for a normal infant. Skinner's wife asked him to invent something to make having another child even more enjoyable than their first child, and he came up with the Aircrib, which no one has ever used for experimentation. (For some details, see the Appendix; also see Ledoux & Cheney, 1987.)

Returning to the operant chamber, regardless of size it is in essence a controllable mimic of the larger environment, a mimic that researchers can simplify in ways that afford them nearly complete control over most aspects of the larger environment. They exert most of this control merely through excluding most of the larger environment from the chamber. These chambers sit in light and sound proof housings, and contain their own lighting, their own sound speakers, and so on with respect to any other sense modality that a particular experiment might require.

More importantly, these chambers contain three essential items. (1) They contain whatever stimuli the experiment requires for evoking or eliciting responses. (2) They contain whatever response manipulanda the experiment requires to ensure that arbitrary though standardized responses are possible and recordable when the stimuli evoke or elicit them. And (3) they contain access ports, along with audio, visual, and other sense–modality stimulus displays, for the delivery of whatever type of reinforcer or punisher the experiment requires. If we must locate the chamber in a noisy environment, a speaker in the chamber may play "white noise" to mask extraneous sounds. We may also control the temperature, humidity, and other conditions in the chamber as any of them might affect experimental outcomes. Controlling all these variables, at least by keeping them constant, not only protects the subject and the results from any extraneous or systematic stimuli that the experimenter or the apparatus might produce, but also provides a relatively stimulus–clean environmental background against which the independent variables of the study can operate. But what collects the data that we require for analyzing the environment–behavior relationships that are developing, maintaining, or disintegrating within the chamber walls under the experimental contingencies?

We collect the data through data recorders and displays of one sort or another that the researchers connect to the operant experimental chambers, along with computerized equipment and interfaces that not only manage the experimental conditions but also manage the recorders and displays. Our focus here is with but one important type of recorder and the graphs it automatically displays, namely the cumulative recorder and its cumulative graphs.

The Cumulative Recorder, Past and Present

In the past cumulative recorders were devices that automatically, through electrical and mechanical means, produced paper cumulative records of behavior from the outputs of operant chamber interfaces. Such devices are now collector's items (see Skinner, 1976; Poling, 1979). Computers running commercially available programs now take the usual outputs from operant chamber interfaces and organize them into the same kinds of cumulative records. Furthermore, these computer programs usually provide additional data arranging and display options many of which are actually quite helpful to the contingencies supporting scientific discovery.

Cumulative recorders as invented. After Skinner invented it, and for the many decades that various companies produced it commercially—but before it became a part of computer programs—the cumulative recorder shared some characteristics with many other scientific data recorders. They all used a roll of paper unrolling at a fixed rate under a couple of ink pens.

With the cumulative recorder, one pen recorded along the bottom edge of the unrolling paper. For this reason some people called it the "lower pen," although due to its main function many others called it the "event pen." You see, this pen could only travel one small step up (its onset) after which it could only travel a small step back down (its offset) and most researchers used this pen to record, by its onset or offset, various experimenter–defined events such as a change from one experimental phase or condition to another. As the paper unrolled, this pen marked a horizontal line at two slightly different levels according to whether it had moved up or moved down its one available step.

On the other hand, people called the upper cumulative recorder pen the "response pen." This pen marked one tiny step up for every response that occurs. For the brand of cumulative recorders with which I am familiar, the width of the paper on the roll allowed the pen to step up a maximum of 400 of these tiny response steps before the pen reset to near the bottom of the paper and started up again with further responses. Of course the downward pen line usually traced only the resetting of the pen and not any data. (I say "usually," because very rapidly occurring responses can leave a couple of response–step jags on this line even while it was tracing the pen reset to the bottom edge of the paper.) This means that every data line (i.e., every non–reset line) actually stacks on top of the previous data line, because these lines show the data of responses *as they accumulate*, which is both why we call these data recorders *cumulative* recorders, and why we call the records that they produce *cumulative records*. Unlike the ever alternating up and down data lines in the non–cumulatie graphs with which we are all familiar, the *cumulative* data line *never* goes down; it only resets to keep the pen tracing the data on the paper (whereas otherwise it would go off the paper edge after the first 400 responses). To get a broader view of any and all changes occurring in the behavior, researchers would sometimes cut the paper records along the reset lines and then tape the data records onto a laboratory wall, starting in the lower left corner and, by attaching the bottom of the latest data line to the top of the previous data line, proceeding in a generally diagonal direction toward the upper right corner. Why is this direction generally diagonal? The answer has to do with another name that people use for the upper cumulative recorder pen.

Instead of calling the upper pen the "response pen," some people call it the "rate pen," because the line it draws relates directly to the rate at which the responses are occurring (i.e., the number of responses divided by the amount of time over which those responses occurred). The tiny steps that increment to record each response are generally smaller than the width of the line that the ink pen draws, so these tiny steps are not easily visible to the naked eye. Recall, however, that the paper is *always* unrolling, so this pen is *also* always drawing a horizontal line. Thus, in the line that the rate pen actually traces, we observe two motions simultaneously. We observe (a) the always horizontal motion as time passes (i.e., the always horizontal motion as the paper unrolls under the pen) and at the same time we observe (b) the vertical motion of the tiny upward steps that accumulate, one for each and every response that occurs. As Figure 8–1 shows, the line of the rate pen combines these two motions, resulting in an *angled* line, a diagonal line rising, from left to right, that we call the *response slope* line, and the angle of this slope indicates the rate of occurrence for the behavior (with a steeper slope showing a higher rate and a shallower slope showing a lower rate—see Figure 8–2). We can, if course, calculate the exact rate in responses per second (or even responses per minute) if that is needed, because we know the speed of the paper (the horizontal distance that the paper travels in a set amount of time) and we can directly measure the number of responses in the vertical distance of pen travel. Many cumulative records even include a reference graphic with typical slopes of known responses per second, slopes with which we can compare any section of an actual cumulative record for a reasonable estimate of the response rate in that section. Also, the angle of the response rate line can trace a general curve, which can tell us—depending on its direction—whether the behavior is increasing in rate or decreasing in rate, as you can see in Figure 8–3.

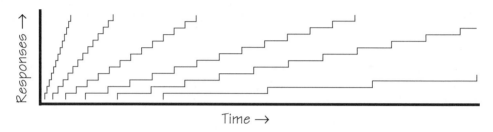

Figure 8–1. Six stylized cumulative records greatly enlarged to show their step–like character but without any pen–reset lines.

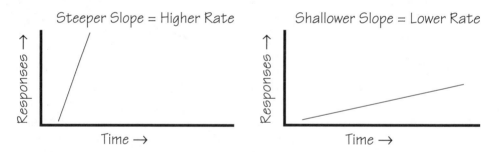

Figure 8–2. Stylized cumulative records showing the relationship between slope and response rate.

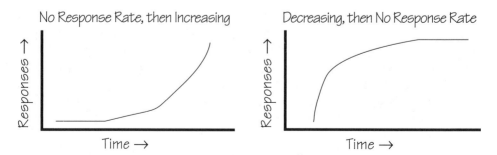

Figure 8–3. Stylized cumulative records showing changing slopes/rates.

Cumulative records include one further type of indicator. When a response produces a reinforcer, the response pen also makes a diagonal tick mark, which some also call a pip or hatch mark, off the response slope line at the same step that records the response occurrence. When reinforced responses occur among unreinforced responses, these pips are readily identifiable, as you can see in Figure 8–4. Again, however, because the ink line is thicker than the space between response steps, if you have a run of reinforcers, in which every response is earning a reinforcer, then all the reinforcer hatch marks run together as a thick and solid black line. You saw many of these thick lines in the figures for our data on the responses produced under the dual reinforcer sources back in Chapter 6.

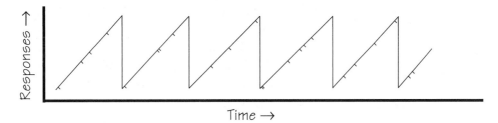

Figure 8–4. Stylized cumulative record showing reinforcement hatch marks and un–erased reset lines.

In those Chapter 6 figures, I simply erased most of every pen reset line. Many researchers follow the more common practice of re–spacing the data slopes to shift them closer together after erasing the pen reset lines. See Figure 8–5 for an example, which shows the result of taking the same data slopes that are in Figure 8–4 and re–spacing them after erasing the reset lines. (This cumulative record mimics the kinds of records that some vr schedules produce.)

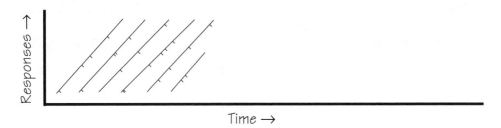

Figure 8–5. The same stylized cumulative record as in Figure 8–4 after erasing the reset lines and re–spacing the data.

On the other hand, Figure 8–6 shows what cumulative really means. Figure 8–6 shows those same data that we see two different ways, in Figure 8–4 and Figure 8–5, stacked on top of each other. At 400 responses per slope (on paper roughly seven inches wide) we would need access to a cumulative recorder that uses a roll of paper about three feet wide. To my knowledge no such behemoth exists. But laboratory walls make very workable substitutes.

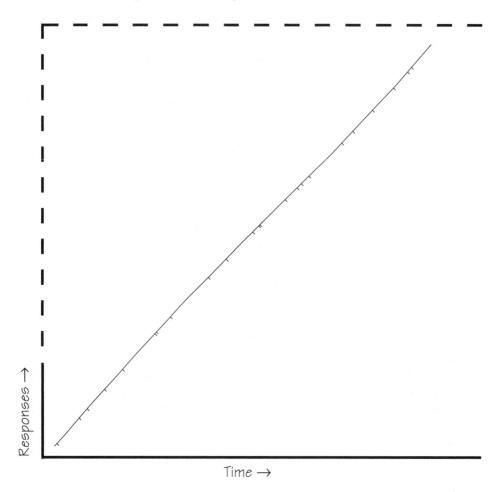

Figure 8–6. The same stylized record as in Figure 8–4 and Figure 8–5 with all the pieces stacked up on each other in ascending order, demonstrating the meaning of "cumulative."

The computerized cumulative recorder. Today (see Ledoux, 2013) researchers use computerized cumulative recorders and their records. These feature some other characteristics that we have not yet covered. When we view the computer screen, we see data records that the computer processes and sends *without pens* to the screen, so we refer merely to the *event line,* and the *response line* or rate line, without reference to pens. Also, the computer sends the data to the screen according to the particular parameters that the experimenter sets for the program (e.g., "Display at _400_ responses vertically and _1_ cm/minute horizontally," although the terms and adjustment ranges may differ depending on which computer program the experimenter uses). We can set such parameters both before the experiment and after it, during data analysis. A particular benefit of computerized cumulative recorders is that they allow contingencies to induce the experimenter to change parameters, which enhances the control over the experimenter's analysis behavior that the data exert. The computer simply recalculates and displays the data according to the newly set parameters. For example, one might change from "Display at _400_ responses vertically and _1_ cm/minute horizontally" to "Display at _1,600_ responses vertically and _1_ cm/minute horizontally"; this would show more data on the screen, perhaps along the lines of Figure 8–6. Also, while the response steps are generally so tiny that they are not readily visible to the naked eye, one can take advantage of the digital resolutions available with a computer–generated cumulative record and, zooming in on a selection of these response increments, see the actual step–up for each individual response.

Either of those changes, and so many other possible changes, might reveal some important details in the data record. Indeed, future cumulative graphs that computer data files produce may even include symbols indicating, at the reinforcer hatch mark, (a) whether the reinforcer was of the contingent or non–contingent type, (b) whether it was of the added or subtracted type, and (c) whether it was of the primary (i.e., unconditioned) or secondary (i.e., conditioned) type. Related symbols may also indicate of the occurrence and characteristics of various punishers, as well as the occurrence and characteristics of various functional antecedent stimuli.

Methodology review and preview. In this chapter we mainly covered some laboratory methodology considerations. In the next chapter we will cover some aspects of practical methodology, aspects more relevant to applied research for culturally beneficial interventions in non–laboratory settings.

Let's end this chapter with our ongoing activity. Keep considering what roles these research methodologies, and those in the next chapter, play in the application of pertinent behaviorological principles and concepts to managing the behavior components of the solutions to global problems. This will help make possible the timely solutions to the non–behavior–related components that traditional natural scientists are addressing.♣

References (with some annotations)

Harris, M. (1974). *Cow, Pigs, Wars, and Witches: The Riddles of Culture*. New York: Random House.

Harris, M. (1977). *Cannibals and Kings: The Origins of Culture.* New York: Random House. (See Vargas, 1985.)

Harris, M. (1979). *Cultural Materialism: The Struggle for a Science of Culture.* New York: Random House.

Ledoux, S. F. (2013). Human multiple operant research equipment. *Journal of Behaviorology,* 16 (2), 3–9.

Ledoux, S. F. & Cheney, C. D. (1987). *Grandpa Fred's Baby Tender or Why and How we Built our Aircribs.* Canton, NY: ABCs. The Appendix contains this book's Foreword (Skinner, 1987).

Poling, A. (1979). The ubiquity of the cumulative record: A quote from Skinner and a frequency count. *Journal of the Experimental Analysis of Behavior, 31,* (1), 126.

Skinner, B. F. (1972). *Cumulative Record: A Selection of Papers (Third Edition).* New York: Appleton–Century–Crofts. The B. F. Skinner Foundation (www.bfskinner.org) in Cambridge, MA, republished this book as "... *Definitive Edition,*" in 1999. Several of the papers in this collection contain material concerning Skinner's invention of various pieces of equipment. The paper, "A case history in scientific method"—on pages 108–131 in the *Definitive Edition*—is particularly relevant to the invention of the *Cumulative Recorder.*

Skinner, B. F. (1976). Farewell my LOVELY! *Journal of the Experimental Analysis of Behavior, 25,* (2), 218.

Skinner, B. F. (1976). *Particulars of My Life.* New York: Knopf. This is the first part of Skinner's autobiography.

Skinner, B. F. (1979). *The Shaping of a Behaviorist.* New York: Knopf. This second part of Skinner's autobiography includes some interesting material about his equipment inventions.

Skinner, B. F. (1983). *A Matter of Consequences.* New York: Knopf. This is the third and last part of Skinner's autobiography.

Skinner, B. F. (1987). The first baby tender. In S. F. Ledoux & C. D. Cheney. *Grandpa Fred's Baby Tender or why and how we built our aircribs* (pp. iii–v). Canton, NY: ABCs. The Appendix contains this paper.

Vargas, E. A. (1985). Cultural contingencies: A review of Marvin Harris's *Cannibals and Kings. Journal of the Experimental Analysis of Behavior, 43,* 419–428.ᘓ

Chapter 9
Some Practical Methodologies

$\mathcal{W}$e considered experimental methodology in Chapter 8. After looking at the history and benefits of methodology, we turned to the relationship between methodology and confidence in research conclusions. Next we discussed the relationship between methodology and the inevitable variability that appears in experimental results. (Can you summarize these two relationships?) Then we described some of the basic experimental methodologies that we called single–subject designs. This is the type of experimental design that behaviorologists and their predecessors used in their *laboratories*. Beyond describing why and how we use these designs, and how these designs exert experimental control, we also detailed some of the laboratory equipment—experimental chambers and data recorders—common to using single–subject designs in the laboratory.

While further considering experimental methodology, in this chapter we move beyond the laboratory versions of single–subject designs. Here we consider some details of single–subject experimental designs that are appropriate to designing, testing, and implementing interventions for solving simple and complex practical problems in applied settings outside the laboratory.

The ABAB/Reversal Design in Applied Settings

The first several decades of natural behavior science occurred from the 1930s through the 1950s. During this time Skinner, his colleagues in the natural science of behavior, and their students (i.e., the predecessors of today's behaviorologists) gradually built their behaviorological science mostly through the kinds of basic laboratory methodologies that we covered in Chapter 8. By the end of this time, they called their natural science "TEAB" (The Experimental Analysis of Behavior; see Sknner, 1957) to distinguish it from the research and practice of the other—mostly traditional (i.e., agential)—psychologists with whom they generally had to share their work units.

Difficulties Beyond the Laboratory

As the number of those early behaviorological scientists grew, and under the demands of some serious behavior–related problems in society, many of these scientists branched out into an increasing number of application areas. But they faced difficulties, because these areas, extending far beyond the confines of laboratory spaces, required methodologies more appropriate to applied settings than to laboratories. For instance, non–laboratory settings lack the kind of rather thorough access to controlling many independent variables that is available in the laboratory, even if only by holding these variables steady.

Could they develop the needed practical methodologies, ones that would help them get hold of enough independent variables in these application areas so that they could devise and test interventions? And would these interventions produce outcomes that society would consider safe, effective, and worthwhile for needy clients? Also, in applied settings, we must approach confidence and variability differently from how we handle these in the laboratory.

 Confidence in applied settings. Although the automatic data recording of the laboratory boosts our confidence the most, the necessary equipment is often unavailable outside the laboratory. Thus we must rely on other methods, the most common involving two or more observers recording data on the same behaviors. But will the data records from different observers be even close? Just how reliable are different observers at recording data on the same behaviors?

 To begin answering such questions, we can turn to measures of *interobserver agreement.* Even though observers can make observational records several different ways (e.g., a total count of all instances of an event during a complete measurement period, a count of event instances during each of several smaller sampling intervals across the full measurement period, a "yes" or "no" on a trial–by–trial basis, and so on) in each case most applied researchers currently use one or another method to calculate the *percentage of agreement* between the observers, with higher percentages supporting higher confidence levels in the research outcomes.

 For example a common method that we can use to calculate the percentage of agreement, between different observers who have made total counts during a complete measurement period, uses this formula: Divide the smaller count by the larger count, and multiply by 100. So, let's say that two observers each counted the total number of student hand–raises during a complete 50–minute class. One counted 18 hand–raises while the other counted 20 hand–raises. The percentage of agreement would be 90%. While we would consider 90% agreement better than 60% agreement, the 90% can still share several problematic possibilities. For one, both observers could be wrong, or perhaps they are both "right." What if the one had actually counted the *number of students* who had raised hands in the class period while the other had counted the *number of hands that went into the air* during the class period, with two students having raised both hands once each; this would lead precisely to these counts, with a resolution requiring a clarification of the definition of exactly what to count, which should have happened before counting began.

 Other complications concern the variety of equations that we use for different kinds of observational records. However, such topics reside more appropriately in an advanced textbook, perhaps one that focuses on applied methodologies (e.g., see Sidman, 1960, or Johnson & Pennypacker, 1980).

 Variability in applied settings. In the laboratory we reduce variability by holding additional independent variables steady so that they have less effect on the behavior dependent variable under study. Given the relative lack of access to many applied–setting variables (some of which can sometimes be so obscure as to remain unrecognized throughout a study) and the increasing economic cost of efforts to hold them steady, we must generally put up with the variability that remains after budget constraints confront us. While this is acceptable if the intervention solves the problems we were trying to address, we nevertheless always remain on the lookout for better, and affordable, practices that hold more independent variables constant.

To address all such methodological difficulties, those early behaviorological practitioners successfully adapted single–subject experimental designs to solving problems in applied areas. They began with the ABAB single–subject design, the design with which they were familiar in the laboratory. To see how they adapted the ABAB single–subject design, we will describe each of its four basic experimental phases, and then give an example using all of these phases.

The Four Phases of the ABAB Design

As we touch on each phase, recall these four intertwined aspects of *all* single–subject designs, aspects that always remain pertinent. (1) Phases represent periods of time during which different conditions exist, different contingencies, different independent variables. (2) We try to continue each phase until the behavior data shows some level of relative stability (i.e., so that, in spite of the remaining variability, changes in behavior that become apparent, in visual data comparisons across phases, help evoke appropriate experimenter behavior such as altering the contingencies to start the next phase). (3) Regarding the visual data comparisons across phases, each subject serves as his, her, or its own experimental control across all phases. And (4) we gain adequate confidence in our conclusions from this experimental control when the data comparisons across the phases basically show the behavior changing when, and only when, the independent variables change.

The first phase—Baseline. In an ABAB–design experiment, we use the term *Baseline* to tact the first experimental phase, the first A phase in "ABAB." This phase starts when we begin to observe, measure, and record the behavior as it happens, and presumably as it has been happening. This gives us a measure of the behavior without any experimental changes, the results of which we can later compare with the results from other phases that contained contingency changes. Also, we take every possible step to prevent our observing, measuring, and recording from affecting the behavior.

The second phase—Treatment (or Intervention). The first B phase, which follows the first A phase in "ABAB," we call the *Treatment,* or *Intervention,* phase. In this second experimental phase, we begin applying some sort of intervention. This may or may not involve a practice or technique that researchers have previously tested; researchers often use this design to test an intervention for reliability, validity, safety, and efficacy. In any case, we start this phase when we change the value of *one* independent variable. We usually predict both that this is the variable most functionally related to the behavior, often an excess or deficit, that we are trying to improve, and that changing this variable will result in that improvement. Alternatively, this may be a different variable that we predict will interfere with, or counteract, the variables of which the behavior of concern is currently a function. If a relevant behavior change is a function of instituting this contingency change, we will see that function in our data when we compare the results from different phases.

However, that one change in behavior, when contingencies change, is by itself actually inconclusive; for all we know, some other, probably unsuspected, independent variable has also changed, and this change was responsible for any behavior change we might observe. This is why we revert back to baseline contingencies in the next phase. As always, we continue to observe, measure, and record the behavior during this phase.

Also, the particular contingency change (i.e., the particular independent–variable change) for this phase may have its own label as an intervention practice or technique. Applied researchers have validated many such standard interventions over the decades; we will introduce a couple in our ABAB and multiple–baseline examples, and others in later chapters.

The third phase—Reversal. The second A phase, which follows the first B phase in "ABAB," we call the *Reversal* phase. In this third experimental phase, we stop applying the intervention technique; we change the conditions as much as possible back to the way they were during the original baseline data collection. That is, this phase starts when we reverse the conditions back to the original A–phase conditions. Some researchers name this whole design after this phase, calling it a Reversal design, or an ABAB–Reversal design. If a desired behavior change was indeed a function of instituting the intervention in the second phase, then stopping that intervention should change the behavior again and in the other direction, and we should see that change in our data when we compare the results from different phases. If the desired behavior change was not a function of instituting the intervention in the second phase but was rather a function of a change in some unknown independent variable, then stopping the official intervention should not change the improved behavior, because the unknown independent variable will likely still be in operation maintaining the improved behavior, and we would see that maintenance in our data when we compare the results from different phases. Again, we continue to observe, measure, and record the behavior during this phase.

The fourth phase—Reinstatement of Treatment (or Reinstatement of Intervention). The second B phase, which follows the second A phase in "ABAB," we call the *Reinstatement of Treatment,* or *Reinstatement of Intervention,* phase. We start this phase when we reinstate the independent–variable value that was in operation during the original treatment phase. The experimental conditions revert back to the B (intervention) phase conditions. If a relevant behavior change was a function of instituting this contingency change in the first B phase, then our data will show that function restored in this second B phase when we compare the results from different phases. As usual, we continue to observe, measure, and record the behavior during this phase.

In the applied settings where we use this design, this last phase reinstates the intervention for two related reasons. We restore the intervention for its value as a re–verification of the sometimes tentative answers to our control question (i.e., "Does the behavior change when and only when, the contingencies change?"). And we also restore the intervention for its beneficial effects for the client. The point of the study is to help the client. Verifying that the intervention works, by starting it then stopping it, helps the client in an important way. But really, helping the client by reinstating the intervention is pretty important as well, as it rebuilds, maintains, and even extends the improved behavior. Figure 9–1 shows some stylized data that an ABAB (i.e., Reversal) design might produce from an intervention concerned with increasing the behavior of concern.

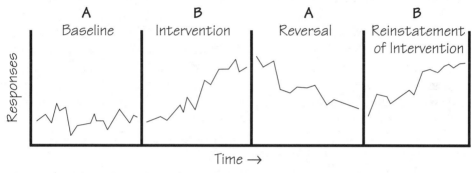

Figure 9–1. Stylized data across the four phases of an ABAB design for increasing a behavior.

The variables in an applied situation are likely to be even more complex than our description of the standard ABAB design might imply. Consequently some researchers extend the design until they are satisfied that conclusions from their data are valid and reliable; these kinds of extensions can add many phases to the basic design, such as ABABAB and ABABABAB. That increasing design complexity begins to approach the complexity possible in our other type of single–subject design, the multiple–baseline design. Before going there, however, here is an example of the ABAB design from an actual study.

Applied–Setting ABAB Design Example

Back in 1975, David Anderson and I worked on a study at the University of Queensland in Brisbane, Australia. That was when and where I began my college teaching career. I give you this example not only because it contains some uncommon features worthy of your attention but also because usually these features only appear in different studies, and that would necessitate giving you two or even three lengthy examples to cover the same ground. With some comments on the frequency of these features today, the features center on (a) a connection between the inadequately informed views of practicum–site directors and methodologically sound studies (something that still occurs too often), (b) an inherent problem with Reversal designs if you succeed "too well," so to speak (something that can still happen, but happens rarely), and (c) the intervention taking place in the client's home, with the client's parent in a therapeutic role (something that is much more common today).

As a fourth–year honours student, David completed most of the work, and most of the writing for the original report, as part of a project for a course on clinical interventions. Out of respect for his writing and the location of the study, the alternate spelling of behaviour appears throughout this example.

David's practicum–site director assigned him to work with a developmentally delayed 6–year–old boy whom we will call Damian (not his real name). Damian's past conditioning history left him with the behaviour problem of repeated and severe tantrums during which he would throw anything within his reach. Thus his home resembled a fortress, with locked doors, and windows closed and barred. At his school Damian had hurt other children with his throwing behaviour, so the school had excluded him.

Damian's mother, a single parent, continually tried to stop his throwing by taking objects from him and by shouting at him. However, these interactions were not regular but rather showed characteristics of an intermittent reinforcement schedule, which thus likely raised the rate of throwing behaviour. Beyond these interactions, his mother gave Damian little attention. Also, she had not used, and felt strongly that she would not use, punishment with him.

Usually for a practicum assignment, the student only needs to try to help a client. In this case, however, far more was at stake. Damian's school required proof of his improvement for readmission, and David's practicum–site director was not only an open and ardent adherent of a mystical epistemology regarding behaviour, an epistemology that still tries to compete with the natural science that is responsible for operant conditioning applications, but David had also observed him getting upset, along with other indicators of his experiencing an aversive emotional reaction, when the topic of operant conditioning arose. Aware that David's clinical interventions course provided students with explicit practical experience with operant conditioning applications, this practicum–site director assigned Damian to him so that David would experience this site director's view that operant conditioning is not applicable in real, non-laboratory cases. So in this case, David not only needed to help a client but he also had to assure that his work was both successful *and* methodologically sound. Then he could report it properly to this practicum–site director as well as to Damian's parent and school so that they would know how to deal with Damian's throwing behaviour in the future were it ever to occur again.

In both those tasks, David succeeded admirably. Figure 9–2, which mostly reproduces the figure from the original report (and which met the standards of the time and place; see Anderson & Ledoux, 1979) shows the daily number of Damian's throwing instances across the four phases of the study.

David implemented a typical ABAB design, using an intervention technique that we call *time out*. At the time this involved, for each throwing instance,

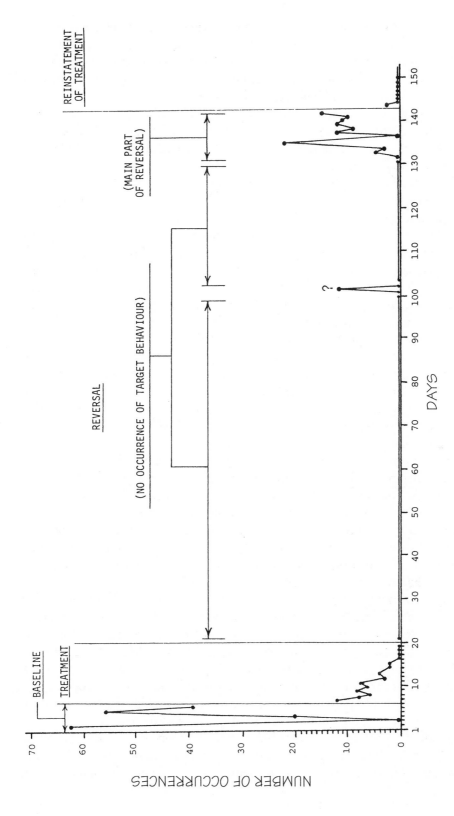

Figure 9–2. Number of occurrences per day of object throwing across four ABAB phases.

Damian spending a few minutes in a room where conditions disallowed the occurrence of added reinforcement. David trained Damian's mother both to use the time–out technique and to mark every throwing instance on a record sheet. Together they defined the target behaviour, an instance of throwing, as "taking up an object, drawing the hand back, and flinging the object ahead." David also recorded throwing instances himself, during reliability checks that lasted at least an hour, first on a daily basis, and then on a weekly basis, throughout the study; interobserver agreement between David and Damian's mother was never less than 90%.

While the Baseline phase was only five days long, throwing instances ranged from zero on one day to over 60 on another day. Given that a project goal was to reduce these instances to less than five per day, such baseline data could be construed as stable enough to evoke proceeding on to the Treatment phase. Subsequent events showed this to have been reasonable. Also, the zero–instances day was a day on which his mother gave Damian a prescribed medication that not only affected his throwing behaviour but also reduced virtually *all* of his behaviour to zero; compared to medication, operant conditioning often provides a more selective, and so beneficial, effect.

During the Treatment phase, Damian's mother implemented the time–out procedure whenever throwing occurred, and also—as instructed—provided reinforcing interactions with Damian during many time periods when his behaviour was more appropriate. As the data show, over the first 11 days of the Treatment phase, throwing instances gradually reduced to zero, which also produced a very happy mother, and later—after the Reinstatement of Treatment phase—a very happy school.

After three further days of zero throwing, David and Damian's mother agreed to begin the Reversal phase by reverting back to baseline conditions, which meant that she would again respond to instances of throwing, not with time out but with plenty of attention from shouting at him, and so on. While uncomfortable with bringing throwing back, Damian's mother understood that he would be unable to return to school if she and David could not document the procedures that brought about the throwing reduction in a reproducible way; they had to stick to sound methodology.

However, during the Reversal phase, they had to deal with the problem of too much success. No throwing instances occurred and so none could be reinforced the way they had been during baseline, so an actual reversal was long in coming. Damian went for 82 days, nearly 3 months, without throwing anything. On day 102 a possible reversal began, but his mother was ill, too ill to respond to this throwing, too ill even to record the throwing instances, which she later estimated at perhaps a dozen throws. The next day she felt better, but no throwing occurred then, nor on any of the next 30 days.

Finally, on day 133, the target behaviour again occurred, followed by the same shouting–attention consequences that had occurred during baseline. Over the next 9 days, the daily count of throwing instances gradually increased, with the exception of day 135, which spiked to 22 instances, followed by a medicated day on which virtually all of Damian's behaviour dropped to zero.

With the Reversal–phase count reaching higher than the count on the first day of the Treatment phase, David and Damian's mother agreed to begin the Reinstatement of Treatment phase. So, starting on day 143, time out again followed any instance of throwing behaviour, while reinforcing interactions with Damian occurred during time periods when his behaviour was more appropriate. On the first day of this Reinstatement of Treatment phase, Damian threw two objects, and went to time out twice. The next day throwing behaviour returned to zero and stayed at that level. A follow–up check showed no throwing instances occurring over the following month.

Damian's school allowed him to return, because the methodology convincingly demonstrated how to deal with throwing behaviour should it recur. The data also showed one way to achieve success in managing environment–behaviour relations through operant conditioning processes or techniques, either by design, which seems definitively preferable, or intuitively. As a result of the practicum–site director declining to acknowledge the study outcomes, David concluded (private communication) that methodologically sound evidence has little chance of upsetting the contingencies responsible for ardent adherence to mystical epistemologies regarding behaviour.

The Multiple–Baseline Design

During the ensuing decades after early behaviorological practitioners adapted ABAB single–subject designs to applied settings, they faced increasingly complex applied situations requiring more sophisticated single–subject experimental designs. For example, what if your intervention involved incompatible responses that prevented a child's further self–destructive behavior, or what if your social–contacts scheduling intervention resulted in an increase in the academic behaviors that were moving a student off probation and onto the Dean's list? In either case, you would not want to use a Reversal design as that would bring the problem behaviors back; however, you may still require methodologically sound verification of the effectiveness of your intervention.

The result of such concerns was the development and implementation of several *multiple–baseline* single–subject designs. We call these designs multiple–baseline designs, because *all previous phases comprise the "baseline" for any particular phase.* That is, each phase becomes part of a growing baseline for every next phase. We compare any particular phase with all previous phases when we are answering our control question. (I am sure you can state that question word for word by now.) For example, if a multiple–baseline study consists of five separate phases, the first four phases comprise the baseline with which we compare the last, fifth phase. What phases comprise the baseline for the third phase?

Multiple–Baseline Designs in General

To begin, we consider details of the multiple–baseline experimental design in general terms. This includes giving both stylized and real examples, with both of these using what may be the most common of the main multiple–baseline forms, the Behavior–Based Multiple Baseline. Then we will consider these three other multiple–baseline forms: the Client–Based Multiple Baseline, the Setting–Based Multiple Baseline, and the Technique–Based Multiple Baseline. Again, as we touch on any phase in these various multiple–baseline designs, recall those four intertwined aspects of all single–subject designs, aspects that always remain pertinent. (1) Phases represent periods of time during which different independent variables are in operation. (2) We try to continue each phase until the behavior data shows some level of relative stability so that, in spite of the remaining variability, visual data comparisons help evoke appropriate experimenter behavior including properly arranging the next phase. (3) Through these comparisons each subject serves as his, her, or its own experimental control across all phases. And (4) we gain confidence in our conclusions from that experimental control when (and the more often, the better) comparisons of the data across the phases show the behavior changing when, and only when, the independent variables change.

The Baseline in multiple–baseline designs. As with the ABAB single–subject design, we begin our coverage of the multiple–baseline single–subject design by revisiting the Baseline phase, which serves the same function in these designs as it serves in Reversal designs. The Baseline phase provides us with a measure of the behavior of concern *without* any experimental changes, a measure that we can later compare with the results from other phases that contain various contingency changes.

The other phases in multiple–baseline designs. A multiple–baseline design proceeds through several phases beyond the initial Baseline phase. As in the ABAB design, the next phase is one or another intervention, but that is where the similarity with the ABAB–Reversal design ends. Instead, the multiple–baseline design takes on the form of ABCDE–N, where N represents the letters for all remaining phases. Typically, we use the term Treatment, numbered consecutively, for each phase after the Baseline phase. Thus we could have Treatment phase 1, Treatment phase 2, and Treatment phase 3, using the abbreviations T1, T2, T3, and so on.

Due to ethical or other concerns, establishing an intervention in the second phase *and then interrupting that intervention* to revert back to baseline conditions in the third phase, even for necessary and justifiable methodological reasons, may not be possible or allowable. For example, your intervention may have succeeded in greatly reducing the occurrence of a child's severe head–banging, and we should and would deem as unethical any reverting to conditions that would or could bring back that head–banging for any reason including methodological reasons. Or, your intervention may have succeeded in eliminating the occurrence of a child's operant vomiting behavior; beyond any ethical concerns, you might not want to revert to conditions that could bring back that vomiting, because cleaning up the mess each time makes you ill as well. Under circumstances like these, the multiple–baseline single–subject design becomes an incredibly necessary and valuable methodology.

In the classic multiple–baseline design, the experimental conditions never reverse; instead, one—and *only* one—variable *changes* in each phase, and the data track any effects on the behavior of concern across those phases. Again, as with the ABAB design, under the multiple–baseline design, we accept the data as demonstrating experimental control when they show that behavior changes when, and only when, the experimental conditions change in each phase. You will find applied behaviorologists emphasizing the multiple–baseline design in research to develop and test intervention procedures and practices that benefit clients and the culture in non–laboratory settings. Let's look more closely at how multiple baselines work.

The Behavior–Based Multiple–Baseline Design

The most common, or most easily conceptualized, basis for a multiple–baseline design is the *Behavior–Based* Multiple Baseline. We say "Behavior–Based," because it is a multiple–baseline form for managing multiple behaviors in a methodologically sound manner but without reversals. Let's say your job is to help a five–year–old child, whom we will call Laddie, and whose contingencies have established problem behaviors of throwing objects that cause different levels of damage. Laddie throws cooking pots, wet sponges, and dog biscuits (which the family dog loves, but that does not help you). Now, your plan invokes the standard, and well researched "time–out" intervention (see Latham, 1994) to reduce the occurrences of throwing behavior. (Excessively briefly, time out in this case could involve, for each throwing instance, Laddie spending one to two minutes sitting at the bottom of the stairs that go from the formal living room up to the bedrooms without any interactions with any added reinforcing stimuli; after all, the full name of this technique is *time out from added reinforcement.)*

Using stylized data, which Figure 9–3 contains, let's run through the process that you might follow to carry out your mission using a Behavior–Based Multiple–Baseline design. As always your first step would involve keeping an initial baseline record of all throwing instances, perhaps with additional observations that might help you answer related questions such as which objects cause the most damage, what consequences follow the throwing, and which objects most often fly through the air with Laddie's help. Before baseline recording ends, you may have determined that throwing pots is the most dangerous but happens the least often (perhaps due to the greater energy that heaving them takes, or to any of a dozen other possibilities; remember, these realistic examples are always more complex than we make them). Throwing dog biscuits, while not dangerous, happens the most often. Wet–sponge throwing, while also not dangerous, ranks as intermediate in terms of frequency. Consequently, you will work on pot throwing first, then on sponge throwing, and finally biscuit throwing. You plot your baseline data as the first phase (labeled A in Figure 9–3) which shows some remarkably stable results that here are merely a function of using perfectly stylized, pretend data. Remember, Figure 9–3 uses only stylized data so that you, the reader, can see clearly what is happening; I have actually never seen data from real observations of real behavioral events look as smooth as stylized data. I suppose that it could happen, theoretically…

Anyway, the first intervention phase (labeled B in Figure 9–3) shows results not only from continuing to record all three types of throwing behavior, but also from putting the pot throwing behavior under the time–out contingency. Note that you only changed the contingency on pot throwing; no contingency changed for sponge throwing or biscuit throwing. Consequently, you see the data showing a decrease in pot throwing but no changes in sponge throwing or biscuit throwing, both of which continue at high baseline levels.

The baseline for pot throwing is the original Baseline phase (labeled A in Figure 9–3). However, you compare the pot throwing intervention with the results from all phases to answer the control question.

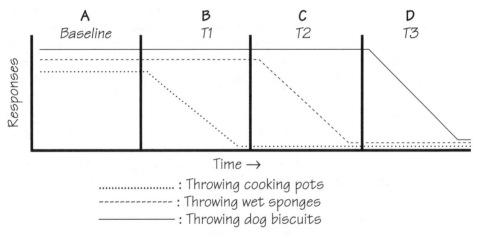

Figure 9–3. Stylized data across the four phases of a Behavior–Based Multiple–Baseline design.

Then, in the second intervention phase (labeled C in Figure 9–3) you again plot results from continuing to record all three types of throwing behavior but, as usual, you again only change *one* contingency, this time the contingency on sponge throwing. You change from the no–time–out contingency on sponge throwing to putting sponge throwing under the time–out contingency, while you *leave* pot throwing under the time–out contingency, as it was during the last phase, and you *leave* biscuit throwing under the no–time–out contingency, also as it was during the last phase. Again, you only *changed* the contingency on sponge throwing; remember, you only change one contingency per phase, and that one change defines a phase. Consequently, you see the data in c showing a decrease in sponge throwing but no change in pot throwing data (which remains at its decreased level from the continuation of the time–out contingency started in the last phase) and no change in biscuit throwing (which continues at its baseline level, because no contingencies for it have changed).

The original Baseline phase plus the first intervention phase (labeled A and B in Figure 9–3) comprise the baseline for sponge throwing. To answer the control question, though, you again compare the sponge throwing intervention (labeled c in Figure 9–3) with the results from all phases.

Finally, in the third intervention phase (labeled D in Figure 9–3) you once again plot results from continuing to record all three types of throwing behavior and, as usual, you only change *one* contingency. This time you change the contingency on biscuit throwing. You change from the no–time–out contingency on biscuit throwing to putting biscuit throwing under the time–out contingency, while you *leave* both pot throwing and sponge throwing under the time–out contingency, as they were during the last (i.e., c) phase. Consequently, you see the data in D showing a decrease in biscuit throwing but no change in either pot–throwing data or sponge–throwing data, both of which remain at a decreased level from the *continuation* of the time–out contingency already in place before this phase began.

The original Baseline phase (labeled A in Figure 9–3) plus the first *two* intervention phases (labeled B and c in Figure 9–3) comprise the baseline for biscuit throwing. And again, you compare the biscuit throwing intervention with the results from all phases to answer the control question.

Study Figure 9–3 carefully. You can clearly see how the data, from each phase and across all phases, combine to answer our control question: The time–out intervention affects each throwing behavior when (i.e., every time) the contingencies change, and only when the contingencies change. Such data from a multiple–baseline study, with any behaviors and any intervention, would confirm our confidence in the intervention as an effective technique worth considering for solving other problems.

Applied–Setting Multiple–Baseline Design Example

Before going into the other forms of the multiple–baseline design, we consider an example from an actual study using one of the multiple–baseline forms. The value of this example, which involves the Behavior–Based form, also resides in two of its other characteristics. One characteristic pertains to the experimenter's evoked exploratory responses that often prove necessary when working with clients in applied settings where other variables can throw you curves. The other characteristic pertains to a generalization effect wherein the intervention in early phases can have a helpful but methodologically confusing impact on the effectiveness of the intervention in later phases.

The primary experimenter for this study was Sheila Knight, another fourth–year honours student with whom I worked back in 1975 at the

University of Queensland. Sheila completed most of the work, and most of the original report writing, as part of a project for the same course on clinical interventions in which David Anderson was enrolled. And again, out of respect for her writing and the location of the study, the alternate spelling of behaviour appears throughout this example.

Sheila's practicum–site director assigned her to work with a young boy whom we will call Nigel (not his real name). He was a normal 4–year–old except that his past conditioning history had left him with a related set of behaviour problems, the combination of which made his parents feel that professional help was necessary. These problems involved the behaviours of inappropriate food seeking, tantrums, and refusing to comply with parental requests or demands (more politely described in the original study as "ignoring parental demands"). These problems collectively typified the "Brat Syndrome," a term that Sheila came across several times in her literature search.

In due course, Sheila and Nigel's mother agreed to try to manage these problem behaviours through the dual contingencies of reinforcing appropriate behaviour while also ignoring (i.e., extinguishing) or using time out for the inappropriate behaviours. They also managed two additional problems. One involved Nigel wetting or soiling his pants when outside the home, and the other involved his leaving home for hours at a time, a kind of "running away."

They were able to track the running away problem and, as the intervention phases of the study resolved other problems, Sheila and Nigel's mother also observed decreases in running away, both in the number of occurrences and in the length of time he was away per occurrence. Home was becoming a less coercive place and so home was evoking less of those primary outcomes of coercion and punishment that we call escape behaviour (getting away) and avoidance behaviour (staying away).

On the other hand, because the wetting and soiling problem was so immediate, serious and aversive, they were unable to take proper data on it. Pants wetting and soiling were so destroying the relationship between Nigel and his mother that Sheila took the clinically correct executive action of clearing up the problem immediately, because its fallout precluded progress on anything else. Usually the mother spent about ten minutes handling Nigel roughly as she cleaned him up right after he came in wet or soiled, all the while providing a continuous stream of adult attention for him by repeating to him all the reasons why he should not wet or soil his pants.

To change these counterproductive practices, Sheila tutored Nigel's mother in the basics of applying dual contingencies. She was to clean him up with a minimum of fuss, and even in silence. Also, she was to praise him whenever he came into the house relatively clean (i.e., not wet or soiled). In addition to that praise, she was to give him a chocolate biscuit whenever he used the toilet appropriately. (Some parts of the world know these biscuits as "cookies.") At first she expressed concern that Nigel would need chocolate biscuits indefinitely (and some of you readers are thinking, "Is that so bad?") but Sheila correctly reassured her that such an outcome, if it occurred, would be more a function of effective biscuit–company advertising than of biscuit use to strengthen proper toileting. Indeed, within ten days, wetting and soiling had reduced to zero while normal, age–appropriate praise maintained appropriate toilet use without further need for *contingent* chocolate biscuits. Now they could all proceed with the main study.

Sheila implemented a typical Behaviour–Based Multiple–Baseline, using time out for problem behaviour along with various kinds of reinforcement for appropriate behaviour as the main intervention. She trained Nigel's mother in both techniques, and in the use of a data record sheet. Together they defined the target behaviours in measurable ways. Nigel's mother reported that the distractions of two other children, along with some difficult marital stress, reduced the consistency of her data collection; when this led to no data for one of the target behaviours on a particular day, no data point got plotted on the data graph. However, Sheila also recorded the behaviours herself, during reliability checks that lasted at least an hour, first on a daily basis, and then four or five times each week, throughout the study; interobserver agreement between Shela and Nigel's mother was always around 95%. Figure 9–4, which mostly reproduces the figure from the original report (which met the standards of the time and place; see Knight & Ledoux, 1979) shows Nigel's problem behaviours changing across the four phases of the study.

The Baseline phase lasted for 14 days, with the data showing that the least problematic target behaviour was food seeking, next was ignored commands, and tantrums comprised the most problematic. Sheila recommended that they apply the intervention to the problems in that order across three treatment phases, because that would give Nigel's mother practice with easier problems before tackling harder ones. Before they could begin, however, they got to face a surprise circumstance, the kind that can arise at any time in an applied setting. In this case the sharp decrease in food seeking on days seven through nine was due to a fire that burned out the family kitchen.

The first Treatment phase (T1) lasted for 11 days (days 15 to 25). In this phase Nigel's mother again implemented dual contingencies, but only on the food–seeking problem; she merely continued to count the ignoring–commands problem and the tantrums. Also, she was diligent at anticipating Nigel's food needs; whenever possible she quickly fulfilled every reasonable food request that Nigel made. She also put inappropriate food seeking on time out. Under these conditions the food–seeking problem all but disappeared. The mother reported that the increase visible on days 21 and 22, along with the dramatic spikes in commands ignored and tantrum behaviour, resulted from a serious marital conflict that ocurred over those days.

The second Treatment phase (T2) spanned 12 days (days 26 to 37). When this phase began, Nigel's mother added, to her implementation of dual contingencies, (a) praise for compliance with her requests or commands, and

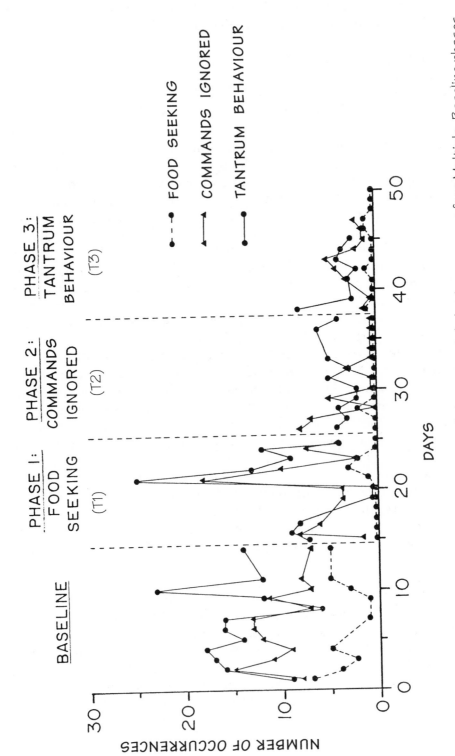

Figure 9–4. Number of occurrences per day of three target behaviours across four Multiple–Baseline phases.

(b) time out for non–compliance. She kept food seeking under time out, and merely continued to count tantrums.

Events did not unfold as serenely as that description implies. When working with a child client, the contingencies one must change often need to affect the parent or teacher more than the child. Sheila had noted in her observations that Nigel's mother was constantly nagging him about (i.e.,paying attention to, and in some cases even evoking) his inappropriate behaviour. So, without taking more formal data (which would have added a study to improve the mother's behaviour to her study to improve Nigel's behaviour) Sheila worked to alter the contingencies controlling nagging so that Nigel's mother could work more effectively on the contingencies to improve his behaviour. This included discussing, modeling, evoking, and reinforcing Nigel's mother's behaviour of giving a command only when she could and would follow it through with praise, or time out, as appropriate.

Apparently Sheila's efforts had the desired effect, because nagging reduced enough that pleasant interactions replaced the nagging, and the reinforcing effects of the mother's part in these interactions conditioned their continuation. The data show that the contingent praise and time out, and less nagging, had succeeded, in the first half of this phase, in producing a mild decrease in commands ignored. However, Sheila and Nigel's mother agreed that a greater reduction was desirable and possible. But what variables were available to make this happen?

Nigel's father had earlier pointed out how pleased Nigel had been when he received a paper star "good boy" badge at school (apparently a rather rare event). So Sheila prepared a bound booklet in which Nigel could paste paper stars, and his mother began giving him not only praise but a paper star when he obeyed one of her commands, with every ten stars earning a special treat (e.g., another chocolate biscuit; he really likes those chocolate biscuits).

Did the stars book help? Nigel demonstrated his enthusiasms for his stars book by eagerly bringing it to the attention of all visitors. And the data for the remainder of this phase showed his "ignoring of commands" dropping to zero.

Now, you would be correct to worry that "ignoring commands" is a non behaviour. But you need not worry, as it was simply the inverse of counting instances of command compliance. At the start of the study, Nigel's mother was far more sensitive to occasions when he failed to comply with her commands than when he complied, which she tended—counterproductively—to ignore, and this made counting instances of "non compliance" easier. But the study designed the intervention contingencies to work on compliance behaviours.

Furthermore, Nigel's ignoring was usually quite active, as a command often evoked a range of obnoxious, refusing–to–comply responses, including hitting and spitting and tantruming. As the new contingencies of this phase exerted control, the reduction of his ignoring commands also reduced all these other problematic responses, including tantrums, which the data showed as gradually decreasing across all treatment phases. His mother also noted that the intensity of his tantrums was decreasing. But tantrums were the focus of the next Treatment phase, so let's look at that phase.

The third, and last, Treatment phase (T3) lasted for 13 days (days 38 to 50). In this phase, because the reduction of evocative stimuli for tantrums was reducing the occurrence of tantrums already, Sheila first instructed Nigel's mother to ignore tantrums. When this proved insufficient to eliminate them, Sheila instructed his mother to add time out for tantrums. She was, of course, to continue her implementation of the dual contingencies for food seeking and command ignoring.

The data for this phase showed that, while the food–seeking rate remained near zero, and tantrums were at a low but not yet zero rate, ignoring commands rose to five on day 43 before falling again. Several interrelated events were producing these results: Nigel's mother had to go to New Zealand due to a death in the family, leaving his father to provide contingency management and data recording, tasks in which he was relatively inexperienced. Under these circumstances, the data may not exactly represent what actually happened. Such circumstances more than adequately exemplify the surprise environmental changes that can occur when working in applied settings. You might think getting any results at all is some sort of miracle; I think, however, that if you give it another thought, you will agree that good results *will* happen, because behaviour is lawful and functionally related to the environmental events that the study manages. Nigel's tantrums provide a good example of how this works.

Due to the relationships between Nigel's tantrums and the other variables we worked with over the first two treatment phases, tantrum behaviours mostly decreased across the entire study. This may partly be due to fewer stimuli evoking tantrums as many common evocatives for tantrum behaviour dropped out when contingency changes reduced other problems.

In summary, in spite of come confusing difficulties, the contingency changes in the interventions in this study succeeded in changing the target behaviours—including the extra behaviours of pants wetting or soiling, and running away—to much more manageable levels in a period of about 50 days. A month after the study ended, at a follow–up meeting just before the academic term ended, Nigel's mother reported that she had only used time out three or four times over the previous 30 days (with no pants wetting or soiling, or running away, occurring either). May all your attempts to help people go at least this well.

Other Multiple–Baseline Forms: Clients, Settings, or Techniques

The Behavior–Based Multiple Baseline is only one of several forms of the multiple–baseline design. Other forms include the Client–Based Multiple Baseline, the Setting–Based Multiple Baseline, and the Technique–Based Multiple Baseline. Instead of multiple behavior problems with one client, you might have three clients who each present the same problem—or, more commonly, each has a different problem. Or, instead of multiple clients, perhaps you have only one client who has only one problem, but that problem happens at home, at school, and at grandma's house. Or, perhaps you have some experimental support for each of several interventions, but your applied field and your clients would benefit if more definitive evidence was available about which was the best practice. You might even face a situation in which many of these elements appeared together (e.g., three clients, each with two problems—not always the same—that occur in a total of four settings, with two intervention techniques available but for both of which the available evidence is weak). Fear not, because in cases like these, one or another—or some combination—of all these multiple–baseline forms would be appropriate.

Let's consider each of those other three multiple–baseline forms in turn. We use Figure 9–5 to exemplify the first two of them, because our stylized data lines remain the same for both our Client–Based, and Setting–Based, Multiple Baselines. The third design, the Technique–Based Multiple Baseline, is a research design beyond the introductory level of this book, so we only touch on it briefly. The data and labels we see in Figure 9–5 replicate the data and labels in Figure 9–3, because we are going to represent the same kinds of events happening in Figure 9–5 as in Figure 9–3. However, for Figure 9–5, we provide a different legend for the data lines. Also, to keep these examples simple, I will again have you using the time–out technique.

The Client–Based Multiple–Baseline design. Beyond the Behavior–Based Multiple Baseline, another form of multiple–baseline design is the *Client–Based* Multiple Baseline. We say "Client–Based," because it is a multiple–baseline form for managing multiple clients in a methodologically sound manner, especially when they present similar problems. Let's say that you have three clients, whom we will call Mary, Billy, and Sue. They all attend the same school but in different classrooms, and each of them presents a similar problem. Each of them occasionally tears pages from books, with Mary tearing out the most pages and Sue tearing out the fewest.

The steps you would take to deal with these clients are nearly the same steps that you took to manage Laddie's multiple throwing behaviors. So, after you take and plot baseline data (the A phase in Figure 9–5) you would put Mary's, and only Mary's, page tearing under time out for the first intervention phase (the B phase in Figure 9–5). Next, you would put Billy's page tearing under time out (the C phase in Figure 9–5) without changing the contingencies for Mary (who is already under the time–out contingency) or Sue (who is still under baseline conditions). Finally, you would put Sue's page tearing under time out (the D phase in Figure 9–5) without changing the contingencies for Mary or Billy (who are both already under the time–out contingency).

Notice that as usual, for each phase, all previous phases serve as baseline, and you visually compare the data for any particular phase with all other phases. Also notice what the data tell you: Your intervention technique was clearly the functional independent variable, because the behavior changed when, and only when, the contingencies changed, and a review of the extensive disciplinary literature will show that these stylized data nevertheless convey the general kind of changes that you will see in the data of real applied research studies (e.g., see the pages of JABA, the *Journal of Applied Behavior Analysis,* which was founded in 1968 with that name, a name that can denote one of many areas of applied behaviorology, although some still take this name as denoting part of "behavior analysis," which is a label that psychology now confusingly controls; recall that behaviorology is not and never was any kind of psychology).

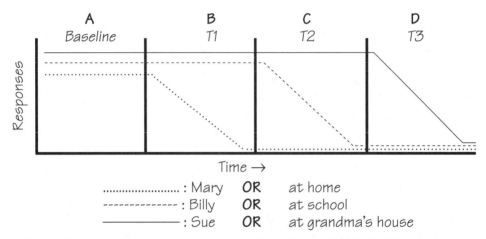

Figure 9–5. Stylized data across the four phases of **EITHER** a Client–Based, **OR** a Setting–Based, Multiple–Baseline design.

The Setting–Based Multiple–Baseline design. Yet another form of multiple–baseline design is the *Setting–Based* Multiple Baseline. We say "Setting–Based," because it is a multiple–baseline form for managing problems that occur across multiple settings in a methodologically sound manner. Let's say that you have only one client, whom we will call Raymond, who has one problem for you to manage, but this problem occurs at home, at school, and at grandma's house. Raymond's problem is that he also tears pages.

The steps you would take to deal with Raymond's problem are nearly the same steps that you took to manage Laddie's multiple throwing behaviors, and the steps you took to deal with Mary's and Bill's and Sue's page tearing (and this is *not* the case because I conveniently structured these examples this way, but because these are the basic steps that one extracts from the research literature). The difference here is that, instead of changing contingencies on each of three behaviors one at a time, or changing contingencies on each of three clients one at a time, here you will be changing the contingencies in each of three different settings in which this problem occurs, one setting at a time. So, after you take and plot page–tearing baseline data for all three settings (the A phase in, again, Figure 9–5) you would put page tearing at home under time out for the first intervention phase (the B phase in Figure 9–5) as the *only* change for this phase. Next, when appropriate, you would put page tearing at school under time out (the C phase in Figure 9–5) without changing the contingencies at home (already under time out) or at grandma's house (still under baseline conditions). And finally, again when appropriate, you would put page tearing at grandma's house under time out (the D phase in Figure 9–5) without changing the contingencies at home or at school (both already under time out).

Notice again that, for each phase, all previous phases serve as baseline. And again notice what the data tell you when you visually compare the data for any particular phase with all other phases: Your intervention technique provides the functional independent variable, because the behavior changed when, and only when, the contingencies changed.

The Technique–Based Multiple–Baseline design. Finally, another form of multiple–baseline design is the *Technique–Based* Multiple Baseline. Sometimes you will see another name for this form, which is "Multi–Element Design." This is a more complicated and specialized research design for evaluating the parameters of one intervention technique, or even for evaluating a set of separate techniques, in a methodologically sound manner. Let's say that the disciplinary literature provides only a little experimental support for each of three particular interventions relevant to your work, and your task is to provide more definitive evidence about which of them, if any, works better than the others. Or perhaps you can implement a particular intervention at several different intensity levels, and your task is to discover any functional differences among the levels.

The steps you could take would be quite similar to the steps in the other examples. Basically—leaving the details for you to consider (as an exercise in managing a real range of realistic complexities)—you would implement only one of several interventions at a time, or only one of several parametric levels of a single intervention at a time. Each change of intervention or parametric level would correspond to a change to the next phase. As usual, when comparing the data, for each phase, all previous phases serve as baseline, and you visually compare the data for any particular phase with all other phases. As long as the behavior of concern changes when, and only when, the variables change, then the levels of the data will answer your research questions.

Practical Methods Summary and Chapter Conclusion

Our coverage of practical methodology focused on single–subject experimental designs that help us develop and test interventions for solving both simple and complicated problems in a range of applied settings outside the laboratory. However, these non–laboratory settings lack the more thorough access that the laboratory provides for controlling potential independent variables while, at the same time, these inherently more complex settings inevitably contain more of these variables. Thus we, and our predecessors, had to develop methodologies appropriate to these settings, methodologies that would allow realistic conclusions regarding the safety and efficacy of interventions in these settings, in spite of the relative lack of access, in these non–laboratory settings, to extraneous independent variables.

To address all such methodological difficulties, we adapted single–subject experimental designs, beginning with the ABAB single–subject design most common in the laboratory, and going on to develop various multiple–baseline single–subject designs. All these designs contain the four pertinent and intertwined characteristics that repetition helps condition: (1) The design phases represent periods of time during which different contingencies operate. (2) Each phase continues until the data show some stability so that the changes in behavior that we see, in our visual comparisons of the data across all phases, help evoke appropriate experimenter behavior (e.g., altering the contingencies to begin the next phase). (3) Across these phase comparisons, each subject serves as his or her or its own experimental control. And (4) we gain adequate confidence in our conclusions from that experimental control when the data comparisons across the phases answer our control question regarding the behavior basically changing when, and only when, the contingencies change.

The development of multiple–baseline designs maintained those design characteristics. In these designs all previous phases serve as baseline for any particular phase. That is, each phase becomes part of a growing baseline for every next phase. And we compare any particular phase with all other phases when we are answering our control question. As a result of these single–subject design methodologies, access to effective interventions in applied settings has greatly expanded beyond the laboratory.

Some Details on the Time–Out Intervention

Before ending this chapter, consider a little more recent information on time out. Back in the late 1970s when our examples of the therapeutic use of time out occurred, the validating research, while extensive, had not examined all the parameters; a data summarizing rule–of–thumb still applied to the length of stay in time out. That rule was "one minute per year of age, up to a maximum of five minutes." Over time, research showed that less and less time, in time out, produced therapeutic results; so today the time–out time "rule" is closer to 30 seconds to one minute *total* time, regardless of age, so long as leaving the time–out setting occurs during appropriate behavior (as leaving during tantrum behavior reinforces the tantrum behavior and so only leads to more, and worse, tantrums).

Chapter Conclusion

Let's end this chapter with our ongoing activity. Keep considering what roles these research methodologies, and those in the last chapter, play in the application of pertinent behaviorological principles and concepts to managing the behavior components of the solutions to global problems. These multiple–baseline designs should prove quite helpful in studies to test possible practices in sustainable living research and other behavioral components of solutions to global problems. This will help make possible the timely solutions to the non–behavior–related components that traditional natural scientists are addressing.✤

References (with some annotations)

Anderson, D. & Ledoux, S. (1979). Parental management of time out and attention in the control of brat–syndrome behaviour. In S. Ledoux (Ed.). *Behaviourism and Mind and Much More* (9–15). Churchill, Victoria, Australia: Gippsland Institute of Advanced Education. In addition to this paper, plus the Knight and Ledoux (1979) paper, and eight other student–authored papers also reporting applied–setting research, this book also contained several professionally authored papers. These included a paper that I wrote for presentation as an invited address to the members of *N'est–ce pas?* (of the College of Nursing, Melbourne, Victoria, Australia) at their June 1977 meeting. My paper was not great, but it did contain my first published use of the term behaviorology, which I had independently coined, in Australia, at about the same time, as I discovered later, that the *Los Horcones* community had coined it in Mexico, and that Ernie Vargas and Julie Vargas and Lawrence Fraley—all at West Virginia University in Morgantown—had coined it in the U.S.A. This was an outcome of our all being under similar contingencies operating around the globe. In 2013, for my behaviorology–major students, I assembled the other eight student–authored papers in this book, along with this Anderson and Ledoux (1979) paper, and the Knight and Ledoux (1979) paper, into a separate volume entitled *Behaviorology Majors Make a Difference.*

Johnston, J. M. & Pennypacker, H. S. (1980). *Strategies and Tactics for Human Behavioral Research.* Hillsdale, NJ: Erlbaum.

Knight, S. & Ledoux, S. (1979). The brat–syndrome and equally fascinating factors. In S. Ledoux (Ed.). *Behaviourism and Mind and Much More* (17–29). Churchill, Victoria, Australia: Gippsland Institute of Advanced Education. See the note in the reference for the Anderson and Ledoux (1979) paper for additional information.

Latham, G. I. (1994). *The Power of Positive Parenting.* Logan, UT: P & T ink.

Ledoux, S. F. (2012). Behaviorism at 100. *American Scientist, 100* (1), 60–65. The unabridged, peer–reviewed version of this paper appeared (2013) under the title, "Behaviorism at 100 unabridged," in *Behaviorology Today, 15* (1), 3–22.

Sidman, M. (1960). *Tactics of Scientific Research.* New York: Basic Books. Authors Cooperative, in Boston MA, republished this book in 1988.

Skinner, B. F. (1957). The experimental analysis of behavior. *American Scientist,* 45 (4), 343–371. In the January 2012 issue (volume 100, number 1) the editor of *American Scientist* repeated much of this article as the first "Centennial Classic" of the journal's centenary year, setting the stage for my update article (Ledoux, 2012) that followed it, and later became Chapter 1.᷈

Chapter 10
Postcedent Processes that Change Behavior

$\mathcal{T}$he discussion of laboratory equipment and methodology in Chapter 8 led directly to our Chapter 9 coverage of methodology in applied settings. We emphasized some details about each of two common applied research designs, the ABAB Reversal design and several types of Multiple–Baseline designs. While we used stylized data to show how these designs basically work, we also got to follow some typical developments in a couple of actual interventions. These interventions, which concerned some behavior problems that stemmed from unhelpful contingencies, employed some postcedent processes that led to changed behaviors. In these examples we followed the actual data documenting the changes in behavior that the interventions produced.

Now, in this chapter, we consider some further details about three of the fundamental postcedent processes that change behavior. These are reinforcement, extinction, and punishment, although we introduce some antecedent factors—establishing and abolishing operations—as well. However, you would be raising a very reasonable question if, given that our behavior formula concerns *antecedents—behaviors—postcedents,* you wanted to know more about *why* we continue to cover postcedent independent variables first.

Today's behaviorological scientists tend to talk about postcedent variables first, and often in more detail than antecedent variables, as a result of two factors that we mentioned in Chapter 5. The one we treated more thoroughly there concerned the difference in accessibility between the energy traces involved in antecedent and postcedent variables. The other, which we treat more thoroughly here, concerns some particular historical circumstances that led to many discoveries about postcedents before researchers turned more focused attention on antecedents. At present the research balance between antecedents and postcedents shows a rough equality.

If these emphases are in balance now, why were they out of balance early in the history of the science? In broad brush strokes, the basic components of this science that we still recognize today were present as early as Skinner's 1938 book, *The Behavior of Organisms.* These consisted of the variables involved in respondent behavior and conditioning, and in operant behavior and conditioning. However, after Pavlov discovered respondent conditioning (i.e., after the contingencies affecting Pavlov led to the first comprehensive descriptions of respondent phenomena, which we call his discovery of them) lots of researchers looked for all the eliciting stimuli that they thought must be responsible for every human behavior. But they could not find eliciting stimuli for a broad range of common and important behaviors, because these behaviors turned out to be operant behaviors, not respondent behaviors. Skinner's discovery of operant conditioning (i.e., the contingencies affecting Skinner that led to the first comprehensive descriptions of operant phenomena, which we call his discovery of them) filled in this explanatory gap.

The contingencies pushing the need to distinguish clearly between operants and respondents, however, evoked Skinner's behavior of emphasizing the role of the postcedent events that he termed reinforcing consequences, because such consequences were of little relevance to respondent behaviors; this emphasis maximized the differences between the two types of conditioning, enabling the distinctions between them to stand out. Skinner and others then researched the types and parameters of such postcedent consequences. While they recognized early on the roles of antecedents, they got around to studying the antecedents in depth more gradually than the postcedents, and they trained their students first in the details of postcedents and then in the details of antecedents.

That sequence has become something of a tradition; for example in both *Science and Human Behavior* (Skinner, 1953) and *The Analysis of Behavior* (Holland & Skinner, 1961) the coverage of postcedent variables precedes the coverage of antecedent variables. This need not have been the case, even considering that *The Analysis of Behavior* is a programmed version of Part II of *Science and Human Behavior*. And most other books and texts also follow the same postcedents–before–antecedents sequence. Since in this book we break new ground in other ways, we have little need to try breaking that pedagogical sequence as well.

Actually, antecedent stimuli turn out to be at least as important as postcedent stimuli. Indeed, we will even find some behaviorologists whose laboratory conditioning compels treating antecedents as *more* important than postcedents. We discuss this in the later chapter on stimulus control. Here, however, we respect the traditional sequence and attend more to postcedent functional variables first. In that regard let's elaborate some details about the postcedent processes of reinforcement, extinction, and punishment.

More about Reinforcement

To put our elaboration of reinforcement into context, we first review the three criteria that determine our labels for various postcedent types. These include not only added and subtracted types, but also unconditioned and conditioned types, of both reinforcers and punishers (although we cover punishers in more detail later). We cover these three criteria in the form of three questions we ask about any particular functional postcedent. The answers to these questions determine the label we use for that particular type of postcedent.

After we look at our three criteria questions as they relate to reinforcement, we mention other types of reinforcers. Then we move on to extinction. Meanwhile, recall that, while we sometimes simplify phrasing by using verbal shortcuts that allude to stimuli reinforcing or punishing people, technically reinforcers and punishers only affect behaviors.

The Three Questions

Of the three questions, the first is always about the behavior of concern while the other two are about the postcedent stimulus. The answer to each question reduces the number of postcedent labels that can apply.

About the change in the behavior. *Does the behavior increase or decrease* (e.g., in rate)? In this context, *increase* includes "maintain," because a postcedent might be functioning merely to increase the rate against the decrease in rate that can accrue merely due to normal, ongoing physiological processes that degrade the physiological structures that mediate the response. We see the result of such an increase balancing that kind of decrease as *maintenance* (i.e., the continuation of a stable rate for that kind of response).

When the occurrence of a postcedent variable results in a decrease in the rate of behavior, we label that postcedent as one or another type of *punisher* (and we will soon consider these types in the section on punishment). On the other hand, when a postcedent variable results in an increase/maintenance of the behavior, we label that postcedent as one or another type of *reinforcer.* But which type? The remaining two questions address the criteria that determine the particular reinforcer type.

About the change in the stimuli. Recall that stimuli are energy changes at receptor cells. Thus we must ask about the type of energy change at those cells. *Does the stimulus energy change involve an increment of energy or a decrement of energy?* An energy increment occurs when a stimulus begins, or when some source presents the stimulus; in this case, and when we already define the energy change as a reinforcer (from our first question) we describe the stimulus as an *added* reinforcer. On the other hand, an energy decrement occurs when the stimulus ends, as when experimental equipment terminates or withdraws the stimulus; in this case, and when we already define the energy change as a reinforcer, we describe the stimulus as a *subtracted* reinforcer. Note that this question shows that *what the stimulus is* (e.g., food, attention, music, money, electric shock, intense noise) is less important than *what happens to the stimulus* (e.g., start or stop). We use *added* and *subtracted* the same way for punishers.

About the history of the stimuli. Our third and last question addresses the historical aspect of stimuli. *Does the reinforcing effect,* from the occurrence of the stimulus (i.e., from its addition or subtraction) *depend on some prior conditioning?* If not (i.e., if no prior conditioning was relevant) then the reinforcing effect is occurring through neural structures (i.e., the structures that are mediating the responses) that genes produced, and we label the reinforcer an **un***conditioned* reinforcer. However, if some prior conditioning (i.e., some earlier *pairing* of this stimulus with other reinforcers, pairing that altered the gene–produced structures) made this stimulus effective as a reinforcer, then we label the reinforcer a *conditioned* reinforcer. As with added and subtracted, unconditioned and conditioned also apply to punishers in the same ways that they apply to reinforcers.

Some postcedent combinations. Note that the various answers to the two questions about stimuli combine to give multi–term labels to postcedents. The possibilities are "added unconditioned," "subtracted unconditioned," "added conditioned," and "subtracted conditioned," each of which can refer to a reinforcer or a punisher (i.e., to reinforcing or punishing stimuli).

A fourth helpful question. We could ask an additional question; the answer to it would not change our *basic* postcedent labeling practices, but the answer would further clarify the status of any particular postcedent. *Does a response produce the functional postcedent that follows it?* While the kind of postcedent, labelled according to the answers to our first three questions, is irrelevant to this fourth question, the answer makes a difference. If the response actually produces the postcedent (which is the more common scenario) then the postcedent falls into the general postcedent class that we call *consequences.* However, if some other chain of functional events produces the postcedent rather than the chain that includes the response it follows, then the postcedent falls into the general postcedent class that we call *coincidental selectors,* a class which can contain both reinforcers and punishers. Coincidental reinforcers produce *superstitious behavior,* and while we described them in Chapter 5, we consider superstitious behavior itself in a later chapter.

Other Types of Reinforcers

Various basic and applied researchers have at times used a number of other terms to describe some aspect of reinforcers, often as a counterpoint to some other reinforcer characteristic. While several of these terms pertain to the topic of a later chapter, we should consider a couple of them here (noting that some folks use these terms to describe punishers as well).

Physical/tangible versus social/verbal reinforcers. We sometimes use the terms *physical* or *tangible* to refer to reinforcers that have more than a fleeting presence. For instance a bag of cookies, a bottle of wine, a beautiful painting, even a 50–dollar bill all have some continuing presence (unless your 50–dollar bill is paying for petrol) and they often leave some sort of record of their occurrence; for example the painting endures and cookies leave crumbs. In contrast a mother's attention, a teacher's praise, a peer group's interactions, or a colleague's compliment all have a fleeting presence; they occur quickly and just as quickly are history, often without leaving any record of their occurrence. And yet these fleeting postcedents can have as potent effects on behavior as their more concrete cousins. In any case we sometimes use the terms *social* or *verbal* when describing these fleeting reinforcers. For both the enduring and fleeting reinforcers, the usual technical terms apply as well (i.e., added or subtracted and unconditioned or conditioned) and circumstances may demand their use before, or along with, using any of these less technical terms.

Coincidental reinforcers. In addition to appropriately attaching any or many of these "type" terms to a reinforcer, we also label the reinforcer as *coincidental* when the preceding, reinforced response did not produce the reinforcer. Some other natural process coincidentally led to the occurrence of the reinforcer at just that time, but the response it followed did not produce it. It happened coincidentally, hence the label *coincidental reinforcer.* Furthermore, we use the term *superstitious behavior* for the behavior that such coincidental reinforcers condition. (We will revisit coincidental reinforcers when we consider superstitious behavior more closely.)

Generalized (conditioned) reinforcers. Recall that an otherwise neutral stimulus (with respect to the behavior of concern) becomes a conditioned reinforcer when it is regularly present when another reinforcer—usually an unconditioned reinforcer—occurs. This pairing, this conditioning, makes the previously neutral stimulus effective as a reinforcer (of the *conditioned* type).

That effectiveness, however, relies on the effectiveness of the unconditioned reinforcer; as long as the unconditioned reinforcer functions effectively, then the conditioned reinforcer also functions effectively. If some process, such as deprivation (of the unconditioned reinforcer) increases unconditioned–reinforcer effectiveness, then conditioned–reinforcer effectiveness increases also. Similarly, if satiation (of the unconditioned reinforcer) reduces unconditioned–reinforcer effectiveness, then conditioned–reinforcer effectiveness suffers also. Conditioned–reinforcer effectiveness relies on the deprivation *level* of the unconditioned reinforcer with which it shared some pairing.

For example, if you usually pair small (i.e., pea–sized) pieces of dog biscuit (which are unconditioned reinforcers) with the *clicks* of your "clicker"—what children used to call a hand cricket—(which thereby become conditioned reinforcers) in your dog–tricks training sessions, then your sessions will be long and productive. (Such "Clicker Training" currently involves the most effective set of practices for training other animals; see O'Heare, 2010, and Pryor, 1999a, 1999b, 2001; also see Johnson, 2004, and Kurland, 2000.) However, if you run out of dog biscuits, and instead you pair cubic–inch chunks of steak with the *clicks* of your hand cricket in a training session, then that session will not be very long or very productive, because your dog will rather quickly become food satiated. As a result the clicks will lose their reinforcing function, and your dog will go bed down and sleep off that great steak meal.

One way to get around the problem of conditioned reinforcers losing effectiveness (e.g., due to the occurrence of unconditioned–reinforcer satiation) is to pair the conditioned reinforcer with several, even many, other reinforcers, so that one or another of those other reinforcers is likely always to be effective due to some functional level of deprivation. Hence the conditioned reinforcer also remains effective. In this circumstance we call the conditioned reinforcer a *generalized* reinforcer. (Technically, it is a *generalized conditioned* reinforcer, but by convention we leave the middle term out.)

As an example of a generalized reinforcer, think of a stimulus, in this case a tangible stimulus, that our culture essentially pairs with just about everything, at one time or another, legal and illegal, so much so that this stimulus has earned a designation as "the root of all evil." What could that be? If you saw dollar signs ($$$) you are right on the money. Also, we treat the stimuli that serve as the "tokens" in a token economy the same way. We pair the tokens with many other reinforcers, making the tokens function as generalized reinforcers. These could be buttons, poker chips, stars, check marks, tickets, or any number of other items.

Establishing and Abolishing Operations

FLASH! We interrupt this postcedent–oriented material to bring you a flash bulletin about an important connection between antecedent and postcedent functional processes. The relation of conditioned–reinforcer effectiveness to the deprivation or satiation level of the relevant unconditioned reinforcer evokes the necessity of some commentary about the broader antecedent processes (which can also be procedures) that we call *establishing operations* and *abolishing operations* (see Michael, 1982). Traditionally, we described these processes as changing both (a) the momentary effectiveness of a reinforcer and (b) the momentary likelihood of any behavior that in the past that reinforcer followed.

However, since the effectiveness of an evocative stimulus actually determines the latter (i.e., the likelihood of any behavior that the reinforcer followed) we must restate what these processes change. They change both (a) the momentary effectiveness of a reinforcer and (b) the momentary effectiveness of any stimulus that in the past evoked the behavior that the reinforcer followed. Let's apply this to establishing operations and abolishing operations more specifically.

We use the term *establishing operation* (EO) for something (a) that momentarily *raises* (establishes) the effectiveness of a reinforcer, and (b) that momentarily *raises* (establishes) the effectiveness of any stimulus that in the past evoked the behavior that the reinforcer followed. *Deprivation* is an example of an establishing operation. We have a deprivation example coming up that involves water (and also relates to extinction).

We use the term *abolishing operation* (AO) for something (a) that momentarily *lowers* (abolishes) the effectiveness of a reinforcer, and (b) that momentarily *lowers* (abolishes) the effectiveness of any stimulus that in the past evoked the behavior that the reinforcer follows. *Satiation* is an example of an abolishing operation. You should take the water–deprivation example that is coming up, and turn it into a water–satiation example; just answer this question: What would instead happen in that example when the child got enough water (i.e., became water satiated)?

Now, understanding that connection between the effectiveness of conditioned reinforcement and the antecedent functional processes of establishing operations and abolishing operations like deprivation and satiation, we return to postcedent processes. What if the reinforcers that have usually followed a behavior stop occurring? What if *no* reinforcers ever again follow the behavior? Then we enter the realm of *extinction.*

More about Extinction

While we mentioned extinction in earlier chapters, we cover several additional aspects of extinction here. Beyond writing extinction contingencies, these include comparisons of the extinction process with the process of forgetting and the procedure of preclusion/prevention. We also consider the role of extinction in intermittent reinforcement schedules. Since both extinction and punishment decrease behavior, we discuss their connections more thoroughly when we move on to punishment after dealing with extinction here.

One type of extinction is *respondent extinction,* and we use this full two–word term, "respondent extinction," when we tact this kind of extinction. (Remember that a tact is a verbal operant that specifies the stimulus that evokes it, which can be anything—concrete or abstract, still or moving, discrete or continuous, and so on—in the internal or external environments; the process of respondent extinction fits that qualification, and so that process itself evokes the "respondent extinction" tact for it.) While respondent extinction is not our focus at this point, it is the kind of extinction in which a respondent conditioned stimulus becomes unable to elicit the conditioned response—and so returns to the status of a neutral stimulus with respect to that response—because it occurs too often without the unconditioned stimulus ever again occurring with it.

However, the type of extinction we focus on here is operant *extinction,* a process which we simply tact with the single term "extinction." This kind of extinction is one of the operant environmental–change processes that brings about the reduction of—and ultimately, if the process continues, the cessation of—the behavior of concern. In this kind of extinction, the reinforcers cease; that is, the reinforcers that have at least occasionally been accompanying or following occurrences of the behavior of concern cease. They no longer accompany or follow occurrences of the behavior of concern and, as a result, a decrease occurs in the rate or relative frequency of the behavior, a decrease that becomes the cessation of the behavior while extinction continues. To connect antecedent and postcedent processes again, without any reinforcers occurring, the effectiveness of the evocative stimulus diminishes, and ultimately it no longer evokes the behavior of concern, leaving it extinguished.

Here is an example of an experience with the extinction process that a student of mine had. Meghan Curry, who had already completed several behaviorology courses, reported this experience in class in 2012. She was at a park one day helping a mother with a rather precocious toddler around two years old. They had been there for a couple of hours and the child was thirsty. That is, as an establishing operation, a couple of hours of water deprivation had produced two related effects. The water deprivation (a) had momentarily raised the effectiveness of water as a reinforcer, and (b) had momentarily raised the effectiveness of any stimulus that evokes any drink–procuring behavior (which, in essence, momentarily raised the likelihood of any behavior that in the past had produced the water reinforcer).

In any case, Meghan was not surprised when, as they passed a water fountain, the child manded a drink saying, "I want some water." (Recall that a mand is a verbal operant under the control of deprivation or aversive stimulation, and the mand statement specifies the relevant reinforcer, in this case water.) However, besides providing water, the water fountain also functioned in a non–standard way; any sideways pressure on the spout would swivel it, sending the stream of water in another direction. Now the child could not both reach the flow–control button and drink (or swivel) at the same time, so Meghan helped him by pushing the button for him. When the water started, though, he would start to swivel the spout toward her; a quick and successful swivel would get her wet. We can hardly resist the implication that getting her wet would be a more potent reinforcer of his behavior than drinking the water, at least for the moment. Perhaps, at some point in the past, getting water on a helper got paired with other reinforcers thus making it a conditioned reinforcer; we seldom have the kind of detailed access to past events that would enable us to specify when and how any such conditioning had occurred. Meghan, however, always released pressure on the button when he started to swivel the spout, which stopped the water flow before she got wet (although water from the first swivelling response came close, perhaps even landing a little splash on her). For lack of any further reinforcements, however, after just three such unsuccessful swivelling attempts, the swivelling extinguished and the child simply drank the water coming from the spout.

All this gets even better. Having observed this sequence, the child's mother asked Meghan how she got him to stop swivelling the spout at her so quickly and just drink. Meghan's immediate reply was "What?..." Her button–off responses had been automatic, that is, under *direct stimulus control* (a topic for a later chapter) of the overt components of the situation. The mother's later inquiry evoked some covert observing responses repeating her timely button releases, and these responses chained to some overt responses reporting those timely button releases to the mother (e.g., saying that she simply released the flow–controlling button each time he began to turn the spout toward her); we will cover such processes in more detail in the chapter on consciousness. On the other hand, the child had been responding both overtly and covertly during the whole episode. While the decrease in his overt swivelling behavior was observable to everyone there, his covert responses—while always public and observable to him—became available to others when he reported, as the swivelling extinguished, that "Oh, when I move the water thing, it [the water] stops." (Precocious? No wonder the mother appreciated Meghan's help.) How would we write contingencies for this example?

Writing Extinction Contingencies

Due to the difference between the *process* of extinction and the *result* of extinction, writing an extinction contingency takes two parts. The first part shows the process while the second part is a sort of ghost version of the first part that indicates the outcome of the process. Let's use our water–spout–swivelling example to demonstrate how to write these two parts.

We first write the reinforcement contingency, but we include in it the change that indicates the ending of reinforcement, which begins the extinction process. Then we write the ghost version of the first contingency showing the result of the extinction process. Here is the "end of reinforcement" contingency under extinction:

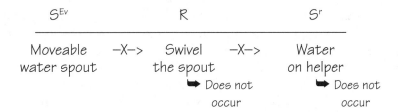

And here is the resulting extinction contingency:

S^{Ev}		R		S^r
Moveable water spout	–X–>	Swivel the spout	–X–>	Water on helper
		➥ Does not occur		➥ Does not occur

The "–X–>" indicates the loss of functional relations in both of the two–term contingencies that make up this three–term contingency of extinction. Together they show us that the moveable water spout loses its evocative function, because no reinforcer occurs in its presence if the response occurs; as a result, the behavior decreases until it no longer occurs. Here are some other important considerations about extinction.

Extinction Versus Forgetting

The extinction and forgetting *processes* differ in an important way. In the extinction process, stimuli initially continue to evoke the previously reinforced behavior, which thus continues to occur; but no further reinforcers follow those occurrences, thereby weakening the evocative function of the antecedent stimuli, with the result that they fail to evoke the behavior, which thus decreases in rate and finally ceases to occur. We saw this happen in our swivelling–water–spout example. However, in the *forgetting* process, *no stimuli evoke the previously reinforced behavior* and so, while no further reinforcers occur because the behavior fails to occur (although they would if it did) the neural structures that otherwise mediate the behavior gradually undergo normal and cumulative physiological changes that degrade the evocable mediation of that behavior until finally the stimuli cannot evoke the behavior, which then cannot occur. For example, if educational contingencies conditioned you to recite a particular poem by rote in high school, and since then essentially no stimuli have ever evoked another recitation of that poem, then the stimulus of this sentence now is unlikely to succeed in evoking the complete rote recitation; it may even prove insufficient to evoke the name of such a poem. The recitation response has become unavailable due to the normal and cumulative physiological changes that degrade the evocable mediation of that behavior.

Extinction Versus Preclusion/Prevention

While forgetting is a *process,* and extinction can be either a process or a *procedure,* preclusion is an intervention procedure that we can apply to help solve some types of behavior problems. We sometimes also call this procedure *prevention* (but I prefer not to use "prevention," because this term can apply to a range of procedures, and that kind of looseness confounds terminology usage). Basically, preclusion takes advantage of the forgetting process. In *preclusion* the practitioner makes environmental changes that preclude any stimuli from evoking the previously reinforced behavior of concern and so the neural structures that otherwise mediate the behavior gradually undergo normal and cumulative physiological changes that degrade the evocable mediation of that behavior until finally the behavior cannot occur (although the use of the procedure seldom continues for this long). Why use this procedure? Because the practitioner may lack access to the usual reinforcers of the behavior of concern, and so may not be able to change the environment in ways that prevent those reinforcers from occurring; hence the practitioner could not employ the extinction process as a procedure. However, with the preclusion procedure, the practitioner need not stop the reinforcers that usually follow the behavior of concern, because the preclusion procedure keeps the behavior from occurring and so producing those reinforcers. Remember, *in extinction stimuli evoke the behavior* (and it occurs) at least initially, but it produces no reinforcers; however, *in preclusion the behavior does not occur,* because no stimuli evoke it (so of course no reinforcers follow it).

For example, let's say that the behavior of concern is a high school sophomore's incessant tossing of crumpled paper sheets in high arcs toward the corner waste basket while the teacher is writing lesson points on the board. Of course, many coercive techniques are available to manage this kind of problem, and some teachers will readily practice them even with less obnoxious problems than this one. But better options are available that avoid the even more obnoxious side effects of coercive practices (which we address in a later chapter). One of those other options is the preclusion procedure. A teacher has precious little access to changing the reinforcers for the tossing behavior, reinforcers often of the *bootleg* kind that come from all over the classroom in the form of peer attention. (Bootleg reinforcers are reinforcers that usually are not a part of, and often interfere with, an intervention.)

A preclusion procedure sidesteps all that. The teacher could, nonchalantly and without missing a beat in her presentation, simply move that waste basket to an out–of–sight location, perhaps under her desk. There the waste basket can no longer invite, I mean, evoke, crumpled paper tossing. Given the short time frame of this example, we can only speculate whether or not precluding *this* evocative stimulus would lead to forgetting the tossing response; I rather doubt that *that* would happen, but this procedure would still solve this teacher's problem (at least until other evocative stimuli occur that lead to another sample of sophomore silliness surfacing).

Extinction and Intermittent Reinforcement Schedules

We have barely touched on reinforcement schedules before this point, but I am sure that you recall from way back in Chapter 1 that these schedules refer to various ways for reinforcers to follow only some responses rather than all responses. While we will attend to those schedules in detail in a later chapter, here our concern is much more limited. Here our concern is simply that extinction relates to these intermittent reinforcement schedules in a couple of ways. One involves a phenomenon that we call the *extinction burst,* and the other involves a phenomenon that we call *resistance to extinction.* We will look at each in turn.

The extinction burst. Data show (see, for example, Ferster & Skinner, 1957) that when a behavior ceases to produce reinforcement, the rate of that behavior momentarily *increases.* We use the term *extinction burst* to describe this momentary increase in rate, and researchers often take advantage of the extinction–burst phenomenon to build intermittent reinforcement schedules.

For example, say you have a laboratory research subject who has been responding steadily with presses on a telegraph key, with each press producing points that are worth money. Yet the study in which he is participating requires stretching his key pressing until you get stable responding on some intermittent reinforcement schedule, say FR–450 (i.e., a fixed ratio schedule on which a reinforcer follows every 450th response). What do you think would happen if you simply changed the contingencies from a reinforcer following every single response to a reinforcer following every 450th response? You are getting the hang of this if you said that you think your subject's behavior would extinguish before you ever got to that 450th response the first time (which, of course, reminds us that just as we only reinforce behavior—not subjects—we also only extinguish behavior; we are not concerned with extinguishing subjects, which is illegal, immoral, and so on).

Instead, to build the ratio up to FR–450, you gradually *stretch the ratio* (a technical term) taking advantage of the extinction–burst phenomenon. When you first stop reinforcing key presses at the beginning of stretching, your subject's pressing rate briefly increases, and in short order the first 15 responses of an extinction burst have occurred in the time he usually took to make five responses. At that point, before the burst rate decreases, you again provide reinforcement, and for a while you continue to reinforce every 15th response. As soon as that FR–15 schedule is somewhat stable, you can stretch the ratio some more, again taking advantage of the extinction–burst phenomenon. This stretching goes on repeatedly until, after enough intermediate steps, you reach the FR–450 that the study requires. If you go too many responses before again providing reinforcement, if you *strain the ratio* (another technical term) rather than merely stretch it, then your subject may—to speak politely, in verbal shortcut terms—become temporarily emotionally unstable and damage the equipment as his or her key pressing extinguishes, and either of these—the damage or the extinction—would unpleasantly delay your research.

Introduction to resistance to extinction. Another interesting phenomenon, relating extinction to intermittent reinforcement schedules, is that even though numerous responses go unreinforced on such schedules, *the responding on these schedules is more difficult to extinguish* when compared to the relative ease with which responding extinguishes if reinforcement had been occurring after every response. Our measure is the number of unreinforced responses that occur during extinction, with more occurring after reinforcement had been intermittent than after it had been continuous. We might sometimes use the verbal–shortcut term *persistence* to describe the extended responding under extinction after intermittent schedules. More commonly we call the phenomenon itself *resistance to extinction.* Of course, persistence is not a cause of resistance to extinction, nor is resistance to extinction a cause of persistence.

The general rule describing resistance to extinction can sound rather counter–intuitive to some people. This rule is that *the more similar schedule responding is to extinction, the more the responding on that schedule resists extinction.* The similarity to extinction is in terms of the number of unreinforced responses occurring *on the schedule,* compared to extinction in which *all* responses go unreinforced. For example, FR–450 resembles extinction more closely than FR–15 (with 449 unreinforced responses for each reinforced response on FR–450 versus only 14 unreinforced responses for each reinforced response on FR–15). Hence the behavior on FR–450 would be more resistant to extinction. The behavior on FR–15 would extinguish first when no more reinforcers followed responding on either of these schedules (i.e., fewer responses would occur in extinction after an FR–15 schedule than after an FR–450 schedule).

Extinction and Punishment

Both extinction and punishment processes bring about reductions in behavior. However, the different ways in which they operate can cause some confusion. For this reason we will consider this topic after we first take a closer look at punishment.

More about Punishment

In many ways the reinforcement and punishment processes demonstrate flip sides of the same behavioral coin. They produce opposite changes in behavior and involve opposing respondent effects. And the effect of a stimulus switches from one side of the coin to the other (e.g., from reinforcer to punisher) simply by "occurring" the opposite way. For instance, a stimulus that reinforces when it starts becomes a punisher when it stops. As an example *receiving* dollars immediately after mowing the neighbor's lawn increases (reinforces) the mowing behavior, whereas *losing* dollars through a fine immediately after jaywalking decreases (punishes) the jaywalking behavior. In this example the dollars *occurred* in opposite ways, but the dollars are still dollars and themselves do not change (with the real changes occurring physiologically in ways that you could, by now, generally describe).

Actually, in anticipation, we should clarify a little about what is happening here in terms of what we will see later in the stimulus–control chapter. A more accurate description of the effect of receiving those dollars (after mowing the lawn) is the strengthening of the function of the neighbor's tall grass as an evocative stimulus for the mowing behavior. Similarly, the effect of losing some dollars (through a fine after, jaywalking) is the weakening of the function of a street clear of traffic as an evocative stimulus for the jaywalking behavior.

In any case, here we look at our three criteria questions as they relate to punishment. Then, after mentioning other types of punishers, we introduce a confounding interrelation between punishment and extinction.

The Three Questions Again

Our three criteria that determine various postcedent types do not apply only to reinforcement. Here is how they apply to punishment. For these three criteria, we ask the same three questions. And again the answers determine the labels we use for punisher–type postcedents.

Of the three questions, the first is always about the behavior of concern while the other two are about the postcedent stimulus. The answer to each question produces part of the complete label for a postcedent stimulus.

About the behavior change. *Does the behavior increase or decrease* (e.g., in rate)? When the postcedent variable produces a decrease in the rate of behavior, we label that postcedent some type of *punisher.* The other two questions address the criteria that determine the particular punisher type.

About the stimulus change. Since stimuli are still energy changes at receptor cells, we must still ask what type of energy change occurs at those cells. *Does the stimulus energy change involve an increment or a decrement of energy?*

Each answer produces the same label for punishers as it does for reinforcers. If the stimulus involves an *increment* of energy, and (from our first question) the behavior decreases, then we describe the stimulus as an *added* punisher. On the other hand, if the stimulus involves a *decrement* of energy, and (from our first question) the behavior decreases, then we describe the stimulus as a *subtracted* punisher. Again note that this question shows that the nature of the stimulus is less important than *what happens* to the stimulus.

About the stimulus history. Our last question still addresses the history or origin of the stimulus. *Does the punishing effect from the occurrence of the stimulus* (i.e., from its addition or subtraction) *depend on some prior conditioning?* If no prior conditioning was relevant, then the punishing effect occurs through neural behavior–mediating structures that genes produced, and we label the punisher an **un**conditioned punisher. However, if some prior conditioning (i.e., some *pairing* of this stimulus with other punishers that altered gene–produced structures) made this stimulus effective as a punisher, then we label the punisher a *conditioned* punisher.

The Postcedent combinations. Again, the various answers to the two questions about stimuli combine to give multi–term labels to postcedents. The possibilities are "added unconditioned," "subtracted unconditioned," "added conditioned," and "subtracted conditioned," each of which can refer to a reinforcer or a punisher (i.e., to reinforcing or punishing stimuli).

Other Types of Punishers

The other terms that some basic and applied researchers use to describe some aspects of reinforcers also pertain to punishers. Thus you will sometimes hear of *physical* or *tangible* punishers. At other times you will hear of *social* or *verbal* punishers. Of course, the use of these does not replace the usual technical terms (i.e., added or subtracted and unconditioned or conditioned). Furthermore, while the pairing of some conditioned punishers can occur with many other punishers, no consensus has yet emerged about a "generalized punisher" term. Similarly, while punishers can occur that the preceding response did not produce, no consensus has yet emerged about a "coincidental punisher" term. We need more basic, laboratory research.

Punishment and Extinction

Another area that needs more research is a certain confounding between punishment and extinction. But this problem presents particular methodological difficulties. Here is basically how this problem presents those difficulties.

Behavior that comes under a punishment contingency already occurs, which means it is already under a reinforcement contingency that may have generated it and in any case now maintains it. (Of course, behaviors are usually under more complex sets of contingencies, but we will ignore that for now, to keep the elucidation of this problem relatively simple.) Thus, with the reinforcement contingency still operating, any punishment contingency decreases the behavior by working against the reinforcement contingency; we describe this outcome as resulting from a sort of *algebraic summation* of the effects of the competing contingencies (see Skinner, 1953, Chapter 14, pp. 218*ff*). However, this means that if the punishment contingency ends, the behavior will rebound due to the continued operation of the reinforcement contingency, which in some cases produces an initial response rate higher than the rate prior to the start of punishment, a phenomenon that some tact as "over recovery."

On the other hand, if the punishment contingency begins and at the same time the reinforcement contingency ends, then the punishment need not compete with the reinforcement. Generally in this situation, the punished behavior not only reduces a little more quickly than when the contingencies are in competition, but it also is less likely to recur should the punishment contingency stop.

The confusion stems from the lack of reoccurrence of the behavior, when the reinforcement contingency ended early on and the punishment contingency now ends. Why does the behavior fail to recur? Is it due to the effectiveness of the punishment? Or is it due to the lack of reinforcement? And that second alternative is but another way of asking if it is due to extinction, because the ending of the reinforcement contingency is, by definition, extinction. Perhaps it is due to both. However, setting up a methodologically sound experiment that could address these possibilities, that could separate these differeng effects, has so far proven elusive, leaving us with more questions for research to answer. Natural sciences will *always* have questions to answer.

Postcedent Change Processes Table

While you may find reviewing our *Postcedent Matrix,* and even our *Postcedent Tree Diagram* (both of which appear at the end of Chapter 5) helpful for the relationships they portray, and the examples that pertain to this chapter's points, we will close the chapter with Table 10–1. This table lets you compare the operant postcedent behavior–change processes that we have covered so far, and see how they are similar and how they differ. (A shorter version of this table first appeared in Fraley, 2008, p. 399.)

However, before closing the chapter, let's lastly continue our ongoing activity. Consider how the various postcedent processes we covered in this chapter, particularly as intervention procedures, pertain to designing successful and timely solutions to the behavior components of global and local problems. Think about the relevance of these processes, as procedures, to generating and maintaining sustainable life–style practices, and to reducing practices that destroy environmental quality.

Table 10–1

Some Operant Postcedent Behavior–Change Processes

[Antecedent–I.V. evoked:] **Response** Dependant Variable	**Postcedent Event** Independent Variable (I.V.)	Effect on Rate of Responding	Name of Behavioral Process
[Antecedent–I.V. evoked:] Response Occurs	Stimulus Starts	↑ (increase)*	Added Reinforcement
	Stimulus Stops	↑ (increase)*	Subtracted Reinforcement
[Antecedent–I.V. evoked:] Response Occurs	No Change in Environment (Reinforcers no longer follow the responses)	↓ (decrease)*	Extinction
Responses *Cannot Occur* Due to **No** [Antecedent I.V.] Evocative Stimuli Occurrence	No Change in Environment (The usual reinforcer could occur, if I.V. evokes response [but does not occur] **& normal physiological degradation occurs)**	↓ (decrease)	Forgetting (the *process* in which **no S^{Ev}'s occur)** [and] Preclusion (the *intervention* that **prevents S^{Ev} occurrence)**
[Antecedent–I.V. evoked:] Response Occurs	Stimulus Starts	↓ (decrease)*	Added Punishment
	Stimulus Stops	↓ (decrease)*	Subtracted Punishment

*Due to change in S^{Ev} effectiveness.

References (with some annotations)

Ferster, C. B. & Skinner, B. F. (1957). *Schedules of Reinforcement.* Englewood Cliffs, NJ: Prentice–Hall. This comprehensive book describes the authors' multitude of laboratory studies researching a wide range of reinforcement schedule parameters; I understand that, for over a decade, it was deservedly one of the most widely cited references in natural–science literature, and we still hold it in high regard.

Fraley, L. E. (2008). *General Behaviorology: The Natural Science of Human Behavior.* Canton, NY: ABCs.

Holland, J. G. & Skinner, B. F. (1961). *The Analysis of Behavior.* New York: McGraw–Hill. This book is the original comprehensively programmed text; the authors successfully applied the laws of behavior that it teaches to its design and use. Indeed it was so successful that McGraw–Hill kept it in print for over 40 years, apparently until—with all the used copies floating around—they could not sell any more new copies. They then gave the copyright to the B. F. Skinner Foundation which will likely have it back in print at some point.

Johnson, M. (2004). *Clicker Training for Birds.* Waltham, MA: Sunshine Books.

Kurland, A. (2000). *Clicker Training for Horses.* Waltham, MA: Sunshine Books.

Michael, J. L. (1982). Distinguishing between discriminative and motivational functions of stimuli. *Journal of the Experimental Analysis of Behavior, 37,* 149–155.

O'Heare, J. (2010). *Changing Problem Behavior.* Ottawa, CANADA: BehaveTech Publishing. While this book openly addresses a wide range of species, Dr. O'Heare has half a dozen other books that more specifically deal with dogs.

Pryor, K. (1991). *Lads Before the Wind: Diary of a Dolphin Trainer.* Waltham, MA: Sunshine Books.

Pryor, K. (1999a). *Don't Shoot the Dog! The New Art of Teaching and Training (Revised Edition).* New York: Bantam Books.

Pryor, K. (1999b). *Clicker Training for Dogs.* Waltham, MA: Sunshine Books.

Pryor, K. (2001). *Clicker Training for Cats.* Waltham, MA: Sunshine Books.

Skinner, B. F. (1938). *The Behavior of Organisms.* New York: Appleton–Century–Crofts. Seventh printing, 1966, with special preface: Englewood Cliffs, NJ: Prentice–Hall. The B. F. Skinner Foundation (www.bfskinner.org) in Cambridge, MA, republished this book in 1991.

Skinner, B. F. (1953). *Science and Human Behavior.* New York: Macmillan. The Free Press, New York, published a paperback edition in 1965.

Reader's Notes

Chapter 11
Varieties of Postcedents Plus Superstitious Behavior

*W*e began to look more closely at postcedent processes in Chapter 10. After discussing some of the reasons why we cover postcedent variables before antecedent variables, we examined various aspects of reinforcement, extinction, and punishment. For reinforcement we emphasized generalized reinforcers and coincidental reinforcers. Then we considered extinction and forgetting, extinction and preclusion, and extinction in relation to reinforcement schedules including extinction bursts and resistance to extinction. Lastly we looked at punishment, not only as compared to reinforcement but also in relation to extinction (e.g., we remain unsure which one is responsible for the decrease in behavior—extinction or punishment—when reinforcement stops and punishment begins). We summarized all of these in a *Postcedent Behavior Change Processes* table at the end of the chapter.

Now, in this chapter, we continue examining postcedent processes, covering other, less–common labels for types of reinforcers (and punishers). Beyond postcedent processes, however, we also acknowledge a wider range of environmental–change options that lead to behavior changes, and we make the detailed connection between coincidental reinforcers and superstitious behavior. Let's begin with the environmental–change options.

More Environmental–Change Options

The accessible variables, changes in which bring about changes in behavior, all reside in the environment (internal or external). The various postcedent processes, including reinforcement, extinction, and punishment, provide but one source of variables that a contingency modifier might access when the need arises to bring about a change in behavior. The other sources…

Hang on. What is a "contingency modifier?" Is that simply some new term for "behavior modifier?" Are we now addressing "behavior modification?" To answer these questions, let's consider an important implication, one that stems from *behavior–changing variables all residing in the environment*. The implication is that the old term, *behavior modifier,* has more problems than merely early on evoking mad–scientist connotations in the popular culture (due to long–standing and pervasive contingencies driving superstitious agential accounts along with the false notion that any control must be of the punitive type). The even bigger problem is that this term never really was all that accurate, and it fell from professional favor years ago for several reasons.

Our reasons for leaving that term out of favor begin with the inaccuracy of its emphasizing just the outcome, just the behavior change, just the modified behavior; hence *"behavior* modifier." This ignores, or at least de–emphasizes, the operating environmental variables in the natural functional history of the behavior change. In their turn these variables include any organism whose participation in the contingencies contributed to the behavior change. However, this old term placed such participants somehow outside of, or above, the laws of nature, including the laws of behavior, by labeling them the *modifiers* (of behavior) as if their part derived from having some agent (e.g., psyche, soul, mind, self, judge, decider, chooser, designer, evaluator, director, and so on) inside the body that told the host body what actions to take to modify the behavior of concern, to make the behavior change, without any traceable natural functional history. We critiqued this kind of causality pattern earlier, and necessarily set it aside.

In reality, such bodies are also behaving organisms and, like all organisms, are limited to operantly and respondently conditioned responses that in one way or another *change the environmental contingencies* on another organism, human or other animal, *and these contingency changes bring about changes in the other organism's behavior.* So perhaps a more accurate label for such people— sans inner agents—could be *environment modifier* or *contingency modifier.* Better still, what about *contingency engineer,* with *contingency engineering* being a general procedural term? (Then, *green contingency engineering* would refer to the efforts of green contingency engineers to help solve global problems.) After all, contingencies are what people modify, as detailed throughout this book. Since all people participate in contingency changes, in modifying contingencies, and so are *all* contingency modifiers, perhaps, for those behaviorologists who work in applied settings outside the college classroom or laboratory, we should just stick to the label *contingency engineers* or, simply, *applied behaviorologists.*

Still, if postcedent processes only provide one source of environmental change variables, what are some other sources of variables that contingencies might compel anyone to access when the need arises for a change in behavior? Some other sources include antecedent processes as well as processes that make more direct changes to body structures. Let's look at each of these in turn.

A Peek at Antecedents

Recall that the present, immediate context in which we are always immersed (part of which is inside the skin, your skin and my skin) always contains numerous stimuli that could function to produce a behavior, and that one antecedent stimulus, or more than one in some sequence or combination, functions in the present to produce *every* behavior as a new behavior. If that *"every* behavior" has not affected you previously, let it sink in a little; it covers all overt and covert behaviors, including verbal behavior, thinking, and consciousness, leaving *nothing* for magical or mystical accounts to explain. The naturalistic implications of that *"every* behavior" for many topics that we have yet to cover is enormous, and we will cover them eventually. While we cannot currently explain every behavior (and perhaps may never have the access or resources to reach that point) we are confident that only natural–science accounts will be able to explain behavior in ways that meet the highest scientific and practical standards, because magical or mystical accounts are not adequate explanations at all; this reminds us once again that, since psychology fundamentally adheres to magical or mystical accounts, mostly of the secular inner–agent variety, behaviorology is not any kind of psychology.

Of course, when we say "stimuli *produce* every behavior," for operant behavior we would say "stimuli *evoke* behavior," and for respondent behavior we would say "stimuli *elicit* behavior." Using related terms in our usual manner, the antecedent stimuli that evoke and elicit behavior can be of either the *unconditioned* variety or the *conditioned* variety, and include eliciting stimuli, evocative stimuli and function–altering stimuli, among others. Also, other environmental–change processes can affect the functional operation of these stimuli. Examples include not only equivalence relations (which we cover in a later chapter) but also establishing and abolishing operations (e.g., deprivation and satiation) which we introduced, in a previous chapter, as changing both the momentary effectiveness of a reinforcer, and the momentary effectiveness of any stimulus that in the past evoked any behavior that resulted in the occurrence of the reinforcer, each of which alters the rate of the behavior of concern.

Any of those antecedent conditions, or processes, or stimuli provide potential access points for making environmental changes that would produce behavior changes. While the operations of these antecedents are the focus of other chapters, here is a simple example. When an open package of cookies on the kitchen counter evokes a child's taking–and–eating–cookies behaviors, a parent is certainly not limited to altering the consequential postcedents of these behaviors as a means of reducing excess or untimely cookie consumption. A much easier intervention involves merely moving the cookie package onto an upper shelf in a cupboard and closing the door. Indeed, keeping the cookie package there in the first place could have prevented that first inappropriate cookie consumption, and many parents have long been under contingencies that generate and maintain top–cupboard storage for cookies. Culturally, some old and misconstrued but witty proverbs describe such contingencies, like "out of sight, out of mind." Now, we could repeat such a saying using "mind" as some sort of standard verbal shortcut, but I think we would be even better off instituting a new saying, something like "out of sight, out of repertoire."

Will the prompt from that kind of antecedent–related saying induce new sayings on your part that relate to other antecedents, *or to any postcedents?* Perhaps a very short postcedent review will help. And please send me any new behaviorologically sound sayings that occur to you, along with your name so that I can credit them to you.

A Quick Review of Postcedents

For this little review, rather than sending you back to Chapter 10, or on to later parts of this chapter, let's simply list the already mentioned or covered postcedent variables which contingencies might compel a contingency engineer (i.e., anyone) to access—knowingly (i.e., with verbal–behavior supplementation) or intuitively (i.e., without verbal–behavior supplementation)—when the need arises to bring about a change in behavior. This list includes all the kinds of reinforcers and punishers, and any postcedent processes (e.g., extinction) and patterns (e.g., schedules) to which they are related.

Some Options Involving Changing Body Structures

Some other sources of variables that a behavior modifier might access, when the need arises to bring about a change in behavior, go way beyond antecedent or postcedent variables. And in this case, the "behavior modifier" label *is* more appropriate than the "contingency modifier" label; these sources of variables address processes that we generally treat independently of the usual contingencies controlling behavior, because these processes are more invasive than the contingency management of behaviorology. The five processes of this sort that we mention here are surgery, drugs, nutrition, disease, and restraints.

Surgery. One kind of body alteration, generally both large scale and permanent, involves surgical practices of different kinds that affect the occurrence of behavior in multiple ways. These practices have occurred around the world and across the millennia. Several even occur in our time. Some behaviors that physiological processes come to constrain often respond rather well to surgical interventions. For example, when an accumulation of certain kinds of injuries or stresses reduces walking to a painful gait limited in both extent and duration, knee surgery often returns the behavioral walking functions to nearly normal.

However, not all surgical practices have such beneficial effects. The practice of severing some portion of the frontal lobes of a brain (e.g., a lobotomy) nearly always yields detrimental effects. On the other hand it so seldom yields solely beneficial effects that such an occurrence gets widely reported as news. Some insist that such practices fit the category of unethical experiments because, among other reasons, no surgeon can know for certain what function the removed portion really serves, so removing it could make things better, or removing it could make things worse. Is taking that risk ethical? Others suggest that such practices should be consigned to museums of medical history.

Perhaps both of those views are right, at least until behaviorology expands to the point when we have developed, tried, and evaluated appropriate but less–invasive procedures. When we can finally say that we really have tried everything and nothing works, can we then try practices like lobotomy? Maybe we will never reach that point, or maybe the contingencies will simply not allow such practices. One such contingency is that "try" usually implies reversible effects, which lobotomy, along with most surgery, disallows.

Some practices of some past cultures inform us about the extremes to which "surgical" practices once carried in their role of producing permanent and large scale changes in behavioral function in multiple ways. In times and places past, a common practice was to cut off the hand of a thief. This practice certainly consequated the earlier theft response. Also, the wrist stump (if the thief survived the "surgery") probably functioned to evoke responses alternative to, perhaps even incompatible with, further thieving responses, which could cost the thief's other hand if not the thief's life. Very likely the wrist stump also regularly elicited emotional responses, as well as various responses that we would label countercoercion (i.e., getting back at those who were related to the "surgery") which the emotional responses could exaggerate, all usually to the thief's detriment. The wrist stump also likely served as an S^{FA} (a function–altering stimulus) that altered the function that the *body with the missing hand* served in evoking merchant responses. The missing hand evoked extra–watchful (of the one–handed shopper) responses from merchants, compared to the more friendly responses that two–handed shoppers evoked.

Drugs. Drugs change behavior by affecting physiological function and structure, including the functioning of the structures that mediate behavior. That is, drugs produce a set of behavior–related changes similar to those that surgery produces in that different drugs change physiological structures in ways that alter either the mediation of responses, or the functioning of antecedent stimuli and processes, or the functioning of postcedent stimuli and processes, or some combination of these contingency components. However, compared to the large scale, usually permanent bodily changes of surgery, drugs operate on the smaller scale of chemical changes that in most cases are temporary.

Since many sources report the effects of drugs, both beneficial and detrimental, to both physiology and behavior, we make the point here by addressing only a couple of the possibilities (and not the most common ones, like the effects that alcohol consumption has on behavior). For example, the complicated motions present when traveling, especially flying, often interfere with behavioral functions by making travelers dizzy or queasy. In that situation, given a body about to fall over, or on the verge of vomiting, stimuli succeed poorly in evoking the usual responses. A common drug remedy that restores some affected behavior–controlling functional relations for a few hours, at some cost of increased drowsiness, involves taking a 50 mg dose of dimenhydrinate (which is available in various over–the–counter forms).

The caffeine in coffee provides another example. You are probably among those motor vehicle drivers for whom caffeine consumption, in coffee or pills, has extended your daily driving range, or at least has increased the safety of your driving, by decreasing the drowsiness that fatigue or repetitive, unreinforcing scenery otherwise induces. Personally I favor alternatives that avoid drugs, while helping solve other problems, such as increased public transport systems.

Nutrition. Another accessible variable that affects behavior concerns nutrition, which is a variable in maintaining both physical and behavioral health. A body that consumes a balanced, nutritionally sound diet is a body better prepared to meet more of the demands that some sets of daily contingencies can impose. For example, consider the nutritional needs to support the stressful contingency demands on the engineer managing the shop floor where workers are crafting and assembling fuel–system components for commercial aircraft. Those nutritional needs differ, possibly substantially, from the nutritional needs to support the contingency demands on a retired engineer engaged in trout fishing several days each week at a beautiful stream located a few hundred yards from the front door of his retirement home. To support their differing daily contingency demands, the shop engineer and the retired engineer each need a different daily nutritional balance to maintain both behavioral and physical health.

Sometimes the need to bring about a change in behavior arises from problems with nutrition. For instance students who arrive at an 8 A.M. class, without first eating breakfast, quickly fall asleep (mostly from low blood sugar rather than from the professor's performance).

That can apply more broadly. If you are taking in the wrong daily nutritional balance to support the contingency demands on your full day's behavior, then your behavior with respect to nutritional intake needs to change. For example, consider that what looks like a healthy, behavior–supporting diet can be an unhealthy behavior–wrecking diet in some circumstances. Let's say you work a desk job that requires only 3,000 calories per day of food intake. Meanwhile your physiology needs the nutrients (e.g., vitamins and minerals) that only a 5,000 calorie per day diet could supply, the kind of diet that a healthy family farmer eats, without gaining excess weight, to support hard work in the fields every day from dawn to dusk. However, if you, with your desk job, eat the 5,000–calorie diet, then over time you would gain quite a bit of excess weight, which is unhealthy and in numerous ways interferes with normal contingency functioning, particularly by limiting behavioral ranges. Typical, daily–life contingencies produce behavior more effectively when the mediating body is not overweight. On the other hand, eating the 3,000 calorie diet prevents you from receiving all the nutrients you require, and that too is unhealthy and in numerous ways interferes with contingency effectiveness. Your physiology can better support behavior mediation when it receives the full complement of nutrients it needs. In such a case, you would likely benefit from taking appropriate supplement doses to get all the needed nutrients without gaining excess weight, which would help maintain physical as well as behavioral health. Prior to dietary changes, however, your contingencies should compel consultations with your physician and nutritionist.

Disease. In a manner rather the opposite of nutrition, disease proves detrimental to behavior, sometimes permanently so. Before reaching that stage, however, a common problem that disease presents to behavior and contingencies is that the physiological fight against disease uses up bodily resources that are then not available for mediating behavior. The body is run down and fatigued in spite of hopefully plentiful rest and relaxation, and we observe a general reduction in the effectiveness of behavior–affecting contingencies. The more specific ways that this happens are as myriad as the number of diseases.

Restraints. The last type of direct changes to body structures includes restraints, which remain variables that some outside behaviorology might access for changing behavior. Restraints *temporarily* constrain behavior in some of the same ways that surgery *permanently* constrains behavior. Restraints, which we also call bondage, simulate surgical bodily changes that prevent contingencies from being effective. For example, while wearing handcuffs, contingencies that might otherwise induce upper–limb aggressive behavior remain temporarily ineffective in a manner similar to the manner in which surgical amputation of the arms makes such contingencies permanently ineffective. Similarly, the coercive contingencies of being "in custody" induce escape behaviors, but those contingencies are far less effective when wearing leg irons, a temporary effect that simulates some of the permanent effects of leg amputation.

Even More Reinforcer Types

Back in Chapter 10, we considered some of the additional terms that basic and applied researchers have at times used to describe some particular aspect of reinforcers. These included physical, tangible, social, verbal, bootleg, and coincidental reinforcers.

Here we consider several more labels, each of which connotes an aspect of some reinforcers that goes beyond the basic, functional characteristics of added or subtracted, and unconditioned or conditioned. This time we cover pseudo, functional, intrinsic, extrinsic, natural and contrived reinforcers. Our comments on these labels apply equally well when we use the labels with respect to punishers. We also review coincidental reinforcers due to their role in our last topic for this chapter, superstitious behavior.

Rewards and Bribery

Misunderstandings, however, about rewards and bribery interfere with analyzing reinforcer types. So let's consider these two topics first.

Rewards. Interchanging the terms *reinforcers* and *rewards* presents problems, because *rewards are not necessarily reinforcers.* Rewards are stimuli that *others* think should reinforce your behavior, perhaps because these stimuli reinforce their behavior. However, a reward has not yet met the definition of reinforcers, at least with respect to the behavior of the organism of concern, possibly you. Remember, we have tested and observed reinforcers being stimuli the occurrence of which, immediately after a response, makes the evocative stimulus for that kind of response more effective across subsequent occasions. Rewards receive no such testing. Besides, if and when a reward meets this definition after testing, then we should call it a reinforcer, not a reward. Also, the concept of rewards supports the false and scientifically irrelevant notion of personal agency. How? A reward is for "you" (as the inner agent inside the particular carbon unit that others tact with your name). Furthermore a reward is *for you* rather than *for your behavior,* whereas reinforcers—as defined—do not reinforce *you*; they only reinforce behavior.

For example, if someone likes not only dark–chocolate peanut–butter cups but also your behavior (i.e., both the candy cups and your behavior reinforces his or her behavior) and follows your behavior with a bag of such candy cups, then you have received a reward. But if you are allergic to peanuts, then those cups are quite unlikely to function as reinforcers for your behavior.

Bribery. One dictionary defines a bribe as "Something… offered or given to someone… to induce him [or her] to act dishonestly" (*The American Heritage Dictionary,* 1982, p. 207). We can broaden that somewhat; a bribe is anything that you offer or give, usually beforehand, to induce someone to act in a manner that is immoral or illegal (Or fattening?) as well as reinforcing for the briber. Reinforcers, on the other hand, occur after desired behavior, and that behavior is usually in the behaver's best interest. So the occurrence of reinforcement, as part of the usual contingencies of life, is not bribery.

An example of bribery would be offering or giving your teenage son the latest, full–featured cell phone if he gives you a third of the illegal drugs he steals from others at school. On the other hand, if you give your toddler a cookie after she washes her hands when she finishes using the potty chair, then you are not bribing her; you are merely conditioning appropriate, life–long toilet and hygiene behavior with cookie reinforcers. Such cookies are neither "rewards" nor "bribes," and using them this way will not make your daughter a cookie junkie for the rest of her life, although other variables could lead to this outcome (e.g., cookie–manufacturer advertising schedules).

Pseudo Versus Functional Reinforcers

With that improved perspective from clarifying the status of rewards and bribes, let's move on to some other reinforcer labels, starting with pseudo versus functional reinforcers, two terms that each have a clear meaning. However, upon further analysis, we find that we really have little need for either term, and using them might even result in confusion.

Functional reinforcers simply refers to stimuli that we have confirmed, through testing and observation, as functioning as reinforcers. At that point the term "functional reinforcers" becomes redundant; we should simply call them *reinforcers.* The term "functional" is not one we need to continue using this way, because reinforcers are, by definition, functional.

On the other hand, some people use the term *pseudo reinforcers* to identify reinforcers that are not functional, which is a kind of oxymoron since reinforcers are, by definition, functional. They use this term to cancel itself, that is, to refer to stimuli that look like they could be reinforcers, that one might want to be reinforcers, that meet some parts of the definition of reinforcers, but that do not meet all parts of the definition. So, since we cannot properly call them reinforcers, some suggest the term "pseudo reinforcers." These pseudo–reinforcing stimuli may follow responses, and the responses may even produce them, but they have not passed tests that would show that their occurrence after a response makes later occurrences of the relevant evocative stimuli more effective in producing such responses. Hence these stimuli are not reinforcers at all; they are more like rewards, and if we need a term for them, the "rewards" term might prove less confusing than the "pseudo" term.

As a practical matter, if reinforcers are part of an environmental intervention, to bring about a change in behavior, we must have tested and observed the functional effectiveness of such reinforcing stimuli beforehand. So they *are* reinforcers and, again, we then call them reinforcers.

Intrinsic Versus Extrinsic Reinforcers

Neither of these two terms, intrinsic and extrinsic, is wrong, and we can certainly continue to use them. However, the difference to which they refer seems gradually and continually to lose relevance as we become more sophisticated regarding the natural, non–agential status of *all* reinforcers on both sides of the difference. We see this difference in how we use these terms.

We can use the term "intrinsic" to describe a reinforcer not as something that benefits an inner agent but as something that a response produces *directly,* without depending on the mediating behavior of another organism. The behavior operates on the internal or external environment, directly producing the stimuli whose occurrence makes the behavior occur more often in the future (or, more precisely, whose occurrence makes the antecedent stimulus more effective at evoking the response when that stimulus occurs again).

For example, when you test fly a just completed paper airplane and it flies out the door and all the way across the street, your construction and test *directly* produce that lengthy flight, and its occurrence strengthens your construction and test responses (or, more precisely, that lengthy flight makes the antecedent stimuli—perhaps a book of paper airplane making instructions on the table when a friend is present who has never seen a good paper airplane or flight—more effective at evoking quite similar construction and test responses when those stimuli recur on a later occasion). The flight is an intrinsic reinforcer.

On the other hand, we can use the term "extrinsic" when describing reinforcers that a response only *indirectly* produces. This indirectness involves the occurrence of the reinforcing environmental change being dependent on the mediating behavior of another organism.

Here is an example. We could describe the door opening as an intrinsic reinforcer when it occurs due to direct responses of turning and pushing the knob. On the other hand, we could describe the door opening as an extrinsic reinforcer when it occurs due to verbal responses to your spouse, with your arms full of groceries, such as, "Please open the door." Sound waves from your mouth seldom, if ever, have enough energy to open a door directly (although some sound–activated servos could complicate the situation). As verbal behavior, however, those sound waves provide an evocative stimulus for another's door–opening response, which *indirectly* gets the door open for you; your spouse's evoked response mediates the reinforcing consequence for your—in this case verbal—response.

Natural Versus Contrived Reinforcers

The usage of these two terms, natural and contrived, leaves us confused, because each is partly right and partly wrong. We will examine both aspects of each term. However, to help us in that task, let's review (from earlier chapters) both our technical, and our non–technical, usages of the term *natural*.

The non–technical use of *natural* pertains to a general notion of nature, as in "mother nature," the "great outdoors," what you see outside your window or outside your town, where you go to "get away from it all," mountains and rivers and skies and trees and wild animals and so on. Supplementing this common, non–technical usage, the technical use of *natural* specifically includes all those general–notion things plus everything else—asphalt and airplanes and cities and cell phones—that is, *all* energy and matter and life forms and behavior, everything real in the universe. As natural scientists the word *natural* (e.g., in such terms as natural variable, natural science, natural scientist) reflects our fundamental approach to *all* real aspects of the universe, an approach that involves dealing only with natural events (i.e., real events, events in nature) as independent and dependent variables, an approach that involves searching for and applying objective and measurable principles, concepts, and laws to all phenomena *of nature*. And that even includes mysticism and superstition (making the study of such behaviors natural topics for behaviorology).

That means that *all* reinforcers are already natural reinforcers. However, since all real things and processes are natural, calling some reinforcers natural implies that other reinforcers are unnatural or non–natural, which is not the case. By definition a reinforcer cannot be non–natural or mystical; in this sense the term "natural reinforcers" is wrong. But here is another sense of the term. What some people have meant by the term "natural reinforcers" is reinforcers that are not part of an intervention that people have arranged. Others have called these "unplanned reinforcers." As an example, consider the seeds that reinforce a bird's fancy foot work that uncovers them on the ground. Or consider the sweet flavor from a bite of apple that reinforces picking it off the tree branch and eating it. However, people are natural parts of the contingencies in intervention's, which leaves little to distinguish "natural" or "unplanned" reinforcers from any other kind. Since the term *intrinsic reinforcers* would seem to apply to them, let's avoid unneeded complications and stick with this term, if we really need an additional descriptor.

What then about *contrived reinforcers?* The contrast between "natural" and "contrived" also implies some impossible non–natural status for contrived reinforcers. In addition "contrived" too easily implies a contriver as an inner agent who arranges these reinforcers independently of any functional natural history; in this sense also, the term is wrong, as we have discredited all inner agents of any stripe. But here is another sense of the term. Some folks use the term "contrived reinforcers" to refer to the reinforcers that a practitioner arranges as part of an intervention. Others have called these "planned reinforcers" or "arbitrary reinforcers." As an example, consider the seeds from a food hopper that reinforce pecks on a disk in a laboratory operant chamber. Or consider the sweet flavor from a bite of apple that reinforces picking it off your desk and eating it after a student, whom you had helped a few moments earlier, had left it there as surreptitious thanks for your help. But again, practitioners are natural parts of intervention contingencies; they are not outside of, or above, natural laws. This leaves little to distinguish "contrived" or "planned" or "arbitrary" reinforcers from any other kind. Since the term *extrinsic reinforcers* would seem to apply to them all, let's avoid unneeded complications and stick with this term, if we really need an additional descriptor.

Some people use the terms "real" for natural, and "artificial" for contrived. However, I am sure that you can now critique these successfully, since the problem, and its solution, for these terms closely approximates the problem and solution for the terms natural/unplanned and contrived/planned. As much a possible, let's exclude all these confounded terms from our technical discourse.

Coincidental Reinforcers

Our last reinforcer–type label concerns *coincidental reinforcers*. We first met the term *coincidental* in Chapter 5 when we discussed "coincidental" versus "accidental." While "accidental" too easily implies no natural functional history for the stimulus of concern, we found "coincidental" implying fewer such problems; thus we retain "coincidental." Here we briefly review coincidental reinforcers due to their central role in our next topic, superstitious behavior.

We label a reinforcer as *coincidental* when the preceding, reinforced response did not produce the reinforcer. Some other natural process coincidentally led to the occurrence of the reinforcer at just that time, but it was not produced by the response it followed. It happened coincidentally, hence the label *coincidental reinforcer.* For example, let's say that every day, on your way home from work, you stop at a favorite grocery for the day's supply of fresh vegetables. The grocery is on the northeast corner of the block while your office is on the southwest corner. Without any discernible pattern, on about half of the past occasions you went around the block clockwise while on the other half you went counterclockwise. About two weeks ago, however, you went counterclockwise every day for that whole week. Then last week you went that way three days out of five. And this week you are back to no discernible pattern. Why did that happen?

The only thing we could discover happened was this. On the Monday of that counterclockwise week, as you neared the southeast corner, a $20 bill floated down in front of you. Of course you snatched it out of the air. You also looked up, looking for who might have lost it, or looking for more $20 bills floating down, or both. On the rather safe assumption that a $20 bill is a generalized reinforcer "for you," its occurrence, following your walking at that location, would have strengthened the behavior of walking along that route. We can rightly call the $20 bill a reinforcer from our observation that you walked counterclockwise every day the rest of that week (which is the way that took you past that location). On the other hand, since no further $20 bills floated down over the next several days (or at any time since this episode occurred) that lack of reoccurrence of this reinforcer gradually led to the extinction of the extra counterclockwise walking that the occurrence of the reinforcer had conditioned. So, after a couple of weeks, you were back in the usual routine of walking clockwise half the time and counterclockwise the other half.

However, the *occurrence* of that $20–bill reinforcer was not functionally related to your counterclockwise walking behavior (i.e., your behavior did not put the $20 bill in the air). And yet, while it only occurred coincidently with respect to your behavior, it did not occur magically. Some set of events comprised the natural functional history of its presence in the air. Perhaps someone left some $20 bills on a night stand near a third–floor window, and a gust of wind blew one of those bills out the window just a moment before you passed below that window. You got it, with the result that travelling counterclockwise occurred more frequently. (By now I think that *you* can improve the phrasing of that account in terms of the effect of the reinforcer *on the stimuli that evoke* travelling counterclockwise. What would you say?)

Now, if the term *coincidental reinforcer* describes the $20 bill, what term describes the counterclockwise travelling behavior? The term we use is *superstitious behavior.* Superstitious behavior is the kind of behavior that coincidental reinforcers condition. Let's consider such behavior more closely.

Superstitious Behavior

Sometimes, using a non–technical term, we describe reinforcers as "fickle." After we have tested them and observed that they have earned the label reinforcer, we also note that they are not picky about what behavior they reinforce, because another principle property of reinforcers is merely temporal; they *follow* responses. Thus, if a stimulus *is* a reinforcer, it will reinforce *any* operant behavior that it follows, and we describe this as reinforcer fickleness. It almost sounds like a polite swear word: "Oh, fickleness!" Yet we may find cause to swear about it. Here is why.

Coincidental Reinforcers and Superstitious Behavior

A curious implication arises from the fickleness of reinforcers. Many people consider humans the smartest species on the planet, in part because, with humans, operant processes can regularly condition a new response in a single reinforcer occurrence. Perhaps they are right about human intellectual prowess. But consider that this reason for placing humans at the top of the intellectual heap ignores the fickleness of reinforcers. That fickleness makes humans a species most susceptible to the occurrence of superstitious behavior, which a single coincidental reinforcer occurrence can condition (e.g., recall the example of the single $20 bill in the air). Thus the implication of the fickleness of reinforcers is that people rather easily fall victim to coincidental reinforcers conditioning superstitious behaviors. The next time that the Halloween holiday rolls around, think about how the occurrence of coincidental reinforcers could have originally conditioned the superstitious behaviors that many well–known cultural superstitions describe, such as "walking under ladders brings bad luck." (Sayings like this, and their origins, provide us with some examples of both "contingency–shaped" behavior, such as the original conditioning of a superstitious behavior, and "rule–governed" behavior, such as the behavior that a current superstition description controls.) Or, consider the origins and continued functioning of the many bothersome, even harmful, agential superstitions that permeate our culture... Can we swear now? Oh, fickleness!

Early Superstitious–Behavior Research

Summarizing some of the early research that analyzed superstitious behavior sets the superstition stage for us. (Skinner reported some of these on pp. 85–87 of his 1953 textbook). In one study the food hopper for each of several operant chambers, each containing a pigeon, provided momentary access to grain every 15 seconds *regardless* of the behavior of any of the birds. This is a *Fixed–Time* schedule, technically a Fixed–Time (FT)–15 seconds schedule, because reinforcement occurs solely by the clock without requiring any response (whereas Fixed Interval schedules require both a time interval and one response). Of course, some response was occurring when the food hopper operated, and the food coincidentally reinforced that response, whatever it was, making it more likely to occur again. As a result, in only a few minutes with this type of schedule operating, observers saw every bird repeating one or another stereotyped response. Examples of stereotyped responses can include head bowing, foot hopping, wing flapping, and turning in circles.

Since none of those various, stereotyped responses produced the reinforcers that followed them, we call the reinforcers coincidental, and we call all the responses *superstitious behavior*. (The experimenters also tried longer times, such as FT–1 minute, but the effect took longer to appear, for reasons discussed in Skinner's textbook.) Basically then, the term *superstitious behavior* tacts behavior that coincidental reinforcers condition.

Some Superstitious–Behavior Examples

Let's consider several more examples of superstitious behavior. These cover the coincidental but still reinforcing effects of a compliment, the coincidental but still reinforcing steps during an elevator ride, and the coincidental but still reinforcing points that follow various player rituals in some court sports.

The compliment. You found the morning temperatures for the first three weeks of October hovering around a cool but pleasant 16° C (i.e., 61° F). Then one day when you awoke in your dorm room, the outside temperature was at freezing (0° C / 32° F). Your first class was at 9 AM, and your winter wardrobe was still at home. To keep warm on your walk up to the classroom building, you had to put on a ratty old sweat shirt that you previously only used for house painting. On your way to class, you saw some friends from another dorm coming out of an 8 AM class on a side path. You waved and heard them say that they liked what you were wearing. Oh, really? Feeling good, being somewhat surprised, and thinking grunge must be back in style, you wore that sweat shirt nearly everyday until taking it home for the Thanksgiving vacation, over which your mother disabused you of your misinformed fashion sense, and made sure you returned to campus with your proper winter wardrobe. What you never found out was that your friends had been addressing the person behind you who was sporting a new, stylish, cashmere cardigan. Your sweat shirt had not produced the compliment, but the coincidental compliment had nonetheless reinforced your sweat shirt–wearing behavior. It was a superstitious behavior (one that continued for several weeks after Halloween).

Elevator buttons. Another superstitious behavior example involves elevator buttons. Given the bad press elevators get in movies—and even in personal experience, if you have ever been stuck in one for any reason—you would probably not be surprised to find that nearly every indicator that a needed elevator ride is progressing toward a proper conclusion turns out to be a conditioned reinforcer. (I say "needed" because, if you are only going up or down a few floors, then taking the stairs is probably better for your health.)

Now I presume that you have taken an elevator ride and so are familiar with the drill. You start by pressing the elevator call button, and it lights up, and the elevator is on its way. But what happens when other people also need to take the elevator? Invariably, one or another also pushes the already–lit call button, and some stand there pushing it several times, even continuously. Of course, none of those extra pushes has any effect on the elevator's arrival. Sooner or later (only a matter of a few seconds generally, and seldom longer than a minute) the elevator arrives; a light comes on and the door whooshes open while a chime sounds. All of those stimuli are conditioned reinforcers, and the button pushing of anyone pushing the call button unnecessarily, when those reinforcers occur, gets coincidently reinforced because, remember, those extra button pushes do not produce the reinforcers; they have no effect on the elevator's arrival. Those extra button pushes are superstitious behaviors, and past encounters with those coincidental reinforcers probably conditioned them in the first place. I suspect that you can well describe what happens when such coincidental reinforcers follow the extra pushing of any of the many other buttons inside the elevator.

Sports rituals. Our last superstitious behavior example involves sports rituals. Perhaps you play one or another court sport (e.g., handball, racquet ball, squash). If so you may have witnessed various superstitious rituals that coincidental reinforcers condition during play.

Let's focus on squash, which I played often and thoroughly enjoyed during the four years that I taught in Australia. The squash ball has little bounce so you must get to it quickly to hit it back to the wall; the long–handled racquet helps with this. But a good player can serve the ball such that it hugs the opposite side wall, forcing you to dig it out of the rear corner to return it, which puts your racquet at risk of damage (a good whack on the wall wrecks a racquet) and a good racquet is not cheap. In competition this might not matter, since the contingencies on winning trump economic concerns (and you have several spare racquets). If you are playing for exercise and enjoyment on a professor's salary, however, you might avoid that risk and lose the point. If you have never played squash, this description may give you some idea of what is going on. Where do superstitious rituals come into play?

Participating in a very active game, you are quickly ready, and eager to serve when your turn comes; but serving before your opponent is ready is rather rude. So, while waiting, some of your energy goes into bouncing the ball, perhaps twice on the floor then once on the wall then once more on the floor; then you serve. If your serve is a good one, right along the far–side wall and into the corner, your opponent may be unable to return the ball and you get the point.

All the particular response components of your service produced that point, and that point further reinforced these components of your serve. However, that point also followed your little, pre–serve, wall and floor ball–bouncing ritual. So that point also coincidentally reinforced that ritual, which we observe in the repetition of that ritual before the next serve, and the one after that, perhaps until your opponent regains the serve. In any case, being a function of coincidental reinforcement, that ritual is an example of superstitious behavior.

We can even see such rituals drift in form. When you again get to serve, the ritual may change. After all, on that very last time you served, the loss of service rather than a point followed the ritual, so we experience no surprise if the ritual changes, drifting into another form (e.g., to bouncing the ball once on the floor then twice on the wall then once more on the floor). Whether that change is due to extinction (no reinforcer) or punishment (a lost point) remains unclear. What is clear is that if a point follows the new ritual form, this new form quickly becomes a new superstitious behavior ritual.

Can you think of other examples of superstitious sports rituals? Consider the antics of major players in various other sports. Generally one can observe pitchers and batters in baseball, and quarterbacks in American football, exhibiting coincidentally reinforced superstitious rituals. Take a break from this book, before starting the next chapter, and watch for such rituals in a game or two of your preferred sport.

However, before going out to the ball game, take a little time as usual for our end of chapter ongoing activity. Consider how we might take into account the variety of postcedents, and the problems that could arise from coincidental reinforcers producing superstitious behavior, when we contribute our share to making timely solutions to the behavior–related components of global and local problems.❧

References (with some annotations)

The American Heritage Dictionary. (1982). Boston, MA: Houghton Mifflin [ISBN 0–395–32944–2].

Skinner, B. F. (1953). *Science and Human Behavior.* New York: Macmillan. The Free Press, New York, published a paperback edition in 1965.℘

Chapter 12
Context, Stimulus Control, and "Rules"

With respect to postcedents, Chapter 11 took us as far as we should go until we bring our coverage of antecedents up to the same level. Having mentioned antecedents in earlier chapters, we previewed them some more in Chapter 11, when we touched on several areas of antecedent environmental–change options that can lead to behavior changes, starting with elicitation, evocation, and establishing and abolishing operations. Additional environmental change areas involved various methods that more directly, and usually more invasively, change body structure in ways that change behavior, including surgery, drugs, nutrition, disease, and restraints. Then we dealt with rewards and bribes as part of covering most remaining terms for "reinforcer types" (e.g., pseudo, functional, intrinsic, extrinsic, natural, and contrived). And we finished the chapter with a discussion and several examples about coincidental reinforcers and the superstitious behaviors that they condition.

Now, in this chapter, we bring our coverage of antecedents up to about the same level as our coverage of postcedents. Our topics begin with the concept of context, which covers all aspects of the setting in which behavior occurs. That quickly takes us into the broad area of stimulus control, where we cover three interrelated processes, (a) *evocation,* which is a less problematic term than the older term "discrimination," (b) generalization, which itself has two types, response generalization and stimulus generalization, and (c) concept formation, which includes the concept of stimulus class. All of these terms may sound like mere jargon now, but not for long.

After a little side trip about seeing ghosts, we finish this chapter with a discussion of "rules" as statements of contingencies. This leads us to differentiate between contingency–shaped behavior and rule–governed behavior, which opens a fascinating world of non–agential complexity for human nature and human behavior, including implications for our concerns about the human behavior changes needed in solving global problems. All later chapters (e.g., on verbal behavior and on consciousness) also support these concerns, with some of these chapters returning to antecedents and postcedents to address directly some contingency–intervention practices regarding evocatives and consequences that could be helpful with these concerns. Right now, though, let's start with the *context* in which behavior occurs.

The Context of Behavior

Behavior always occurs in some context. Even in the vacuum of space, behavior occurs in a context, often in a capsule of some sort (e.g., a part of the International Space Station, a shuttle, or a lunar lander) or just a space suit; without at least a space suit, behavior in space ceases quite quickly and permanently. The context is the local environment—another name for "space suit" is "environment suit"—and this environment always includes not only the part that we call the *external* environment but also the behaving organism and its insides, the internal environment. As you recall, the laws of nature operate on both sides of the skin.

Some of those laws of nature govern behavior, which we summarize in our basic behavior formula, our starting point for analyzing any behavior, this way:

Antecedent functional independent variables

⇓

Behavior dependent variables

⇓

Postcedent functional independent variables

We refer to the *antecedent–variables* part of the formula as the context in which behavior occurs. Having introduced many of the postcedent variables already, this time around we emphasize the role that various aspects of context play in controlling behavior. And this includes some interrelationships between antecedent and postcedent variables.

Based on the foundations provided by the previous chapters and this one, in the first few chapters of *Part II* (which starts with the next chapter) we will alternate, with increasing complexity, between antecedent and postcedent variables. That will prepare us to cover the remaining chapters, with topics including verbal behavior, consciousness, and reality, all of which thoroughly intertwine antecedents and postcedents.

Before we get there, though, we need to take many contextual variables into account; we need to be able to juggle as many as we find are responsible for a behavior. Just as we have talked so far as if reinforcers follow every response (which is not the case, as our chapter on reinforcement schedules will explain) we have also talked as if only one antecedent stimulus is involved in evoking a response, which also is not the case, as this chapter explains. We accomplish this explanation, without insurmountable difficulties—except perhaps the economic one of adequate research funding—through *n–term contingencies* (i.e., contingencies containing as many more terms as we need to conclude the analysis, instead of the mere three or four terms in our previous contingencies). These contingencies not only feature several contextual antecedent stimuli, each of which alters the function of the next stimulus until we reach the stimulus that actually evokes the behavior of concern, but these contingencies also keep the concept of context within science, by setting aside as irrelevant any inner–agent, or other mystical, accounts of context.

Some researchers prefer to describe the n–term–contingency situation as each stimulus operating *on the condition* that another particular previous stimulus is present (i.e., conditional stimuli). At our still introductory level of analysis, however, I find that a quicker valid understanding accrues when we describe the n–term–contingency situation as strings of stimuli in which each stimulus alters the function of the next stimulus (i.e., function–altering stimuli). At a more advanced level of analysis, the context actually consists of complex, in–flux combinations of interacting antecedent functional variables; we manage this complexity by focusing on the generally small number of these functional variables that are mostly responsible for the behavior of concern. Meanwhile, to begin to see the roles that function–altering stimuli play, we must first master some other aspects of *stimulus control.*

Stimulus Control

The term *stimulus control* refers to a broad area of phenomena all relevant to the antecedent control of behavior by stimuli. The two basic areas traditionally included under the stimulus–control rubric are *generalization* and *evocation.*

To boost our understanding, recall that for designing and implementing environmental *postcedent* behavior–changing interventions, we found no *requirement* to involve the physiological level of analysis, because physiology addresses the question of *how* behavior happens while our engineering efforts derive from the science that addresses the question of *why* behavior happens, which includes accessible independent postcedent variables available for changing in interventions. The same applies to antecedent events. Nevertheless, reference to physiological events helps us more fully comprehend what is going on in the natural functional history of behavior with both postcedent and antecedent events.

When we discuss antecedent phenomena, like generalization and evocation, what kind of physiological events help us understand them better? Basically these antecedent phenomena involve effects that we describe as stemming from stimulus–produced behavior mediation either by genetically produced neural structures or by neural structures that have gradually accumulated structural changes through uncountable conditioning episodes across a human's or other animal's history right up to the present moment and continuing until the process that we call life stops. Such episodes, of course, affect postcedent phenomena as well, including the full range of contingencies of reinforcement (e.g., contingencies in which occur—separately or in combinations—processes such as reinforcement, punishment, extinction, generalization, and evocation). While we will examine physiological aspects where appropriate, let's now turn to our first stimulus–control topic, generalization.

Generalization

The phenomena that we describe as *generalization* (which in the past some researchers described as "induction"; see Skinner, 1953, Chapters 6 and 8) concerns a pair of observations that parallel our two generalization types, stimulus generalization and response generalization. After conditioning produces a behavior in the presence of one stimulus, we observe that behavior occurring to some extent in the presence of other similar stimuli (i.e., stimulus generalization). We also observe that, after a stimulus comes to control one response, that stimulus also controls other similar responses to varying extents as well (i.e., response generalization). As an example of the former observation (stimulus generalization) after conditioning produces the behavior of a professional astronomer photographing celestial objects with a large observatory telescope, we often observe the similar stimuli of backyard astronomical equipment also evoking her behavior of photographing celestial objects. And as an example of the latter observation (response generalization) after ballistics–formula stimuli come to control the behavior of an artillery unit commander in national defense forces, we might see those same ballistics–formula stimuli evoking her behavior of testing, for fun, the ballistic potentials of various rifle and pistol cartridge loads on the target range.

We then wonder what makes those extra functions happen, as if we required a new basic process to explain them. But this is a kind of pseudo problem that results from our arbitrarily breaking up the continuous process, or flow, of behavior into countable response units convenient for measurement and analysis, and then trying to explain similar behavior whose occurrence should not have surprised us in the first place had we stuck with behaving as a continuous process. When reinforcement occurs, it affects not only the particular kind of response it follows but also all other related kinds of responses that share elements with it; after all, recall our discussions of the physiological basis of behavior. These kinds of "elements" may actually be the "unit" of behavior rather than the response, but such elements never appear alone; some combination always appears as parts of the response of concern, which makes these elements less suitable as our unit of behavior. The elements that constitute verbal behavior may provide a particularly noteworthy example (as we will see in the verbal behavior chapter later). Nevertheless, we need both those arbitrary response units and a way to talk about the effects of similarity on observed stimulus and response interrelationships. For that we turn to generalization. While typically the processes of generalization *enlarge* the set of evocative stimulus relations, let's consider each of our two types in turn.

Response generalization. Basically, *response generalization involves different responses*. More completely, *response generalization involves the same stimulus evoking **different responses**.* These different responses share two related and often inseparable characteristics. (a) These responses are members of the same *response class*. A class of something involves a bunch of things that have something in common, which means that a response class is a group or set of responses that have something in common; they have the same effect on the environment in common (i.e., they have the same reinforcing consequence). And (b) these responses vary, sometimes substantially, in topography (which means that the responses vary from one another in physical form, for instance, in one or another of their dimensions, such as vector or speed or intensity).

Here is an example about door handles that shows both of these characteristics, response class membership and topography variation. Normally, the horizontal handle on your front door evokes the simple response of grasping it with your whole hand and pushing it downward, because that response produces the reinforcing consequence of the door opening. Of course, other members of the "door opening" response class, although varying in topography, would also produce that consequence, such as pushing with any finger alone, or with a fist, or even with an elbow or a knee and so on. All of these variations in response topography share the similarity that each would get the door open; the door–handle stimulus has an evocative effect on all of them.

The door handle's evocative effect, however, produces a *particular* response topography under appropriate conditions. When you arrive at your door totting four full grocery bags, the door handle evokes only the elbow or knee members of the door–opening response class, even though the handle has never evoked them before, so reinforcers have never followed them before, in this context. However, as elements of other responses, other stimuli have evoked elbow–use and knee–use responses before, and reinforcers did follow on those occasions. We use the term generalization, response generalization, to describe the occurrence of the elbow–use or knee–use response on this occasion.

Similarly, if you were carrying, not grocery bags, but a big, open–topped box from which groceries were threatening to spill, the door handle would not evoke either the elbow or knee response, either of which could destabilize the box. In this case the handle would instead evoke a verbal response perhaps in this form: "I cannot get the door. Would you please get it?" If, as a result of this question (i.e., this mand) the door opens due to your spouse, who was shopping with you, turning the handle, then your verbal response—so utterly different in topography from the other door–opening response class members—proves also to be a door–opening response–class member. Again, we use the term response generalization to describe the occurrence of this verbal response.

Now imagine the generalization process, as part of all our contingency processes (including those we have yet to discuss) extending across the board to all types and levels of behavior phenomena; not only is behavior complex, but we can also account for it with these processes. Humans need no magical processes in their accounts. Let's now turn to *stimulus* generalization.

Stimulus generalization. Basically, while response generalization involves different responses, *stimulus generalization involves different stimuli*. More completely, while response generalization involves the same stimulus evoking different responses, *stimulus generalization involves **different stimuli** evoking the same type of response*. The rule of thumb, from the overall research results, is that, while the evocative stimulus involved in conditioning the response evokes the *most* responding, *the more similar a new evocative stimulus is*—in comparison to the evocative stimulus involved in conditioning the response— *the more responding that new stimulus evokes.* Conversely, the less similar a new evocative stimulus is, the less responding that new stimulus evokes.

Here is an example, using hypothetical data. Let's say you condition responding to occur in the presence of a yellow light; that is, using the evocation–training procedure that we will talk more about shortly, you reinforce responses occurring when the light is on, hence these responses continue, while you ignore any responses when the light is off, hence these responses extinguish. Then you present a series of lights, each light of a different color, sometimes yellow, red, orange, green, or blue, in a randomized order. And you keep track of whether the light on each trial evokes the response or not. You then plot your data. To keep this example simple (and unrealistically perfect), our hypothetical data show that while the yellow light evoked the response on 99% of its presentations, green and orange each evoked the response on 70% of their presentations, and blue and red only evoked the response on 30% of their presentations.

Figure 12–1 plots those hypothetical data points. The pattern of plotted points traces what we call a *generalization gradient.* Such gradients from actual studies, often much like this one, show that the more similar a new evocative stimulus is to the original evocative stimulus (in this case, a yellow light) the more responding that new stimulus evokes (e.g., orange and green). Conversely, the less similar a new evocative stimulus is, the less responding the new stimulus evokes (e.g., red and blue; indeed, red and blue lights may function more like a light that is off, which would evoke no responses).

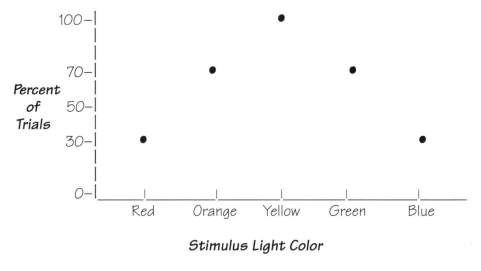

Figure 12–1. The percent of trials on which each stimulus–light color evoked a response (hypothetical).

Now let's turn to our next stimulus–control topic, the process of evocation. Under this topic we can discuss that training procedure we mentioned.

Evocation

While the process of generalization *enlarges* the set of evocative stimulus relations, the process of evocation *shrinks* the set of evocative stimulus relations. The evocation process often reduces the number of stimuli that evoke the response of concern while making the remaining evocative stimuli more effective. Again, the term *evocation* is a less problematic term than the older term "discrimination." Let's consider to what these terms refer before considering the benefits of using "evocation" rather than "discrimination."

The stimulus–evocation process. We first mentioned this process when we increased the precision of our basic behavior formula from A–B–Cs (… consequences) to A–B–Ps (…postcedents) in preparation for moving on to explicit three–term contingencies (e.g., $S^{Ev}\text{–>}R\text{–>}S^R$). The basic evocation process appears quite simple. Due to relevant past conditioning, the S^{Ev}s (*evocative* stimuli) evoke the response. That is evocation. The "relevant past conditioning" mostly involved stimuli following the responses that the S^{Ev}s evoked. Since we then observed these responses occurring more often, we called these stimuli *reinforcers*. We can describe that as the reinforcers strengthening the responses in the presence of the S^{Ev}s, because that is what we see.

However, your improving behaviorological repertoire benefits from more precision, particularly about the relation between the evocative stimuli and the reinforcing stimuli, even if the details are less visible. In terms of the whole contingency, and to respect the work in related natural–science areas by addressing a more complete natural functional history, the reinforcers actually serve a function different from merely "strengthening responses." *The occurrence of reinforcers actually strengthens the relation between the evocative stimuli and the responses that these stimuli evoke;* the reinforcers bring about enduring changes in neural microstructures, the microstructures that future occurrences of the evocative stimuli affect. As a result of these changes, the S^{Ev}s thus *more* effectively induce the neural mediation of the response, which thus occurs more often (the observation of which is, again, what leads us to call these postcedent stimuli *reinforcers*). Also remember that neural mediation includes muscle movement, since muscles contract only when appropriate neurons fire.

All this works with punishers as well, which we say weaken the responses that they follow. However, *the occurrence of punishers actually weakens the relation between the evocative stimuli and the responses that these stimuli evoke;* the punishers bring about enduring changes in neural microstructures, the microstructures that future occurrences of the evocative stimuli affect. As a result of *these* changes, the S^{Ev}s thus *less* effectively induce the neural mediation of the response, which thus occurs less often (the observation of which leads us to call these postcedent stimuli *punishers*).

For instance recall our Chapter 10 example about receiving dollars after mowing the lawn well, but losing dollars in a fine after jaywalking. Here is the effect of those consequences, one a reinforcer and the other a punisher, on the related evocative stimuli. Receiving dollars after mowing *strengthens* the function of the neighbor's tall grass as an evocative stimulus for the mowing behavior, while losing dollars after jaywalking *weakens* the function of a street clear of traffic as an evocative stimulus for the jaywalking behavior.

Let's look even more closely at S^{Ev} and reinforcer functioning. With a relevant conditioning history, S^{Ev}s transfer energy that triggers changes in nervous system structures (on a moment by moment basis). These changes include the neural firings that *are* the mediation of behavior (neural or motor, and involving more substantial bodily energy resources). Meanwhile reinforcers also transfer energy that triggers changes in nervous system structures, but these changes result in altered structures *that are more enduring,* leaving the body different, such that the later occurrence of the relevant S^{Ev}s more easily—but still involving substantial bodily energy resources—evokes the behavior again, which we observe as the behavior occurring more frequently.

Now you can take a turn. That account for reinforcers needs rephrasing to turn it into a similar account for punishers. Try it; then we go on to consider other details about how evocative stimuli come to evoke behavior.

Evocation training. As we just saw, the effects of the energy traces that we call reinforcers *provide the conditioning history* that makes S^{Ev}s evoke behavior. The reinforcing stimuli functionally feed energy back into the organism's nervous system, changing it so that the now different nervous system mediates behavior differently. When the S^{Ev} that had evoked the behavior before confronts the now changed organism again, the nervous system mediates the same appropriate behavior more readily or quickly. (Stimuli called punishers work like those called reinforcers except that the enduring nervous–system structural changes leave the S^{Ev}s *less* effective as evocative stimuli so that the behavior occurs less frequently.) How does this happen? How can we establish a stimulus as an evocative stimulus?

For starters, the environment—the context—always contains numerous potentially functional stimuli. Sticking to publicly observable events, if a response occurs, and a reinforcer regularly follows when a particular stimulus is consistently present, then we see the later appearance of this stimulus as producing, as in evoking, this kind of response (which usually produces the reinforcer). We then call this stimulus an evocative stimulus.

A simple definition of evocative stimulus implies that function along with hinting how that function occurs: An evocative stimulus is a stimulus in whose presence a reinforcer follows the evoked response. (We say "in *whose* presence," because the stimulus can be, and often is, an organism.) So how does a functioning evocative stimulus come about?

Contingencies produced a laboratory procedure for conditioning functioning S^{Ev}s. The procedure usually involves first conditioning some standard response without any added antecedent stimuli, and then presenting a series of trials with two kinds of added antecedent stimuli. On some trials, in the presence of a particular stimulus, reinforcement follows the response, in which case we call this antecedent stimulus an S^{Ev}. On other trials, under the condition of no S^{Ev} present or in the presence of a different stimulus, *no* reinforcement follows the response, in which case we call this antecedent condition or stimulus an S^Δ. (We pronounce this symbol "ess–delta," because "Δ" is the fourth letter of the Greek alphabet, Delta. It corresponds to our English letter "D," which we used early on, in S^D, for discriminative stimulus, before anti–ghost contingencies compelled the move to S^{Ev}.)

Remember, the behavior that extinguishes is the same kind of behavior that continues; the only difference is in the antecedent stimuli. This "same" behavior continues in the presence of the S^{Ev} but extinguishes in the presence of the S^Δ. That is, with the S^{Ev} present, the behavior that reinforcement follows occurs more often. On the other hand, with the S^Δ present, the behavior that nothing follows occurs less often; without reinforcement, this behavior extinguishes. As an example, recall the stimuli in our recent hypothetical generalization–gradient experiment. Our starting steps are pertinent. We would have started by reinforcing (i.e.,conditioning) a standard response with each of our half–dozen subjects, perhaps presses on a telegraph key by each of our college-student subjects. With presses established, we would then have limited the reinforcer occurrences only to presses that occurred in the presence of a light that was illuminated yellow, which would be the S^{Ev}, the training stimulus; in this case, the presence of the non–illuminated light (i.e., the light in the "off" condition) would be the S^Δ. Responding would become well established in the presence of the "on" yellow light, but extinguished in the presence of the "off" light. Evocation training would then be complete.

That example prompts some other points as well. For one, lots of errors occur in this kind of training (i.e., lots of responding occurs under S^Δ until such responding extinguishes); we will consider a method for *errorless evocation* (which others previously called "errorless discrimination") when we tackle the fading procedure in a later chapter. This same chapter will also introduce the procedure of backward chaining, which stems from an inherent characteristic of S^{Ev}s. We call this characteristic the "dual function of a stimulus," and it refers to each S^{Ev} also becoming an S^r because, in a pairing process, it is present when the reinforcer occurs. Thus these antecedent stimuli come to serve two functions, that of an S^{Ev} for the next response and an S^r for a previous response.

Also, the example implies that we can use various kinds of stimuli as training pairs for S^{Ev}s and S^Δs. These antecedent stimuli can merely involve the presence or absence of the same stimulus. Similarly, these antecedent stimuli can be the same stimulus in either an "on" condition or an "off" condition. Or, we can use two different stimuli, such as a green light as the S^{Ev} and a red light as the S^Δ. Different considerations will determine which of these pairs we use in our experiments (e.g., if our research subjects are color–blind, then using colored lights as antecedent stimuli is definitely inadvisable). Now, though, rather than go into that level of detail, let's instead proceed with our problem regarding the term "discrimination."

Problems with the term "discrimination." Another simple, and less common, definition of evocative stimulus begins to imply the problem with the term "discrimination": An evocative stimulus (or, as we used to say, a discriminative stimulus) is a stimulus that discriminates occasions when a reinforcer follows the evoked response. In this case a stimulus is "doing" the discriminating rather than an inner agent, which some folks might see as an improvement. But no such thing happens, in either case; neither stimuli nor inner agents discriminate. Stimuli are just stimuli; they neither change nor discriminate. Any changes happen at the physiological level inside the skin. But traditional cultural conditioning evokes inappropriate inner–agent–account responses when we go inside. If the stimuli are not discriminating, then surely the inner agent has this job. But inner agents lack status, so they cannot discriminate either.

These antecedent stimuli *only* either merely occur before a response happens that a reinforcer follows or, after such conditioning, they evoke the response in the process that we tact as evocation. Unfortunately the discrimination term compounds problems and causes students unnecessary difficulty and confusion. For that reason leaving it to history seems best. But let's consider how it came into use so that we might avoid similar future problems.

Let's review the connection between functional antecedent and postcedent stimuli, using the older discrimination terms that we allowed in a couple early chapters as part of our early terminology history. These include not only reinforcer (a term we still use) but also S^As and S^Ds (the S^{Ev}s that we previously called discriminative stimuli). We changed to S^{Ev}s (for "evocative stimuli") as part of reducing the residual "ghosts in the machine" problem that occurred with a range of behavior–related terms when some inner–agent implications inhering in useful terms exerted stimulus control over term usages. In this case the term "discriminative" came into use because in the laboratory these S^Ds and S^As determined the training–procedure difference between reinforcers occurring or not; these stimuli "discriminated" whether or not reinforcement occurred in laboratory experiments. But philosophical slippage led to some people talking about these discriminative stimuli as involving some sort of "discrimination," in which some ghostly inner agent in the organism was "doing" the discriminating. Perhaps this occurred, the argument went, as some putative higher–order "mental function," a function that some construe as something only inner agents can accomplish, which gives inner agents something to "do," thereby making them seem less redundant than they are.

However, one does not discriminate whether reinforcement will occur or not, because the inner agent "one" who would do that is a non–existent, and irrelevant fictional construct. No inner–agent involvement occurs in these contingency relationships, and we have no need for such involvement. More importantly, we also lack any need for the ghostly implication of these older—and as verbal shortcuts, perhaps still usable—terms, especially since they can mislead people, particularly those who are just getting started in appreciating this science. Furthermore these problems are in addition to the term "discrimination" having other, and unrelated, usages in common language. The single solution for all of these problems involves instead speaking of evocative stimuli and evocation, which can help sensitize us to avoid this kind of mystical error in our thinking, talking, and other disciplinary behavior.

Are S^{Ev}s the only kind of functional antecedent stimulus? No; another common kind of functional antecedent stimulus that we consider here is the function–altering stimulus, the S^{FA}.

Function–Altering Stimuli

As many have noted, the role of evocative stimuli in controlling behavior expands substantially in the presence of *function–altering stimuli*. These are stimuli in the presence of which other stimuli function differently from the way they otherwise would function. Up to now our discussions have usually implied that only one antecedent stimulus is involved in evoking a response. While this does occur, more commonly a few stimuli, even several, participate in evoking a response, because some stimuli function effectively as evocatives only if other stimuli are functional first (or at the same time). For instance, stimulus B might be neutral (i.e., if B occurs alone, nothing else happens; it is an S^N) with respect to our behavior of concern; but if B occurs in the presence of stimulus A, then we may find that B is no longer neutral and instead now functions evocatively. We would designate stimulus A as a "function–altering" stimulus (an S^{FA}), because its presence alters the function of stimulus B, which we would now designate as an S^{Ev}. Several such S^{FA}s might occur, together or one at a time, before a stimulus evokes a response.

To deal with those situations, we expand our analysis from our minimal three–term contingencies to *n–term contingencies* (i.e., contingencies containing as many terms as we need to make a full analysis). These n–term contingencies feature several contextual antecedent stimuli, each of which functions to alter the function of another stimulus until we reach the stimulus that actually evokes the behavior. We will look at two relatively simple examples.

The first example contains only one function–altering stimulus, and it alters the function of another stimulus *from the status of a neutral stimulus to an evocative stimulus*. At work a longtime co–worker asks you where you are taking your spouse to dinner to celebrate your anniversary that day. While I encourage you to consider the possible role of that coworker–provided stimulus with respect to certain covert (i.e., neural) responses—in the class we call remembering—regarding anniversaries, let's move on to a more practical matter. You have less than $20 in your wallet, and paying for dinner with a credit card could dampen the enjoyment of the experience as you and your spouse had agreed to cut back on credit card use. Then, while rummaging in your pockets to check your exact change, you find a stipend check that you had placed there last week. It is for a rather small amount (and, as a verbal shortcut replacing a proper account—with no agency implied—we could say you forgot it). But the amount would more than cover a nice evening out.

How does the check become cash? You usually pass your bank on your way home from work. But usually the bank functions as a neutral stimulus with respect to stopping while travelling home. However, the check alters the function of your bank such that today the bank functions as an evocative stimulus for stopping and cashing the check, which in turn provides the cash you need for the night on the town with your spouse. Here is how we would diagram the example:

$$S^N \searrow$$

S^{FA}		S^{Ev}		R		S^r
Check in pocket	$\rightarrow$	Bank	$\rightarrow$	Cash check	$\rightarrow$	More funds for dinner

How about a slightly more complex example? This second example also contains only one function–altering stimulus, but this case shows another possibility. In this case the function–altering stimulus alters the function of an already evocative stimulus. The S^{FA} alters the function of the S^{Ev} from evoking one behavior to evoking another behavior. A desired video recording on a store shelf generally functions as an evocative stimulus for buying the video. But, for a body conditioned poorly in terms of general social mores, the occurrence of a power outage while that body is at the store alters the function of the video on the shelf. For that body, with the power out, the video no longer functions as an evocative stimulus for buying it; instead the video functions as an evocative stimulus for stealing it (e.g., slipping it into a pocket and slipping out the door). Here is a diagram of the multi–term contingency in this example.

$$S^{Ev}_1 \searrow \text{(for buying)}$$

S^{FA}		S^{Ev}_2 (for stealing)		R		S^r
Power outage	$\rightarrow$	Video on shelf	$\rightarrow$	Stealing [rather than buying]	$\rightarrow$	Video in hand

As with most of our examples, this example is more complicated than I have presented it. Can you point out how? For instance, consider that the power outage, by distracting the clerks and disabling the security cameras, reduces the threat of punishing consequences for this stealing response.

Furthermore, note that (while the verbal behavior chapter has implications for the meaning of the term "meaning") this example also shows us two opposed meanings of the term *responsible,* as in "people are responsible for their actions." The two meanings are the usual agential meaning and a more recently recognizable scientific meaning. Regarding the agential meaning, supposedly the human being involved is responsible for the bad behavior, because the bad behavior occurred upon the mystical inner–person agent telling the body to take the action; thus we should hold the person—inseparable from the body—responsible, and blame and punish him or her. However, scientifically accounting for the bad behavior in this example leaves no room for such agential claims about responsibility. Nevertheless, regarding the scientific meaning, we can say that, as part of the general contingencies operating in society, the behaving body *is* responsible for the evoked action, but only in the sense that the behavior has consequences; the body is responsible in the sense that this body will experience the consequences of the evoked action, and in this case those consequences are supposed to make such actions less likely in the future. Any notion of responsibility, then, inheres in the contingencies rather than in putative person agents.

Separately, we should ask a question. Will agential notions inform the substance of those consequences (the way they inform the typical consequences that our current scientifically uninformed—with respect to behavior—penal system manages)? Or will science inform them (as Fraley discussed at length in his book, *Behaviorological Rehabilitation and the Criminal Justice System,* 2013)? At present we remain more in the grip of the agential notions.

In yet other ways, this example is more complicated than I have presented it. Can you say how else? In any case, it is typical of one of my early warnings to you, namely that *all* human behavior examples are more complex, and have more to their full accounts, than an introductory book can cover and still be merely an introductory book. (Please take advantage of the more advanced books that the bibliography contains, starting with Fraley, 2008. No book is perfect, including this one and Fraley's book; but his book is the best of several books covering more complex behavior in more complex details.)

The connection between postcedents and antecedents. At this point we should review again the answer to another more complex question. What makes these stimulus functions work? What makes S^{FA}s and S^{Ev}s work? What makes the former alter functions? What makes the latter evoke responses?

What makes this all work is the energy feedback into the nervous system that functional postcedents provide (reinforcers or punishers). That energy feedback strengthens or weakens (from reinforcers or punishers, respectively) the relation between each stimulus—function–altering or evocative—and either the stimulus that comes next or the response that the stimulus evokes. The occurrence of those reinforcers or punishers brings about enduring changes in neural structures, relevant to response mediation, that future occurrences of the function–altering and evocative stimuli affect. As a result of these changes, either (a) the function–altering and evocative stimuli *more* effectively induce the neural mediation of the response, which thus occurs more often, as observed with stimuli that we then call reinforcers, or (b) the function–altering and evocative stimuli *less* effectively induce the neural mediation of the response, which thus occurs *less* often, as observed with stimuli that we then call punishers.

Beyond answering questions like those, we should also cover more of the potential functional relations present in multi–term contingencies, more of the large number of multiple terms involved in the context of behavior, such as the number we address when we speak of *n–term* contingencies. For instance one S^{FA} alters the function of an S^N, making it function as an S^{FA} that alters the function of another S^N, making it function as an S^{FA} that alters the function of yet another S^N, and so on, until an S^{FA} alters the function of an S^N, making it function as an S^{Ev} that evokes a response. And as we saw, an S^{FA} can also alter the function of each of any number of S^{Ev}s from S^{Ev}s for behaviors of one kind to S^{Ev}s for behaviors of other kinds, as the last example's diagram showed for one S^{Ev}. (And a single S^{FA} can also, among other things, alter several S^Ns and S^{Ev}s together. And what about multiple S^{FA}s? …)

Those kinds of n–term contingencies help us understand how a combination of operant antecedent stimuli can look so similar from one circumstance to the next, and yet the subtle differences between the circumstances result in a quite different behavior, all lawful but not so easily visible, at least not as easily lawfully visible as respondent eliciting stimuli. We previously described this as a sort of competition among functional antecedent stimuli that helps us distinguish them from respondent antecedent eliciting stimuli (S^{El}s) but, rather than a competition, it is really just more orderliness in the universe. Here is a diagram for one way multiple S^{FA}s can affect multiple S^Ns, with arrows also showing the strengthening effect of the reinforcer occurrence (S^r) on *all* of the earlier stimulus functions:

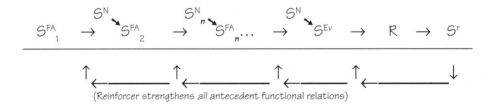

(Reinforcer strengthens *all* antecedent functional relations)

As with similar contingency diagrams, you would read that diagram this way: In the presence of the first S^{FA}, an S^N functions as another S^{FA} in whose presence an *n* number of neutral stimuli function as an *n* number of S^{FA}s until, in the presence of the last S^{FA}, an S^N functions as an S^{Ev} that evokes a response that a reinforcer follows. And the occurrence of the reinforcer strengthens *all* of the previous antecedent functions.

Let's see how those functions work in practice. Here is a diagram similar to our check–cashing basic example of a function–altering stimulus at work, this time with arrows showing the strengthening effect of the S^r on *all* the earlier stimulus functions. What we see is that, as a result of the reinforcer occurrence, a future check in the pocket will function more effectively as an S^{FA} that makes an otherwise behaviorally neutral bank (S^N) function more effectively as an S^{Ev} in evoking the stopping and check–cashing responses:

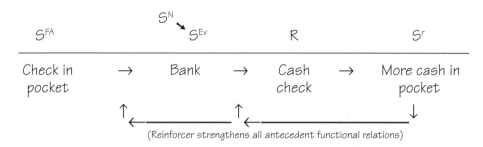

$$S^{FA} \qquad \overset{S^N}{\underset{S^{Ev}}{\searrow}} \qquad R \qquad S^r$$

| Check in pocket | → | Bank | → | Cash check | → | More cash in pocket |

(Reinforcer strengthens all antecedent functional relations)

Consciousness responses and direct stimulus control. In anticipation of the later chapter on details about consciousness, we should at least mention a connection to stimulus control. All the various behaviors of consciousness (e.g., awareness, recognition, comprehension, thinking, problem solving) are strictly covert neural behaviors that data (Fraley, 2008) suggest occur on a single channel, so to speak. That is, apparently one behavior at a time happens with each behavior–capable body part, although behavior can happen with several behavior–capable body parts simultaneously. For consciousness behaviors, the brain comprises the behavior–capable body part. For overt neuro–muscular behaviors, the behavior–capable body parts include mouth, hands, arms, legs, and so on. Overt neuro–muscular behaviors, while sometimes occurring with controlling–stimulus supplementation from neural consciousness behaviors, also often occur on their own channel, in the absence of neural behaviors of consciousness, under a kind of stimulus control that we call *direct stimulus control* (i.e., one need not be at all aware—a consciousness behavior—of S^{FA}s and S^{Ev}s, or even of postcedents like S^rs, for these stimuli to affect behavior). When S^{FA}s and S^{Ev}s control behavior in the absence of stimulus supplementation from neural consciousness behaviors, we speak of *direct stimulus control*.

Here is a quick example. Recall that like most people, you have likely found yourself (no inner agents implied) at some point on a road (perhaps a point where an uncommon stimulus, such as several turkeys crossing the road, has evoked some heavy braking) wondering (another neural consciousness behavior) how you got there, that is, where the last couple of miles went. During those miles your consciousness behavior involved a neural response chain that your next professional speech (or some such stimulus) had evoked and that focused on expanding the outline of that speech (or some such daydream–like response chain), while all your usual driving responses (e.g., steering) had been under the direct stimulus control of the various features of the road (e.g., the center line and gentle curves). Your neuro–muscular driving behaviors were under direct stimulus control. This shows us that stimulus control affects *all* behavior, neuro–muscular and neural alike.

Concept Formation

The notion of a *concept* turns out to be intimately related to some aspects of stimulus control. Since we have covered some essential components of stimulus–control, we can consider the concept of concepts more closely. The term concept often refers to some topic such as a man, a kitchen, a pet, or smaller or larger sets of these (e.g., cats, domesticated animals, even non–pets like wild animals). In each of these and innumerable other cases, you might have difficulty listing even just the necessary characteristics included in the concept, because exceptions that fail to invalidate a particular instance of the concept would confound the attempt to list the necessary characteristics. Yet while we cannot give the kind of explicitly adequate definition that we might want for a particular concept, we still have no trouble "applying" it intuitively (i.e., applying it—as our later section on rules describes—under direct control of the contingencies that produce the concept).

One reason for that difficulty is that concepts have no existence apart from the contingencies covering the stimulus properties and the responses they evoke and consequate. Each concept is *in* the contingencies; it is not the outcome of any inner agent gathering information upon which to conjure the concept. Being in the contingencies, concepts have a physical status that inner agents lack. A related reason resides in the processes through which contingencies produce concepts. What may amaze us is the discovery that a description of the concept–producing process in general is simple enough that some researchers have demonstrated the process with other animal subjects like pigeons. While I generally refrain from non–human examples in this book, I include this example, because the pigeons were under contingencies regarding "applying" the concept of people.

Here is a procedure in which contingencies would bring about a concept of people, a procedure that I will describe as if a professor had just completed the study. While in general this procedure comes from several research articles, for which I provide references later, the actual origin of some details have become unclear over the last three decades as they came from graduate–school courses and laboratory discussions. I recall verifying the legitimacy of these details at the time, but I did not store the sources as my days then were filled with concerns that seemed more important at the time. Furthermore, I claim some poetic license, within scientific constraints, to make the example as interesting as its title, which one of the references (Whaley & Malott, 1971, p. 179) provides. The title is "The Pecking Pigeon People Peeper" experiment.

This procedure began with the professor sending out a dozen students with cameras and enough print film among them for about 1,500 pictures in total. The students were told to walk around campus, in and out of buildings, snapping pictures left and right, up and down, close by and far away; just use up the film as this was not an art project.

When all the pictures were developed and printed, the professor and the students had a long meeting in which they first classified every picture as either a *people* picture or a *no–people* picture, and tossed each picture into the appropriate pile. The *people* pictures somehow had people in them, tall, short, slim, fat, alone, in groups, clothed, near naked, adults, children, close up, far away, and any combination of these in any of numerous positions or activities. None of this applied to the *no–people* pictures, of course. At the end of the meeting, they had two piles of pictures, a people–picture pile and a no–people picture pile. What was the value of two piles?

The answer requires something new. Recall our discussion of response classes. Well, welcome to *stimulus classes*. A class of stimuli is any bunch of stimuli that shares something in common. In the case of these two piles of pictures, every picture differed from every other picture in many, *many* ways. But for each pile, all the pictures in the pile shared something in common; in one pile *all* the pictures had people in them while in the other pile *all* the pictures had no people in them, making each pile a separate stimulus class.

Rounding the numbers to keep the example simple, each pile had about 750 pictures. The students pulled out the 250 pictures on the literal top of each pile; they would use these as training stimuli, leaving some 500 pictures in each pile that they reserved for the crucial last step of the study. The students then proceeded to implement training contingencies with their six pigeon subjects. They would present two photos at the same time to the bird, one *people* picture and one *no–people* picture, alternating position in a randomized fashion to prevent picture–type position from becoming part of the contingencies.

The two different picture types served two different stimulus functions. The *people* pictures served as S^{Ev}s that evoked a peck–the–people–picture response (which produced the reinforcer of a grain of food) while the *no–people* pictures served as S^{Δ}s (so no reinforcer followed responses of pecking them). As a result of these procedures, the pigeons became very accurate (around 95%) at pecking at *people* pictures while not pecking at *no–people* pictures. (Some of their errors could have been speeding errors—pigeons are quite fast—where, under conditions of some food deprivation, the pecking response started as soon as the gate began to rise to expose the stimulus pictures and, if some part of the S^{Δ} picture bore similarities to persons, then that would evoke the response, the momentum of which could lead a pecking response to connect with the incorrect picture.)

Now, so far, this is only a description of a sophisticated evocation training procedure, sophisticated in that the training was based not on two *stimuli* but on two *stimulus classes*. Then things got very interesting.

The next step was to dig out all those reserved pictures, the ones which no pigeon had ever seen before, and present each of them as stimuli to each pigeon subject. Would the birds continue to peck accurately only at *people* pictures, and not at *no–people* pictures, when all of the pictures were novel pictures? Indeed, that is what happened. The birds' accuracy dropped only a couple of percentage points. We would have to say that in general the birds had got the concept of people.

Also, I understand that one particular picture caused a lot of errors for every bird, so it was inserted as a challenge picture several extra times, and the birds kept pecking it even though the human group had classified it as a *no–people* picture. Out of curiosity, someone examined it with a magnifying glass and found, in an upper story window in a background building, a person visible. Think about that.

More importantly, *can we say what happened* such that we can say "the birds got the concept of people?" Yes, we can. The differences between the stimulus classes were differentially evoking the birds' responses. At the same time, the similarity that each stimulus shared with all the other stimuli within each stimulus class induced generalization within each class. Remember, each stimulus class had either all *people* pictures, or all *no–people* pictures, so we observed stimulus generalization occurring *within* each stimulus class (i.e., all class members of each class affected responding the same way) at the same time as we observed differential evocation occurring *between* stimulus classes (i.e., members of one class evoked pecking while members of the other class did not). And such contingencies affect people more often than pigeons. (For more and related information, see Cumming, 1966; Herrnstein & Loveland, 1964; Holland, 1958; Skinner, 1999; Verhave, 1966; and Whaley & Malott, 1971.)

Seeing Ghosts

Stimulus–control concepts also help us with other complex response phenomena. Early in Chapter 5, we said, "We can consider ghosts and other spooky phenomena elsewhere." Well, we have now laid enough antecedent and postcedent foundation to make considering them possible. Ghosts are spooky enough, and seeing them can be a quite real phenomenon, but probably not in the traditional manner that superstitious cultural contingencies condition. As clearly an exercise in interpretation, and taking into account what we know about contingency–produced behavior, we can speculate about some realistic circumstances that could typically produce the behaviors that we label "seeing ghosts." (Trying to get research dollars, to help fund studies with laboratory analogs of these interpretations, could be the real ghost story.)

Where to begin? Reports of ghosts were more common prior to science winning increasingly more battles with superstition in the twentieth century. Indeed, in that century, science and technology took off literally like a rocket. The changes in computer (and other gadget) technology that people currently witness in their lifetime, while amazing, may never equal the changes that people born in the 1890s witnessed during their lifetimes compared to those who came before them. These changes included going from foot or horse or horse–drawn buggy as the most common means of transportation, for several past millennia, to people walking in space and driving a horseless buggy on the moon; such changes included going from hand–carried letters as the most common means of communication, for several past millennia, to cell phones, the internet, and email. (For a fascinating tale elaborating on these and other categories of extensive change, *as well as the relative lack of change in dealing with behavior scientifically in that same period,* see Joe Wyatt's novel, *The Millennium Man,* 1997.) So let's consider seeing ghosts in the late 1800s, before the twentieth century.

Think about the circumstances for people living in the late 1800s. Many people lived within multi–generation extended families, often in the same big house, often on a family farm or business, and often for their whole lives. Thus, very likely, complex but mostly unvarying and repeating patterns of stimuli would evoke similar patterns of responding, all melding into a substantial familiarity and routine. When family members died, however, the stimuli that their presence had provided were removed from the equation. Not so, though, with all the stimuli that, for years to decades—depending on your age—had inevitably occurred along with the presence of those deceased family members. Being paired with this presence, these stimuli not only inevitably participated in the stimulus complexes evoking everyone's behavior while they were alive, but also these stimuli *continued* to participate in evoking everyone's behavior. Since some of the behaviors that these stimuli shared in evoking involved seeing the now deceased person, no one should be surprised that these stimuli might on occasion again evoke the behavior of seeing the deceased. Now though, with that person dead, traditional superstitious contingencies induce describing the seeing behavior as "seeing a ghost," whereas scientific contingencies induce describing the seeing behavior as the product of the related evocative stimuli.

Here is a possible example. You live in those times, and in circumstances similar to the ones we described. More precisely, you are now in your late fifties, your mother is in her late seventies and your grandmother—your mother's mother—just died at nearly one–hundred. Your grandmother had always run the household in your experience, although she had "allowed" your mother to handle an increasing number of tasks over the last three decades, just as your mother had been "allowing" you to handle some routine tasks in the last couple of decades. The family house is, of course, filled with many other folks of all ages and various, often overlapping relationships (brothers, sisters, cousins, aunts, uncles, cousins, spouses, and I must mention cousins). All played various roles; all had a share of indoor or outdoor chores, and all affected everyone else's behavior.

Now, ever since you can remember, you knew bedtime had arrived for everyone when you saw a flickering light on the stairway; that was your grandmother making her way, with candle in hand, up the stairs to her room, and checking on everyone else as she went. Nearly every night, for as long as you can remember, you saw the flickering light on the stairway and you would then see her as she passed your door across from the top of the stairs.

But now she is gone. Or is she? Your mother never adopted that nightly check–everyone routine. Yet occasionally at night you see a light flickering on the stairway, and about once each month since her passing, *you see your grandmother* where she used to disappear from view at the top of the stairs, and then, just as in so many times before, she is gone from view.

One need not posit a "haunted house" as the only explanation of your apparitions, your seeing behaviors. Other explanations are more plausible, more parsimonious, even if still interpretive or speculative. What we know about the why and how of behavior supplies at least a reasonable and scientifically grounded, interpretive alternative to the haunted house notion, and may dovetail well with related accounts from other natural sciences.

The stimuli that your grandmother's physical presence and interactions had provided, as part of everyone's contingencies, are now gone. However, *all the other stimuli* that had been present, when your grandmother had been present and interacting and going about the routines that her contingencies compelled, *are still present.* Past contingencies had made these stimuli part of ongoing contingencies, as function–altering stimuli and evocative stimuli and consequential stimuli; after all, life goes on.

Since some of these stimuli had been involved when your *grandmother's presence* evoked seeing her in the past (remember, seeing is a behavior under the control of the same laws of nature that control all behavior) these stimuli continue to evoke the behavior of seeing your grandmother even in her absence. Now though, the stimuli of the presence of *only a vague light* on the stairway evoke your grandmother–seeing response. That vague light may only be someone else going by candlelight back to bed after visiting the newly installed, indoor flush toilet downstairs at about the time at night when everyone has gone to bed, which is when you had so often in the past seen your grandmother disappearing at the top of the stairs. Unsurprisingly, that vague light again evokes the response of seeing your grandmother.

You are not seeing a ghost. Without implying any ghostly inner agents in you, you are simply again behaving the seeing, under different evocative stimuli, that you behaved in the past under the evocative stimulus of your grandmother's presence.

You will need to think more about those possibilities, since space limitations here allowed only a speculative, bare–bones interpretation. (Pun intended, I guess. As I write this, Halloween is only a couple of weeks away. While we can address the subtle impact of such contextual, and multiple–control variables on verbal behavior in general, and humor in particular, in a later chapter, Halloween, the topic of ghosts, and that "bare–bones" phrase provide an actual example. Perhaps we will see it again.)

Now, however, let's move on to a topic that behaviorologists have dealt with since Skinner well elucidated it back in 1966 (see Skinner, 1969). This is a contingency–control topic that we loosely summarize with the term "rules."

Rules as Contingency Statements

The rules that we discuss here address a range of phenomena that not only subsume the usual notion of rules, such as school rules, but also enable us to deal with more of the more complicated aspects of human behavior. Here the term *rules* refers to *statements of contingencies* (i.e., statements that explicitly or implicitly include at least a behavior and its contingent consequences, and often its evocative stimuli as well). These statements are verbal behavior, behavior that arises in humans through conditioning that other humans mediate.

Rules are present stimuli that are a part of contingencies by supplementing other stimuli in the contingencies. They sometimes serve this function by bridging the gap between behavior and the delayed consequences of defective contingencies (i.e., contingencies in which the selectors are too far removed in time to be directly effective). For example, good grades are contingent upon thorough studying, but they occur far too long after the studying to function as reinforcers of that studying. A rule like, "More thorough study produces better grades," repeated in the present, can supplement other contingencies on study behavior, leading to the kind of studying that gets good grades.

Through the conditioning of verbal behavior, such rules occur when circumstances evoke verbal descriptions of contingencies (i.e., rules) that we have experienced. The rules then function *as stimuli* (which some researchers have called *contingency–specifying stimuli,* or S^{CS}s). Here we seldom use the term contingency–specifying stimuli, because these stimuli become a part of other contingencies as function–altering stimuli or evocative stimuli, and we use these terms instead. But these contingency–specifying stimuli lead us to differentiate between two kinds of operant behavior, "contingency–shaped behavior" and "rule–governed behavior." As we consider each of these in turn, note that contingencies control *both* of these kinds of behavior. However, only the "rule–governed behavior" involves contingencies that include rules. Again, we differentiate between them, because treating them separately helps us better deal with complex human behavior.

Contingency–Shaped Behavior

We call behavior *contingency shaped* when contingencies condition that behavior without involving stimuli that equate with statements of contingencies. In common language we might say that the behavior occurs through direct personal experience. For instance, we keep a safe distance from bee hives because, in the past, getting close resulted in getting stung. Such contingencies thus directly determine the distance–keeping movements, so we call that behavior contingency–shaped. Let's trace relevant variations of *this* example beyond contingency–shaped behavior and through (a) rule–governed behavior, (b) instructions, and (c) other verbal supplemental stimuli.

Rule–Governed Behavior

The behavior that some contingencies compel can be dangerous for our health (e.g., honey reinforces behavior requiring more than mere closeness to hives). Must we all experience such contingencies directly for those contingencies to produce appropriate behavior? Do the natural laws of behavior include any process to save at least some, perhaps most, of us from having to experience the bad effects of dangerous outcomes? Yes. We need not each experience all contingencies directly, because the laws of behavior encompass rule–governed behavior. The process involves a contingency evoking a verbal–behavior description of the contingency, and that description, which we call a rule, then affects (i.e., supplements the controls on) behavior that we term *rule–governed behavior.* This process saves us from all having to experience directly every contingency of daily life (not just dangerous ones). Indeed, this is a major foundation for the cultural practice of education.

We call behavior *rule–governed* when the contingencies governing the behavior include rules, statements of contingencies (i.e., contingency–specifying stimuli) that function as either function–altering stimuli or as evocative stimuli. That is, the behavior occurs in part due to rules rather than solely through direct personal experience. For instance, while simple contingencies can directly condition keeping a safe distance from bee hives, you probably must get stung as part of that conditioning. However, observing someone else, even another animal, get stung when close to a hive can evoke a rule, a verbal–description behavior of the contingencies, a rule like "getting close to bees can get you stung," a rule that can generalize to future settings (i.e., similar future settings evoke the rule). Later, the buzzing of bees may evoke restating that rule, which then helps compel safe–distance–keeping movements, movements that we then call rule–governed behavior.

Instructions

Those kinds of rules can become *instructions* for others who have neither experienced some contingency nor seen someone else experience that contingency. For these others the instructions become part of contingencies that lead to rule–governed behavior.

While we usually phrase rules in a rather general manner, we often phrase instructions in a more situation–specific fashion, although an instruction can simply be someone else stating or restating a rule in the presence of others for their benefit. Still, instructions are rules, statements of contingencies, that the contingencies of someone else have induced, and which they or others then later pass on to still others whose behavior is then appropriate without their coming directly under the control of the original contingencies. For instance perhaps your grandfather had once got too close to bees and got stung. When his daughter (your mother) was getting too close to bees as a child, your grandfather instructed her with the rule, "getting close to bees can get you stung." Or perhaps he phrased the rule as the instruction, "Stay away from *those* bees or you will get stung." Either way, the important point is that his statement controlled the distance–keeping behavior of your mother. Now, your grandfather stayed away from bees, and *never again* got stung, due to the contingencies of his personal experience perhaps with rules serving supplementary functions. Your mother, on the other hand, *has never been stung*, but that is due to her distance–keeping behavior being rule–governed, that is, being under contingencies that include a rule about keeping a safe distance from bees, a rule that possibly occurred as an instruction from her father. And when you were little, and getting too close to bees, she passed this instruction on to you. Perhaps you also have never been stung and will never get stung.

A Role for Supplemental Verbal Stimuli

The verbal stimuli of rules and instructions, as statements of contingencies, often supplement the other variables in contingencies, making the contingencies more successful in various ways. While rules can certainly function this way on the overt level of some stimulus that evokes writing down the rule or speaking or reading it out loud, rules commonly function this way through the covert, purely neural, verbal behavior of consciousness that we call thinking. Such behaviors are real events and their occurrence then participates in the current contingency as a supplemental stimulus altering the function of some other stimuli. In this case the behavior occurs under additional variables besides direct personal experience. For instance, the buzzing of bees may evoke the covert thinking response, "getting close to bees can get you stung," a response the elements of which your mother first conditioned decades ago when she saw you about to touch a bee on a garden flower, and so instructed you. This thought, from a past instruction, then supplements, as a supplemental verbal stimulus, the current contingency in which the annoying bee buzzing was already arresting forward motion toward a hive such that the now combined stimuli instead induce motion that puts greater distance between you and the hive, preventing you from getting stung.

A Rule–Governed Behavior Shortcoming

In spite of the help that rules provide, rule–governed behavior has a major shortcoming. Once conditioned, the rules can last longer than their benefits, because the contingencies can change faster than the rules that we derive from the contingencies. For example if the safe distance from bee hives in *current contingencies* has increased, over what the rule implies, due to the hive harboring genetically altered bees that arose from interbreeding with certain strains for whom stimuli (e.g., people or other animals) at greater distances still evoke (or elicit) attack responses, then the previously safe distance that *current rules* help evoke is now an unsafe distance, and behavior that the rule still governs will get you stung. Perhaps we must come under the control of a rule about rules: Keep comparing rules with the actual contingencies they purport to describe, or get stung (i.e., get hurt in whatever way the contingency involves).

A Residual Confusion about Contingencies and Rules

Before we finish the chapter, let's revisit a residual confusion. Rules are *not* separate phenomena from contingencies; *rules are parts of some contingencies.* As evoked statements of contingency components, rules constitute natural phenomena that are part of various contingencies. A rule can even be a part of a contingency that involves contingency–shaped behavior. Again, the differentiation between contingency–shaped behavior and rule–governed behavior occurs, because it helps us deal with some complicated aspects of human behavior. Rules, which are not magical, or agential or "cognitive," are still natural parts of other contingencies that also generate, maintain, and reduce as well as shape behavior. Here is an example of a rule, a simple statement of a contingency, a contingency–specifying stimulus, simply being part of a contingency, in this case serving as a function–altering stimulus that changes an evocative stimulus for one response into an evocative stimulus for another response. This example describes something that has never happened to you before, and you have not discussed such events with anyone before, because you come from a nice small town where events like those in this example are extremely rare; but you are in the big city now, enjoying some spare days before presenting data as a scientist at a holiday–time sustainable–living convention. Thus, this example is an example of a contingency–shaped behavior that encompasses a "rule" (although some authors prefer the term "contingency–conditioned" over "contingency–shaped," because the example depicts no clearly separate shaping steps).

Imagine you are walking on a poorly lighted street late on one of those cold winter–holiday nights, and you see someone else walking from the other direction. This other walker is fighting the biting breeze with a cute ski mask that looks like it has white rings around the eye openings making the mask look like one of those cute old–time cartoon dogs or racoons. But just as you are about to pay a compliment on the ski mask, the other walker waves an illegal, black–market gun at you and says, "Give me your wallet or I'll shoot you." Now, every non–criminal citizen like you has a right to keep and bear arms for many legitimate reasons including self defense as in this case, but not everyone exercises their right, and that may include you. Where people are allowed to exercise that right, even if few actually exercise it, the allowance itself serves as an S^{FA} that, for a potential robber, changes many potential victims from S^{Ev}s to S^{Δ}s for robbery. Think about that (and check the FBI statistics for verification). For our example, however, in this particular big city, exercising that right is currently disallowed.

So, those words, "Give me your wallet or I'll shoot you," occurring under those circumstances, constitutes a contingency–specifying stimulus (i.e., a rule) that serves as a function–altering stimulus that changes the "cute" ski mask from an evocative stimulus for a compliment into an evocative stimulus for compliance. While at first you hesitate, which evokes some further abusive and potentially very dangerous (for you) posturing responses on the part of the robber, you finally comply by handing over your wallet. (You may also notice that the "white rings" were just overly large eye openings.) Your compliance reduces the threat as the thief removes your cash, throws your wallet behind you, and disappears up an alley. Again, this is an example of a rule, as a contingency–specifying stimulus, participating in a contingency as an function–altering stimulus. (Due to its situation–specific nature, we could call the robber's statement an instruction, but we call it a rule here, because its implication of severe consequences matters more in our example than its specificity.) Here is the diagram for this contingency.

$$S^{Ev}_1 \text{ (for compliment)}$$
$$S^{FA \, (CS)} \qquad S^{Ev}_2 \text{ (for compliance)} \qquad R \qquad\qquad S^{r-}$$

"Give wallet or else"	→	[Cute] Ski mask	→	Compliance (rather than compliment)	→	Threat reduced

Note that in this diagram, the S^{FA} appears as $S^{FA \, (CS)}$. The S^{FA} in this example "specifies a contingency"; it is a function–altering stimulus of the contingency–specifying type, hence $S^{FA \, (CS)}$. However, only rare circumstances, like an example in a book, evoke inclusion of the "^{CS}."

Now, is that an example of contingency–shaped behavior, or an example of rule–governed behavior? If, as described, this is the first robbery you experienced, we would describe your compliance as an example of contingency–shaped behavior, even though a rule was involved in the contingency controlling your behavior. Consider, though, that generalization from this instance could lead to a verbal description of the contingency that leads to more effective behavior in these circumstances. This rule could be, "When a robber confronts you in that city, comply right away." Whether the rule arises from having survived a robbery, or from talking about your—or someone else—experiencing a robbery, is immaterial. The rule is now a verbal S^{CS} supplementing other contingencies. Now, if a robber again confronts you and you comply more quickly, due to this rule, we call *this* compliance *rule–governed behavior*.

As usual, before moving on to the next chapter, take a little time for our end of chapter ongoing activity. Consider how the various parts of stimulus control, contingency–shaped behavior, and rule–governed behavior figure into our share of making timely solutions to the behavior–related components of global and local problems. The interrelations between rule–governed behavior and the development of some needed, green cultural practices will likely prove particularly pertinent.♣

References (with some annotations)

Cumming, W. W. (1966). A bird's eye glimpse of man and machines. In R. Ulrich, T. Stachnik, & J. Mabry (Eds.). *Control of Human Behavior* (246–256). Glenview, IL: Scott Foresman.

Fraley, L. E. (2008). *General Behaviorology: The Natural Science of Human Behavior.* Canton, NY: ABCs.

Fraley, L. E. (2013). *Behaviorological Rehabilitation and the Criminal Justice System.* Canton, NY: ABCs.

Herrnstein, R. J. & Loveland, D. H. (1964). Complex visual concept in the pigeon. *Science, 146,* 549–551.

Holland, J. G.. (1958). Human vigilance. *Science, 128,* 61–67.

Skinner, B. F. (1953). *Science and Human Behavior.* New York: Macmillan. The Free Press, New York, published a paperback edition in 1965.

Skinner, B. F. (1969). *Contingencies of Reinforcement: A Theoretical Analysis.* New York: Appleton–Century–Crofts. The B. F. Skinner Foundation (www.bfskinner.org) in Cambridge, MA, republished this book in 2013.

Skinner, B. F. (1999). Pigeons in a pelican. In B. F. Skinner. *Cumulative Record—Definitive Edition* (630–647). Cambridge, MA: The B. F. Skinner Foundation.

Verhave, T. (1966). The pigeon as a quality control inspector. In R. Ulrich, T. Stachnik, & J. Mabry (Eds.). *Control of Human Behavior* (242–246). Glenview, IL: Scott Foresman.

Whaley, D. L. & Malott, R. W. (1971). *Elementary Principles of Behavior.* Englewood Cliffs, NJ: Prentice–Hall.

Wyatt, W. J. (1997). *The Millennium Man.* Hurricane, WV: Third Millennium Press. Given the time period over which he wrote this novel, the author, a friend and colleague, often used the term "behavior analysis" as the name for the natural science of behavior to show that it is no kind of psychology. Today, however, with the psychology discipline claiming behavior analysis as part of itself, this could confuse readers who therefore should substitute "behaviorology" for these usages in their reading.✧

PART II
Advanced Developments &
Answers to Long–Standing Questions

Chapters 13 – 24

WARNING

Without the conditioning preparation that derives from mastering the chapters of Part I, much material in Part II, especially starting with Chapter 19, can be needlessly difficult intellectually, emotionally, or both. This happens, because the extensions of the naturalism of human nature and human behavior that occur in the more complex contents of Part II can easily exceed the level of conditioned naturalism that occurs in the reader's repertoire merely through past general education and cultural experience. *Part I provides a minimum supplement or antidote. So, to get the most out of all Part II chapters,* **cover Part I thoroughly before going on to the chapters of Part II!**

PART II

$\mathcal{P}$art II directly extends earlier chapters to introduce some topics that arise from the natural combinations of the basic natural–science assumptions, methods, principles, concepts, and applications of behaviorology that we covered in Part I. These combinations produce processes that operate in more complex ways and so produce more complex behaviors and changes in behavior. As always, these basics and combinations are not any kind of psychology.

That greater complexity allows us to begin offering natural–science answers to some ancient human questions. These natural–science answers provide more realistic and helpful alternatives to the traditional, culturally conditioned and currently available mystical answers to these questions, regardless of whether the mysticism stems from theological or secular sources.

Some of the topics of those ancient questions appear in several chapter titles, starting mainly with Chapter 19. Along with the chapter numbers, and presuming that your have heeded the *WARNING,* here are the Part II topics, with italics for some of the ancient questions that we begin to address:

13 Arranging consequences—differential reinforcement and shaping procedures;

14 Arranging evocatives—backward chaining and fading procedures;

15 Basic schedules of reinforcement;

16 Aversive control problems and alternatives;

17 Some applied behaviorological research considerations;

18 The stimulus equivalence relations horizon;

19 On attitudes, *values, rights, ethics, morals,* and beliefs;

20 *Language* is verbal behavior;

21 Accounting for *consciousness;*

22 Cultural concerns of *life, personhood,* and *death;*

23 The unexpected *nature of reality* and robotics;

24 Evolutions and Epilogue.

WANTED!

Courageous *&* Bold
Faculty, Deans, *&* Provosts

of college *Natural–Science* units to implement Doctoral, Master, *&* Bachelor programs in the *natural science of human behavior,* Behaviorology,

emphasizing basic experimental research *&* culturally relevant applications to understand human nature better *&* to make a better world for individuals *&* humanity.

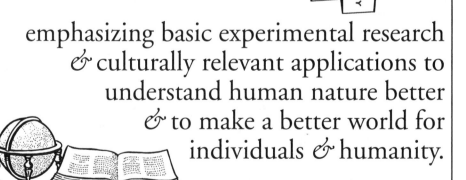

Contact Board Members of TIBI *(The International Behaviorology Institute)* at www.behaviorology.org

Chapter 13
Arranging Consequences—
Differential Reinforcement and
Shaping Procedures

*I*n some earlier chapters we considered many properties and characteristics of our subject matter, behavior. We also described all manner of functional postcedent variables along with brief forays into the interrelations of these postcedent variables with some antecedent variables. And finally, in Chapter 12, we covered the range of functional antecedent variables in some detail.

Those details included the phenomena under stimulus control, which we extended to contingency–shaped and rule–governed behavior. Under stimulus control we started with response generalization and stimulus generalization. Then we covered other topics including the process of evocation, evocation training, evocative stimuli, function–altering stimuli, and some of the ways that all of these interrelate (e.g., concept formation). We stressed the fundamental nature of behavior, whether neuro–muscular behavior or just neural behavior, whether walking or talking or emoting or thinking or remembering or any other behaviors, whether operant behavior or respondent behavior; in all cases behavior lacks any explanatory need for the mysticism or superstition of any inner agent. Instead we focused on the fundamental nature of behavior both (a) as the always–new product of current stimuli, and (b) as real events that serve stimulus functions for other behaviors, including the way that the verbal responses that describe contingencies function as rules that alter subsequent behavior. Under such rules we considered the roles both of instructions and of other verbal supplemental stimuli. We even had some fun with ghosts, or at least with one scientific, interpretative accounting for them.

Being the last chapter in Part I, Chapter 12 completed our basic coverage of behaviorological concepts, methods, and principles. Now, in Part II, we introduce some of the combinations of those basics, combinations that provide us with processes that operate in more complex ways and so produce more complex behaviors and more complex changes in behavior. And we consider how to harness these processes as intervention procedures to begin engineering environments to interact with physiology (i.e., bodies) to produce behaviors to prescription. Of course contingencies have always compelled humans to engage in such engineering practices with respect to behavior, but with results that mystical foundations limited. Perhaps now, upon the foundation of behaviorology, the natural science of behavior, we can increasingly provide interventions with less damage from misunderstandings and misuse, and so with more benefits from the improved effectiveness that typically accrues when our efforts deal with a subject matter as a natural science.

We begin expanding those efforts with this chapter. Then, across the next several chapters, we maintain balance by bouncing back and forth between postcedent and antecedent processes or procedures. For our first bounce, this chapter returns to postcedents by considering two potent postcedent behavior–change procedures, *differential reinforcement,* and *the method of successive approximation* that we more commonly call *shaping.* Each of these arises from the combination of the two simpler postcedent processes of reinforcement and extinction. We begin with differential reinforcement.

Differential Reinforcement

A rather simple combination of reinforcement and extinction leads to the procedure that we call *differential reinforcement.* We use this name, because the procedure involves reinforcing some responses differently from other responses. This procedure makes the behavior of concern occur more often, so long as it already happens at least occasionally. If this behavior never occurs, then we cannot reinforce it, and so we would need some other procedure to make it occur more often.

To describe how differential reinforcement works, we must invoke a concept that we previously introduced, and that is the concept of *response class.* A response class refers to a group of responses that have in common the same effect on the environment. The classic notion is that the members of a response class all produce the same kind of reinforcement. Consider, for example, the many different responses which could produce an open door. These could include turning the handle with a hand or any one of four fingers (which already constitutes five different members of the door–opening response class) or a thumb or an elbow or a knee or even a mouth—no, not that way. (Well, I suppose you *could* open a door that way.) But I am referring to opening the door with the words, "Please open the door" when an appropriately conditioned audience is present. Regardless of the differences in topography, all of these responses are members of a door–opening response class, because they all produce an open door. So, how does the differential–reinforcement procedure work? How does it involve the concept of response class?

In differential reinforcement we treat response–class members differentially. That is, we consequate response–class members differently. To be explicit, we *reinforce* the response–class member that constitutes the behavior of concern while we extinguish all the other response–class members. We call this *differential reinforcement,* which we define as the procedure of reinforcing one currently occurring member of a response class while not reinforcing (i.e., extinguishing) all other members of that response class. Consider some examples.

Differential Reinforcement Examples

Very likely we are all interested in helping others. Sometimes we do this just by talking. If we understand differential reinforcement, we are likely to be of greater help, naturally and automatically, even in these ordinary, everyday encounters. Here is what can happen.

Helping a friend. Say you are talking with a friend who makes a range of verbal responses related to the personal effects of some hard times. These responses include comments that indicate some comprehension of the problems he is experiencing and some reasonable steps toward solutions, plus comments that reasonably clarify his feelings about himself and any involved others. On the other hand, his responses also include comments that indicate some pessimism, some confusion, some apprehension, and some mis–comprehension regarding his problem–related behavior (including feelings).

Now any of your friend's verbal responses could evoke verbal responses on your part that, at least as attention, could function as reinforcers for your friend's responses (i.e., your friend's responses are all members of the same response class). However, I doubt that you would want to reinforce *all* of those responses. Indeed, I am sure you would not want to reinforce certain ones. What would be helpful? Since these responses are already occurring, we bring differential reinforcement to the rescue.

You would be helping your friend even if your main reactions involved following his problem–comprehension responses, and his emotion–clarifying responses, with some warm and empathic verbal and postural responses of understanding and agreement on your part. Meanwhile, you would also be helping your friend if you followed his pessimistic, confusing, apprehensive and mis–comprehending responses with verbal and postural responses of relative indifference or silence, which essentially means ignoring (i.e., extinguishing) these problematic responses. With these steps you are applying differential reinforcement. You might even witness some relatively immediate verbal indication of improvement on your friend's part. If your friend's problems were severe, he would likely need more help, perhaps even from an applied behaviorologist. The general lines of this example also apply at the professional level (see Truax, 1966). Some have claimed that behavior–focused procedures only manage some symptoms while ignoring the more deeply rooted problems that they assume must exist and that supposedly cause the symptoms, problems such as those that the machinations of some miscreant inner agent putatively cause, and that these "real," deep–rooted problems will surely manifest again through new symptoms in a sort of "symptom substitution" scheme. However, such claims are grounded in mystical assumptions, and credible substantiation of these claims has not occurred.

Nevertheless, the reader should be aware that trying to help oneself or others with procedures grounded in natural–science data violates the draconian laws that are on the books in some nation states. This particularly applies in states that maintain no–longer–appropriate legal sanctions favoring the guild interests of those like psychologists whose discipline cannot qualify as natural science. Check your state statutes and be careful to obey the laws, even while we all work to change them to better reflect reality. Meanwhile, let's consider a more complicated example of differential reinforcement.

Helping a professor. A human resource problem (i.e., not enough faculty) once left a colleague teaching an overbooked *Introduction to Behaviorology* course (BEHG 101). The course roster showed 160 students in a large lecture hall that had a broken sound system. The professor could project her voice adequately. However, the acoustics of the room made it difficult for the 40 students in the left rear quadrant of the room (from the professor's perspective) to hear well unless the speaker was standing well to the left of the podium, near the edge of the blackboard; from this location everyone could hear her well.

Due to the large class size, the professor had switched from her usual interactive classroom activities (which involved applications of the very science she was teaching for conditioning new behaviorological repertoires) to the lecture method (which became somewhat obsolete long ago with the invention of the printing press, and more obsolete more recently with the invention of programmed instruction; see Holland & Skinner, 1961, for an excellent example). She lectured for a two–hour period each week, and the students then met in groups of 40 for the other hour for review, questions, and quizzes. Each quadrant met with one of the four graduate teaching assistants (TAs) who were working with her on the course.

Now, the students in that left–rear quadrant were an unhappy lot, because they could not hear the professor very well. When they complained to their TA, she suggested that they talk to the professor about possible solutions to this problem. Since the professor was a bit famous, these young freshmen could not really see themselves following this advice. But another possible solution occurred to the TA when she began to review the last lecture with them. In that lecture the professor had described differential reinforcement, and… Wait. What's that? Since the students had been unable to hear the professor clearly, they really had little clue about what differential reinforcement was. So the TA reviewed differential reinforcement for them very thoroughly, and that was when they hatched the solution plot. Here is what happened.

Together, the TA and the students defined the professor's lecturing response class in a way that seemed conducive to changing it; in this case they defined it *positionally.* That is, they defined each of seven locations, from which lecturing could occur, as members of the response class of lecturing. These seven positions were at the podium, just right and just left of the podium, middle right and middle left of the podium, and far right and far left of the podium. Then they arranged to reinforce only the one member of this response class that enabled everyone in the room to hear the professor; this member involved lecturing from the far–left location (from the professor's perspective; i.e., on the professor's far left). They would ignore lecturing from all the other locations; they would extinguish all other response–class members.

All the TAs then coordinated their groups in a cooperative demonstration of differential reinforcement. Of course they needed an effective reinforcer. Since it was a state university that could only pay its professors poorly, they considered that throwing paper airplanes made from dollar bills might work, but as students they too lacked dollar bills. Then they thought of attention, and they laid their plans. At the next lecture, when the professor arrived and at any time she was anywhere except near that left edge of the blackboard, everyone would cast their eyes downward toward their desk, and move their pens slowly as if disinterested and doodling. However, whenever the professor was near that left edge of the blackboard, a position from which everyone could hear her, all heads would snap up, every eye wide open with interest, and every pen taking copious notes on everything the professor said.

Would that kind of attention reinforce the professor's behavior of lecturing from that location? They thought it could work, because that response at least occasionally occurred already. The professor's routine involved starting at the right edge of the blackboard (i.e., the left edge from the student's perspective) and putting a topic on the board, and turning around to comment on that topic at length before putting up the next topic. Gradually she would work her way through the topics and across to the left side of the room. What happened?

The first time she got to the left edge, on her third or fourth topic, she had put the topic of superstitious behavior on the board and then turned around and saw 160 bright–eyed, smiling and most–attentive faces. This was certainly an improvement; she had earlier noted the apparent gloomy mood of the class, and had wondered what caused it and, more importantly, how to fix it. Apparently a great deal of interest in superstitious behavior existed. So instead of the couple of minutes that she usually spent at that location, and on that topic, before returning to the other side of the room and beginning on another topic, she spent just over ten minutes on that left side talking about superstitious behavior.

By the way, the ten minutes on that topic provided an unplanned example of what kind of behavior? If you said, "superstitious behavior," then you are correct. The attention indeed reinforced lecturing from that location as planned—witness the increase from the usual three minutes to over ten minutes there—but the attention also *coincidentally* reinforced talking about that particular topic as well. As I am sure you recall, we use the term *superstitious behavior* to describe behavior that coincidental reinforcement conditions.

Well, this sequence played out a couple more times, with the students' heads returning to the disinterested position as soon as the professor moved from that sweet spot on the left side of the room. But each time that her routine took her back there, the heads popped up again and she spent a longer time there, finally lecturing for the last 35 or 40 minutes of the period from that spot while she erased the old topic from the board behind her and wrote the new topic in the same place. The students had successfully differentially reinforced the member of her lecturing response class that involved her lecturing from that particular spot on the left side of the room, a position from which everyone could hear her. The term we use for such outcomes is *response differentiation.*

At the end of the class, with a TA's help, the students thanked the professor for speaking from the acoustically helpful spot, and for this real demonstration of differential reinforcement. After getting over her initial surprise—and at least partly humorous threat to flunk them all—she agreed. It was a real, and realistic, application of the differential reinforcement procedure. The students also found that, like so many of the better professors, she was indeed very human and approachable. They need not have gone to such lengths to solve the classroom sound problem; she made the necessary phone call (as this was in the days long before the invention of email) and the problem got fixed right away.

While I have changed some details for pedagogical reasons, a professor of my acquaintance reported in a textbook the kind of experience that this example describes (see Whaley & Malott, 1971, p. 65–66). However, what if a different professor had taught this class, a new professor who stayed near the podium in front of the students, a professor who never ventured over to the side of the room where all the students could hear the professor well? Then the students *could not use* this simple differential–reinforcement procedure, because the target behavior of lecturing from that side of the room never occurs. Are any other procedures available to manage environmental variables that could produce the behavior of lecturing from that side? Yes, at least one is available, and the short name we use for it is *shaping*. Let's now turn to the shaping procedure, the procedure that you use when the target behavior never happens.

Shaping

Before changing our classroom sound–problem from an example of simple differential reinforcement to an example of *shaping,* let's consider what shaping involves. As with differential reinforcement, the shaping procedure involves a combination of the reinforcement and extinction processes. But differential reinforcement, which we sometimes call *simple* differential reinforcement, and shaping involve these two processes in slightly different ways. With simple differential reinforcement, extinction and reinforcement occur according to response–class membership; reinforcement follows the member that comprises the target behavior while extinction follows the other members. However, in shaping, both the response–class membership, and the *criteria* for which response–class member gets reinforced, regularly change. In essence, *shaping is the repeated use of differential reinforcement for conditioning a behavior that initially is not occurring.* Let's see how that plays out.

If the target behavior *is* occurring, then we need not start a shaping procedure. We apply simple differential reinforcement to strengthen a target behavior that initially occurs at least occasionally, because we *can* reinforce it since it is already occurring. But when the target behavior never occurs, we cannot follow it with a reinforcer. Then, we need to apply the shaping procedure, which does not require an already occurring target behavior.

The full name for the shaping procedure is *the method of successive approximation,* because it is the method of reinforcing responses that ever more closely approximate the desired, or target, behavior. You must, of course, start with a behavior that is currently occurring, because you can only reinforce a behavior that is occurring. But which one? At a minimum, after you have identified your target behavior, the initial occurring response that you start reinforcing should share some characteristic or dimension with your target behavior. Then, along this dimension, you reinforce those particular response versions that ever more closely approximate the target behavior.

However, to really describe shaping and how it works, we need to clarify that notion of "response versions." It comes from the concept of *variation in behavior.* When particular instances of a behavior happen, instances that we call responses, these instances are never absolutely identical. Generally, trained observers, and often untrained observers as well, can readily note the difference, the variation, between responses (i.e., the difference is usually of sufficient magnitude to evoke recognition and other relevant responses from observers—no inner agents are ever necessary or involved).

If we select a current, initial behavior that shares some dimension with the target behavior then, along this dimension, some of the small variations in the response instances will be slightly less like the target behavior while others are slightly more like the target behavior. We can shape the current behavior a little toward the target behavior by reinforcing those slightly more similar variations while extinguishing the slightly less similar variations. This is a use, our first use, of differential reinforcement in shaping our target behavior. How do we move on to additional uses of differential reinforcement?

As a result of that first use of differential reinforcement, the center of the variation range of the current—but now somewhat different—behavior shifts toward the target behavior. We then change the criteria for reinforcement to the side of the range of the newly current behavior that is still, and now even more, similar to the target behavior. This is not the behavior we reinforced under our first use of differential reinforcement; we consider this a new behavior, and we then use the differential–reinforcement procedure *again* to reinforce the slightly more similar variations that are on this side of the range while extinguishing the slightly less similar variations that are on the other end of the range. This is *another* use of differential reinforcement.

We then repeat that process of both changing the criteria for reinforcement, and applying differential reinforcement, again and again until the resulting behavior *is* our target behavior. We have then shaped this target behavior (which, remember, had not been occurring) using the method of successive approximation that involves the *repeated* use of the differential–reinforcement procedure on the newly shaped variations of each newly current behavior, each of which more closely approximates the target behavior along some dimension.

All that helps us see why, taking behavior variation into account, one of the easiest ways to define shaping is as *the **repeated use** of the differential–reinforcement procedure to condition a series of new response variations toward an initially non– occurring target behavior.* Also, note that the change in the criteria, regarding which responses get reinforced, can happen either with every reinforcement, as is often the case for laboratory studies, or only after responding has stabilized in each differential reinforcement use, as is often the case for applied interventions. That is, the change from the use of differential reinforcement with a member of one particular response class, to the next use with a member of a different response class, depends on criteria that reinforcement providers change either after *each* reinforcement occurs following a response or after *several* reinforced responses occur that establish some stability for each intermediate behavior between the initial and the target behaviors. To see shaping at work, let's start by changing our classroom sound–problem from an example of the simple differential–reinforcement procedure to an example of shaping in which the criterion for which responses get reinforced changes after several reinforcers stabilize responding.

Shaping Examples

Reconsider the question that ended our original classroom sound–problem example: What if a different professor was teaching this class, a new professor who stayed near the podium in front of the students, a professor who never ventured over to the side of the room where all the students could hear the professor well, a professor with whom the students could not use the simple differential–reinforcement procedure, because the target behavior of standing on that side never occurs? In this case we bring *shaping* to the rescue.

Students shaping professor behavior. We pick up the example when the whole class is in the classroom for the professor's next lecture. The whole class is then about to implement the intervention that the TA and students from the acoustically challenged left–rear quadrant have designed. To improve the consistency of applying the criteria changes that move from one differential reinforcement contingency to another, the sound–deprived students' TA is sitting down in the first row right in front of the professor, and everyone is keeping an eye on her for the cues regarding when to provide the reinforcer and when not to, which includes when to wait, because a criterion change requires a closer approximation before reinforcement occurs. When her head goes up all wide–eyed and smiling, then all heads go up all wide–eyed and smiling.

Also, having previously seen this new professor lecture from the center of the podium as well as while leaning to the left of, and to the right of, the podium (and only from these positions) the students and TA have defined the professor's initial lecturing response class as consisting of these three members (i.e., lecturing from the center, left, and right of the podium). Since the left–of– podium response–class members most closely approximate the target behavior of lecturing from the edge of the blackboard at the professor's far left, these are the initial responses that they reinforce when these responses occur.

Here is how this two–hour lecture class might go, based on the reasonable assumption that the successful reinforcer of the beaming–and–smiling–faces in our previous and *real* example would continue to function as a reinforcer for the behavior of this new professor. At first, each time the professor leaned to the left, the attention from all the eager faces beaming and smiling would occur. While some (perhaps much) back and forth movement of the professor might occur, along with a similar amount of heads–up and heads–down movement of the students, we would still see the reinforcing effect appear in the increasing time that the professor spends leaning or, more likely, standing on the left of the podium while lecturing.

A change in the criterion for reinforcement would become appropriate when standing to the left of the podium had become somewhat stable. At that point, the leading TA would wait for some additional leftward movement before going heads up with the reinforcer. Why could this additional leftward movement happen? Because when the left–of–podium standing responses stabilized, variation was still present. *Some variation is always present.* The professor now occasionally lectured a little farther to the left of this new position, as well as a little back to the right of this new position. Then, using differential reinforcement *again,* a new lecturing position would stabilize, and another reinforcement–criterion change would become appropriate.

Using differential reinforcement repeatedly, each time on slightly farther left positions than each new position that had just stabilized after several reinforcements, the reinforcements that the class delivered gradually shape the positional lecturing responses across the room, from the room center over to that left–edge position where everyone can hear the professor speak. This was a behavior that initially was not occurring. Yet now it occurs, thanks to the particular balance of reinforcement and extinction in the shaping procedure.

Note that the class had to work with whatever initial response was available. If the professor had not leaned to the right or left near the podium, but had only *looked* toward the right or left, then the class would have had to start by reinforcing looking to the left, because that would have been the initial response that most closely approximated the target behavior. I am sure you can imagine some the intermediate steps that shaping would require to go from that point to the target behavior.

People shaping pigeon behavior. Now, let's consider one last example of shaping. This time the criteria for which responses get reinforced change after each reinforcement. That is, after each reinforcement we will require additional response changes that more closely approximate the target behavior before another reinforcement occurs. This example also shows an increase in complexity in that shaping occurs with respect to the behavior of a literal flock of subjects, pigeon subjects, rather than a single individual, through the effects of an intermittent reinforcement schedule, because while any member of the flock could get the reinforcer, only one member actually gets any particular reinforcer. Here is how that worked.

This shaping example occurred in France. My wife and our son both play English Handbells. Some time back, she signed up herself and our son, Miles, who was 18–years–old at the time, to participate in the first ever *massed* English Handbell concerts in France. Over 100 ringers (who all paid their own way) and some 500 instruments were on the week–long tour with several concerts. I went along both for the music and to accompany our daughter, Susannah, who was nine–years–old at the time.

This shaping example occurred on the day of the last concert, which was at the *Eglise de la Madeleine* in Paris. During the afternoon, while the ringers were inside the church for a pre–concert practice, Susannah and I passed the time outside in the church yard. Unfortunately, very little was happening that might engage her there, although about a dozen pigeons flocked around elsewhere in the yard. They were initially rather unapproachable, always about ten feet away, and always keeping that distance if we moved toward them or away from them.

Then, recalling the small bread roll that I had put in my coat pocket from lunch, in case hunger struck one of us during the afternoon, I asked Susannah if she would like the pigeons to be more friendly. When she responded positively, I suggested that we use the roll to shape some pigeon approach behavior. Before describing how we shaped that behavior, ask yourself what steps you would take; we have already covered everything you need.

After turning the bread roll into lots of half–inch pieces, we sat on a bench and observed the pigeons, noting the ten–foot distance that they kept away from us. Then I suggested to Susannah that she toss a bread piece half way between us and the pigeons. As expected, the birds' excitement level increased when she complied, but they would not come close enough to get the piece. Briefly, however, they came closer by one or two feet. After retrieving that piece, she tossed it just within the edge of the milling flock's perimeter. Of course, they went for it, but the result was also that the flock's perimeter had moved closer to us. And this movement reinforced our bread tossing; remember, all of us—you and I and the pigeons—are all behaving organisms with behavior under control of the same fundamental natural laws. Our responses of tossing bread toward the flock of birds when their movements were slightly closer to us reinforced those movements, compelling even closer–to–us movements, while these same closer–to–us movements reinforced our bread tossing.

For several minutes we repeated that kind of shaping interaction, with the criteria for the next reinforcer changing each time to require movements that made the perimeter closer to us. By the end of only those several minutes, the pigeons covered Susannah from head to knee (as she was still sitting on the bench). Luckily they were rather polite pigeons, as they covered her only with themselves; they dropped nothing unpleasant on her. When we had distributed all the pieces of bread, the pigeons lost interest in us. However, the perimeter of their continued movements nearby had shrunk to about five feet, and they were still maintaining that closer distance when the time came for us to go inside about a half hour later.

So we see that shaping is a fun as well as a potent process. Yet all that is happening is that reinforcers are following some behaviors while other behaviors are extinguishing. Now, reinforcement and extinction (each alone, or together as in differential reinforcement) are natural processes that can occur as helpful procedures. And shaping also is a natural process, one that again we can use as a procedure to help improve problem behaviors.

Due to potential dangers, however, please resist trying to shape wild pigeons. While pigeons lack teeth (which makes them far safer as wild shaping subjects than squirrels) they can peck. Instead, I recommend shaping the usual range of fun performances, and healthful routines like exercise, in your domesticated pets. They make great initial shaping subjects. The best pet training protocols, which are applications of behaviorology to generating and maintaining behaviors of other animals, involve "clicker training," and a number of good references are available (see O'Heare, 2010, and Pryor, 1991, 1999a, 1999b, 2001; also see Johnson, 2004, and Kurland, 2000).

Professionals shaping client behavior. Before leaving the differential–reinforcement and shaping topics, let's revisit, and go beyond, the reference I gave you as an example of the application of these procedures at the professional level. That reference was Truax, 1966, an article addressing the inherent directiveness of Rogerian "non–directive" psychotherapy. Truax showed that Rogerian therapy was indeed directive, which accounts for its helpfulness, because the experienced therapist's feedback reinforces appropriate client responses in ways similar to our earlier helping–a–friend example. A truly random (i.e., non–directive) feedback pattern would likely reinforce more of a client's more common negative comments, which could easliy make the client's problems worse. This topic was one of the facets of the "Rogers/Skinner debates." The Truax article and these debates are part of the history that behaviorology and psychology shared, before the attempts of proto–behaviorologists to change psychology into a natural science had received adamant refusal, which led to the organizing of behaviorology as a discipline separate from and independent of psychology.

The debates between Carl Rogers (the developer of Rogerian psychotherapy) and B. F. Skinner (the original behaviorological scientist) are legendary with respect to the incommensurability between natural sciences like behaviorology and non–natural–science disciplines like psychology. Rogers represented so-called humanistic psychology, a psychology that presented itself as proud of seeing science (i.e., natural science, philosophically and methodologically) as largely irrelevant to human nature and human behavior. Such a position makes this psychology school perhaps the clearest version of psychology's general anti–science allegiance to the mysticism of inner agents that conveys secular-religion status on the psychology discipline. Other humanists, however, openly acknowledge the relevance of natural science to human nature and human behavior. We see this in the (American) Humanist Society's naming of B. F. Skinner as the *Humanist of the Year* for 1972 (Bjork, 1993, p. 220) after the publication of his book, *Beyond Freedom and Dignity* (1971).

From Consequences to Evocatives

The procedures we tackled in this chapter involved arranging *consequences* in particular ways to produce particular outcomes. The consequence–related procedures that we covered were differential reinforcement and shaping.

Procedures that involve arranging *evocative stimuli* in particular ways to produce particular outcomes are also possible. In the next chapter, we will tackle two of these, the backward–chaining procedure and the fading procedure.

Lastly, as usual, recognize the value of the two applied behaviorology procedures that this chapter described; think about how differential reinforcement and shaping might be integral parts of solutions that help reduce overpopulation and build sustainable lifestyles. In this way the natural science of behavior (i.e., behaviorology) contributes to the solutions to the behavior–related components of global problems, thereby helping to make possible, in a timely manner, the solutions to the non–behavior–related components that traditional natural scientists are addressing.❧

References (with some annotations)

Bjork, D. W. (1993). *B. F. Skinner: A Life.* New York: Basic Books.

Holland, J. G. & Skinner, B. F. (1961). *The Analysis of Behavior.* New York: McGraw–Hill. This book is the original comprehensively programmed text; the authors successfully applied the laws of behavior that it teaches to its design and use. It is still in use, because it works so well.

Johnson, M. (2004). *Clicker Training for Birds.* Waltham, MA: Sunshine Books.

Kurland, A. (2000). *Clicker Training for Horses.* Waltham, MA: Sunshine Books.

O'Heare, J. (2010). *Changing Problem Behavior.* Ottawa, CANADA: BehaveTech Publishing. While this book openly addresses a wide range of species, Dr. O'Heare has half a dozen other books that more specifically deal with dogs.

Pryor, K. (1991). *Lads Before the Wind: Diary of a Dolphin Trainer.* Waltham, MA: Sunshine Books.

Pryor, K. (1999a). *Don't Shoot the Dog! The New Art of Teaching and Training (Revised Edition).* New York: Bantam Books.

Pryor, K. (1999b). *Clicker Training for Dogs.* Waltham, MA: Sunshine Books.

Pryor, K. (2001). *Clicker Training for Cats.* Waltham, MA: Sunshine Books.

Skinner, B. F. (1971). *Beyond Freedom and Dignity.* New York: Knopf.

Truax, C. B. (1966). Reinforcement and non–reinforcement in Rogerian psychotherapy. *Journal of Abnormal and Social Psychology, 71,* 1–9.

Whaley, D. L. & Malott, R. W. (1971). *Elementary Principles of Behavior.* Englewood Cliffs, NJ: Prentice–Hall. In this book these authors also discuss the Traux (1966) article example; see pages 69–71.☙

Chapter 14
Arranging Evocatives— Backward Chaining and Fading Procedures

$\mathcal{W}$e commit these early chapters of Part II to considering, in some detail, various combinations of basic behaviorological concepts, methods, and principles, combinations that compile into more sophisticated processes and procedures. Recall that "procedures" is simply our term for processes that not only happen through the natural mediation of other behaving organisms, including other people, but that also happen as a function of contingencies that include reinforcers involving behavioral outcomes that benefit various individuals and sometimes society. At the professional level, we may use the term "interventions" for these procedures, these process combinations.

In Chapter 13 we began with combinations focused on arranging *consequences* in particular ways to produce particular outcomes. We began with two potent *postcedent* environment–change procedures, the *differential reinforcement* procedure, and *the method of successive approximation* that we more commonly call the *shaping* procedure. Both of these combine the two simple postcedent processes of reinforcement and extinction in ways that bring about beneficial behavior changes. In elaborating these two procedures, we also revisited the two concepts of response class and variation in behavior.

Now, in this chapter, we continue our coverage of combinations of behaviorological basics. This time, however, we focus on two potent *antecedent* environment–change procedures for arranging *evocative* stimuli in particular ways that produce particular behavior–change outcomes that are also of considerable benefit. This time we focus on the procedures that we call the *backward–chaining* procedure and the *fading* procedure.

Backward Chaining

The term "chaining" describes the process of responses linking together through the stimuli with which they share functional relations. This is how behaviors connect, or *chain,* to each other, something we have touched on in various previous chapters, something that happens both to neural behaviors (e.g., thinking, daydreaming, and other consciousness behaviors) as well as to neuro–muscular behaviors (e.g., putting on pants or playing a solo piece for cello or any other musical instrument). The chaining process leads to chains of stimuli and responses that we simply call stimulus–response chains.

Some stimulus–response chains are rather fleeting and might occur only once in any particular or exact sequence while other chains are more fixed and repeatable. Current contingencies often establish the components of fleeting chains (e.g., the contingencies of an approaching holiday, and its traditional complex recipes, that compel thinking about shopping for the necessary ingredients even during driving, while your driving responses are under direct stimulus control). On the other hand, prior contingencies often establish the components of fixed and repeatable chains (e.g., the contingencies that result in the musical compositions that control the building of the playing chains).

Some fleeting chains, particularly chains of neural responses (e.g., thinking about shopping for those recipe ingredients) happen as each response, as a real event, functions as the stimulus that evokes the next response. However, in many chains of neuro–muscular responses, each response produces a distinct stimulus that both reinforces that response and evokes the next response. We call this phenomenon, of a stimulus both reinforcing one response and evoking another response, the *dual function of a stimulus*. One interpretation of musical improvisation is that at least some parts or aspects of improvisation involve this kind of fleeting chain. Each playing response or musical phrase produces a stimulus sound that both reinforces the response or phrase that produced it and also shares in evoking the next response or phrase, which produces a stimulus sound that both reinforces the response or phrase that produced it and also shares in evoking the next response or phrase, and so on. Other contingencies control the length of the improvisation chain, including both the contingencies surrounding a music pattern, and—at least in traditional jazz—the contingencies determining the next player's improvisation time.

Under contingencies that drive applying helpful interventions, our concern here stresses chains in which the stimuli and responses explicitly link through the dual functions of stimuli. Let's examine this phenomenon more closely before turning to chaining procedures.

Dual Function of a Stimulus

The dual–stimulus–function phenomenon arises from the process that produces many evocative stimuli. Recall that in this process these stimuli are regularly present when another, reinforcing stimulus follows a response; as a result the presence of these stimuli comes to evoke the response, and we then call these stimuli *evocative stimuli*. In addition, and as an inherent part of this process, many of these stimuli–becoming–evocative–stimuli are also pairing with (i.e., occurring at the same time as) the reinforcers, which also makes these stimuli themselves conditioned reinforcers. Each such stimulus then serves two functions, the functions of both evocative and reinforcing stimuli, which we tact as the dual function of a stimulus. These stimuli now function both as a conditioned reinforcing stimulus (for a previous behavior) and as an evocative stimulus (for the next behavior). Indeed, we define *dual function of a stimulus* as a stimulus serving two functions as a result of that stimulus being regularly present when a reinforcer follows a response, which makes the stimulus come to function both as a conditioned reinforcing stimulus for the previous behavior and as an evocative stimulus for the next behavior.

We can diagram a stimulus–response chain by showing how each stimulus, in serving those two functions, links responses together into a chain. Here is one way to make such a diagram (indicating functions in temporal order):

$$\ldots \rightarrow S^{r/Ev} \rightarrow R \rightarrow S^{r/Ev} \rightarrow R \rightarrow S^{r/Ev} \rightarrow R \rightarrow S^{r/Ev} \rightarrow R \rightarrow S^{r/Ev} \rightarrow \ldots$$

Both functions appear in the single term "$S^{r/Ev}$." As an S^r each stimulus functions to reinforce the previous response while as an S^{Ev} each stimulus also functions to evoke the next response.

In building a chain, stimuli and responses can link together in either direction. They can link *forward,* from the start to the end of the chain, or backward, from the end to the start of the chain. Each direction can be helpful in some cases, although any evoked link only leads to later links.

The most valuable applied possibilities, however, seem to surround the procedure of building stimulus–response chains "backwards." Before covering this backward–chaining procedure, let's first consider "forward" chaining.

Forward Chaining

As a procedure, forward chaining links stimuli and responses into a chain by starting at the beginning of the chain and building links toward the end of the chain. This kind of chaining works well for simple or short chains with a small number of quick or discrete responses. For example, educational contingencies usually build the short chains that we call word spelling through forward chaining. We usually reinforce starting with the first letter of a word that someone is spelling, and each letter shares in evoking the next letter, and then the next letter, until all the remaining letters are present in the correct sequence (a process that also benefits from both equivalence relations and intraverbal relations, each of which we cover in their respective chapters).

Longer or more complicated chains, however, benefit from starting with the end of the chain and building links backward, toward the beginning of the chain. Let's now look at backward chaining.

The Backward–Chaining Procedure

As a procedure backward chaining has provided many benefits in a range of application and intervention areas. This kind of chaining works well when both the stimuli and the responses in the chain are explicit and directly accessible. Again, as a procedure, backward chaining links stimuli and responses into a chain by starting at the end of the chain and building links toward the beginning of the chain. Here is how that works.

You start with the final response, which is usually one that the main, often unconditioned, reinforcer follows. Then, using standard evocation training, you reinforce that response only in the presence of a particular stimulus. As a result this stimulus becomes an evocative stimulus for the response, and it also becomes a conditioned reinforcing stimulus that can condition another response by following the other response. When it follows the other response, which it reinforces, it also evokes the original response. Together, these two responses are now linked by that stimulus occurring between them.

Repeating that whole procedure builds another response onto the chain, which then has three linked responses. Repeating the procedure again adds another link. For an open chain, the more often you repeat the procedure, the more links you accumulate.

Let's break that procedure down just a little more. You have a response (R_A) that produces a reinforcer (that here is an unconditioned reinforcer, S^R) in the presence of a stimulus (S_1). This makes S_1 (as S_1^{Ev}) evoke R_A. The reinforcer's occurrence in the presence of S_1^{Ev} also makes S_1^{Ev} become a conditioned reinforcer, S_1^r. You arrange a prior response (R_B) to precede and produce S_1^r, which reinforces R_B *and* evokes R_A due to the dual functions that such stimuli serve. You then arrange for that S_1 to follow the prior R_B *only* in the presence of another stimulus, S_2. Soon S_2 (as S_2^{Ev}) evokes R_B which produces the reinforcing S_1^r that also evokes (as S_1^{Ev}) R_A, which produces the unconditioned reinforcer. This is a two–response chain under dual stimulus controls. Here is what this sequence looks like, with the dual function visible as a "dual" stimulus in two forms—one above the other—having the same number:

$$S_1^{Ev} \rightarrow R_A \rightarrow S^R$$
$$S_2^{Ev} \rightarrow R_B \rightarrow S_1^r$$

Observe how this diagram (and the next one) reads left to right as usual in terms of time flow, and *bottom to top* in terms of response progression, but *right to left* in terms of chain–construction procedural steps. Of course, you can add any number of additional stimulus–response links to the chain by repeating the procedure. If a chain inherently has a limited number of responses (i.e., a *closed chain*) then that number automatically limits how often you repeat the procedure; an *open chain* means that other variables control the limit. We will cover both types of chains in our examples. However, here is a generic diagram of an open chain (as long a chain as I can fit across the page) in which the responses are lettered, with the *last* response being letter A, and each dual stimulus is numbered, with the *last* $S^{r/Ev}$ being number 1:

$$S_1^{Ev} \rightarrow R_A \rightarrow S^R$$
$$S_2^{Ev} \rightarrow R_B \rightarrow S_1^r$$
$$S_3^{Ev} \rightarrow R_C \rightarrow S_2^r$$
$$S_4^{Ev} \rightarrow R_D \rightarrow S_3^r$$
$$S_5^{Ev} \rightarrow R_E \rightarrow S_4^r$$
$$...S_5^r$$

What would the next steps be, with respect to $S_5^{r/Ev}$? Also, note that the end of this chain, from the $S_2^{r/Ev}$ to the S^R, is identical to the two–response chain that we diagramed earlier. Both diagrams exemplify the definition of backward chaining. In the definition, *backward chaining* is the procedure of taking advantage of the dual function of a stimulus to build chains of responses by linking them with stimuli; you strengthen the function of the evocative stimulus for a response by reinforcing the response in its presence, and then using this stimulus, which has also become a conditioned reinforcer, to strengthen another response before it, and repeating this process from end to beginning until the chain is complete. Each stimulus functions both as an evocative stimulus for the next behavior and as a conditioned reinforcing stimulus for each added "previous" behavior. As we describe the examples, recognize that the ones that involve "practice" stress that such practice includes always proceeding *to the end of the chain* regardless of where in the chain a practice run begins. This supports success so much that we will say it again.

Let's now consider some examples of backward chaining. You might find drawing your own diagrams of the parts of each example beneficial, or at least compare each example to our earlier diagrams to see how they fit.

Some Backward–Chaining Examples

These examples include both open and closed chains, both humans and (in one example) other animals, both children and adults. Applying a couple of them can even improve your own performances.

Barnabus the rat. Our first example describes the 14–response chain that made Barnabus the rat famous half a century ago. People thought his performance was unbelievable, and they thought the training for it must have been nearly impossible. Perhaps some praise, or at least appreciation for the effort, should go to the people who were a natural part of Barnabus's contingencies and whom we call his conditioners. Then again, perhaps that is inappropriate, not only because they, like everyone else, lack inner agents who earn the praise by telling their host bodies the right things to do, but also because the task is actually not quite as hard as the public presumes; it was not easy, but neither was it nearly impossible, as any good scientifically informed and clicker–expert animal trainer will tell you and show you.

To grasp Barnabus's act, imagine a four–foot tall, four–story house that would make any doll envious. It was open across the front so people could observe the performance. The inside contained not only floors and rooms but also some connecting structures and other items that served as evocative and— through dual–stimulus functions—reinforcing stimuli for the Barnabus act. These included a light, spiral stairs, drawbridge, ladder, chain, pedal car, tunnel, straight stairs, tube, elevator, string (attached to a flag and the elevator), buzzer, response lever, and food pellets. Note that some of the appropriate responses likely required some skillful shaping on the part of Barnabus's conditioners (e.g., raising the flag, which lowered the elevator, and pedaling the car).

The act begins with Barnabus in the house waiting near the spiral staircase. The stimulus that evokes the first response is a light. When it comes on, Barnabus is off, up the staircase. He continues through the whole stimulus–response chain, ending up eating a small food pellet near the bottom of the staircase, where the light is now off. When it goes on, he will again be off.

Let's list the components of the stimulus–response chain that constitutes Barnabus's act. Note that the first response on the list is the last response that his conditioners conditioned, with the last response on the list being the first one that they conditioned. And remember that the procedure involves Barnabus always running through *all* responses, from the current new response to the last response at the end of the chain, each time they are conditioning or practicing a new stimulus–response link at the then start of the chain. The list shows most stimuli as $S^{r/Ev}$s; their appearance/occurrence reinforces the previous response and evokes the next response. Here is the chain:

S^{Ev}:   The light goes on (perhaps due to a timer set to give Barnabus a rest between runs) evoking…

R: Climbing the spiral staircase which leads to a platform on the next level where…

$S^{r/Ev}$:   *The presence of* a drawbridge evokes…

R: Pushing down the bridge and crossing it, which leads to another platform where…

$S^{r/Ev}$:   (The presence of) a ladder evokes…

R: Climbing the ladder, which leads to where…

$S^{r/Ev}$:   (…) A chain evokes…

R: Pulling the chain, which makes…

$S^{r/Ev}$:   A car appear that evokes…

R: Climbing into the car in which…

$S^{r/Ev}$:   Pedals evoke…

R: Pedaling, which takes the car through a tunnel, which ends at…

$S^{r/Ev}$:   The bottom of some stairs, which evokes…

R: Climbing the stairs, which leads to…

$S^{r/Ev}$:   A tube, which evokes…

R: Running through the tube, which leads to…

$S^{r/Ev}$:   An elevator, which evokes…

R: Entering the elevator, where…

$S^{r/Ev}$:   A string (attached to a flag and the elevator) evokes…

R: Pulling the string, which raises the flag while lowering the elevator down through all floors, which leads to…

$S^{r/Ev}$:   An open access on the ground floor that evokes…

R: Exiting the elevator, during which his steps trip a pressure switch in the floor setting off…

$S^{r/Ev}$:   A buzzer, which evokes…

R: Pressing a nearby response lever, which releases a pellet of food that, when it hits the food tray, produces…

S$^{r/Ev}$: A sound, which evokes...

R: Stepping to the food tray in which the now visible...

S$^{r/Ev}$: Food pellet evokes...

R: Eating the food pellet, which produces...

S^R: The unconditioned reinforcing effects from food consumption, with the pellet tray conveniently close to the light and the spiral staircase where, when the light comes on again, the chain begins again.

Barnabus may not need a standing ovation, but observing his thrilling performance, over and over again, could easily evoke one. Note that we have broken down the end of his performance into his last three stimulus–response components (from the buzzer to the food eating) although his conditioners would likely have conditioned these three as a unit.

On the other end of the chain, which is certainly long enough, the complexity of the doll house could have enabled an even longer chain, giving this chain the character of an open chain. Regardless of the length (i.e., the number of stimulus–response links) Barnabus's conditioners followed the procedure we previously described to condition each response in relation to the relevant evocative and reinforcing stimuli. Your describing each of the procedural steps that they took for each link can be tedious but beneficial.

For the record my first exposure to this example was, as a student, from its appearance in Whaley and Malott (1971, pp. 298–300). They cited Pierrel and Sherman (1963) as their source. As with many examples in this book, I have for decades shared the examples in this chapter with my students. Let's consider another example that the Whaley and Malott book inspired. This is a human example, which can generalize to help anyone around young children.

Putting on pants. Have you ever tried to teach a toddler to put on his pants by himself? With your lifetime of conditioning experience, putting on pants is now quite routine; for the toddler, however, it can be a major undertaking. Depending on the method, the outcome can lead to very strained or very pleasant relations. To make the latter reality, use backward chaining. Here is how backward chaining applies to this closed chain (i.e., a chain with definite beginning and ending points).

You start, of course, with the last component of the chain. So you put the pants on the child and pull them *almost* all the way up, say, to just below the child's rump. Then you provide a verbal evocative stimulus for the child's response, perhaps by saying, with plenty of enthusiasm, "Pull up your pants." You might need to put the child's hands on the top of the pants when you say that; you might even need to mold the response by taking the child's hands while they are grasping the pants and helping pull the pants up; supplementing the usual reinforcer (i.e., the pants being on) with more added attention and verbal reinforcers can be helpful here. If the interaction seems fun (for both of you) then you can remove the pants and restart and run through the process a couple more times, or you can let progress accumulate with each normal occasion for putting on pants.

When saying "Pull up your pants" effectively evokes the child's pull–up response from this initial position, you are both ready for the next chain component. Next time, you pull the pants up only to the above–knee level and then say, "Pull up your pants." That will evoke the pulling up response which will put the pants at the below–rump level, a position that now *already functions* as the evocative stimulus for pulling the pants the rest of the way up.

The remaining chain components accrue similarly. When each new position effectively evokes the child's response of pulling his pants up such that previous positions then also evoke further pulling until the pants are on, then you move on to the next chain component, then the next, until you can hand the child his pants and simply say, "Put on your pants," and the pants get put on. A typical set of chain components includes the pants at below–rump level, at above–knee level, at above–ankle level, on only one ankle, and simply handing the pants to the child, or even letting the child remove the pants from a drawer. Rather than being a bother for both of you, backward chaining makes the task both successful and fun.

Whaley and Malott (1971, p. 300) inspired that putting–on–pants example through several examples that they cast in terms of teaching developmentally delayed children various skills. They cited Breland (1965) and Caldwell (1965) as their two main sources. In every case the examples followed the backward–chaining procedure that we previously described to condition each response in relation to the relevant evocative and reinforcing stimuli. Let's consider another human example, one which can help any player of a musical instrument.

Performing a piece of music without the sheet music. Have the contingencies operating in your past ever produced musical–instrument playing beyond the beginner's level? If so, are you ever under contingencies to perform a solo piece in a setting where the presence of sheet music is inappropriate (e.g., a concert setting or a competition)? In preparing to perform this kind of closed chain, would forward chaining help most, or would backward chaining help most? In both cases you first analyze your piece for the boundaries of its inherent sections and sub–sections, not only for the usual musical reasons but also because these become your chain components. Let's explore both forward and backward chaining in these circumstances, especially in terms of the likely outcomes that might occur during and after the performance.

In the simpler forward–chaining procedure, you begin to practice your piece by playing through its first chain component several times, repeating a couple of times any little parts that give you any trouble. Then you play through the second chain component several times the same way, followed by playing to this point from the beginning. You then repeat this whole procedure for all remaining chain components.

Seemingly all prepared, and a little nervous, you start to play your piece in the concert hall. The beginning is easy; after all, you have practiced it more often than any other part, an inherent characteristic of the forward–chaining procedure. As you get farther into the piece, however, little errors creep in, but as a professional you push on. Still, the piece is getting more and more difficult. Why? Because the farther you go, the more you get into less and less practiced material. These contingencies force far more focus on technical aspects of playing just to get the notes right, while the artistic aspects gradually all but disappear. You are far more nervous near the end of your piece than when you started, and the audience seems more tempted to throw tomatoes than roses. Such are possible outcomes of forward chaining.

Even though it seems more complicated, let's try backward chaining instead. In this procedure, you begin to practice your piece by playing through its *last* chain component several times, (always repeating a couple of times any little parts that give you any trouble). Next you play through the second–to–last chain component several times, each time playing through the last component as well; you always play through *to the end of the piece,* every time you play any chain component. Then you play through the third–to–last chain component several times, as usual playing to the end of the whole piece. And you repeat this whole procedure for each and every remaining chain component.

This time, seemingly all prepared, and a little nervous, you start to play your piece in the concert hall. The beginning is easy; after all, you were practicing it off stage just before your stage call. This time, however, as you get farther into the piece, your backward–chaining practice pays off as your performance gets ever better, because the farther you get, the better you play, as you have practiced all later parts more than earlier parts, an inherent characteristic of the backward–chaining procedure. Now *these* contingencies induce far more focus on the artistic aspects of playing, because your practice procedure already polished the technical aspects. You are far less nervous near the end than when you started, and you are playing better than at the beginning. Possibly more importantly the audience seems more appreciative of your technical and artistic prowess, and more prepared to throw roses, and perhaps even money.

Again, the best outcome in this example followed the backward–chaining procedure that we previously described to condition each response in relation to the relevant evocative and reinforcing stimuli. I first discovered that many musicians already followed this procedure when I asked my spouse, Nelly Case (a professor of music at the Crane School of Music at SUNY–Potsdam) about it. She pointed out that many voice and other instrumental performance faculty long ago came under practical contingencies to practice pieces through the backward–chaining procedure, and to pass this procedure on to their students, although they may not have known this scientific name for it.

Is music playing the only activity to benefit from backward chaining? As a sample to show that this procedure can benefit many more activity areas, let's briefly consider one last human example, one which can help anyone with a public–speaking engagement.

Delivering a speech without a text (or an ear piece). Have the contingencies operating in your past ever required delivering a speech that at least appears extemporaneous (i.e., no written text, and no ear piece/

teleprompter)? As was the case with the last example, in preparing to perform this kind of open chain, backward chaining would be very helpful, certainly more helpful than forward chaining, and for similar outcome reasons. (Can you explain why we could call this example an open chain?)

Of course, your speech will not be extemporaneous; it will be well rehearsed. You begin to practice your speech by reading the last sentence, as the last chain component, several times (although some use paragraphs as chain components, but that is more difficult). Ultimately you are trying to recite the sentence without reading it or, in more technical terms, until the first word of the sentence evokes its full recitation (with more of those controlling contributions from intraverbal relations which, as we mentioned earlier, we will cover in the verbal behavior chapter). Next you read the sentence before that several times, again trying to recite it without reading it, but each time *always* also reciting—without reading—all remaining sentences through to the end of the speech. That is, you practice until hearing each sentence without reading it reinforces reciting that sentence *and evokes reciting the next sentence* (i.e., in simple terms, each sentence leads you into the next sentence). You repeat these steps for each and every remaining sentence *back* through the speech until you can start the first sentence, which you have practiced the least, and then chain through all the sentences to the end, which you know so well, as you have practiced it the most by always reciting the speech through to the end. Before long the first word of the speech is sufficient to evoke the full recitation of the whole speech. Then, normal contingencies induce other characteristics of good public speaking to occur that can endear you to your audience.

My first exposure to this example occurred through personal experiences, the contingencies of which definitively demonstrated the increased reinforcing value of backward–chaining practice. Your contingencies will predictably generate similar outcomes in many activity areas.

However, continually recognize that when applying backward chaining with practice repetitions, such practice must include always proceeding *to the end of the chain* on each practice run regardless of where in the chain a practice run begins. This key element of the procedure maximizes the benefits of backward chaining. Now, let's move on and consider the procedure that we call "fading."

The Fading Procedure

Some procedures share the characteristic of *gradual change,* which evokes part of our descriptive responses regarding these procedures. We have already seen the kinds of gradual change that the *shaping* procedure brings about. Shaping starts with one response and gradually, as we apply the procedure, changes that response along some dimension (e.g., topography) to a different response.

Fading also involves gradual change and, like shaping, can occur either as a process or as a procedure (i.e., an intervention). While shaping gradually changes characteristics of responses, fading gradually changes characteristics of stimuli, leading to different effects of evocative stimuli. The fading procedure starts with at least two stimuli that share one or more characteristics. Each of these characteristics can be similar or dissimilar across the stimuli. We refer to such similarities and dissimilarities as dimensions of the stimuli. Fading involves gradual changes of, or in, a stimulus (or dimension of a stimulus) that result in the *transfer of stimulus control* from one stimulus (or dimension of a stimulus) to another stimulus (or dimension of a stimulus). In some cases, the procedure actually involves literally fading out a stimulus.

To keep our description of fading simple, it may imply that the whole stimulus gradually changes. However, far more commonly, only one or another dimension of the stimulus gradually changes. Before fading begins one stimulus exerts control over (i.e., evokes) the response while the other stimulus is merely present along with this evocative stimulus. As the fading procedure unfolds, we make small changes in the evocative stimulus while it repeatedly evokes the response and the other stimulus remains unchanged.

At that point two possible outcomes exist. In one outcome the changes make the evocative stimulus into a different—sometimes very different—*yet still effective evocative stimulus,* while the other stimulus simply remains unchanged; as we will see in an example, this is a valuable outcome, especially if the final evocative stimulus form could not originally evoke the response. In the other outcome, the changes make the evocative stimulus into a different—sometimes very different—stimulus, *but one that no longer serves an evocative function,* while the other, unchanged stimulus actually becomes the evocative stimulus; as we will see in a different example, this too is a valuable outcome, especially if the unchanged stimulus could originally not clearly evoke the response. In both of these cases, whether changing the evocative stimulus form or making the unchanged stimulus evocative, the control that a stimulus exerted at the start of fading gradually changed into control that a different stimulus exerts at the end of fading. The point of the fading procedure is to get evocative stimulus control to transfer from one stimulus, or dimension of a stimulus, to another stimulus, or dimension of a stimulus.

Notice also that something "extra" often happens in fading. In the usual evocation training process that makes a stimulus function as an evocative stimulus, errors are common; we call each unreinforced response that occurs in the presence of only the non–functional stimulus (i.e., in the presence of the S^Δ, as we described in the last chapter of Part I) an error. In fading, however, the stimulus that evokes the response at the end of fading usually comes to evoke the response errorlessly, that is, without the errors that occur during regular evocation training. Fading can produce *errorless evocation.* (In the past we called this errorless "discrimination," a term with inner–agent implications—Who is doing the discriminating?—such that we rarely use it anymore, as we discussed in a previous chapter). Let's see if such errorless evocation, in our two examples of fading, evokes neural (i.e., non–agential) recognition responses on your part.

Two Fading Examples

Here are two examples of fading. In both, gradually changing one stimulus results in a stimulus–control transfer. In the first example, the transfer is to the changed form of the stimulus that underwent gradual change. In the second example, the transfer is to the unchanged stimulus. As a matter of record, Whaley and Malott also inspired both of these examples (1971, pp. 195–201).

Changing the evocative–stimulus form. Let's say you needed to teach a young child to respond to her name by pointing it out from among at least two possibilities. Through the fading procedure, you succeeded. Here is an example of how that could have happened.

You began by putting the child's name, Laurie, in white letters on a black card, and you put another name, Debbie, in white letters on another black card. Then you presented the cards to Laurie along with the evocative stimulus, "Point to your name." Unfortunately, she merely took the cards and waved them around, or ran around the room with them. However, you had earlier interactions with Laurie, and these showed that she likes small fish–cracker snacks. So you made giving her a fish cracker contingent upon her pointing to a card. After a little shaping, every time you placed the cards before her and recited the evocative stimulus, she pointed to one of the cards, and you provided a fish cracker. (You should diagram these contingencies as we describe them; that would help clarify them, and further expand your repertoire.)

So Laurie was now at least pointing to the cards. However, even though you often reversed the position of the cards, roughly half of Laurie's responses were still errors with respect to her name, because she seemed to be pointing to whichever of the cards was closest to her; the names were not exerting any stimulus control over her responding.

At this point you invoked the fading procedure. You took the white letters of her name off the black card, and moved them to a white card of the same color as the letters; this card duplicated every aspect of the other name's black card except color. In addition you began providing a fish–cracker reinforcer for only some of her correct responses, which now involved pointing to the white card with her white–lettered name on it. Within just a few trials, she was consistently pointing to the card with her name on it, although the overall whiteness was the stimulus controlling her responses, not her name.

Next, you instituted a series of gradual changes. Every time she pointed to the correct card on 40 consecutive trials, you moved the letters of her name from the current card to a new card of a *slightly* darker shade. And you found that she continued to make correct responses after every such shade change.

After 11 or 12 gradually darkening shade changes, the card with her name, *Laurie,* on it—in white letters, as always—was the same shade of black as the card with the other name, *Debbie,* on it, which was the card–name combination that you had never changed. More importantly she still continued to point to her name card. Fading succeeded.

The control that the white–name–on–white–card stimulus originally exerted, on Laurie's pointing behavior, transferred to the white–name–on–black–card stimulus that ended the sequence of gradual changes. The only difference between the two cards at the end of this sequence was the difference between the names, *Laurie* and *Debbie.* Now, instead of whiteness controlling her pointing responses, her name *Laurie* (in white letters on a black card) controlled her pointing responses.

Also, throughout all the changes to ever darker shades, Laurie's responding was errorless. When the card and letters were both white, the whiteness was the stimulus controlling her responses. As the color of these cards gradually changed to darker shades, no errors happened even as the stimulus control was shifting from the overall card color to the pattern of the letters of her name. The final *Laurie* stimulus (as white letters on a black card) came to evoke her pointing responses errorlessly, an example of errorless evocation.

In this name–card fading example, at least one early version of which really happened (Whaley & Welt, 1967) stimulus control transferred to the changed form of the stimulus that underwent gradual change. Now let's look at a fading example in which stimulus control transferred to the *unchanged* stimulus.

Making the unchanged stimulus evocative. Let's say you needed to test the hearing of children who lack language behavior. Through the fading procedure, you succeeded. Here is an example of how that could have happened.

Late one summer a dozen children turned up, in the kindergarten and first–grade cohort of your local school, showing not only little or no language (i.e. verbal) behavior but also a range of other behavior deficits and excesses, some severe. The school reported that typically only one or two such children show up each year in the entering cohort. Out of concern both for these children and to discover why the number was larger than usual (and harboring a reasonable fear of some connection with one of the civil or industrial hazardous waste dumps near your city, which would require an additional level of intervention) the school called the behaviorology (not the psychology) department at your university for help, and the department chair directed the call to you.

Under control of the "little or no language behavior" part of the report, and considering that the natural–science analysis of verbal behavior (AKA language) clarifies the vital role of hearing in verbal–behavior conditioning, one of your first questions would have been "What do the children's hearing tests indicate?" However, since hearing tests at the time required at least minimal language skills to make the test steps work, you already knew that the answer would have been, "What tests? These kids lack language so we cannot give them hearing tests." You knew that you would have to find another, and non–verbal, method to test these children's hearing. And you also knew that the longer you took to solve this problem, the sooner someone else might move, with little data, and put as yet unjustified diagnostic labels, such as "autistic," on these children, which would likely lead to inappropriate treatments (e.g., unnecessary drugs or psychological therapies or institutionalization).

Essentially, you wanted to find a procedure that gets the children to "tell" you, somehow, that "I hear a sound" and "I no longer hear a sound," which are the responses that various verbally instructed methods produce in standard hearing tests. Your recent work with the fading procedure induced your application of fading to help solve this problem.

With the cooperation of the school, you converted a small meeting room into a temporary experimental chamber, and you set up, along one side, a false wall with some stimuli, manipulanda, and reinforcer sources. You would manage the stimuli and maintain the supply of reinforcers while sitting in the space behind the false wall. The manipulanda were two levers protruding from the wall that you set a few feet apart so that a child could not operate both at the same time. Below each lever you made a small hole in the false wall through which your reinforcer dispensers could drop various reinforcers (e.g., raisins, chocolate chips, peanuts, and various other consumables, or trinkets) into a small wall–mounted tray. Above each lever you installed a light. Turning a rheostat that you placed behind the false wall would dim both lights together in fixed increments. Three feet above the level of the lights, and between them, you also mounted a small speaker on the false wall. You wired the speaker so that it sounded a simple tone whenever the left light was on, but was silent whenever the right light was on. Only one light was ever on at a time, and the tone was a mid–volume, mid–range tone; your hearing–science colleagues said that if a child could hear at all, he or she would very likely hear this tone.

Let's consider the children's interactions with this apparatus. First, however, here is a diagram of how the stimulus set–up of this apparatus worked:

Left lever S^{Ev}: Light on (& Tone **on)**   Right lever S^{Ev}: Light on (& Tone **off)**
(Left lever S^{Δ}: Light off [& Tone off])   (Right lever S^{Δ}: Light off [& Tone on])

	HIGH WALL SPEAKER	
LEFT LIGHT		RIGHT LIGHT
LEFT LEVER		RIGHT LEVER
S^R TRAY		S^R TRAY

To begin, you demonstrated this apparatus with each child, one at a time. You made a few presses on the left lever with the left light lit (while the light on the right was off) and at each press a reinforcer fell into the tray. After giving these to the child, you put the child's hand on the lever, and off the child went, eagerly working the lever, which collected reinforcers. After a moment, that left light went out as the right light came on. Following some unproductive left–lever–press errors, the child moved over and began to operate the right lever under the now lit right light, which now produced reinforcers. Each time the current evocative light went off, the other light came on at the same time and evoked switching levers and pressing some more. To avoid satiation, you changed the reinforcement schedule from CRF, on which every lever–press response produced a reinforcer, to a VR–8 (variable ratio–8) schedule, on which an average of eight responses produced a reinforcer.

Within a matter of minutes, the behavior of each child came under control of whichever light was on, switching quickly when the current light went off. But so far this was only a nice demonstration of stimulus evocation, with a lit light as the evocative stimulus. How did this help to get each child to "tell" you that "I hear the sound" and "I no longer hear the sound"? Here is how.

To get each child to "tell" you that, you implemented a fading procedure. Recall that whenever the left light is on, a sound plays over the speaker, and whenever the right light is on, no sound plays over the speaker, as our S^{Ev}/S^{Δ} diagram showed. So you started adjusting the rheostat to make the lights dimmer; light intensity provided the dimension along which fading operated. And the children continued to respond through every decrease in light. Typically, you gradually faded the lights across half a dozen rheostat settings until the lights were not only completely off, but you physically removed them from the wall (which is not actually a necessary step in the fading procedure). Yet each child still continued to respond appropriately, pressing the levers according to whether the sound was playing over the wall speaker or not. After fading out the lights, here is how the remaining stimulus set–up worked:

Left lever S^{Ev}: Tone **on**   Right lever S^{Ev}: Tone **off**
(Left lever S^{Δ}: Tone off)   (Right lever S^{Δ}: Tone on)

HIGH WALL
SPEAKER

LEFT   RIGHT
LEVER   LEVER

S^R TRAY S^R TRAY

The children were finally telling you, "I hear the sound" and "I no longer hear the sound." As the lights gradually dimmed, literally fading out to off, the evocative stimulus control that they had exerted gradually transferred to the wall speaker. The lights gradually stopped evoking presses on a particular lever as the presence and absence of the speaker sounds came to evoke the children's responses. And this happened errorlessly.

Throughout the fading changes that made the lights dimmer, the children's responding was errorless. At the start, the lights were the stimuli evoking their responses, and once conditioned, these responses continued virtually without error, even though by the end of fading only the presence and absence of the speaker sounds evoked their responses. The children made essentially no response errors as the stimulus control shifted from the lights to the wall speaker. The speaker–sounds stimuli came to evoke their responses errorlessly, another example of errorless evocation.

Did all your effort make any difference. You bet! As a last step, you shaped the wearing of standard earphones. Your hearing–professional colleagues then proceeded to run a standard hearing test with each child, which enabled your physician colleagues to make a proper diagnosis of each case where the test revealed hearing difficulties, and to move forward with appropriate corrective strategies. The school then placed all the children whose hearing improved into special language–delayed classes to "catch them up" as much and as quickly as possible. What about the couple of children who tested with normal hearing but still showed severe behavior deficits or excessses? This is about the number that this school typically experienced turning up in each kindergarten or first–grade cohort. The school asked your behaviorology department team to make further evaluations of these children and to arrange for interventions appropriate for each child's particular condition. As for why a dozen children lacking language repertoires entered the school that year, we may never know; unfortunately little evidence ever supported any particular account.

While in the earlier name–card fading example, stimulus control transferred to the changed form of the stimulus that underwent gradual change, in this hearing–test fading example, stimulus control transferred to the unchanged stimulus. Versions of this example have also really happened. The outcome of the Meyerson and Michael intervention (1964) contributed to hearing–science practitioners developing a standard method using fading as part of a procedure to test the hearing of anyone lacking facility in the particular language of the hearing–test provider.

Timeliness of Works Cited, and Chapter Conclusion

You probably noticed that nearly all of the works cited in this chapter occurred in the 1960s. Certainly other, more recent references are available. These, however, keep us in contact with some of the classic research that has occurred in this science over the last 100 years; for this reason we often provide older references among our best recent ones.

Lastly, as usual, recognize the relevance of the behaviorological application procedures that this chapter described. Think about how we can make them—how can we make backward chaining and fading—contribute to the natural–science–of–behavior (i.e., behaviorology) work regarding the solutions to the behavior–related components of global problems, thereby helping to make possible, in a timely manner, the solutions to the non–behavior–related components that traditional natural scientists are addressing.

References (with some annotations)

Breland, M. (1965). Foundation of teaching by positive reinforcement. In G. J. Bensberg (Ed.). *Teaching the Mentally Retarded* (pp. 127–141). Atlanta, GA: Southern Regional Education Board.

Caldwell, C. (1965). Teaching in the cottage setting. In G. J. Bensberg (Ed.). *Teaching the Mentally Retarded* (pp. 159–163). Atlanta, GA: Southern Regional Education Board.

Meyerson, L. & Michael, J. (1964). Hearing by operant conditioning procedures. *Proceedings of the International Congress on Education of the Deaf,* 238–242.

Pierrel, R. & Sherman, J. G. (1963, Feb). Barnabus, the rat with college training. *Brown Alumni Monthly,* 8–12.

Whaley, D. L. & Malott, R. W. (1971). *Elementary Principles of Behavior.* Englewood Cliffs, NJ: Prentice–Hall.

Whaley, D. L. & Welt, K. (1967). Use of ancillary cues and fading technique in name discrimination training in retardates. *Michigan Mental Health Research, 1,* 29–30.

Reader's Notes

Chapter 15
Basic Schedules of Reinforcement

After covering, in Chapter 13, the *postcedent* processes of differential reinforcement and shaping, we turned in Chapter 14 to the *antecedent* processes of backward chaining and fading. Now, in this chapter, we return to postcedent processes by considering schedules of reinforcement. All of these processes also work as procedures (i.e., interventions).

In our examples of reinforcement in past chapters, we generally stated or implied that reinforcers occurred after each response, which gives us the reinforcement schedule that we traditionally call "CRF," for "continuous reinforcement." We sometimes pointed out that reinforcers could occur after only *some* responses rather than after *all* responses, but we resisted adding that increased complexity until we could focus on it, as in this chapter.

Here we consider the basics of the schedule topic, the topic of intermittent reinforcement, in some systematic detail. Several all–natural (as no other kinds exist) contingency arrangements manage the occurrence of reinforcers after some, rather than all, responses. These comprise the intermittent–reinforcement schedules. We begin by covering the four fundamental reinforcement schedules, followed first by a selection of other schedules, and then by the building of schedule performance as part of a range of related schedule considerations.

Four Fundamental Schedules

The term *schedules of reinforcement* refers to the contingency patterns of reinforcer occurrence, with most schedules involving reinforcers occurring occasionally rather than after every response. These schedules provide a range of exacting contingency determinants for the behavior patterns of humans and other animals, in both individual and social circumstances. These schedules also replace a host of putative inner–agent "motivational" causes of behavior.

We consider four schedules as fundamental for reasons that include their early discovery and their roles in compound schedules. We differentiate these four in terms of either the number of responses since the last reinforcer occurred, which we call *ratio* schedules, or the amount of time—plus a contingent response—since the last reinforcer occurred, which we call *interval* schedules. The values of either schedule type can be fixed or variable, thereby defining the four fundamental intermittent schedules of reinforcement: fixed ratio (FR), variable ratio (VR), fixed interval (FI), and variable interval (VI).

As we describe each of the four basic schedules, notice that each produces its own characteristic response–rate pattern. For simplicity here, we will analyze each of the patterns from these four schedules in terms of just two characteristics of the response rate *under established (i.e., stable) responding.* One characteristic concerns whether the response–rate pattern appears generally high or low while the other characteristic concerns whether the response–rate pattern appears generally steady or unsteady. From observations of these response–rate characteristics of a behavior of concern, you can deduce the general schedule type likely controlling the behavior. Keep checking Figure 15–1 for these characteristics as we consider each of the four fundamental schedules.

Ratio Schedules

We define ratio reinforcement schedules as those on which a reinforcer is contingent *only* on some number of responses occurring since the last reinforcer. This number of responses per reinforcer (i.e., the ratio of responses to reinforcers) can have a set value or a changing value.

Fixed–ratio schedules. If the value of this number remains constant, then we call the ratio schedule a *fixed–ratio* (FR) schedule. Let's set the responses–to–reinforcer ratio at 100 to 1 (although it could be any value). In this case, a reinforcer follows every hundredth response, 100 responses for each reinforcer. We denote this schedule by writing FR–100 (and, in a manner similar for all reinforcement schedules, we read FR–100 as "*ef–ar* one–hundred").

The value, of course, could be a higher value or a lower value. It could even be 1 (i.e., a ratio of 1 to 1, one response per reinforcer). This would constitute an FR–1, a schedule we already know as CRF (continuous reinforcement).

We note several characteristics of behavior on fixed–ratio schedules (see Figure 15–1). Most obviously, the cumulative record shows a step–like pattern. Also, the overall response rate, while not steady, is typically relatively high. We refer to such a behavior pattern as a "break and run" pattern; once stimuli evoke the first response of a ratio, we observe the organism running through the remaining responses until a reinforcer follows the last response of the run.

After the reinforcer occurs, however, a pause in responding occurs before stimuli evoke the start of another run. We have traditionally called this pause the *post–reinforcement pause,* although some folks prefer to call it the "preratio pause." In either case, these pauses on fixed–ratio schedules have generated considerable research interest and we will have plenty to say about them.

What brings about those basic response pattern characteristics of a high and unsteady rate? The answer to this question can involve some complex variables as well as some simple ones that match the scope of this book.

As for the high rate, recall that reinforcer occurrence depends solely on responding. Thus, more reinforcers occur when more responding occurs, which means that the faster responses follow each other (i.e., the higher the rate) the faster reinforcers accumulate. This part of the schedule contingency could induce continually rising rates; but such rates rapidly produce counter–controlling contingencies, like fatigue, which oppose continually rising rates and lead, by algebraic summation, to the observed, relatively high rate.

As for the unsteadiness of the rate, for starters recall that when responses produce no reinforcers, in the process of extinction, the response rate takes a hit, as the saying goes. Now, while we look at the fixed–ratio cumulative record—in Figure 15–1—and consider the fixed–ratio contingencies, would any responding right after reinforcer occurrence ever produce reinforcement? Since the answer is no, we should experience no surprise when no responding occurs right after reinforcement. However, since further reinforcement remains contingent on completing the next response run, we still experience no surprise when, after a pause, the next run commences. Other more complex variables actually have a greater hand in determining the post–reinforcement pause and its duration, and thus the unsteadiness of the fixed–ratio response rate; we will soon touch on some of them briefly.

The example of fixed–ratio schedules that most people easily recognize, possibly because they see it as making people work too hard, involves piece–rate pay (a pay method that FR contingencies produced long before this science discovered the schedule). Regardless of what you are producing, your pay is neither by salary nor by the hour; your pay, your reinforcer, accrues after completing a fixed amount of product, such as sewing five shirts or filling a bushel basket with apples.

Variable–ratio schedules. If the value of the number of responses per reinforcer changes, varying around some average, then we call the ratio schedule a *variable–ratio* (VR) schedule. Let's again set the responses–to–reinforcer ratio at 100 to 1. In this case, however, a reinforcer does not follow every hundredth response but instead follows every *average* of 100 responses. We denote this schedule as a VR–100. That is, on a "VR–100" schedule, a reinforcer would occur, not after *every* hundredth response—which would be an "FR–100" schedule—but after every set of responses, with the sets *averaging* 100 responses; for instance using one of several available methods to arrange this VR schedule, ten reinforcers would occur in 1,000 responses with each reinforcer following a set comprised of between one response and 200 responses.

A few behavior characteristics stand out on variable–ratio schedules (see Figure 15–1). With the cumulative record showing only rare pauses, the overall response rate maintains a fairly constant steadiness at a relatively high rate. What brings about these response–rate characteristics? Regarding the high rate, again reinforcer occurrence depends solely on responding. Thus, the higher the rate, the faster reinforcers accumulate (until counter–controlling contingencies—fatigue is again a good example—oppose continually rising rates leading, by algebraic summation, to the observed, relatively high rate). As for the steadiness of the rate, recall that occasionally the variation in ratio sizes includes the occurrence of two (and sometimes even three) reinforced responses in a row, or at least very close together. *And such occurrences have happened in the past* as the contingencies in the current VR schedule established stable responding; contingencies arranged this way induce steady responding. That is, pauses or other delays under these contingencies reduce the amount of reinforcement and so become less likely to occur, which leaves a relatively steady response–rate pattern.

Sometimes people describe this pattern as a sort of intense *persistence* in responding, but recall our previous discussion about description versus explanation: No "things" called intensity or persistence cause the continuously steady, high–rate behavior pattern. Instead and at best, the intensity and persistence labels provide verbal shortcuts for the functional contingencies that make up the variable–ratio schedule responsible for the pattern; at worst, these labels comprise aspects of an inner agential explanation that is unneeded in accounting for the behavior.

Outside the laboratory VR schedules are common. The example of variable–ratio schedules that perhaps too many people easily recognize involves games of chance such as gambling. The relatively rapid and steady response pattern that VR schedules produce readily evokes images of the behavior of players on traditional casino "one–armed bandit" slot machines. Hence we often refer to the VR schedule as the "gambler's" schedule. Considering the pattern of behavior observed in the laboratory on the VR schedule, let's examine the behavior of the gambling slot–machine player that you observe. The player's behavior involves a repeating pattern of *put coin in slot and pull handle down* then *put coin in slot and pull handle down,* and so on, again and again, at a high and steady rate, with only occasional interruptions from collecting the monetary reinforcer. And this behavior pattern often continues, either on into the night (until closing, if the casino ever closes) or until the player runs quite short of funds (i.e., "loses his or her shirt" as the saying goes). Again, for such reasons, we call the VR schedule the "gambler's schedule."

Contingencies, like those in VR schedules, produced gambling centuries before science discovered and analyzed this schedule. Back then, as now, the laws of nature, including the laws of behavior, affected people, controlling all behavior. So even back then, as now, the VR schedule–induced response patterns compelled purveyors of games of chance intuitively to arrange VR schedules for control of the behavior of their players. However, today as back then, VR schedule effects—not the "gambling habits" of fictitious inner agents—are responsible for the behavior that often reduces individual citizen wealth while swelling government treasury coffers from lotteries and gambling taxes.

Interval Schedules

While ratio schedules depend solely on the fixed or variable number of responses to determine the occurrence of a reinforcer, we define interval reinforcement schedules as those on which a reinforcer is contingent *only on one* response occurring *after* an interval of time passes since the last reinforcer. During the interval no reinforcer is available; a reinforcer becomes available only at the end of the interval, when the interval times out. However, a merely available reinforcer remains undelivered; it only occurs after a response occurs. Thus, any responding during the interval remains irrelevant. That is, regardless of how many, or how few, responses happen during the time interval since the last reinforcer occurred, only the *first* response that occurs after the interval has ended produces the reinforcer. Even if this response is the only response to occur since the last reinforcer (i.e., even if no responses happened in the interval) or even if this response is but one of a series of responses in a high–rate sequence of responses (some of which may even occur after it) *only this one response produces the reinforcer.* Although the reinforcer is contingent upon this one response, the reinforcer may coincidentally affect any other occurring responses. Meanwhile, the length of the interval since the last reinforcer, usually stated in minutes or seconds, can have a set value or a changing value.

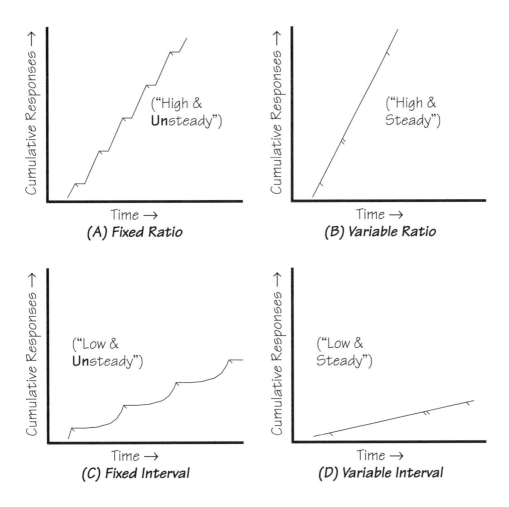

Figure 15–1. Stylized cumulative records showing a typical behavior
pattern for each of the basic reinforcement schedules:
(A) Fixed Ratio,   (B) Variable Ratio,
(C) Fixed Interval,   & (D) Variable Interval.

Fixed–interval schedules. If the value of the time interval since the last reinforcer remains constant, then we call the interval schedule a *fixed–interval* (FI) schedule. Let's set the interval at 100 seconds (although it could be any value, from a few seconds to many minutes). In this case, a reinforcer follows the first response that occurs after the 100–second interval times out (i.e., after each 100–second interval). We denote this schedule by writing FI–100 seconds. That is, on an "FI–100 seconds" schedule, a reinforcer would follow the first response to occur after each 100–second interval, with each interval starting with the delivery of the previous reinforcer.

Here, with fixed–interval schedules, we again note several characteristics of behavior (see Figure 15–1). Most obviously, the cumulative record usually shows a scalloped pattern, and the overall response rate, while not steady, is typically relatively low. What brings about those response–rate characteristics?

We can view the generally low rate as resulting from a sort of competition between contingencies. This kind of competition applies as well to variable–interval schedules, so we will repeat some details then. Here are the principal points. The reinforcement–maximizing aspect of the contingencies tends to exert a lot of control over responding. One way to maximize reinforcement (i.e., one way that collects every reinforcer as soon as it becomes available) involves a high rate of responding such that, as soon as an interval times out, a response occurs that collects the reinforcer; however, this wastes lots of energetic effort. The alternative involves very low–rate responding in which *every* response collects a reinforcer with no wasted effort; however, the collecting response might occur long after the reinforcer became available, a wasted time period for the schedule, which means that some potential reinforcers go uncollected, thereby not maximizing reinforcement. The algebraic summation of the effects of these opposing contingencies results in a generally low overall response rate that neither wastes too much effort nor misses many available reinforcers.

To see better how that works, consider that essentially in 600 seconds (ten minutes) six reinforcers could occur on an FI–100 second schedule. While FI responding actually looks different, responding that occurs once per second would collect every one of these six reinforcers, but at the cost–burden of over 590 unneeded responses, because only six of these responses actually produce reinforcers. Alternatively, responding that occurred rather minimally, say, at only once every 120 seconds (two minutes) would only collect five of the six possible reinforcers (83%), because each 100–second interval *only starts* after reinforcer occurrence; in this case each interval starts 20 seconds "late," and 600 seconds only has room for five such 120–second periods. However, a relatively low but not minimal rate of, say, one response every ten seconds avoids about 90% of the wasted response efforts while collecting all six of the reinforcers when they become available. (Yes, you should and can check my math.)

As for the unsteady, scalloped–looking FI rate pattern, our fixed–ratio concern about extinction crops up here also. Responding right after the reinforcer occurrence has never produced reinforcement, so we experience no surprise when no responding occurs right after reinforcement. However, while responding in the interval also produces nothing, the end of the interval still gradually approaches as time passes, and this induces a gradual increase in response rate such that by the time the interval ends, responding has reached a high rate for at least a short period. Since the interval times out during this high–rate series of responses, one of these responses produces the reinforcer immediately, while the coincidental aspect of this reinforcement—with respect to the other responses—affects many responses in the series, making that kind of series more likely, at least as the end of each interval approaches. This slow start, then a gradually increasing middle rate, and finally a fast–rate that leads up to reinforcement, all show up in the cumulative graph as a scallop.

An example of a fixed–interval schedule scallop that many people can easily recognize involves behavior while waiting for a bus. You take the bus twice a day, to and from work. (Yes, I have been on this schedule also.) A bus arrives at your stop every 20 minutes but, as happens all too often, you arrive just in time to see one pulling away; thus you have 20 minutes until the next bus arrives. So you take a seat, dig a good book (perhaps this book) out of your bag, and begin to read. Occasionally you look up, checking to see if the bus is visible. It would still be a couple of blocks away, which is good, as you need time both to put your book away and to dig out the required exact change. However, your looking–up–and–checking responses occur not on a regular basis but on a basis of gradually shrinking periods. For the first ten minutes or so, checking happens only every four, then two, minutes. For the next seven minutes or so, checking happens only every minute, then about every 40 seconds. In the last few minutes before the bus arrives, checking happens every 15 second, then every ten seconds. Finally the reinforcing bus appears. If we plotted your checking responses cumulatively, we would see that they formed a typical fixed–interval scallop, and this scallop would fairly closely repeat each time you arrived at the bus stop just in time to see a bus pulling away. Now, this example describes single intervals, each up to 20 minutes in duration, that typically occur about eight hours apart on each working day, which differs from the sequentially repeating intervals of an FI schedule. Nevertheless, the example and regular FIs share enough contingency similarities that they suit our pedagogical point about the kinds of contingencies that produce scalloped response–rate patterns. Also note that while for simplicity we cast this example in terms of *"you,"* the schedule contingencies account for your responding, not any inner agential "you."

Variable–interval schedules. If the value of the time interval since the last reinforcer changes, varying around some average, then we call the interval schedule a *variable–interval* (VI) schedule. Let's again set the interval at 100 seconds, but it is not fixed at 100 seconds; rather the interval varies around an *average* of 100 seconds. We denote this schedule by writing VI–100 seconds. In this case, a reinforcer does not follow the first response that occurs after every 100–second interval, because the intervals differ in length from one another. Instead, the reinforcer follows the first response that occurs after each different–length interval times out, with intervals varying in length from a couple of seconds to several minutes, but averaging 100 seconds. That is, on a VI–100 seconds schedule, reinforcers follow the first response to occur after each interval ends, with the intervals averaging 100–seconds. Thus, using one of several available methods to arrange this VI schedule, ten reinforcers could become available in 1,000 seconds, with the intervals ranging in duration from, say, two seconds to four minutes (i.e., 240 seconds). As usual each interval starts with the occurrence of the previous reinforcer.

Also as usual we note certain characteristics of behavior with variable–interval schedules (see Figure 15–1). Most obviously, the cumulative record typically shows a relatively steady pattern with a relatively low overall response rate. What brings about those response–rate characteristics?

We can again view the generally low rate as resulting from a kind of competition between contingencies. Recall that reinforcer occurrence under interval schedules depends solely on a single response happening *after* each interval times out. Responding during the interval is irrelevant. However, collecting all possible reinforcers *is* relevant. Yet nothing obviously indicates the passage or end of the interval; if something did, that indication would evoke the one needed response that collects the reinforcer. Instead, one way to maximize reinforcement involves constant, essentially high rate responding such that, as soon as the interval times out, whenever it times out, a response is occurring that collects the reinforcer; however, this involves a lot of wasted energy, something that nature generally abhors. The alternative involves very low–rate responding, such that every response collects a reinforcer; however, the collecting response might occur long after the reinforcer became available, which means that some potential reinforcers go uncollected. The combination of these opposing contingencies results in a generally low overall response rate that neither wastes too much energy nor misses many available reinforcers.

As for the steadiness of the rate, recall that occasionally the variation in interval sizes includes some short intervals that allow the occurrence of two (and sometimes even three) reinforced responses in a row, or at least very close together. Such occurrences will have happened in the past while the variable–interval schedule was establishing the stable behavior pattern of the current organism. Contingencies arranged this way induce steady responding. That is, scallops or other pauses or delays under these contingencies reduce the amount of reinforcement and so become less likely to occur, which leaves a relatively steady response–rate pattern.

An example of a variable–interval schedule that people can easily recognize involves hunting. Humans and other animals engage in various activities that provide food, and one such activity is hunting. One type of hunting involves waiting in a particular place and checking the different available paths for prey (and catching it) when it comes along. The length of the interval, during which checking–the–paths–for–prey responses occur, may average several hours (e.g., a VI–5 hour schedule) but any particular prey–waiting time can range from a few minutes to longer than daylight. Multiple prey may appear on a path within minutes of each other, or no prey may appear all day long. Since prey may wander by at any moment, the checking–the–paths–for–prey responses occur in a steady pattern at a relatively low rate with respect to each available path, which is a typical variable–interval schedule–induced behavior pattern.

Response–Rate Patterns Predict Schedule Types

Recall that each of the four basic schedules produces its own characteristic response–rate pattern (see Figure 15-1). From these patterns you can predict, through deduction, which basic schedule is likely responsible for a pattern of behavior that you observe, at least as part of your starting point for analyzing a behavior of concern.

We analyzed each of the patterns from these four schedules in terms of two characteristics of the response rate under established responding (i.e., stable responding, after performance on the schedule has stabilized, as opposed to responding while schedule performance is under construction). Was the pattern response rate generally high or low? And was the pattern response rate generally steady or unsteady (e.g., stepped or scalloped)? Can you already see how to deduce the general schedule type likely controlling the behavior from observations of these response–rate characteristics? What schedule characteristics correlate with which response–rate pattern characteristics?

When the answers to those questions begin to evoke correct responses, you have another reason to call the VR schedule the gambler's schedule. Why? Because we can cast the predicting of basic schedule types from response–rate patterns in terms of the gambler's schedule. You begin by gambling that you can recall the rate characteristics of the gambler's schedule (i.e., a high and steady rate). Then, you sort all the schedule contingency parts (i.e., ratio or interval, and fixed or variable) by all the behavior–rate pattern characteristics (steady/unsteady rate, and high/low rate). This gives you several conclusions: Ratios tend toward higher rates (while intervals tend toward lower rates) and fixed values tend toward unsteady rates (while variable values tend toward steady rates). As a result of this exercise, and addressing just the gambler's VR schedule contingencies and rate characteristics, we see that the VR schedule characteristic of high rate correlates with the ratio contingency while the characteristic of steady rate correlates with the variable–value contingency. Variable ratio correlates with steady and high rate, respectively.

Observing the rate characteristics of a behavior of concern leads to deduction of the likely schedule controlling the responding. If you observe a relatively high and unsteady rate, then you deduce a fixed–ratio schedule (fixed, from the unsteadiness of the rate, and ratio, from the highness of the rate) as it is the schedule that most commonly shows the high but unsteady rate characteristics. Gambling that you can stay focused on the rate characteristics of the gambler's schedule (high and steady) go ahead and deduce several other examples that you can imagine observing.

VR Schedules, Comet Discoveries, and Helping Others

Before moving on to other reinforcement–schedule types, let's touch on one of innumerable examples of reinforcement–schedule impacts on science. Over a period of more than 40 years, astronomer Alan Hale has sighted over 500 comets, one of which was a major comet that he co–discovered with Thomas Bopp in 1995. That comet, Comet Hale–Bopp, was his 199th sighting and, arriving in 1996, was also one of the really spectacular comets in the last 100 years although, given my high northern latitude, I personally enjoyed Comet Hyakutake more, which arrived a year later in 1997. Hyakutake's arrival also coincidentally correlated with my promotion to full–professor, which happened shortly after its arrival. The occasional occurrence of this kind of coincidental correlation in the past has, over the centuries, induced the superstition that comet arrivals always foretell the coming of notable events, good or bad.

Meanwhile Alan Hale continued searching for comets. Shortly after his 500th sighting, David Levy published an article (Levy, 2012) on Hale's achievement in *Sky and Telescope* magazine. Reading this article can give a clear sense of the natural, gradual building of the reinforcement schedule affecting Hale's behavior, which led to his spending so many nights at his eyepiece, the behavior that produced the occasional comet sighting. Of course, as we have considered before regarding our use of real examples, many other variables affect this behavior, including other related reinforcers (such as big reinforcers like the attention from co–discovering Comet Hale–Bopp, and little reinforcers such as other beautiful deep–sky sights while comet searching). Yet in general the high and steady rate implied in the article, along with the sighting reinforcers being contingent on the amount of responding rather than, so far as we know, some aspect of time passage, leads to the conclusion that the behavior is largely a product of a variable–ratio reinforcement schedule.

Perhaps more importantly Hale's 500 comet sightings also lead to an example of the contributions behaviorology can make in support of other natural scientists, as we first described near the end of Chapter 1. Perhaps due to a lack of acquaintance with behaviorology and its discoveries of the natural laws governing behavior, the author of that article essentially attributed Hale's 500–comet–sighting achievement to something he called *determination* (p. 26). Now, our concern with this attribution is not a criticism of the article author; after all, any lack of acquaintance on his part with behaviorology likely derives from the lack of opportunities for exposure to this natural science, which in turn accounts for the call, in Chapter 1, for expansion of behaviorology programs, departments, and degrees. Rather, our concern is a criticism of the general culture, a culture in which superstition is still a more potent reinforcer than science for the behaviors of so many of its members. That is, our concern is that this type of attribution has problems and so requires some comment.

Attributing extensive behaviors, occurring with apparently little tendency to fall for numerous available distractions, to "determination" typifies the fairly common kind of account that many commentators offer in our culture. Now, *determination* may represent a properly understood verbal shortcut for a more complex and accurate scientific account. However, more often "determination" alludes to a characteristic of some inner agent that then magically tells the host body to keep at the behavior of concern. Accepting the risk of being wrong in this case, but for educational reasons, I suggest that—possibly for many of the article's readers—the latter applies here. For many of these readers, "determination" confirms a culturally presumed but irrelevant and redundant inner–agent account. As such, "determination" is a scientifically incorrect explanation typical of a superstition–steeped culture, the kind of scientifically inadequate account that we exposed and rejected in earlier chapters.

After all, scientifically, and just like everyone else (including you and me) Alan Hale lacks inner agents of any sort (e.g., self, soul, mind, or psyche) including one with some excess amount of "determination" to cause him to stick with a 500–comet–sighting program. Furthermore, the circularity of "determination" causing the searching for 500 comets stands out clearly. You can easily recite the question–answer sequence exposing this circularity (i.e.: Why did he search...? ... How do you know that he had great...? ...).

That kind of commentary, by authors who are more behaviorologically informed, on more of the regularly appearing superstition–based cultural examples of behavior causes, can increase the benefit that a wider knowledge of behaviorology can bring to humanity. It would especially help other science authors avoid crediting mystical causes. And it will help provide the general public not only with a greater and scientifically more accurate understanding of their own behavior but also with how knowing more about the *variables of which behavior is a function* can help them contribute to solutions—in their own best interest—to global and local problems.

Some Other Reinforcement Schedules

Beyond the fundamental schedules, additional reinforcement schedules also affect behavior. Some of these schedules, which appear in the laboratory or in applied settings, feature relatively simple contingencies. We find other, more complicated schedules mostly in the laboratory where we refer to the group of them as compound schedules.

Extensive details about these schedules remain outside the purview of this book. Here we provide only brief descriptions of some of them.

Other Simple Laboratory and Applied Schedules

Laboratory preparations sometimes involve *fixed–time* and *variable–time* schedules. These schedules contain the simplest of schedule contingencies, perhaps even simpler than a CRF schedule. In the CRF schedule, the reinforcer remains contingent upon responding; not so in the time schedules. Applied settings involve a wider range of additional simple schedules. These include *fixed–duration* and *variable–duration* schedules along with schedules we call *DRH* and *DRL* schedules (differential reinforcement of high, or low, rate). Let's look briefly at each of these, beginning with the time schedules.

Fixed–time and variable–time schedules. We first ran across fixed–time (FT) schedules back when we covered coincidental reinforcers and superstitious behavior. These are schedules that contain no response requirement; this clearly distinguishes them from ratio and interval schedules, both of which have response requirements.

Instead, with time schedules, reinforcers merely occur as a function of the timing out of some period of time. On fixed–time schedules, the period has a set length, while on variable–time (VT) schedules, the period has a length that varies around some average. The early research on superstitious behavior featured an FT–15 second schedule.

Fixed–duration and variable–duration schedules. Unlike time schedules, which lack a response requirement, duration schedules require responding that occurs repeatedly or continuously. These schedules, which generally occur in applied settings, feature reinforcers that follow responding that continues across an interval of some duration. On fixed–duration (FD) schedules, the time period has a set length, while on variable–duration (VD) schedules, the time period has a length that varies around some average.

Here are some examples. A child's work on math–problem sheets might be on an FD–3 minute schedule. On this schedule a reinforcer occurs at the end of each three–minute period during which the child works continuously on the math sheets (i.e., the child works for the duration of a three–minute period). For another example a young child may recently have begun violin lessons and so is not yet under control of the reinforcers that playing and practicing can supply. So the child's behavior of practicing the violin could be on a VD–5 minute schedule. On this schedule a reinforcer occurs at the end of each period, averaging five–minutes long, during which the child plays continuously (i.e., plays for the duration of the period); however, each period varies in duration around the average of five minutes. Some periods might last only half a minute or a minute, while other periods could last six or eight or ten minutes, with still other periods irregularly lasting durations between and among all these values.

DRL schedules. Two other schedules that you typically find operating in applied settings are the *differential reinforcement of low rate* (DRL) schedule and the *differential reinforcement of high rate* (DRH) schedule. These schedules feature reinforcement that occurs contingent on a particular *rate* at which a certain type of response happens. For these schedules we look for, or specify, not only how many of these responses must occur but also the amount of time in which this number of responses must occur. As usual for reinforcement schedules, responses other than the behavior of concern, even if they occur during the time periods of the schedule, exert no effects on schedule contingencies. Let's look at DRL schedules before DRH schedules.

Researchers use various DRL–schedule methods to generate and maintain a low rate of responding. Here we focus on one that involves an applied–setting intervention closely deriving from Ferster and Skinner's (1957) schedule work. In this method the number of responses of concern that must occur in the time period is *zero* and, if zero responses occur in the time period, then the reinforcer follows the next response of concern. However, every response of concern that happens during the time period restarts the timer for another time period. For instance, on a DRL–10 second schedule, if no response of concern occurs during the ten–second period, then the next response of concern produces the reinforcer. However, each response of concern that happens during the ten–second period restarts the timer for another ten–second period.

DRH schedules. These schedules are even more complex than DRL schedules. In DRH schedules the point is to generate and maintain a high rate of responding. So the number of responses of concern that must occur in the time period is specified at some *greater than zero* value. If that number of responses of concern, or more, occurs in the time period, then the reinforcer follows the first response of concern that occurs after the time period ends. However, if fewer responses of concern than the specified number happen during the time period then, at the end of the period, the timer restarts for another time period.

Compound Research Schedules

For a taste of research complexity, let's list some of the common compound schedules. Researchers often combine or otherwise rearrange the elements of the basic schedules to conduct studies with more complex schedules. We will touch on five compound schedules, the chained, tandem, multiple, mixed, and concurrent schedules. In most of these compound schedules, only one schedule component of the compound schedule is available at any given moment. A technical example will accompany only the chained schedule; this can provide a taste of what is involved. However, further details on these compound schedules goes beyond the purview of an introductory text (see Ferster & Skinner, 1957; or Fraley, 2008, Chapter 17).

In *chained* schedules the researcher chains together two or more other, usually basic, schedules in a row, with each link in the chain starting with a change in the evocative stimuli; completing the requirements of *all* the links produces the reinforcer. For example, on a chain FR–5 FI–5 second VR–10 schedule, the the reinforcer would occur after five responses under S_1^{Ev}, then one response after a five–second interval (during which any other responding was irrelevant) under S_2^{Ev}, and an average of 10 more responses under S_3^{Ev} (which could be a return to S_1^{Ev}).

In *tandem* schedules the contingencies are the same as under chained schedules, with one notable exception. In tandem schedules, no evocative stimuli are present for the different schedule components.

In *multiple* schedules the simple–schedule components alternate, with each component under control of a different evocative stimulus. Reinforcers are contingent on meeting the requirements of either schedule component.

In *mixed* schedules the contingencies are the same as under multiple schedules, again with one notable exception. In mixed schedules, no evocative stimuli are present for the different schedule components.

In *concurrent* schedules, unlike the previous schedules, two or more schedules are available at the same time. The researcher may arrange an FR–10 under S_1^{Ev} and a VR–12 under S_2^{Ev}, with the manipulandum for each concurrent component always available to the subject. We would denote this compound schedule as CONC FR–10 VR–12.

Some researchers have spent much of their careers testing parametric variations of one or another compound reinforcement schedule and reporting their results in the *Journal of the Experimental Analysis of Behavior* (JEAB) over the last 50 years. In the process they have uncovered some of the subtle schedule nuances that show up as rather robust changes in behavior patterns. Let's turn now to some other, perhaps even more interesting, schedule considerations.

Other Schedule Considerations

We focus on two principal areas that cover a range of related schedule considerations. One pertains to schedule characteristics, such as the post–reinforcement pause and resistance to extinction. The other, which we tackle first, pertains to the building of schedule performances.

Building Schedule Performance

Three areas require comment regarding the building of schedule performances. These include (a) recognizing the difference between later, stable performance on a schedule, and the earlier, constant but gradual changes in contingencies that change behavior while building up a schedule, (b) the role of the extinction burst in schedule construction, and (c) the importance of stretching—while not straining—the schedule.

Schedule construction versus performance stability. We previously pointed out that our cumulative graphs of various schedule performances show the behavior pattern of performances that have stabilized on their respective schedules. Depending on the subject species, this stable responding may have taken the experimenter from days to months to establish. Behavior that occurs under the contingencies of schedule building in this period may differ markedly from the stable behavior that occurs on the established schedule later. Also much schedule–induced behavior of concern outside the laboratory may exist only under changing, schedule building contingencies.

Hence some research questions involve the behavior that occurs under the contingencies of schedule building. While researchers addressing such questions make very clear that their data graphs show response rates under schedule construction, wherever no such clarifying statements occur, we assume that the behavior under study involves final, stable performances on the stated schedule.

The extinction burst. One of the most important processes that enables the building of schedule performance to begin is a process that we call the *extinction burst.* We regularly observe that when reinforcement stops occurring for responding that previously received reinforcement, especially on a CRF schedule, that kind of responding briefly increases in rate. This brief rate increase is what we call the extinction burst.

A mundane example concerns virtually everyone's typical reaction to a vending machine that fails to deliver the selected goods. You put in your money, push the button for your selection, and wait. Something is supposed to happen—the machine is supposed to deliver your goods—every time (i.e., a CRF schedule). If nothing happens, I doubt you think or say, "Oh, well; life is uncertain. The universe operates magically. Maybe next time the machine will deliver my cookies," and walk calmly away. Instead, some other things happen. For many people, the earlier response of putting in money extinguishes instantly; at least I have never witnessed someone continue to feed money into a malfunctioning machine, although one machine serviceman informed me that this does happen. Meanwhile, the response of button pushing typically increases briefly in rate, like an extinction burst; my machine–serviceman informant even reports that this extinction–burst button pushing sometimes breaks the buttons. Additional extinction–induced and emotionally exaggerated behavior also occurs, such as loud cursing and kicking the machine. The brief increase in rate is the salient part that we tact as the extinction burst. Why is it important? Because researchers employ extinction bursts to build the size or duration of reinforcement–schedule ratios. Here is a better example, one that helps us see other values of the extinction burst in schedule building.

Let's say that, as part of your job researching astronaut behavior under FR schedules that could end up requiring high fixed ratios, you wanted to compare performance on FR–600 and FR–1,200. You start each astronaut subject out on CRF. Would you predict success in changing their reinforcement schedule from CRF directly to FR–600? I am sure that you at least said "No" (and will keep covert those other invidious comments). Indeed, moving directly from CRF to this high FR schedule would not be a good procedure because it carries a virtual guarantee of the behavior extinguishing before—probably long before—the first FR–600 reinforcer follows the initial six–hundredth response.

However, as soon as you stopped reinforcing every response, which essentially implements an extinction contingency, some in–extinction responses would occur, very likely at a rate higher than the previous rate under CRF. While perhaps altogether 50 or more responses might occur, at a gradually decreasing rate, before extinction set in, you would not wait that long. Instead you might reinforce the tenth response in this extinction burst, and then continue for some time to reinforce every tenth response. These contingencies would relatively easily move your subject off the CRF schedule and onto an FR–10 schedule. Your next intermittent–reinforcement schedule–building step pertains to our next topic.

Stretching versus straining the schedule. Your subject now responds under FR–10 and for a while you would continue to reinforce every tenth response. However, as soon as responding on that FR–10 schedule approaches stability, you change the schedule requirement again, perhaps to FR–25. Then, after a little stability, you move to FR–45, and later to FR–85, then on to FR–160, and so on. With each FR–schedule increment, you are *stretching* the ratio further. And you can continue stretching the ratio across more steps until, after enough intermediate steps, you reach the FR–600, as your study requires. You let performance under this schedule stabilize fully so that you can construct a comprehensive picture of an FR–600 performance to compare later with an FR–1,200 performance. Then you want to go on to establish and stabilize at FR–1,200 so that you can also construct a comprehensive picture of this schedule. With it you can compare the performance on these two schedules.

However, what if you get carried away, and try to go directly from FR–600 to FR–1,200? After all, your pattern of ratio–size increases had already included a doubling jump of nearly 300 responses, so you might think that a doubling jump of 600 responses, from FR–600 to FR–1,200, was reasonable. But any sizable jump is risky; that jump from FR–300 to FR–600 was risky, although it worked out. If the 600–response jump ends up requiring too many responses before another reinforcement, then the jump fails and extinction begins. We call this *straining* the ratio rather than merely stretching it, and you would need to back up to some earlier ratio that still works, and then stretch the ratio again in smaller steps until you bring your subject up to the required FR–1,200.

For another example, consider some professional gamblers (e.g., poker players and pool hustlers). They come under a particular contingency that has been shaping their behavior since long before behavior science discovered the ratio–stretching process. This contingency involves letting the victim win while gradually stretching the victim's schedule ratio. This process keeps their "pigeon" victim playing—and losing—the game for longer periods. In the beginning, the professional gambler arranges for the victim's playing to produce small wins frequently on low–ratio VR schedules; gradually, however, the victims lose more than they gain in their wins, which now occur only occasionally, and usually with little at stake, as the card sharks or hustlers gradually stretch the ratio, switching to higher–ratio VR schedules. Whether they stretch the ratio through skillful playing or cheating is beside the point (although skillful playing seems more common); in either case the high and steady response rate induced by the VR schedule countercontrols the effects of other contingencies that might otherwise induce a quicker exit by the victim from the games.

Generally however, straining, rather than just stretching, ratios and intervals in schedule construction causes delays in our research along with delays in providing any application benefits derived from the research. The contingencies of such aversive delays, however, may not always suffice to induce only stretching the schedule during construction. Thus discussions like this one function as rules that supplement these contingencies. Meanwhile let's turn our attention to other schedule characteristics.

Schedule Characteristics Considerations

When considering other schedule characteristics, three areas demand comment. These include (a) the limited hold, (b) the post–reinforcement pause, and (c) resistance to extinction.

The limited hold. A somewhat common extension of schedule contingencies involves an arrangement that we call a *limited hold.* Applicable to interval schedules, this refers to a constraint on how long a reinforcer remains available. Interval schedules make the next reinforcer available at the end of the fixed–length or variable–length interval; as soon as the reinforcer is available, the next response collects it, regardless of how much time passes before this response occurs. However, a limited hold makes the reinforcer available for only the amount of time specified in the limited hold. A limited hold puts a limit on how long the schedule holds an available reinforcer. If no reinforcer-collecting response occurs in the limited–hold interval, then that reinforcer is lost and the next no–reinforcer–available interval begins.

Recall our bus–arrival example that featured intervals similar to an FI–20 minute reinforcement schedule. Let's say that you were the only one waiting for the bus, but your book provided so many reinforcers that only the arrival of the bus and the opening of its doors finally evoked your behavior of putting away your book prior to rising and boarding the bus. However, what if, to maintain the schedules, bus–company policy allowed drivers to keep the bus doors open only for five seconds? If no passenger climbed aboard in that time, then the doors closed and the bus moved on. (Rude? Yes. But it makes the example.) Since these seconds elapsed before you even got your book stashed, the bus doors closed and the bus moved on before you even stood up. This bus–arrival schedule would then contain a 5–second limited–hold (i.e., an FI–20 minute, LH–5 second schedule) limiting you to five seconds in which to move enough that your movement evokes the driver's behavior of keeping the door open until you board.

The continued reading, which replaced watching responses and delayed your boarding movement, not only cost you your ride that time, but also started another 20–minute waiting interval. We leave you on the bench contemplating limited holds, and turn to another kind of pausing, the temporary decrease in rate that we sometimes observe after the occurrence of a reinforcer.

Post–reinforcement pause. We call this pausing, this temporary rate decrease to zero after reinforcer occurrence, the *post–reinforcement pause* (PRP). While we previously discussed it only under fixed–ratio schedules, where it stands out most obviously, it not only occurs occasionally under variable–ratio and variable–interval schedules, but it also occurs extensively under fixed–interval responding, as the first part of each fixed–interval scallop.

At first glance the PRP looks like the result of a simple interaction between reinforcement and extinction contingencies, and this may still be part of why post–reinforcement pauses occur. After all, on FR and FI schedules (except very low ratio FR schedules, and FI schedules with intervals of only a second or three) two reinforced responses in a row never occur; some number of responses or some amount of time must occur before another response produces a reinforcer. This leaves the functional effectiveness of the evocative stimuli, right after reinforcement, at the lowest level, and responding fails to occur right away. Of course, evocative stimuli remain present, and physiologically retain some functional effectiveness from the occurrence of past reinforcements. As a result they evoke the responding that leads to the next reinforcer.

However, researchers have found that pausing is a function of more complex interactions of several experimentally identifiable independent variables, including ratio size and reinforcer magnitude (see Schlinger et al., 2008, although be weary—as is the case with some other references as well—of the unclarified agential accounts or verbal shortcuts that the authors use in this paper). Experimenters have conducted most of the pausing–related research using FR schedules, and consensus remains somewhat out of reach with the exception of the unneeded and redundant status of accounts with agential components; they reject these.

In his schedules–of–reinforcement chapter, Lawrence Fraley (2008, Chapter 17) bypasses the zero–rate portion of the fixed–ratio post–reinforcement pauses in favor of focusing on a fine–grained analysis of the small scallop that we can sometimes observe in the transition from the break to the run on these schedules. While this detailed analysis extends beyond our present purview, researchers can apply it to the common, larger–scale scallops that occur after the zero–rate post–reinforcement pauses on fixed–interval schedules. We leave to the self–correcting nature of natural science to provide us with a thorough account in due time. Stay tuned.

Resistance to extinction. No chapter on schedules of reinforcement could be complete without some comment on *resistance to extinction.* This refers to an interesting phenomenon that occurs with intermittent–reinforcement schedules; even though numerous responses go unreinforced, *the responding on these schedules seems more difficult to extinguish* than is the case when reinforcers occur after every response.

The general and counter–intuitive sounding rule describing resistance to extinction holds that *the more similar a schedule is to extinction, the more responding on that schedule resists extinction.* The similarity to extinction derives from comparing the number of unreinforced responses occurring regularly on the schedule with the reality that, on extinction, *all* responses go unreinforced. For example, FR–1,000 resembles extinction more closely than FR–10, with 999 unreinforced responses for each reinforced response on FR–1,000 (i.e., 99.9%)

versus only nine unreinforced responses for each reinforced response on FR–10 (i.e., 90%). In this case we find that far more individual responses occur in extinction after FR–1,000 than after FR–10; the behavior that was on FR–1,000 resists extinction more than the behavior that was on FR–10.

For more than 50 years, many researchers, calling this the *partial–reinforcement effect* (PRE) have tried to discern what could make the occurrence of fewer reinforcers result in a greater number of responses. They have offered many suggestions, most of which arose during the period in which historical contingencies forced natural scientists of behavior to cohabit academic departments with psychologists (see Ledoux, 2002). Thus we are not surprised to find that many of these suggestions have included inner–agent accounts. Of those that adhered to naturalism, the "response–unit" account may have the most empirical support. This account suggests re–conceptualizing a response unit as the complete behavior requirement that produces the reinforcer so that, for instance, one response is the response unit on a CRF schedule, and 100 responses comprise the response unit on an FR–100 schedule. That is, the "unit" that produces reinforcement is a multiple of the unit that usually evokes our tact of "a response." While this suggestion may go a long way to accounting for the PRE, other suggestions remain in the research hopper. In any case the level of detail that we require for dealing with any of these suggestions, which should take physiology into account, extends beyond the scope of this volume.

Conclusion

Overall, schedule research has repeatedly led to several general conclusions. The most applicable include these three: (a) Many features of behavior emerge as the effects of the contingencies comprising particular reinforcement schedules. (b) Schedules with only subtle contingency differences often produce distinctly different response patterns. And, perhaps most importantly at the present time, (c) the direct effects of schedules of reinforcement reduce a wide range of putative inner–agent emotional and "motivational" causes of behavior to misleading redundancies.

Before closing this chapter, let me mention an additional schedule effect, one that we call *adjunctive behavior*. This refers to behavior that appears to observers as an intrusion on other ongoing behavior. Recognizing and analyzing this behavior helps us explain some behavior that otherwise appears quite out of place. Nevertheless it results from contingencies, particularly those occurring in relation to other schedule effects and behaviors. To follow up on it, as its complex analysis goes beyond our scope here, see Chapter 18 of Fraley, 2008.

Let's close this chapter by revisiting our ongoing activity. We simply cannot underestimate the pertinence of reinforcement schedules, and their effects, for the role they play both in various individual contributions to global problems, and to applications related to managing the behavior components of the solutions to global problems in timely ways. Consider how they will help make possible the kinds of repertoire changes in sustainable lifestyles that will enable the success of solutions to the non–behavior–related components that traditional natural scientists are addressing.⚜

References (with some annotations)

Ferster, C. B. & Skinner, B. F. (1957). *Schedules of Reinforcement.* Englewood Cliffs, NJ: Prentice–Hall. This comprehensive book describes the authors' multitude of laboratory studies researching a wide range of reinforcement schedule parameters; I understand that, for over a decade, it was deservedly one of the most widely cited references in the natural–science literature, and we still hold it in high regard.

Fraley, L. E. (2008). *General Behaviorology: The Natural Science of Human Behavior.* Canton, NY: ABCs.

Ledoux, S. F. (2002). An introduction to the origins, status, and mission of behaviorology: An established science with developed applications and a new name. In S. F. Ledoux. *Origins and Components of Behaviorology— Second Edition* (pp. 3–24). Canton, NY: ABCs. This paper also appeared (2004) in *Behaviorology Today, 7* (1), 27–41.

Levy, D. H. (2012, June). Alan Hale and his 500 comets. *Sky and Telescope,* 24–26.

Schlinger, H. D., Derenne, A., & Baron, A. (2008). What 50 years of research tell us about pausing under ratio schedules of reinforcement. *The Behavior Analyst, 31* (1), 39–60.❧

Chapter 16
Aversive Control Problems and Alternatives

Chapter 15 revisited reinforcement not only by addressing the range of schedules on which reinforcement can occur as it controls behavior, but also by addressing some of the numerous and subtle effects on behavior that various schedules provide. This chapter flips the consequence coin to introduce some concerns about how *coercion* operates in controlling behavior. Coercion includes punishment, the threat of punishment, and even subtracted reinforcement. These three constitute the major components of aversive postcedent control, which is often the opposite—in both effects on behavior and emotional effects—of reinforcing postcedent control, particularly added reinforcement.

Control by both coercion and added reinforcers has *always* been with us. For example, over 3,000 years ago in the Near East, the Hittites sold some of their captives into slavery. In contrast to that aversive consequence, the Hittite Kings settled other captives on vacant land, gave them "supplies to start farms, and granted a three–year remission on taxation" (Dise, 2009).

In spite of both kinds of control having been with us for so long, our human experiences evoke differing levels of reaction to control by aversives and control by added reinforcers. Our verbal behavior about control overemphasizes aversive control, pre–scientifically the most obvious kind of control, as if it were the *only* kind of control. The literature of freedom and dignity (which B. F. Skinner discusses; see Skinner, 1955–56, 1971) emphasizes ending that kind of control but largely ignores, even denies, the vital effects on our existence of added–reinforcement control, an ignorance that can lead to disastrous global outcomes. Since we have emphasized added–reinforcement control in several chapters, we now cover coercion control.

Behavior control, which is ever present and always operating, comes in two flavors, which we call positive control and coercive control. These phenomena follow natural laws of behavior the way physical phenomena follow natural laws of physics. To enhance successfully dealing with ourselves and the world around us, these laws of behavior require thorough investigation, with results shared with everyone as a major step that countercontrols any contingencies inducing the misuse of these laws. As we saw through previous chapters, we have made some progress along these lines. Still, we have much more to discover about the laws of behavior, particularly the laws involving coercion.

So, taking into account the effects and implications of what we have already discovered, here we cover the basics of aversive control, its problems, and its alternatives, in three respective sections. Starting with escape and avoidance, we move on to eight coercion traps, and we finish with some alternative ways of dealing with the kinds of behaviors that too easily otherwise evoke punishment and other coercive methods of control. First, however, let's briefly review punishment, some of its problems, and why we so often and so easily respond to the behavior of others with punishment.

Punishment Review

Recall that punishment, as a postcedent process or procedure, occurs either as the addition, or the subtraction, of stimuli. The stimulus addition or subtraction functionally results in a reduction in the ongoing rate of the behavior that the stimuli followed. The rate reduces because the energy trace from these postcedent stimuli alters nervous–system structure such that the antecedent stimuli that usually evoke the behavior function less effectively; they no longer as readily evoke the punished behavior. For example, say a burned out light evokes your response of grabbing a new bulb and moving quickly to replace the burned out light without first unplugging the lamp. Then, an electric flow starts as your fingers contact both the metal base of the new light bulb and the lamp socket, a shocking experience that we call added punishment. As a result of that punishment, a burned out light will no longer as readily evoke the same behavior. The next time a light burns out, it will evoke slower movements, or more careful movements, or unplugging the lamp first, or all of these.

Unfortunately you see added reinforcement happening too rarely when you look around. Instead, nearly everywhere, you see the common kinds of coercion. You see subtracted reinforcement, in which some behavior produces the end of punishment (as ending hand contact with the lamp produces the end of the shock). Or you see the threat of punishment. Or you see punishment. When an adult offends, the offended takes a swing; when children act out, a parent threatens them or spanks them or both; when a pupil misbehaves, legal substitutes for paddling occur; and when countries disagree, war often follows. *We cannot recommend any of these as satisfactory solutions.* When contingencies drive secular–law violations, feelings of guilt—if the necessary emotional conditioning has occurred—and fines or jail follow. When contingencies drive religious–law violations, feelings of sin—if the necessary emotional conditioning has occurred—and penances, excommunication, or—some folks say—eternal damnation follow. The data on coercion and its problems compel my support for any contingencies that make you find such practices in need of improvement if not outright replacement.

Despite its wide use, punishment has many problems; here we mention only a few. We describe the use of punishment as *incredibly risky.* In the long run, it drives wedges between people (e.g., between parent and child, and between teacher and student). It induces disabling emotional anxieties and exaggerations of some subsequent operant reactions. It compels those on the receiving end *to get away* from the punisher (including any person providing the punishment), *to stay away* from the punisher, *and to get even* with the punisher. In technical terms we refer to these last three major effects of punishment—getting away, staying away, and getting even—as *escape, avoidance,* and *countercoercion.*

We often consider countercoercion as counteraggression. Coercion is a form of aggression, and an attack induces those under attack to counterattack. The punisher usually bears the full force of the counterattack but, if that is not possible, then others or even objects bear the force of the counterattack, which often takes an exaggerated form due to emotional components. These effects can even flow through a community like a wave; a boss reprimands and threatens to fire an employee who goes home and yells at the spouse who yells at a child who yells at the cat that goes out and screeches at the neighbors who yell at the police officer over the phone about the screeching cat—and everyone lives happily… Oops; sorry, wrong ending. We soon address emotional components, along with escape and avoidance, more fully.

Meanwhile, given such problems, we must wonder why we and other people so often and so easily respond to the behavior of others with punishment. Perhaps the biggest reason is that punishment *is* an effective procedure, when it occurs in certain ways, and especially in the short term. If a punisher of only necessary *intensity* is *contingent* on a particular behavior and *immediately* and *consistently* follows this behavior every time it occurs, then the punishment rapidly produces a sizeable decrease in the punished–behavior rate.

Due to that effectiveness, the immediate consequences of punishment affect its use far more than the long–range consequences affect its use. The immediate consequences of using punishment *reinforce its use,* but the long–range consequences are disabling. While the long–range effects of punishment involve those problems of emotional effects and escape, avoidance, and countercoercion (among others), one immediate effect of punishment involves the quick and substantial reduction in the rate of the punished behavior. *This rate reduction reinforces the behavior of applying punishment,* which we see later as increases in the rate of punishment–application behavior, and which thus accounts substantially for why we so often and so easily respond to the undesirable behavior of others with punishment. Other kinds of reinforcers can follow punishment use as well, such as when a manager praises a foreman for keeping staff noses to the grind stone through punishment (an added reinforcement for punishment use) or when a principal stops criticizing a teacher when the teacher "gets tough" with the students through punishment (a subtracted reinforcement for punishment use). These also increase the rate of punishment application, even though available alternatives, to which we will soon turn, would more effectively deal with the behaviors of concern.

If you are on the receiving end of punishment, however, the stimuli evoke those behaviors of escape, avoidance, and countercoercion. Since the traditional use of the escape and avoidance terms breeds some confusion, let's consider escape and avoidance in more detail, before considering coercion traps and some alternatives to punishment.

Escape and Avoidance Behavior

Two preliminary considerations surround our coverage of escape and avoidance. We acknowledge a couple of common, easier–to–use, but slightly less accurate terms. And we describe some respondent outcomes of operant consequences.

As a verbal shortcut, people often use the term *aversive stimuli,* or simply *aversives,* for stimuli that meet the definition of punisher. However, the clarity and accuracy of these terms falters somewhat, because the terms apply to both operant and respondent stimuli. You will see this when we not only call a punisher, like a spanking (an added punisher) or a fine (a subtracted punisher) an aversive, but we also call a stimulus that elicits negative emotional reactions, like the stink of a broken sewer pipe, an aversive. Sometimes the same aversive stimulus has both operant and respondent effects; for instance the stimuli from whacking your thumb with a hammer not only function as operant consequences (e.g., punishing sloppy swinging) but also function to elicit negative emotional reactions (e.g., anger) with subsequent exaggerated effects (e.g., you throw the hammer with force rather than merely drop it).

That is not surprising, is it? Operant consequences not only alter response rates, but they also elicit respondent effects. Added reinforcers not only increase response rates, but they also elicit positive emotional respondent reactions. On the other hand, while subtracted reinforcers increase response rates, and added or subtracted punishers decrease response rates, *all three* also involve negative emotional respondent reactions. When you add the difference between *unconditioned* reinforcers and punishers, and *conditioned* reinforcers and punishers, to this mix of operant and respondent/emotional components, then you begin to reveal the complexities of escape and avoidance.

Escape

We call a response an *escape* response when the result of that response is the termination of *ongoing* aversive stimulation. The presence of an occurring aversive stimulus evokes a response that ends ongoing exposure to that aversive stimulus. This is subtracted reinforcement, and the escape response occurs again in the future when that aversive evocative stimulus occurs again. Note that the stimulus that evokes the response by starting is the same stimulus that the response stops, thereby affording relief from the aversive stimulation. Does punishment play a part?

Yes. The ongoing aversive stimulation that the evoked response escapes usually happens in the first place as the occurrence of a punisher. These aversive stimuli function in multiple ways. When such stimuli start, they serve both as postcedents punishing the last response (possibly coincidentally) and as stimuli evoking the next response (i.e., another kind of dual function of a stimulus). However, they also elicit negative emotional responses, and an effect of the occurrence of these emotional responses is to change the body in ways that exaggerate subsequent responses, often including the escape response.

Those comments on escape behavior apply equally well, as they should, to escape from both unconditioned *and conditioned* aversive stimuli. (Double check by reading them.) Note however, that we traditionally limit our use of the term escape to responses that terminate *unconditioned* aversive stimuli that are ongoing (i.e., that someone is currently experiencing). This usage leaves room for something we call "avoidance."

Avoidance

In contrast to escape, which has the single outcome of terminating ongoing aversive stimulation, we use the term *avoidance* when the response has two outcomes: (a) The response *prevents* or *delays* the occurrence of the *unconditioned* aversive stimulus, so we call it an avoidance response. (b) And the same response simultaneously also terminates the *conditioned* aversive stimulus, so we would call it an escape response except that it already has a name, "avoidance." These conditioned aversive stimuli became conditioned through pairing with unconditioned aversive stimuli. And the "avoidance" response works, not through some magic with the non–occurring *unconditioned* aversive stimulus, but due to *escaping* the *conditioned* aversive stimulus. So in terms of actual functioning, we really only have escape behavior, but using the two terms these ways supports our understanding of, and work with, these phenomena. What ways again? These ways: We say "escape" when the response ends the unconditioned aversive stimuli, and we say "avoidance" when the response both ends the conditioned aversive stimuli and prevents/delays the unconditioned aversive stimuli.

Let's repeat that common but less accurate way to differentiate between "escape" and "avoidance." Use the term *escape* for a response that simply ends unconditioned aversive stimuli *that you are currently experiencing,* and use the term *avoidance* for a response that simply prevents or delays the unconditioned aversive stimuli *that you are not currently experiencing.*

Improving the accuracy, avoidance prevents or delays the occurrence of unconditioned aversive stimulation while also escaping conditioned aversive stimulation. In other words, in escape you are experiencing the unconditioned aversive stimulus, and the response terminates it; meanwhile, in avoidance you are not experiencing the unconditioned aversive stimulus, but you *are* experiencing the conditioned aversive stimulus, and the response prevents the former and terminates the latter.

Escape/Avoidance Example

Whaley and Malott (1971) summarized a study that Hefferline, Keenan, and Hartford had published in *Science* in 1956. Even after over 50 years, this study bears re–description. While some aspects of the study addressed concerns over past conditioning history, the study provides an early experimental example of escape and avoidance with humans. In addition it provides an example of how, for humans, stimuli affect responses without either of these evoking human consciousness behaviors (e.g., awareness). Let's say this in everyday language, which requires ignoring the inherent but unsupportable agential implications: We need not be aware of responses or stimuli for them, respectively, to occur or to affect (as in evoke or consequate) our further behavior.

Hefferline and colleagues arranged for 12 human subjects to participate in their study. Each subject took a turn in the comfortable seat in the test room and, for the 80 to 90 minutes of the experiment, listened to music. Noise occasionally sounded on top of the music. When the noise was happening, a tiny but measurable thumb twitch—too small to evoke awareness responses—turned off the noise for 15 seconds (i.e., escape conditioning). The same kind of response, when the noise was not happening, delayed the onset of noise for 15 seconds (i.e., avoidance conditioning).

The researchers arranged their study as a single–subject experimental design that followed an ABCA pattern of phases. The A phase began each subject's experimental session and lasted for five to ten minutes; during this phase music occurred without noise. This constituted a baseline phase for measuring the thumb–twitch behavior of concern prior to any conditioning. The B phase came next, lasted for about an hour, and included the escape and avoidance conditioning. The C phase lasted for ten minutes after the hour–long B phase, and involved an extinction contingency; during this extinction the noise sounded continuously, but responding had no effect on it. The second A phase followed the extinction phase for a few minutes and, as a reversal phase, returned to the baseline conditions of music playing continuously without any interfering noise.

To measure the behavior of concern, the experimenters attached three pairs of electrodes to the subjects, although *only* the wires attached to the thumb and edge of the left hand operated; the experimenters then observed and recorded the electrical voltage of thumb twitches through these wires. Thumb twitches showing between one and three microvolts constituted the behavior of concern. Neither the subjects nor the experimenters could directly observe this imperceptible level of thumb twitch.

The researchers set up four conditions, one for each group of three human subjects. The four conditions differed mostly in that each group received different instructions. The instructions to Group 1 were merely to listen through the earphones. The instructions for Group 2 added that a small, invisible response would turn off, or postpone, the noise, and that their task was to discover that response and use it. The instructions for Group 3 added even more, specifying that a tiny twitch of the left–thumb would affect the noise. The instructions for Group 4 were the same as for Group 3, but Group 4 got to watch a meter, for the first half hour, that indicated when a correct response had occurred.

Each subject also participated in an interview after the experimental session. Experimenter questions evoked each subject's verbal–reporting behavior about any motor behavior, and any neural behaviors of consciousness, that various aspects of the study had evoked.

Let's consider the outcomes for these groups; as we proceed, consider how fully lawful each outcome is. For Groups 1 and 2, the escape and avoidance conditioning proved effective; tiny twitches occurred that escaped or avoided the noise. However, while Group 1 members thought that their behavior had no effect on the noise, two members in Group 2 reported giving up the search for the right response half way through the experimental hour, and the third insisted that he had discovered and used the required noise–reducing response, which he described as involving hand–rowing movements, ankle wriggling, jaw displacement, exhaling, and waiting; we would describe such a pattern as superstitious behavior. The escape and avoidance conditioning failed for two members of Group 3; their "small" instruction–evoked thumb–twitch responses were not small enough; indeed they were so large that they interfered with responses tiny enough to work. The conditioning succeeded for the third member of Group 3, but only because he misunderstood the instructions and tried to increase pressure gradually on an imaginary button, which allowed some reinforceable responses to occur. Unsurprisingly, given the presence of the feedback meter, the escape and avoidance conditioning succeeded for all the Group 4 members; more importantly they continued responding appropriately after the meter became unavailable. Apparently, after overt feedback of covert processes brings these processes under contingency control, the overt feedback becomes unnecessary. We will see more of this phenomenon when we later look at Progressive Neural Emotional Therapy (PNET).

In summary, this study provides an early experimental example of escape and avoidance with humans. It also shows that once overt feedback helps bring covert processes under contingency control, the overt feedback becomes unnecessary. Equally importantly, the study exemplifies how, for humans, stimuli affect responses without these stimuli or responses necessarily evoking human consciousness behaviors. Using everyday language (while ignoring the inherent but unsupportable agential implications) for responses to occur, we need not be conscious of them, nor need we be conscious of stimuli for them to evoke or consequate our further behavior; recall our example of daydreaming taking up the "single" neural consciousness channel while still driving safely. The driving responses occur not due to any evoked awareness of road features, but under the direct stimulus control that those road features provide.

Escape/Avoidance Summary

Technically escape responses are those that terminate ongoing, unconditioned or conditioned, aversive stimulation. However, since escape responses that terminate conditioned aversive stimuli also often prevent unconditioned aversive stimuli from occurring, or at least delay their occurrence, we tact them differently, calling them avoidance responses. Nevertheless, this is merely a convenient terminological convention; avoidance still involves the subtracted reinforcement of escape from ongoing conditioned aversive stimuli.

With careful analysis of the unconditioned and conditioned stimuli involved, you can see both escape and "avoidance," along with countercoercion, as likely outcomes of what Glenn Latham (1998, 1999) has labeled the eight coercion traps. Let's look at these traps now, as they function as an establishing operation for alternatives to coercion.

Eight Coercion Traps

Dr. Glenn Latham spent decades working with parents at home and teachers at school. He was discovering the particulars of some parent and teacher behavior that drive much of the misbehavior of children, and he helped design many of the best practices that help change these parent and teacher behaviors so that they lead to more appropriate child behavior at home and school (see Latham, 1994, 1998, 1999). Sadly, his discoveries are not yet nearly as widely applied as they could be; perhaps your reading this book will help in this arena.

When addressing the best steps parents and teachers can take to improve matters, Latham pointed to what he called "traps" that ensnare people in coercive practices, and he delineated what he found to be the eight most common of them. Here we also call them coercion traps. While Latham cast them in terms of children, parents, and teachers, their description applies equally well to peers, friends, relatives, employees, managers, officers, soldiers, public servants, and diplomats. If I left *anyone* out, count them as included.

Coercion traps can befall anyone. Each one increases the level of coercion, which then leads to more getting away, more staying away, and more getting even. What are alternatives to these traps? We will get to alternatives after we touch on each trap in turn.

(1) The Criticism Trap
The criticism trap involves verbally berating people when their performance is inadequate. In focusing on negative performance aspects, the criticism usually fails to describe, let alone teach, what an adequate performance involves and how to reach it. For example, you may have heard (or said) "You have been cleaning this messy room for an hour but little has changed…" I doubt that any child has ever responded to such criticism by saying, "Oh, thank you for clarifying my inadequate effort. I will take more specific steps immediately, and quickly, to get this place cleaned up to your complete satisfaction." (Consider what you could say instead to the child that might be more helpful.)

Criticism also elicits negative emotions that exacerbate subsequent responses, and these responses are more likely to include various forms of escape, avoidance, and countercoercion. Indeed, this applies to every one of these eight criticism traps, as our regular repetition will show.

(2) *The Arguing Trap*

Arguing is neither debating nor discussing, either of which can come to a reasonable, even helpful, outcome. The arguing trap often begins with two people verbally brawling in an un–winnable game of one–upmanship, but it easily escalates further by involving additional players. Tempers fray and flair up as the verbal weapons elicit negative emotional responses, all of which contributes to the usual, arousal–affected outcomes of leaving, avoiding, and hurting back. As coercion rises, positions polarize and nothing gets solved. While arguing provides each party with lots of attention–type added reinforcers—which is part of why it happens—it fails to build useful repertoires. Not merely a most ineffective method for managing behavior, arguing takes a prize for counterproductivity. Because you surely have already heard too many arguments, I will not give an example, and I doubt I need to suggest that you stay away from the arguing trap.

(3) *The Logic Trap*

When a question arises, and you are carefully staying away from arguing, logic becomes very tempting. And if the moment is calm, without the usual looming deadline hanging in the air and heightening emotions, then logic may help elucidate why something has happened or why it will happen or how you can improve it.

Instead, however, we often grasp at logic merely to make our own wisdom attractive to another person. We are likely only telling someone something that they already know, or we are giving logical instructions based on traditional lore, or common sense, or even reason. All these attempts usually provide attention for inappropriate behavior. And all these attempts to manage behavior through logic usually backfire by affecting others as aversive stimulation (which elicits negative emotions and their effects while evoking the three standard coercion and punishment outcomes, namely escape, avoidance, and counterattack).

For example if, in the middle of a school week, a young teen's friends arrive after dinner, and he is to go out with them, the question occurs regarding how late he may stay out. Avoiding an argument, you instead start listing a dozen various logical reasons why he should not stay out late, but before you get even half way, he—possibly politely—cuts you off saying, "We're in a hurry. Just what time?" Logic will not work here; other variables are at play (see Latham, 1994). For you to keep avoiding the aversiveness of an argument, you better not say too early a time, so instead you say a later time, a time later than the time you think you should have said. That is, the aversiveness of an argument about the time evokes your stating a time later than the time that other variables had been compelling you to specify, and the removal of the threat of that argument reinforces stating that later time. Now, though, you might ask: How many months must pass before you would have considered your young teen to be even chronologically ready for the time that you stated? Look where the logic trap got you.

(4) The Questioning Trap

If you need information for problem solving, asking questions is not only appropriate but may also be necessary. If you arrive home to find the sitter taking an open bottle of window–washing solution away from your young child and saying that the child had been drinking it, asking the child *why* she was drinking it would only give attention to inappropriate behavior while also wasting time that would be better spent getting medical treatment. On the way to treatment, an appropriate question to ask would be "How much of the bottle did she drink?" That information might be of use to the ER physician.

The questioning trap is not about questions needed for problem solving; the questioning trap involves asking questions about inappropriate behavior, questions like the "why" question. You may get an answer, but it will likely displease you. For example, "Why is your homework incomplete?" "Because I hate the topic." Such answers usually evoke quite negative reactions, both emotional and operant, from the questioner, and a mutually coercive downward spiral begins, a spiral that includes the three standard outcomes of coercion and punishment: escape, avoidance, and countercoercion.

Prevent all that; never ask questions about misbehavior. Here are three more reasons not to ask questions about misbehavior. The questions are threatening, and so tend to evoke lying, evasion, and defensiveness. Asking such questions also provides lots of attention for the misbehavior, which makes it worse. And besides, usually the ones who ask such questions seldom want answers; they want appropriate behavior, and an answer does not equal appropriate behavior.

Anticipating our later discussion of coercion alternatives, while you should never ask questions about misbehavior, you should ask questions about future behavior. Extending our logic–trap example, and presuming that you have earlier clarified with your young teen that you expect him home by 10 P.M. on such nights, you should now ask "At what time do I expect you back home?" Ignoring all the verbal–behavioral "noise" that might occur, you calmly repeat your question as necessary and, when he finally answers "10 P.M.," you can say, "Thank you. I'm glad you understand that." Rather than you ordering your young teen home at a certain time, which is inherently aversive, *your young teen has told you* what time he should be home, which is far less aversive for everyone. Furthermore, if you have treated questions this way for the last decade, the exchange becomes rather routine and pleasantly lacks the otherwise usual verbal–behavioral noise. Is this not a desirable alternative to any coercive practice? It is a very predictable outcome (again, see Latham, 1994, 1998, 1999).

(5) The Sarcasm Trap

One of the most hurtful coercion traps is the sarcasm trap, which includes pointed teasing. Sarcasm occurs as a sort of shock treatment that supposedly helps manage behavior, but it fails miserably at this task. Sometimes disguised as humor, sarcasm seldom evokes laughs from the receiver, but its coercive effects are clear; these effects include the usual suite of negative emotions building walls between people, along with the standard outcomes of escape, avoidance, and countercoercion. We have all experienced sarcasm; we may have even indulged in it. A common and confusingly hurtful sarcastic comment on a student's poor grade is that, "At least they cannot accuse you of cheating." Imagine your feelings, and likely counterreactions, were someone to say that to or about you.

(6) The Despair/Pleading Trap

This "despair/pleading" trap features two names, because "despair" implies the emotional component of the problem while "pleading" alludes to the operant component. This trap implies that the person whose behavior needs managing is at fault for these problems, and this implication begins the downward spiral of coercion. All the negative emotions and their effects ensue along with the three standard operant coercion outcomes of escape, avoidance, and counterattack. After a history of failed encounters regarding improvements, the circumstance of little or no progress on some often minor issue elicits the despair and evokes the pleading. These typically take forms such as saying, in strained emotional tones, "What am I *ever* going to do with you?... Everything I try gets nowhere!... After working my fingers to the bone day after day to pay your way, this is your best effort?..." And so on. I hope you have never heard such things except in comedy routines, although this trap is not very funny.

Now, what if the other person actually responded with an answer? What if they offered to help, to tell you what to do to fix things? Then he or she had better be well out of reach, because the despairing and pleading person's behavior could quickly devolve from the verbal realm to the physical realm. This kind of offer to help is most unlikely to occur, and even more unlikely to be accurate. Just imagine the likely coercive reaction on the part of the despairing pleader if, in reply to "What am I going to do with you?" the other person said, "Well, I can make some suggestions. I have spent some time in the library and online studying the literature of behaviorology regarding alternatives to negative management practices with children and others. With these in hand, I am sure I could tell you how to increase your competence in this area." If you are on the receiving end of the despair/pleading trap, I must recommend against taking this approach; it may allude to reasonable alternatives, but it is a polite yet still in–your–face form of subtle counterattack. It is not one of the reasonable alternatives, not in this form, and not with someone caught in the despair/pleading trap.

(7) The Threat Trap

We fall into the threat trap when we claim that something bad will happen if some performance fails to shape up. The "something bad" seldom involves something "reasonable" but instead is either extreme or absurd. Nevertheless the threat gets the negative emotions going, along with their effects. Often emotions are already running high, and the threat spikes them further. Escape, avoidance, and countercoercion are not far behind. For example, emotionally exaggerated threats commonly concern curfews: "If you are not back by midnight, I will ground you for five years!" Such threats are possibly more inappropriate than the behavior evoking the threat. And, as usual with any of these coercion traps, falling into the threat trap is counterproductive; the threat is useless verbal behavior that emotions like anger exaggerate, evoking subsequent regrets. Furthermore the threat is quite often unenforceable, which calls the threat–maker's competence into question. Even worse than these problems, threats are but a short step away from force. Indeed the abuse, inherent in some verbal threat behavior, pushes this behavior into the force–trap category.

(8) The Force Trap

The force trap involves using verbal force (e.g., shouting) or physical force (e.g., grabbing, hitting, kicking, shaking, shoving, squeezing) to "manage" another's behavior. None of these forms of coercion and punishment is appropriate, some exceed legality, and all fail to teach more desirable behavior. Furthermore, all of these coercion and punishment forms elicit a full range of long–lasting negative emotions and effects while also evoking extreme forms of escape, avoidance, and counterattack, including dropping out, escalating violence, and the ultimate escape, suicide.

Imagine what happens in and to society when our institutions and agencies institutionalize these or worse kinds of force as standard operating methods. Our challenge, which Sidman (2001) shows we have the means to meet, concerns applying appropriate scientific behavior controls in ways that reduce not only the use of interpersonal force, but also the use of institutionalized force, at all levels of social interaction, while at the same time institutionalizing recognized and capable positive, alternative controls, the same ones that will also enable us to solve global as well as local problems.

Trap Conclusion

In addition to all the other problems, all of those eight coercion traps undermine the relationships of the people involved, driving wedges between them. However, the world will not end if on occasion you fall into one or another of these traps. Indeed, the data that Latham and other researchers have collected shows that you can substantially reduce, perhaps even minimize, the negative effects of coercion if you maintain a positive–interactions to negative–interactions ratio of eight or more to one (e.g., see Latham, 1994, 1998). That is, react positively at least eight times for every one time that you react negatively. This may seem difficult at first, but gets easier with practice.

For example, after Latham trained teachers and staff at one school to implement positive practices, and to collect data tracking their progress, they reported moving from more negative interactions than positive interactions to a positives–to–negatives ratio of over 100 to one before the end of the term. What about those positive interactions? They range from a wink and a smile to some simple verbal praise or more elaborate acknowledgement of quality performance; they take little time; and they involve alternatives to aversive controls. Let's turn to that topic now.

Control by Alternatives to Aversive Controls

Recall that, along with control in all other natural–science subject matters, an absence of behavior control is not an option. Natural variables, including the behavior of other people, control all behavior. However, while coercive controls like punishment leave us feeling constrained, the alternatives like added reinforcement leave us feeling free. The less coercion we experience, the more we experience that kind of freedom. Given the problems of coercion, including those we have discussed, alternatives to coercion are vital to our present well–being and future survival (read Sidman, 2001).

We will cover alternative controls to coercive control first in broad, general terms. These we will follow with a sampling of some basic, specific alternative controls. To start these alternative controls, here is a story that addresses one non–coercive alternative, and exemplifies the rule that behavior usually responds better to positive (i.e., non–aversive) controls, like added reinforcement, than to negative (i.e., aversive) controls, like punishment or subtracted reinforcement.

A Story about Better Controls than Coercive Controls

This short children's story (which is adapted from Ledoux, 2002) introduces the positive–control alternatives to coercive control. The title is *Jamie's Lesson* and, under literary license, we need not take offense at the inner agents that the simple story phrasing might coincidently imply.

"Jamie's Lesson." *Let's start with some questions. You go to school? You see some kids behaving meanly? And you see others behaving nicely? Well, this is a story about Jamie, and about an early lesson that she had on helping others behave nicely.*

In the middle of winter, with a cold sun in the bright blue sky, and a thin glaze of ice on the ground, Jamie's classmates played in the grounds surrounding the school. However their teacher, Mr. Glenn, saw Jamie off to one side, sniffling. Going over to her, he asked, "Jamie? Are you okay?"

"I don't like Freddy!" she replied adamantly. "He's so mean. He called me clumsy, just because I slipped on the ice."

"I can understand why you are upset," Mr. Glenn calmly said. "We feel unhappy when other people call us names."

"And everyone laughed, too," Jamie added.

"We feel even worse when others give attention to name calling," Mr. Glenn continued pleasantly. "We have talked in class about a helpful way to handle these things. Can you recall what we said?"

With a little hesitation, Jamie replied, "We said we should ignore misbehavior, and pay attention when people behave nicely." After a pause, she continued. "But Freddy never behaves nicely!"

"Well," Mr. Glenn said, "at times like these, seeing any good behaviors can be difficult. But tell me just one good behavior from Freddy recently."

"Well," Jamie said, deep in thought. Then, beaming, she said, "yesterday I saw him go right over to a little kid who fell off the slide, to see if he was okay. Oh, and this morning he helped pick up a box of spilled pencils—and he wasn't even the one who spilled them. That was nice of him."

"Wow!" said Mr. Glenn. "That's great. That's two good behaviors! Did you happen to tell him you thought that was nice of him?"

"...Oops," said Jamie.

"You can still tell him, if you want to," said Mr. Glenn. "That can still help him become better at nice behavior."

"That would be good," Jamie replied. "I will!" And off she went to tell him. *You too can catch people being good. Just once each day, notice someone's good behavior, and tell him or her that it was nice. Each of us paying such compliments every day makes a much better world.*

That story shows one alternative to coercive control. We will consider others, both general and specific. While we may describe many of these alternatives in terms of parents or teachers managing a child's behavior, all of these alternatives also apply, with appropriate adaptation, to friends, spouses, employers, employees, government agencies, international relations, and solutions to global problems.

Some General Alternatives to Coercion

The category of general alternatives to aversive control spans the range of non–coercive antecedent and postcedent processes and procedures. Think for a moment about those that we have already covered, and how the control that each exerts usually occurs non–aversively. They include evocative and function–altering stimuli, added reinforcement and differential reinforcement, shaping and fading, reinforcement schedules and backward chaining, verbal rules and instructional controls, establishing operations and certain uses of extinction, and even some respondent unconditioned and conditioned stimuli.

All those independent variables control behavior in generally non–coercive ways. Let's look more closely at a few of the many specific ways to manage behavior without coercion.

Some Specific Alternatives to Coercion

Our specific alternatives to coercion focus on some different ways to deal with three categories of behavior. These categories cover appropriate behaviors, inconsequential behaviors, and inappropriate behaviors.

A new distinction for us separates *inconsequential* behaviors from *inappropriate* behaviors. We would label most of the annoying behaviors of children and others as inconsequential, even if that covers most of a person's behavior. This is essentially harmless behavior. It continues, and becomes even more annoying, when it gets attention, especially adult attention, especially adult attention that it need not receive. On the other hand, inappropriate behavior consists of behavior that causes emotional harm or physical damage to the behaving person or others, or property damage. This is behavior that we cannot ignore; but we must manage it carefully, because some methods make it worse. Before looking at ways to manage each of these, let's consider better ways to generate and maintain appropriate behavior.

Appropriate behavior. Unfortunately, two inconsistent traditional cultural notions, (a) that behavior lacks causes, and (b) that only coercion affects behavior, have led to the widespread practice of ignoring most appropriate behavior. However, words like "ignoring" merely provide a verbal shortcut for the technical term *extinction*. "Ignoring" also implies an agent that "does" the ignoring, but since we know better, we can ignore this problem (all agential puns aside). Meanwhile, what happens to any behavior, including appropriate behavior, on extinction? Yes, its rate gradually reduces. For appropriate behavior this outcome ranks high in undesirability.

To reverse that outcome, we must stop ignoring appropriate behavior. We must stop saying, agentially, "We should not pay any attention to that behavior, because that is the way people should behave," as if their behaving appropriately or not merely depends on the type, good or bad, of mystical person agent inside them. Instead, we must catch people being good by following their good behavior with reinforcement. We must remember that stimuli evoke all behavior, including appropriate behavior, and these stimuli get their effectiveness from the consequences of the behavior, particularly attention reinforcers. The best way to generate appropriate behavior is to follow it with added reinforcers, with intermittent reinforcers then maintaining the behavior.

For example, as a child grows up, most parents prefer reasonably uncluttered bedrooms. Some of the reasons for messy children's rooms concern incomplete or missing contingencies that otherwise condition room–cleaning behavior. This could mean that, while some present contingencies would *maintain* room cleaning, no present contingencies *generate* room cleaning. Certainly, using coercive methods would solve nothing; they fail to teach how to clean rooms, but they do not fail to engender a host of negative effects.

Besides, parents want clean rooms not just for the present moment, but clean rooms now are supposed to portend clean rooms as an adult. However, the state of a child's bedroom poorly predicts cleanliness as an adult. In the long run, the parents' behavior better predicts the child's later adult behavior. Even in the same family, the state of children's rooms can range from that of a junk yard to that of the bridal suite in a five–star hotel. But the later adult behavior of these children tends to duplicate not their childhood behavior but the behavior of the adults that were around during their formative years. And this parental modeling need not pertain solely to how the parents keep their own room; it can also pertain to modeling how to clean a room, a non–coercive practice that actually teaches children how to keep rooms clean.

The earlier such parental modeling occurs, the better. In line with the principle that behavior responds better to added reinforcers than to coercive consequences, parents can begin this modeling when a child is about two years old. Parents can start prompting "picking–up–after–themselves" behavior by modeling it as they (the parents) straighten up the child's room, paying attention to the child's participation, even shaping it, and making that participation fun for the child in what seems like a game because the parents also provide many reinforcing stimuli as enjoyable compliments and contingent praise. As a result, the quality of the room's condition improves, and that is easier to maintain through adolescent years with the more occasional though still appreciated compliment and praise. Some might even say that the child has a room–cleaning "habit," but by now you can easily say both why that is not accurate and what accounts for room cleaning.

Inconsequential behavior. The easiest behaviors to manage are those we call inconsequential behaviors. These include age–typical behaviors like dressing down, mild sibling rivalry, meaningless verbal blows, and jousting, or rough–housing that often leads to laughing, not crying. These behaviors threaten neither the basic quality of life nor the safety of limb and property. How should you respond to such junk behaviors? (Hint: The verbal noise, which sometimes follows a parent's compliments of a child's good behavior, is inconsequential behavior, so just ignore it.)

You treat those kinds of inconsequential behaviors in the opposite way that you deal with appropriate behavior. While you attend to appropriate behavior, attending to junk behavior simply gets you more of it, which is counterproductive. Instead, basically, you *ignore* inconsequential behaviors; you simply put them, or leave them, on extinction. Also, rather than count instances in extinction, you can more easily time how long a current incident takes to stop when ignored. While later stimuli might re–evoke the annoying behavior, Dr. Latham's data shows that about 80% of current inconsequential behaviors, if ignored, end within 90 seconds. Try it; actively ignore the next annoying, inconsequential behavior; turn away from it while timing it; engage in something else during those 90 seconds. For example next time your child bothers you by tapping a fork against a glass at dinner, just surreptitiously time it while continuing other normal dinner–time behaviors. You will be pleased with the outcome.

Nevertheless, those may seem like the longest 90 seconds you have ever experienced, but that time interval is far shorter than the time required to undo the problems that coercion–based practices bring when they are used, in place of calm extinction, for managing inconsequential behavior. The calm extinction not only reduces the problem behavior but also models for children how they should deal with the problem behavior of others, a most valuable lesson for both right then, and adulthood later. Remember, indeed ponder, the proverb, "Patience rather than anger escapes 100 days of sorrow." Inappropriate behavior, however, requires a more complex strategy. Let's consider that next.

Inappropriate behavior. Unlike appropriate behavior, which we prefer and which benefits from plenty of added reinforcement, and unlike inconsequential behavior, which we can consider relatively benign although annoying and which benefits from active ignoring, inappropriate behavior threatens the basic quality of life and the safety of limb and property. We cannot ignore it; we would not want to reinforce it; and responding to it coercively not only produces the usual negative effects but also still pays a sort of attention to it that only makes it worse, especially if the inappropriately behaving person is under attention deprivation, receiving little attention for better behaviors. So, what steps might mitigate inappropriate behavior?

For one way to manage inappropriate behavior, we turn to a well–researched (e.g., the University of Kansas Achievement Place model) and long–implemented (e.g., for decades now, at Boys and Girls Town) multi–step practice that stops inappropriate behavior and redirects it toward reinforceable appropriate behavior. We call this practice the *Teaching Interaction Strategy,* or the *Corrective Teaching Procedure,* because it stops the misbehavior and redirects it by teaching (i.e., conditioning) an appropriate, alternative behavior.

The Teaching Interaction Strategy has six steps that occur in a calm, empathetic and non–threatening manner. You start by saying something positive. An easy and valuable enhancement involves making eye contact with the misbehaving person, and holding that contact briefly, in silence if possible; this kind of contact often has a chilling effect on the misbehavior. In the next step, you describe the problem behavior in just a few words. For step three you describe a better, alternative, appropriate behavior, also in as few words as possible. Then, you state a clear reason why the new behavior is more desirable than the inappropriate behavior. In step five, you practice the desired behavior, at least verbally. Finally, you follow the appropriate practice behavior with positive feedback; while the other steps occur in a matter–of–fact manner, you can show more enthusiasm in reinforcing the appropriate practice behavior.

Let's list those six steps together, and then consider an example. Here are the six steps (from Latham, 1994 and 1998, which contain much more discussion of these steps, and many more examples, from both child care and education):

(1) Say something positive;
(2) Briefly describe the problem behavior;
(3) Describe a better, alternative behavior;
(4) Give a reason why the new behavior is more desirable;
(5) Practice the desired behavior;
(6) Provide positive feedback for the better behavior practice.

Here is an example of those six steps in action: Some of your child's friends stopped at your house on the way home from school one afternoon, and you know from past experience that they sometimes seriously squabble. A commotion in the back of the kitchen induces directing your gaze that way. Just barely, you overhear Tom call Gerry a name unfit to repeat and, as you get close, you see Gerry punch Tom in chest. Quickly putting your arms between them and separating them, you look each one in the eye for a second or two in turn while saying nothing. Then, you apply the Teaching Interaction Strategy with Gerry first because, even though both behaviors are inappropriate, Gerry's hitting behavior is more serious than Tom's name–calling behavior. Here is one way you could apply the strategy, with the steps numbered and summarized.

(1) Speak positively: "Gerry, you are usually quite pleasant to have around."

(2) State the problem: "But when Tom called you that name just now, you reacted by hitting him."

(3) Give an alternative: "One better way to respond to name calling is to ignore it and simply walk away."

(4) Say why the alternative is better: "By ignoring it, you stay out of trouble while also reducing that misbehavior by preventing it from getting further attention; Tom will probably forget it, and it won't affect your friendship."

(5) Practice the alternative: "So Gerry, when someone calls you a name, how will you deal with it next time?" (Reasonable answers would be: "I will ignore it" or "I will just walk away.")

(6) Provide positive feedback: "That's good, Gerry. I know how hard that will be at the time, but in the long run, that will be easier on you too."

Of course you may need to ignore some behavioral noise during that strategy. You may even need calmly to repeat parts of it. Still, you now have the six steps, and the example of dealing with Gerry's inappropriate behavior of punching Tom. You can see the importance of dealing with such problems instructively, without emotion but with direction. You may have thought of some other, more common—and coercive—ways to handle such a situation, but let's put those more damaging ways behind us. Since I cannot provide a laboratory or practicum setting in which you might practice the six steps, spend a few minutes outlining, for your own practice, what you might say when implementing the Teaching Interaction Strategy again, this time with Tom, because his name–calling behavior was also inappropriate.

Coercion Alternatives Summary and Extension

Again, independent variables control all behavior at all times. But different types of control affect us in different ways. Emotionally, coercive control constrains us while added–reinforcement control frees us. The less coercion we experience, the less loss of freedom we experience. In other words the more we experience the alternatives to coercion, the greater is our freedom experience.

You can even find alternatives to coercion for some of the most coercive institutions in our culture. Fraley has described coercion alternatives in detail for two of these institutions. In *Dignified Dying—A Behaviorological Thanatology* (2012) he presents the applied behaviorology behind better ways to treat our terminally ill neighbors, friends, loved ones, and selves, along with their survivors. And in *Behaviorological Rehabilitation and the Criminal Justice System* (2013) he elaborates on the applied behaviorology behind improvements in managing penal incarceration in ways that help inmates as well as society.

Conclusion

Overall, however, we need far more detail to do even minimal justice to the topic of "aversive control, problems and alternatives." This chapter could not adequately introduce all the needed basics even if it was 50 pages longer. Happily, other books treat this topic thoroughly. You will find reading one or another of them quite advantageous. An excellent starting point is Murray Sidman's *Coercion and its Fallout* (Sidman, 2001). Note, however, that Sidman originally wrote his book before 1989, which was before the term *behaviorology* was in general use, so he refers to "behavior analysis" in his book, a name that psychology has recently claimed even though, back then, this name could tact the natural science of behavior. Sidman's book relays no kind of psychology (which leaves readers to reduce confusion by substituting "behaviorology" when "behavior analysis" appears in this text). Instead, Sidman's book vastly expands this chapter. These topics deserve that kind of coverage, given our long–standing overuse and misunderstanding of coercion and the negative impact it has on efforts to solve global and local problems at all levels of society, from your household to the relations your country has with other nations.

That thought can be part of closing this chapter with our ongoing activity. We must not underestimate the negative impacts that coercion and punishment, and their effects, have on behavior, negative impacts that would prevent the timely solution of global problems. On the other hand, and more importantly, think about how the alternatives to coercion will help make possible the kinds of repertoire changes in support of sustainable lifestyles that will enable the solutions to those problems to succeed. This success could begin with the practice of providing reinforcers for recycling, rather than charging people money for recycling, or increasing the response effort for recycling, or merely coercing people to recycle by just punishing them when they throw recyclables in the trash. Not only could we go far from this start, but we must. Indeed, we are not exceeding the data when we acknowledge that avoiding another multi–thousand–year, mystically overrun dark age, and possibly humanity's very survival, depends on taking the realities of behaviorological science and interventions into account when designing physical, chemical, and biological solutions to global problems. We all face the problems together, and with a natural–science team effort we are far more likely to solve them successfully.♣

References (with some annotations)

Dise, R. L. Jr. (2009). *Ancient Empires before Alexander* (Course 3150, Lecture 9). Chantilly, VA: The Teaching Company.

Fraley, L. E. (2012). *Dignified Dying—A Behaviorological Thanatology.* Canton, NY: ABCs.

Fraley, L. E. (2013). *Behaviorological Rehabilitation and the Criminal Justice System.* Canton, NY: ABCs.

Hefferline, R. F., Keenan, B., & Hartford, R. A. (1956). Escape and avoidance conditioning in human subjects without their observation of the responses. *Science, 130,* 1338–1339.

Latham, G. I. (1994). *The Power of Positive Parenting.* Logan, UT: P & T ink. This book is Dr. Latham's most comprehensive work applying behaviorology to child care (and education).

Latham, G. I. (1998). *Keys to Classroom Management.* Logan, UT: P & T ink. This book applies behaviorology to education.

Latham, G. I. (1999). *Parenting with Love.* Salt Lake City, UT: Bookcraft. This book applies behaviorology to child care, and much of it parallels the author's two–part, two–hour video entitled "The Making of a Stable Family." Dr. Latham was a good friend and colleague, and one of the four founders of *The International Behaviorology Institute* (see "In Memoriam" at www.behaviorology.org).

Ledoux, S. F. (2002). A parable of past scribes and present possibilities. *Behaviorology Today, 5* (1), 60–64. While opening with the "Jamie's Lesson" story, the principle point of this article concerned the educational abuse of children, and the financial abuse of taxpayers, that results from ignoring the "Project Follow Through" research outcomes (see Watkins, 1997).

Sidman, M. (2001). *Coercion and its Fallout—Revised Edition.* Boston, MA: Authors Cooperative. Be aware that Sidman originally wrote this book before 1989, which was before the term *behaviorology* was in general use, so he refers to "behavior analysis" in his book, a name that psychology has recently claimed even though, back then, this name could tact the natural science of behavior. Since this book does not relay any kind of psychology, readers can avoid confusion by substituting "behaviorology" when "behavior analysis" appears in this text.

Skinner, B. F. (1955–56). Freedom and the control of men. In B. F. Skinner. (1972). *Cumulative Record: A Selection of Papers Third Edition* (pp. 3–18). New York: Appleton–Century–Crofts. The B. F. Skinner Foundation (www.bfskinner.org) in Cambridge MA, republished this book as "... *Definitive Edition*," in 1999.

Skinner, B. F. (1971). *Beyond Freedom and Dignity.* New York: Knopf.

Watkins, C. L. (1997). *Project Follow Through: A Case Study of Contingencies Influencing Instructional Practices of the Educational Establishment.* Cambridge, MA: Cambridge Center for Behavioral Studies.

Whaley, D. L. & Malott, R. W. (1971). *Elementary Principles of Behavior.* Englewood Cliffs, NJ: Prentice–Hall.

Chapter 17
Some Applied Behaviorological Research Considerations

*L*ast time, in Chapter 16, we turned our attention to what we might see as the dark side of the behavior–control reality, the coercive parts of the laws of behavior, the aversive controls that humans used against each other long before our natural science of behavior discovered these controls. Besides reviewing punishment and its role in aversive control, we also considered escape and avoidance phenomena, eight coercion traps, and some general and specific alternative controls to aversive controls.

Furthermore we recognized that, because all behavior is controlled either through positive controls or through coercive controls, making knowledge of *all* the laws of behavior, both positive and negative, available to everyone constitutes perhaps the most valid resistance to the misuse of these laws, for instance, to abuse or suppress people, or for purely personal gain. For this reason assuring that as many people as possible understand at least the basics of these laws provides yet another reason for founding and expanding programs and departments of behaviorology in our colleges and universities. And this reason may equal in importance our earlier reason, in which humanity requires behaviorology now, not only comprehensively in the repertoires of fully trained behaviorologists—to teach and help others—but also, due to the substantive behavior components of global problems and solutions, at least minimally in the repertoires of the traditional natural scientists who are working in the team efforts to solve these problems. Let's now consider some additional topics that can support these efforts.

In this chapter we consider some specialized areas of ongoing applied behaviorology research. Of the many topics that we could include here, we will consider three that portend impacts and implications well beyond their own borders, with emphasis on the last one. We will cover (a) the General Level of Reinforcement (GLR), (b) Progressive Neural Emotional Therapy (PNET), and (c) developing behaviorological practices from behaviorological principles.

General Level of Reinforcement (GLR)

As a testable concept worthy of further research, the *General Level of Reinforcement* (GLR) stands poised to evoke a scientific redefining both of our overall understanding of therapeutic interventions for helping people and of our particular understanding of some specific intervention areas and practices. Here are some highlights of this concept.

A GLR Pioneer

Joseph Cautela published on the concept of the GLR in 1984, and he expanded on this concept when he delivered the B. F. Skinner Memorial Lecture at the fifth annual convention of The International Behaviorology Association (TIBA) in 1993. His remarks then appeared in the journal *Behaviorology* in 1994, where he defined the GLR as "the number, strength, and duration of reinforcers per unit time" (p. 2). He also gave a formula for the GLR, although it is a conceptual formula more than a calculable formula. Cautela then explored how we can apply the GLR concept to help us better deal with the world around us, and he made various research suggestions. Today, in conjunction with our physiology colleagues, and using the more recently developed digital data–collection hardware and software—including smart phones—for measuring many of the physiological variables related to reinforcement, researchers might be able to elaborate and apply a more calculable GLR formula.

GLR Implications for Helping Others

Meanwhile let's consider some therapeutic intervention implications of the GLR. Dealing with some problem behaviors requires a behaviorological practitioner to change the antecedent environmental conditions that are evoking the behaviors. Dealing with other problem behaviors, however, calls for changing the postcedent environmental conditions to increase or reduce the reinforcers that are affecting the behaviors. Remember, this means *immediate* reinforcers; delayed "reinforcers" operate through other processes (e.g., rules or instructions) and so are likely less relevant and possibly less effective.

Any reducing of reinforcers, however, lowers the GLR and typically elicits aversive emotional reactions, while evoking escape (e.g., leaving the setting or therapy) and avoidance (e.g., staying away from the setting or therapy) and countercoercion (e.g., violent responses toward the practitioner and sabotage of the therapy). Virtually all helping professionals have observed such responses.

To reduce such difficulties when an intervention requires a reduction of some immediate reinforcers, the practitioner must assure an increase in other reinforcers, contingent on more appropriate behavior, that maintains or, better, increases the GLR. Indeed most savvy practitioners begin by increasing the reinforcers for a client's adaptive behavior as an early component of any intervention. Such practitioners also assure that these new reinforcers are immediate, not delayed, consequences. For example, when your physician or nutritionist (see Guarneri, 2006, 2012) takes you off all those foods that you love but which are ruining your health, the delayed "reinforcer" of better health substitutes poorly for the loss of those favorite food reinforcers. When combined with whatever medical diagnoses your physician gives you, the reduction in GLR through the loss of these food reinforcers can be quite depressing. Instead, these helping professionals turn to other more immediate reinforcers to take up the slack, and preferably before the slack occurs. Our next example covers some ways to accomplish that.

GLR–Related Examples

Some cases of depression result from very clear decreases in the GLR. Say you take a job that requires moving to a city that you have never even visited. You are pleased to have a job, but few of the stimulus complexes in the new city evoke relevant responses; the streets, neighborhoods, bookstores, bars, shopping malls, doctors, dentists, hairdressers, and so on, are completely unfamiliar in that conditioning has yet to establish these stimuli as evocative of your responses. In this sense much of your past behavior repertoire is mostly useless. For example a bookstore–visiting response would produce reinforcers in your old city but not—yet—in the new one. The move has made your repertoire irrelevant, and it only slowly changes from that status during the period in which the new environment gradually conditions appropriate responses to new evocative stimuli, a period that can stretch out for months or more, depending on variables like the time commitments that the new job requires. Thus for some time, few reinforcers are forthcoming, an unsurprising depression–inducing decrease in GLR, which for many shows a severity that evokes the clinical "depression" label. The label might get you only some talk "therapy." More likely, a prescription for an anti–depressant drug follows this label, and popping these pills can make you feel a little better. But these "therapies" mostly leave untouched the environmental source of your depression, the decreased GLR from the lack of evoked relevant responding.

In cases like that, and perhaps to some extent in all depression cases, what is often depressed is not so much "the client" (or some putative agent inside the client) but the client's GLR. Then the success of an intervention, in "reducing the depression," turns on changing the client's environment, particularly so that it evokes an increasing amount of reasonable client behaviors that produce appropriate reinforcers, thereby restoring, even expanding, the client's GLR.

For another example, what if a therapy takes on the maladaptive behavior of cigarette smoking? Then every missed cigarette is a GLR decrement. Failure on the therapist's part to understand this, to recognize it, and to assure that it is at least counterbalanced, lays a foundation for a failure of the therapy. Therapy does not fail because some obnoxious, hostile, resistant inner agent inhabits the client's body (an example of blaming the victim); instead therapy fails because some variables functionally related to the problem behavior remain unchanged. This holds for all problem behaviors, from overeating, to excess alcohol consumption, to wasting environmental resources, to name but a few.

What are some ways to raise the GLR? Cautela (1994) discusses many; here are some of them. In addition to the various covert conditioning procedures (e.g., covert reinforcement and creative fantasizing) with which Cautela's name is closely associated (Cautela, 1970; Cautela & Baron, 1977; Cautela & Kearney, 1986) one can provide reinforcing verbal feedback after each of many appropriate client responses. For example, using the linguistic level of the client, and so not taking offense at the otherwise useless agential implications, one can say many things—at an appropriate rate and time—like, "you did well," "that's a good idea," "you are really coming along," "that took courage," "you look good today" (Cautela, 1994, p. 13). One can also increase the GLR by training the client's significant others and associates—workmates, roommates, relatives, friends, family—to provide more reinforcement for the client's adaptive behavior (i.e., to give more attention, compliments, praise, and interaction as appropriate). One can increase *and also maintain* the GLR by conditioning the client's participation in additional, preferably legal, reinforcing activities (e.g., by reinforcing a client's further extensions into hobbies such as coin, stamp, antique, or other collecting, or pursuits such as sports, gardening, writing, photography, or other artwork creation). Furthermore, training the client in self–control techniques, assertive behavior, and the self–control triad (Cautela, 1983) can increase and maintain the GLR as well.

Before moving on to Progressive Neural Emotional Therapy (PNET), here is how Cautela concluded his 1994 General Level of Reinforcement paper. In his experience once you fully consider the concept of the GLR, you see implications everywhere, "in street violence and world wars, in low morale (low reinforcement) and low production" (p. 14), and even in the way the GLR might affect resistance to extinction.

Progressive Neural Emotional Therapy (PNET)

We can define *Progressive Neural Emotional Therapy* (PNET) as the refined and standardized practice of what people once inadequately described as a kind of relaxation training and which we now might better describe as successive muscular–emotional re–conditioning. This practice reduces the negative effects of anxiety and stress, which are emotional and physical factors that we have long associated with numerous neuro–emotional maladaptive behaviors.

Many types of therapies boast a broad array of tactics to calm and relax clients as part of treating their problems. Past researchers examined which components helped and which did not. The outcomes, however, were inconsistent. Different client–problem combinations responded to different combinations of components, leaving the meaning and value of "relaxation training" unclear in general as well as for particular clients. Some factors in that inconsistency include both the range of non–scientific descriptions of the relaxation process and the range of different relaxation–training protocols.

For some decades now, other researchers have worked to develop a *standardized* set of relaxation–training protocols (e.g., see Ferreira, 2012). These efforts resulted in PNET, which grounds a standardized relaxation–therapy protocol in natural–science laws of behavior. By *standardizing* the principal protocols, PNET reduces or eliminates anxiety as a response to stress *for essentially all clients,* including those without clinical problems, because every client get the full training; in this sense all clients need, and get, all parts of the procedure. Most improvement occurs during the intervention training procedures, with further improvement occurring after the training for those clients who continue to practice the PNET procedures. Indeed, the research indicates that we could all benefit from some PNET training. As part of such training, PNET conditions the client in the skill of providing the relaxation cues, the stimuli that evoke relaxation, as a function of the usual stimuli that induce stress or anxiety. In other words the occurrence of stress or anxiety outside the therapy setting then successfully evokes the client's responses of providing the cues that evoke relaxation, and the client relaxes.

Behaviorological and physiological factors explicitly interface in the production of those beneficial outcomes. Decades ago, Skinner recognized the inescapable connections between physiology and behaviorology when he stated that analyzing behavior "is essentially a statement of the facts to be explained by studying the nervous system. It tells the physiologist what to look for" (Skinner, 1969, p. 283).

Behaviorological therapists have used the PNET passive cue–controlled progressive–relaxation protocol "across a wide range of populations varying in diagnoses, gender, age, ethnicity, medical conditions, physical disabilities, and intellectual impairments" (Ferreira, 2012, p. 5). And PNET is a particularly valuable tool for addressing not only functional behavior problems but also physical health problems, as we will see.

The Standardized PNET Protocol

The PNET protocol, through the successive steps of the relaxation experience, de–conditions the problematic muscular–emotional responses to stress while conditioning, or reconditioning, muscular–emotional responses describable as calm and relaxed, which then become the usual behaved reactions to the restructured environmental stimuli. This process works along the same lines as our Chapter 16 thumb–twitch example. In PNET the therapist provides the reinforcing feedback that establishes the evocative control of the verbal "relax" stimulus over the relatively unobservable relaxation responses. Similarly, in the thumb–twitch example, the readings on a biofeedback meter supplemented the reinforcing consequences of both the end of noise and the continuation of the music in conditioning the otherwise unobservable escape and avoidance thumb–twitch responses for all the subjects with access to the feedback meter. Importantly, after the meter became unavailable, the thumb–twitch responses successfully continued. In both this case and in the case of PNET relaxation training, after the overt feedback of covert processes enables bringing these processes under contingency control, the overt feedback becomes unnecessary, and the contingencies control the covert responses. This phenomenon also extends to the effects of an optional guided–imagery component—the Imagery Phase—of the PNET protocol.

The PNET protocol consists of three main phases, which we call the *Preparatory Phase,* the *Induction Phase,* and the *Transfer Phase,* with the optional *Imagery Phase* occurring at a particular point *during* the Induction Phase. The therapist includes appropriate parts of each phase at every PNET session. And each client participates in the number of sessions, on a daily to weekly schedule, needed to condition the occurrence of cue–controlled relaxation both during and outside the therapy session. Let's look briefly at each protocol phase.

The Preparatory Phase. In the *Preparatory Phase* of PNET, the behaviorological therapist first interviews the client, and conducts a functional assessment, as part of verifying whether PNET, another procedure, or some combination of intervention practices would best benefit the client. After identifying and defining the behaviors of concern, examining client medication regimens, listing client medical conditions, clarifying available reinforcer parameters, and delineating likely therapeutic interactions and outcomes, the therapist introduces the client to the PNET training room and equipment. The uncluttered, quiet, softly lighted room, featuring pleasant temperatures and clean air, contains a full–size reclining chair for the client, a therapist chair, and some device on which to play recorded surf sounds at the rate of six surf sounds per minute. After the client settles comfortably into the reclined chair, the therapist describes what he or she might experience during the session. You will find more complete details on the contents of this and the other PNET protocol phases in Ferreira's 2012 paper.

The Induction Phase. The *Induction Phase* of PNET consists of ten steps. Most of these steps involve progressing through the relaxation of various muscle groups (e.g., [left/right] hand, arm, shoulder, mouth, neck, chest, abdomen, back, leg, foot) while verbally reinforcing the observable relaxing responses, and pairing them with the surf sounds at the same time as the therapist says "… relax." The therapist says this repeatedly as part of the always slow, quiet and calm evocative instruction of "let your [muscle group] … relax." Later in each session, which typically lasts 35 to 45 minutes, the instructions evolve to "feel your [multiple muscle groups] … relax" (e.g., "feel your hands and your arms, … relax"). The optional Imagery Phase would occur between the single–muscle–group instructions and the multiple–muscle–group instructions.

The Transfer Phase. For the *Transfer Phase* of PNET, the therapist provides the client with a PNET training recording (e.g., a CD) and instructs the client in its use to mimic the session conditions for PNET practice at home. With practice, the client's own verbal mand, "Relax," will automatically evoke client–body mediation of the relaxation response. This mand will successfully come to function regularly as a countercontrol to the usual effects of anxiety–eliciting or stress–inducing stimuli. When this happens, the Transfer Phase concludes as the client no longer needs PNET sessions with the therapist.

The optional Imagery Phase and its added value. To best serve some client's needs, PNET contains a optional *Imagery Phase.* The therapist introduces this phase after the completion of at least three consecutive, fully successful, regular PNET sessions. In the Imagery Phase, the therapist guides the client's imagery during relaxation by providing evocative verbal stimuli for visualized imaginary scenes in terms of the kinds, intensities, durations, and possible developments of the scenes that relate the scenes to therapeutic outcomes. The imagery can involve any function–altering behavior patterns including experiences, emotions, ordinary scenes, fantasy scenes, or some combination of such scenes. Together the client and therapist prepare two or three appropriate imagery scenarios in advance.

That preparation takes into account that experimental data show such imagery feeding back into, and affecting, physiological systems in beneficial ways (see Ferreira & Duncan, 2002). For example, Ferreira (2012) describes an imagery set that altered the imagining client's measured hand temperatures and number of hand warts; he speaks of an image depicting a beach setting "in such a way as to elicit sensations of touch (sand, blanket, water), temperature (cool breeze, warm water, hot sun), taste (salt water, soft drink, ice cream), smell (ocean air, hot dogs on a grill, sunscreen lotion), and [sight] (sunset, ocean view, people)" (p. 9). With this beach–setting Imagery–Phase scenario, which included imagery like the client imagining his warts melting away under the hot sun, PNET was able to help the client's medical condition. Ferreira continues, "in this case a client with previous PNET training and with multiple warts on both hands, for which no traditional medical treatment had worked, was rid of his warts after seven PNET sessions over a 12–week period. This study also collected data on hand temperatures during the Imagery Phase, which involved imagining hands in the hot sun; records of measured hand temperatures showed an increase" (p. 9). Let's look at some other PNET examples.

More PNET examples. Ferreira (2012) provides several more examples. In one he describes the benefits of PNET for an elderly man diagnosed with cancer:

> E. F. was a strong, athletic 82–year–old man when he received a diagnosis of "tumor, urothelial carcinoma *in situ*," a medical descriptor for bladder cancer. Tumors identified as carcinoma *in situ* are likely to increase in invasiveness and malignancy, and have been described as treacherous with unpredictable outcomes. During his initial interview..., he appeared upset as he remarked "Of course the diagnosis was devastating. Here I was an 82–year–old man who listed excellent physical health as the second highest priority in life suddenly the owner of a malignant tumor." It was apparent that he was experiencing considerable stress and, as he put it, "this is a most anxious time for me."
>
> PNET with guided imagery sessions began with some urgency because his oncologist had scheduled a cystectomy in three weeks. It was during this interim that all PNET sessions were conducted. E. F. had successfully completed seven PNET sessions before the pre–surgical evaluation was completed. The second biopsy revealed "no definite evidence of metastatic disease" and that the "bladder appeared normal" (p. 4).

Another author (Johnson, 2012) also describes a health–related PNET outcome. Johnson's research concerned measuring any effects of PNET on tardive dyskinesia, which is a movement disorder that neuroleptic medications often induce. The three participants in his subject–based multiple–baseline study all showed tardive dyskinesia reductions after PNET. The reduction for two of his subjects brought the rate down to 25% to 35% of the baseline rate.

From examples like these, think about the extensive, health–related, preventative–medicine possibilities that PNET could provide if everyone routinely received half a dozen PNET training sessions before they finished high school, including the transfer training to continue PNET on their own regularly or as needed. Imagine how many developing health problems PNET might nip in the bud. Perhaps other practices (e.g., yoga, meditation) can also produce some outcomes like the ones PNET produces. But do these other practices have PNET's scientific grounding? This kind of grounding more reasonably justifies widespread dissemination and application. Let's turn now to another topic that also has similar far–reaching implications.

Developing Behaviorological Therapies

So far we have seen the value of the General Level of Reinforcement (GLR), and the benefits of Progressive Neural Emotional Therapy (PNET). Now let's discuss the overall process of developing behaviorological therapies. For this topic we will use the control of smoking behavior as an extended example.

Therapies and interventions stemming from different sources address problems with different and unequal methods. Interventions with behaviorological science as the source thoroughly analyze and directly control, change, and eliminate the independent variables that are in explicit functional relations with the presenting problems (i.e., the "causes" of the problems). However, therapies that originate in disciplines not fully respectful of natural science deal with problems without thorough analysis of, and direct control of, such variables, in many cases because these therapies are attending to mystical or other unreal variables instead. Or, these other approaches to therapy focus on variables that may relate, if at all, only indirectly to the relevant functional variables (although sometimes with a measure of adventitious success; see Fraley & Ledoux, 2002, Chapter 3). By countercontrolling problem–producing variables, the benefits of the behaviorological approach include an inherently greater likelihood of clinical success. We will surmise how to improve success with a range of presenting problems by examining in detail the reasons for the success of a particular behaviorological–therapy example. This successful example addresses the problem behavior of smoking cigarettes. (See Ledoux, 2002a, for details that go beyond those contained in this section.) Let me repeat, however, that *our primary point is not to expound on therapies for smoking control but to use such a therapy to exemplify a general behaviorological approach to developing therapeutic interventions.*

Scientific Therapy Steps

Designing a scientifically informed therapy, such as a behaviorological intervention, takes several steps to keep it consistent with the functional realities of both the problem behavior and the desirable alternatives to this behavior. Of the two most fundamental of these steps, the first involves thoroughly analyzing the behavior of concern to discover and rank as many as possible of the variables—historical and present, antecedent and postcedent—of which it is currently a function. The second step involves arranging environmental changes as helpful, separate, reproducible independent–variable techniques that address the problem–"causing" independent variables in ways that countercontrol as many of them as possible, beginning with those that contribute the most to the occurrence of the problem behavior. In other steps the therapy builds in extra supports that the laws of behavior can provide, such as processes that gradually shape up the alternative desired behavior while the therapist introduces the techniques that countercontrol the problematic independent variables. As success builds, the design includes gradually adding processes that fade out client reliance on the therapist and therapy process while also conditioning the client skills that take over the therapeutic role. Depending on the nature of the problem, the therapy may include other steps, perhaps ones that build mutual supports, through scheduled and unscheduled group interactions, among clients who share similar problems and solutions.

Those are some of the general steps that behaviorologists take in designing any therapy or intervention. Keep thinking about them so that you can spot their appearance and application as we go through our extended example on smoking behavior. Why examine smoking behavior as the example?

Many therapies to combat smoking exist. Some simply take an intuitive approach, which provides a contrasting background against which you can more easily see what makes a behaviorologically designed therapy different, and why the differences are important. The most basic contrast concerns the failure of scientifically uninformed "therapies" to analyze smoking as a behavior functionally related to accessible environmental independent variables. Some such therapies completely ignore the behavior status of the smoking problem while others address important but secondary issues such as nicotine addiction. I call this a secondary issue, because the data from our smoking–control example (see Morrow, Gmiender, Sachs, & Burgess, 1973) showed that 90% of those who completed the full therapeutic program were still not smoking one year later even though no special pharmaceutical techniques to address nicotine addiction were even available at the time. We will elaborate on these results later. Meanwhile some folks push currently available pharmaceutical aids for nicotine addiction as sole or primary solutions for smoking problems. How vital these aids might be in a scientifically grounded stop–smoking program remains an unanswered experimental question. I suspect that they will prove helpful but not vital, because smoking is mostly a function of other variables.

In any case we also emphasize smoking control as our example due to the tremendous pressures that smoking puts on both public–health concerns and health–care costs. After you see, by way of the stop–smoking example, how the behaviorological steps to therapy design work, and succeed, you will be able to conceive how the steps can apply to any number of problematic behaviors, including those that are problematic because they damage the life–supporting processes on the space rock that we all call home. Remember, however, that this section *primarily* concerns the value of behaviorology informing therapy and intervention design; it only secondarily concerns how to stop smoking even though this example takes the lion's share of the space.

A Comprehensive Stop–Smoking Therapy Example

We can sort anti–smoking therapies according to a particular criterion: Either they essentially involve guessing what steps to have clients take to try to stop smoking while also adding *new* variables to control smoking behavior (most commonly, "The Record" of time since the last cigarette) or they involve behaviorologically analyzing the variables explicitly responsible for the smoking behavior and then designing steps to countercontrol those variables directly. Joseph Morrow, Susan Gmiender, Lewis Sachs, and Helene Burgess took the *behaviorological* approach, although this term was not yet available for them to use. In 1973 they reported their evaluation of a such a therapy (i.e., Morrow *et al,* 1973; henceforth we will refer to their procedures and report simply as the Morrow Study).

The Morrow–Study therapy differs from most non–behavioral smoking–control therapies in several ways. Primarily this therapy treats smoking as a behavior rather than, as was then the fashion, as a pathological symptom perhaps indicative of problems with the client's putative inner agential self. Also, as we already noted, this therapy demonstrates a high clinical success rate. And this therapy focuses on directly countercontrolling the explicitly analyzed functional variables producing smoking while refusing to use the added variable of the client's "record" of how long they have gone without ever again smoking any cigarettes, a variable on which many other smoking therapies tend ultimately to rely to prevent smoking.

Today, decades later, savvy therapists continue using updated versions of the Morrow–Study smoking–control therapy. We mostly feature the original Morrow–study therapy here—as old as it may be—because it was the earliest *comprehensive* behaviorological–science smoking intervention. Again, however, smoking control remains secondary; rather, we examine this early comprehensive *and successful* therapy to appreciate deriving scientific practices from scientific principles, a general process that fundamentally enhances the success of most efforts to improve the human condition.

Let's cover the Morrow–Study therapy by discussing its three main characteristics. These are (a) the therapy success rate, (b) the therapy process, and (c) the therapy procedures.

The Therapy Success Rate

Perusal of the smoking–control literature reveals that the phrase "smoking–control therapy" covers many different intervention types. Most of the non–pharmaceutical using therapies (and even some of them) invoke aversive techniques as the main method to bring the client's cigarette smoking down to a low daily rate. Such aversive techniques, which directly address only one or two of the variables functionally related to smoking, include using nausea–inducing drugs, horror stories about the diseases to which smokers are more susceptible than non–smokers, gory visual comparisons of healthy and smoke–damaged lung tissues, ash trays shaped like open lungs, or emotional appeals about children wanting to imitate smoking parents. These therapies typically schedule the aversive techniques, individually or in groups, for the first week or two of the therapy program, which may be the full extent of the therapy.

Regardless of all that variety, nearly all of these therapies initially succeed. By the scheduled end of the aversive–technique period, the daily smoking rate of nearly all clients has dropped to zero or nearly zero.

However, at the end of the scheduled therapy intervention, after that initial success, clients are on their own. To prepare clients to stay off cigarettes after their initial success, most therapies leave clients with only one fundamental controlling variable. This added variable, which does not directly address variables producing smoking, takes the form of their "Record" of how long they have gone without smoking a cigarette. Unfortunately, clients see a single post–quitting cigarette as breaking the record, and the smoking problem resumes.

Some therapists thus see the record of smoke–free days as a powerful but fragile variable. "How long have I gone without smoking? Six days!" Later: "Six weeks!" Later still (maybe): "Six months!" And even later (rarely): "Six years!" The longer one goes without smoking, the more control this record variable seems to exert. The problem with this variable, however, is its fragility. Sooner or later, for most clients, too many stressors inevitably pile up on a particular day (e.g., your spouse gets upset over undone household chores, and your boss increases your work, and a broken pipe floods the basement, or the family's teenager crashes the car). This results in the otherwise successful quitter smoking a single cigarette, and then, usually, more cigarettes. "How long since my last cigarette? Just six minutes? My record [of weeks or months or years] is wrecked. I might as well have another cigarette!" The unbroken record, so powerful while unbroken, is erased.

Behaviorologically designed therapy arrangements, however, discard the notion that clients must never have another cigarette. In the Morrow Study, the "quitter's procedure" allows occasional cigarettes but only under conditions that lead to even less smoking, not more. Besides, in terms of health, any single cigarette probably causes less damage than taking a long walk on a summer afternoon in a smog–polluted city. Without the quitter's procedure, however, a single post–quitting cigarette erases the record and so can induce a return to smoking. So relying on the record as a therapy technique actually undercuts continued success.

The stressor–produced breaking of clients' records also produces other adverse effects. Clients' self–esteem aside, not only have they returned to smoking but also their *efforts* to quit have ultimately gone unreinforced, which makes quitting yet again even more difficult to attempt; the behavior of quitting is on extinction. The joke that "it's easy to quit smoking; I've done it hundreds of times" is not funny to smokers who have tried, and failed, to quit.

In contrast to those problems, the behaviorological smoking–control therapy featured in the Morrow Study shows a high, long–term clinical success rate. Their study involved 55 clients (25 women and 30 men). Ages ranged from 28 to 65 years. Initial smoking age ranged from 13 to 28 years. The number of years that these clients had smoked ranged from 9 to 45 years. The daily rate of cigarette smoking at the start of the study ranged from 15 to 60 (three–quarters of a pack per day to three packs per day). As with clients in other smoking–control therapies, these clients all ceased smoking, as scheduled, by the end of the initial week of therapy. Overall, of the 55 clients starting the program, 34 (or 62%) were not smoking one year later. However, the therapy consisted of two sequential components; one was individual–based (during the first week) and the other was group–based (during three months of weekly meetings). So the Morrow Study tracked long–term success rates separately (a) for those who did not complete both components and (b) for those who did complete both components. The results showed that 35 clients (64% of the 55 starting clients) completed the first component but only part, if any, of the second; of these, only 16 clients (46%) were still not smoking one year later. The results also showed that 20 clients (36% of the 55 starting clients) completed *both* therapy components; of these, 18 clients *(90%)* were still not smoking one year later.

Therapies seldom achieve that kind of clinical success rate, regardless of the presenting problems they handle. To what might the success rate of this therapy be due? The Morrow Study implicitly suggested, and here we explicitly suggest, that the success is due to the comprehensive, designed connection, based on a behaviorological analysis, between the therapy techniques and the variables functionally related to smoking. We discuss these connections by detailing the behaviorological process and procedures of this therapy, including how to get more clients to complete both therapy components and so increase their chances for long–term success.

Based on their results, the Morrow group made two recommendations for adjustments to improve their therapy design, adjustments which later therapists adopted (e.g., see Ledoux, 2002a). One adjustment concerned the number of meetings after the daily sessions. Interested in reducing the recidivism rate found in their study, the Morrow group recommended that clients continue attendance at group meetings on a monthly basis for nine more months. This adjustment would extend therapy duration to a full year, while increasing the interval between meetings through an additional step (daily to weekly *to monthly).* Such a schedule more gradually reduces—compared to the original Morrow study—the therapist's part in the contingencies generating and maintaining clients' efforts to quit smoking. This gradual reduction in reliance on the therapist (akin to a fading procedure) contributes to clients' continued success when the therapy ends and they proceed independently.

The other adjustment concerned special fees. These were fees that could also address the recurring concern about recidivism. The Morrow team suggested that therapists require each client to pay a special fee that would be refunded if, and only if, the client had attended virtually all therapy meetings *regardless* of whether or not the client had quit smoking. They suggested that making the refund contingent on attendance, rather than on success, might reduce the chance of clients quitting therapy, which should increase the likelihood of their quitting smoking. Given the money that quitting smoking saves (even after paying the therapist's regular fees) they suggested setting this special fee at $50 (in 1970 dollars). In addition, some clients see the refund as a nice bonus for their successful efforts. We retain these adjustments in the therapy process and procedures that we report here.

The Therapy Process

The process of this therapy contains two sequential components. The first involves daily individual sessions and the second involves weekly (and later, monthly) group sessions. (Detailed descriptions of all the self–control techniques, which we merely mention by name in this section, appear in the "Therapy Procedures" section.)

Individual daily sessions. In this therapy the first component emphasizes contingencies on the individual. The client and therapist meet Monday through Friday for five daily sessions, each one hour long. Each of these daily sessions has some different activities, which focus especially on the introduction of various self–control techniques (i.e., techniques that, after the therapist verbally conditions them, the client's environment evokes without the therapist present; no inner self–agent is involved). These daily sessions also have some activities in common, including these six, which occur each day: The therapist collects the record of the number of cigarettes the client smoked in the previous 24 hours and praises drops in smoking rate. The therapist supervises while the client engages in the technique called *satiation smoking* (a maximum of three "satiation cigarettes" per session). The therapist and client discuss the previously introduced techniques and any difficulties the client may have experienced using them. The therapist teaches the client the new techniques for the session. The therapist provides any instructions relevant to the client's efforts between then and the next session (e.g., any restrictions on smoking between sessions, and the need to continue to keep an accurate record of how many cigarettes get smoked between sessions). And the therapist responds with empathy to a client's doubts about his or her ability to succeed in quitting and, regarding any difficulty the client is experiencing, the therapist expresses confidence, draws comparisons with similar clients who have successfully quit smoking, and engages in other helpful verbal exchanges. Clients may smoke between some daily sessions, always provided, however, that they engage in all relevant techniques that the therapist had previously introduced, and that they continue accurately counting all the cigarettes that they smoked.

On *Monday* the therapist first stresses the importance of client cooperation if the client is to reach zero smoking by the end of the first week; the therapist's role is to aid the client in reducing smoking so that the smoking rate is zero between the Thursday and Friday sessions. After that the client will go without smoking from Friday's individual session to Monday's first group session, which is the first step of fading out the therapist. Then the therapist teaches the client the technique that we call *satiation smoking.* The client smokes three satiation cigarettes during the session, using the client's normal brand of cigarettes. The therapist intersperses these satiation cigarettes with teaching the three techniques that we call *pure activity, anti–social chair,* and *difficult to obtain.* The therapist also explains the point of these and later–introduced techniques in countercontrolling common causes of smoking and so ultimately eliminating smoking. Using all the techniques that the therapist has introduced, clients may smoke between the Monday and Tuesday sessions.

On *Tuesday* the client again smokes three satiation cigarettes, the first two of the client's normal brand. The third is a strong, domestic, unfiltered brand provided by the therapist; if the client normally smokes this brand, the therapist provides an even stronger tasting imported brand. Between these satiation cigarettes, the therapist reviews previously introduced techniques, and introduces more techniques. These include the techniques that we call *alternative behaviors, review emotional responses,* the filter–related part of *change cigarettes* (which is a two–step technique; the therapist will introduce the second step, the brand–related part, at the next session), and *damage cigarettes.* Again, using *all* the techniques that the therapist has introduced, clients may smoke between the Tuesday and Wednesday sessions.

The *Wednesday* session again includes three satiation cigarettes, but only one is of the client's normal brand; the other two are the strong, unfiltered brand. Between these, the therapist reviews previously introduced techniques, and introduces several more. The first two are for immediate use, as usual. These include the technique that we call *rehearsal of difficult times,* and the second, brand–related part of the two–step *change–cigarettes* technique. The therapist also introduces, for *later* use, the technique that we call the *quitter's procedure.* The therapist stresses the importance of getting the smoking rate down to the lowest possible level between this session and the next so that the client can more easily reach the zero rate between the Thursday and Friday sessions. To help reduce the rate to the lowest possible level, the client should postpone smoking as long as possible, and as usual use all the techniques that the therapist has introduced while also smoking any cigarettes as satiation cigarettes. Finally, the client is to destroy or give away all of her or his remaining cigarettes just before the Thursday session.

On *Thursday* the client smokes only two strong, unfiltered–brand satiation cigarettes. For most of this session, the therapist and the client discuss the importance of going without cigarettes for the next 24 hours, and again review all the techniques. As part of that review, the therapist advises the client on managing any problems that she or he has experienced in the application of these techniques. As an aid to abstaining, the client can look forward to smoking a satiation cigarette at the Friday session. And from now on the client is to use the *quitter's procedure* technique for any cigarette smoking.

On *Friday*, the last individual session, the therapist allows the client to smoke a strong, unfiltered–brand satiation cigarette. Having abstained for 24 hours already, most clients experience great discomfort over this cigarette, and they need not complete it. The client and therapist review what the client has been through, and discuss what the client might experience in the adjustment to no more smoking except through the quitter's procedure. In the rest of this session, the therapist describes to the client the scheduling, purposes, and benefits of the upcoming weekly—and later, monthly—group sessions.

Virtually all clients achieve zero smoking between the Thursday and Friday sessions. Something akin to a shaping process usually occurs as they extend this one–day success first to three days, by abstaining from the Friday individual session to the first group session scheduled for the following Monday, and then to a weekly—and later a monthly—basis between subsequent group sessions.

Group sessions. The second therapy component emphasizes group contingencies. The therapy continues with eleven weekly group meetings spread over three months, followed by nine monthly group meetings, which brings the duration of the complete therapy program to one year. Each group meeting is of two hours duration in the evening. The group is comprised of clients who have completed the individual therapy sessions. At any particular meeting, some clients may be attending for the first time, others for the last time, and most somewhere between these points.

At each group meeting, similar activities take place. In an informal atmosphere of healthy beverages and snacks (No smoking!) clients introduce themselves, state how long they have been "on" the quitter's procedure, and describe any problems they are encountering or what topics they would like to see the group discuss. Together they discuss events in their lives related to no longer smoking, the difficult situations they described facing, and how to handle these. Thus each client benefits from the experience of the others. When they exhaust therapy–related subjects, they are free to discuss any other topics of mutual interest. In these ways they bond together into a mutual support group of non–smoking friends (the first and often the only such friends some clients have) who have had similar quit–smoking experiences. For some clients these friendships and social contacts endure on their own after the end of formally scheduled meetings.

The therapist keeps the group focused and provides verbal reinforcement when applicable. Should a client need additional help, the therapist works out an appropriate individual program with that client.

What's the point? All those process details behind that successful smoking therapy, and the procedure details that we cover next, can readily distract us from our primary concern. To reiterate it, our primary point is not to emphasize successful smoking–control therapy, as valuable as that might be, but to use such therapy to exemplify the general behaviorological approach to developing therapeutic interventions. In this approach we locate and analyze the independent variables responsible for a behavior of concern, and arrange change techniques that directly counter those variables, using established principles as the foundation for successful intervention practices. The smoking–therapy procedures of the Morrow study exemplify this approach.

A summary table. Table 17–1 tracks the daily sessions at which the therapist introduces the various anti–smoking procedures. This table also specifies the general type of variables that each anti–smoking procedure addresses.

The Therapy Procedures

The Morrow–Study therapy employs a particular set of procedures to produce appropriate responses. Most of these procedures involve self–control techniques that can be found in the literature (e.g., Skinner, 1953, Chapter 15, gives a general discussion while Stuart & Davis, 1972, focus specifically on overeating). Our coverage here acknowledges some inherent technique overlaps, and includes some refinements to the techniques and their names.

The Morrow Study specifically designed each technique to counter one or another of the variables, or variable components, explicitly controlling smoking. These variables fall under the four categories of (a) *antecedent stimulus considerations,* which include respondent unconditioned and conditioned emotions as well as operant evocative stimuli, (b) *response considerations,* (c) *postcedent stimulus considerations,* which include unconditioned and conditioned reinforcement, and (d) *the combination of variables* that applies when clients are "on their own" during and after the group–based component of the therapy (i.e., the "quitter's procedure"). Let's look at the techniques in each of these categories in turn.

Antecedent stimulus variables. Four techniques address antecedent stimulus variables. One addresses unconditioned emotional respondents. Another addresses conditioned emotional respondents. And two address two different sources of evocative stimuli. While maintaining our rejection of inner agents, we invoke descriptions here that contain verbal shortcuts mimicking some of the vernacular terms and phrases that we might use with clients.

We use the term *satiation smoking* for the technique that addresses unconditioned respondent emotional variables. This is the aversive respondent procedure in this therapy. In this technique, which offsets the unconditioned pleasant respondent effects that smoking elicits, and which also serves a satiation function, the client *inhales on a cigarette every six seconds.* As the rate of smoking outside the sessions decreases, satiation smoking becomes increasingly uncomfortable. The aversive effects are temporary, and reinforcement occurs when clients discard an uncomfortable satiation cigarette.

We use the term *review emotional responses* for the technique that addresses conditioned respondent emotional variables. In this technique, which offsets the conditioned pleasant respondent effects that smoking elicits, clients list (on a card that they carry with them) all their main reasons that carry an emotional impact for quitting smoking. Therapists instruct clients to review the list several times each day during the week of individual sessions, and as part of the quitter's procedure. In reviewing the list they should focus on each reason until it elicits the relevant emotion. A list item could involve feeling the pleasure that

Table 17–1

Names of techniques by the session/day of their introduction and under the general type of variables that they countercontrol

SESSION	VARIABLES		
	Antecedents	Responses	Postcedents
Monday	❧ Satiation smoking		❧ Pure activity
	❧ Difficult to obtain		❧ Anti–social chair
Tuesday	❧ Review emotional responses	❧ Alternative behaviors	❧ Change cigarettes (filter part)
			❧ Damage cigarettes
Wednesday	❧ Rehearsal of difficult times		❧ Change cigarettes (brand part)
			❧ Quitter's procedure (introduction only)
Thursday			❧ (Quitter's procedure: begin use)

a positive reason for ending smoking elicits (e.g., pleasure at being able to smell flowers again) or feeling the fear that a negative reason for ending smoking elicits (e.g., fear that the client's toddler might play with the cigarette lighter). As time passes clients remove from the list any reason the review of which fails to elicit an appropriate, strong emotional reaction. Clients also add to the list reasons newly discovered to elicit appropriate emotional reactions (e.g., hearing the infamous saying, "kiss a smoker, lick an ashtray").

Two other techniques address sources of evocative stimuli. Both techniques help bring about the initial cessation of smoking and then decrease in relevance.

We use the term *difficult to obtain* for one of those two evocative–stimulus focused techniques. In this technique, which offsets the evocative power of client–owned smoking materials, clients put their cigarettes and matches in places that are separated in space as much as reasonably possible. The new locations necessitate additional response effort, such as bending over, to retrieve an item. This makes satisfying an urge to smoke more difficult (see Ledoux, 1973, for research support of the response status of urges to smoke). For example, at home the cigarettes might be kept in a footlocker in an upstairs bedroom (with the TV located in the downstairs living room) while the matches might be kept on a high shelf in the basement. Should an urge to smoke occur during a TV program, the increase in time and effort, that obtaining the needed items requires, is likely to preclude the occurrence of smoking at that time. The difficult–to–obtain technique helps because, when matches or, especially, cigarettes are nearby, they all too easily evoke smoking. For example a smoker reaching into a pocket or purse for paper and pen to write down a telephone number would, upon encountering a pack of cigarettes, likely light up.

We use the term *rehearsal of difficult times* for the other evocative–stimulus focused technique. In this technique, which offsets the evocative power of smoking materials *owned by others,* the client makes special arrangements with everyone from whom he or she might receive cigarettes. For one day (the day immediately after this technique is introduced) all these people are to offer cigarettes unexpectedly to the client. The client is never to accept an offered cigarette but instead is to practice saying "no" to the offers. Clients who experience an excessive urge to smoke are to get their own cigarettes under already introduced techniques. Clients need to rehearse difficult times, due to the minimal likelihood of saying "no," when someone offers a cigarette, especially given the price of cigarettes. Some clients report *never* having said "no" under such circumstances. Under conditions of deprivation during and after stop–smoking therapy, saying "no" is likely to be all the more difficult. Rehearsing difficult times enables the client, under controlled if theatrical conditions, to practice saying "no." Then, having already said "no" at least under some relevant stimulus circumstances, clients find saying "no" easier when they are later faced with a real need to say "no."

Response considerations. We use the term *alternative behaviors* for the technique that addresses the problems associated with the response status of smoking. In this technique, which offsets the response vacuum created when smoking behavior stops, the therapist helps the client construct a list of alternative behaviors in which the client is increasingly to engage as smoking decreases. For this list many clients emphasize past activities, such as hobbies or skills, in which they no longer engage but to which they would like to return. The cessation of smoking may even make some preferred activities affordable. The Morrow Study provided these details:

> [We ask clients] to include at least three activities in each of three categories: physical activities (such as jogging, bike riding, gardening, walking, etc.); quiet activities (i.e., crossword or jigsaw puzzles, sewing, arts and crafts, carpentry, reading, etc.); and oral activities (such as sucking, chewing or biting on lemon drops, life–savers, gum, celery, fruits, various non–toxic objects, etc.) (Appendix A, item IV).

The alternative–behaviors technique is needed, because behavior does not occur in a vacuum; it occurs in time. For smokers, smoking behavior fills part of each day. As therapy produces less smoking, more of each day becomes behaviorally empty. If we make no plans to fill this time by design with beneficial activities, then other, often troublesome, activities fill it through other normal behavioral process like generalization. Extra eating is a common but undesirable replacement, as generalization occurs from one consumption response class to another. Many people who have quit smoking have reported an increase in eating, and consequently weight. The alternative–behaviors technique fills this otherwise empty time with behaviors that are less likely to cause problems like weight gain.

The alternative–behaviors technique can also help maintain, and even increase, a client's GLR. For example, I worked with one couple who enjoyed entertaining dinner guests. They stopped when their doctor, who wanted them to quit their heavy smoking, suggested how badly their house must smell to their non–smoking guests from all the stale tobacco ash and smoke. Upon successfully quitting, they used the funds previously expended on smoking to strip and redecorate the entire interior of their home. Then they and their guests enjoyed these dinners even more (and they only invited non–smokers).

Postcedent stimulus variables. Four overlapping therapy techniques address *the reinforcing capacity of smoking*. All work to weaken that capacity. The first two focus on countercontrolling *conditioned*–reinforcement value. The other two focus on countercontrolling ***un****conditioned*–reinforcement value. We need these four techniques to help counter any continued generating and maintaining of smoking due to its reinforcement value.

We use the term *pure activity* for one of the two techniques that offset the *conditioned*–reinforcement value of smoking. Normally a smoker engages in other reinforcing activities while smoking. This pairs the reinforcing aspects of these activities with smoking, which establishes some of the conditioned–reinforcement value of smoking. The pure–activity technique countercontrols much of this normal, automatic pairing of various physical–activity reinforcers with smoking. In this technique, when clients smoke, they *only* smoke; they engage in no other activities while they smoke. This means, for instance, no television viewing, no stereo listening, no eating or drinking, no reading, no driving, no studying, and so on. This technique thus reduces some of the conditioned–reinforcement value of smoking, because it stops the pairing of smoking with the reinforcers from various physical activities.

We use the term *anti–social chair* (which we sometimes call the "smoking spot") for the other technique that offsets the *conditioned*–reinforcement value of smoking. Engaging simultaneously in smoking and *social* activities, a common occurrence, also pairs the *social*–reinforcer components of many activities with smoking. This again increases and maintains some of the conditioned–reinforcement value of smoking. The anti–social chair technique extends the pure–activity technique by countercontrolling much of this normal, automatic pairing of social reinforcers with smoking. In this technique the therapist and client discuss locations, suitably isolated from normal social environments, that the client can use as smoking spots where he or she will be *alone* when smoking. Consistent with the pure–activity technique, these are locations where the client can keep a chair to be used exclusively for smoking. (The overlap of the pure–activity technique and anti–social chair technique presents no problems.) The client selects one spot for use at home and one spot for use at work. The client is not to smoke anywhere else. When away from these places, a public toilet is a suitable, anti–social smoking spot (where legal). Some clients put an uncomfortable chair in spots like a spare room, bathroom, stairwell, basement, attic, or garage. This technique thus further reduces the conditioned–reinforcement value of smoking, because it stops the pairing of smoking with social reinforcers.

The Morrow Study kept the two techniques that are focused on the *unconditioned*–reinforcement value together. They referred to the combination as *reduction of reinforcement value* because both shared in reducing this reinforcement value. Here we describe these two techniques separately.

We use the term *change cigarettes* for one of the two techniques that offsets the *unconditioned*–reinforcement value of smoking. In this technique clients change their cigarettes in two steps. The first step is filter–related. At the scheduled session, clients who smoke filtered cigarettes remove the filters, and use only unfiltered cigarettes thereafter, making the cigarettes somewhat aversive (or "distasteful," as clients describe them). Clients who smoke unfiltered cigarettes change to a filtered brand. The second step is brand–related. At the next session all clients change to a strong, even imported, unfiltered brand, making the cigarettes even more distasteful. These changes in the cigarettes offset some of the unconditioned–reinforcement value of smoking by making smoking increasingly aversive.

We use the term *damage cigarettes* for the other technique that offsets the *unconditioned*–reinforcement value of smoking. In this technique clients damage their cigarettes either by poking half–a–dozen pin holes in them, or by waterlogging them, or both. Damaging the cigarettes in any of these ways before smoking them makes them distasteful. Thus such damage to the cigarettes also offsets some of the unconditioned–reinforcement value of smoking by also increasing the aversiveness of smoking.

The combination of variables. We use the term *quitter's procedure* for the therapy technique that addresses the combination of variables that we invoke near the end of the one–week, individual–based therapy component. During this week some previously introduced techniques stand alone (e.g., the alternative–behaviors technique that continues whether or not other techniques are in use) while others become irrelevant to the client's efforts (e.g., the difficult–to–obtain technique, since the client no longer owns cigarettes). The quitter's procedure combines all remaining, relevant techniques in a way that avoids clients having to rely on an "it's been x amount of time since the last cigarette" record that verbally implies falsely to them that they can never again have another cigarette. Instead the quitter's procedure allows future cigarettes but only under specific, rate–reducing controls.

The Morrow Study designed the quitter's procedure so that successful clients need not fear falling for the temptation to have a post–quitting cigarette.

> [We developed this technique for clients] in response to the expressed fear... that they would again return to smoking if and when they ever had a cigarette after they quit, and their expressed inability to stop smoking if that meant they never, under any circumstances, could have another cigarette. [We told them] that they could have a cigarette, after a period of zero smoking, which would satisfy their urge to smoke while at the same time making the smoking response so aversive that the probability of emitting the smoking response would not be increased... (Appendix A, item VIII)

Therapists describe the quitter's procedure to clients as a series of steps. These become successively more difficult so as to counter increasingly strong urges to smoke. In this procedure, when stimuli evoke an urge to smoke, clients face a series of what, in lay language, we may conveniently (i.e., without the otherwise proper objections to inner agents) call choice, or decision, points.

In the presence of some likely unrecognized stimulus that is evoking smoking, clients initially observe and report to themselves (as a "public of one"; see Ledoux, 2002b) an urge to smoke. Knowing that some environmental stimuli are evoking smoking, clients leave the immediate environment (i.e., the room or other location) in which the urge to smoke occurred, because the first step in weakening such control is to move away from that environment. Once in another setting, they consider—the first decision point—whether or not to smoke. Simply changing environments is often enough to render ineffective whatever stimulus started evoking smoking.

If that evocative effect still favors smoking, then clients next review each item on their emotional responses list until it elicits the relevant emotion. Having reviewed the list, they again consider—the second decision point—

whether or not to smoke. Reviewing this list often renders ineffective whatever stimulus is still evoking smoking.

If the evocative effect on smoking continues, then clients review the final sequence of the quitter's procedure; they review their required actions *if* the urge to smoke still persists: (a) They must buy an expensive pack of specified, strong tasting, unfiltered cigarettes. (b) They must trash 19 of the cigarettes from the pack right then at the store (because the urge to smoke involves only one cigarette, and keeping the other 19 would only provide them with 19 more evocative–stimulus "temptations"). And (c) they must take the remaining cigarette to an agreed upon smoking spot where they must put pin holes into the cigarette and/or waterlog it before, as a pure activity, smoking that cigarette as a satiation cigarette! Having reviewed the required actions of the final sequence of the quitter's procedure, they again consider—the third decision point—whether or not to smoke. Reviewing this final sequence, and experiencing the aversive emotional reaction to it, often renders ineffective whatever strong stimuli are still evoking smoking.

However, if the contingencies still compelling smoking retain some effectiveness, then clients walk, if possible, to a store where they *might* buy a pack of cigarettes, since they no longer own or "bum" any. Just prior to paying for a pack, however, they review the quitter's—procedure steps one last time and consider—the fourth decision point—whether or not contingencies compelling smoking are really that strong. This second full review of the remaining quitter's–procedure steps, before actually paying for a pack, often succeeds at the last moment to deter smoking. If this proves insufficient against the current smoking compelling contingencies, then clients follow through with the rest of the quitter's–procedure steps, finishing with a satiation cigarette.

By completing the quitter's procedure, the client has satisfied the urge to smoke but has not reinforced smoking. Later we will discuss *why* clients follow these stringent steps.

One other factor bears on the continued successful application of the quitter's procedure. As part of the discussions on this procedure, therapists also repeatedly remind clients that smoking a post–quitting cigarette in any way other than through the use of the quitter's procedure carries a high probability of producing a return to smoking.

Clients need the quitter's procedure, because virtually everyone who has smoked and quit can expect to experience an occasional urge to smoke. Presumably one or another as yet unaddressed smoking–evoking variable has momentarily raised the probability of smoking. Some of these variables are essentially unaddressable. For example, only a rare client, such as the president of a college campus, would have the authority to ban, say, cigarette vending machines from the work place. Yet such a measure would be required to minimize the smoking–evoking stimulus–control effects of these machines in this environment. Nevertheless, such measures are simply beyond normal reach. Under these circumstances the quitter's procedure seems vital to continued success in therapy compliance because it provides the client a way to "satisfy" urges to smoke while minimizing the reinforcers from smoking.

The quitter's procedure also increases the understanding of those who have never smoked regarding the extent of the problems faced by smokers who are under contingencies (i.e., who "want") to quit. The rigorous rate–reducing controls of the quitter's procedure surprise many of these non–smokers who see such controls as extreme. They ask "How could anyone ever follow that procedure? How could they throw away the money tied up in 19 cigarettes, or take those other steps? Why would they put themselves through all that?" Smokers who quit through using the quitter's procedure provide the answer: They experience, in various ways, that *the quitter's procedure works,* so they follow it, finding this procedure to be an invaluable aid to successfully quitting smoking. Actually most smokers find the prospect of *never again* having another cigarette not only far worse than the quitter's procedure but also a virtually impassible barrier to successful quitting. Those who have used the quitter's procedure to stop smoking have shown, *through their compliance with it,* that using this procedure is indeed easier than never ever again having another cigarette.

Therapy Example and Chapter Conclusion

This smoking–control therapy *provides an example* of a therapy that addresses its concerns comprehensively and is theory based in that it derives its practices from established behaviorological principles. Its substantial clinical success rate exemplifies the value of behaviorologically analyzing explicit functionally controlling variables *and designing intervention steps that countercontrol them.* This benefits clients, in this case clients with smoking problems. Extension of these design steps to many other kinds of concerns will likely enable people with those concerns to attain similar success rates.

Potential candidate problems for behaviorologically designed interventions range from over–indulging in food or drink to wrecking planetary resources. As we finish this chapter, pause and ponder, in the spirit of our usual ongoing activity, how behaviorology can help traditional natural scientists solve that last candidate problem by taking into account the lessons that this chapter provides not only regarding the scientific design of interventions but also the value of PNET and the GLR in managing contingent environment–behavior changes.❦

References (with some annotations)

Cautela, J. R. (1970). Covert reinforcement. *Behavior Therapy, 1,* 33–50.

Cautela, J. R. (1983). The self–control triad. *Behavior Modification, 7* (3), 299–315.

Cautela, J. R. (1984). General level of reinforcement. *Behavior Therapy and Experimental Psychiatry, 15* (2), 109–114.

Cautela, J. R. (1994). General level of reinforcement II: Further elaborations. *Behaviorology, 2* (1), 1–16.

Cautela, J. R. & Baron, G. (1977). Covert conditioning: A theoretical analysis. *Behavior Modification, 1,* 351–368.

Cautela, J. R. & Kearney, A. J. (1986). *The Covert Conditioning Handbook.* New York: Springer.

Ferreira, J. B. (2012). Progressive neural emotional therapy (PNET): A behaviorological analysis. *Behaviorology Today, 15* (2), 3–9.

Ferreira, J. B. & Duncan, B. R. (2002). Biofeedback–assisted hypnotherapy for warts in an adult with developmental disabilities. *Alternative Therapies in Health and Medicine, 8* (3), 141–144.

Fraley, L. E. & Ledoux, S. F. (2002). Origins, status, and mission of behaviorology. In S. F. Ledoux. *Origins and Components of Behaviorology—Second Edition* (pp. 33–169). Canton, NY: ABCs.

Guarneri, M. (2006). *The Heart Speaks: A cardiologist reveals the secret language of healing.* New York: Touchstone

Guarneri, M. (2012). *The Science of Natural Healing* (Course 1986). Chantilly, VA: The Teaching Company.

Johnson, P. R. (2012). A behaviorological approach to management of neuroleptic–induced tardive dyskinesia: Progressive neural emotional therapy (PNET). *Behaviorology Today, 15* (2), 11–25.

Ledoux, S. F. (1973). *The Experimental Analysis of Coverants.* M.A. thesis, California State University, Sacramento.

Ledoux, S. F. (2002a). Successful smoking control as an example of a comprehensive behaviorological therapy. In S. F. Ledoux. *Origins and Components of Behaviorology—Second Edition* (pp. 243–258). Canton, NY: ABCs. This paper documents and extends most of the contents in the original 1973 Morrow Study (see Morrow, Gmiender, Sachs, & Burgess, 1973). This paper also appeared (2011) in *Behaviorology Today, 14* (1), 3–13.

Ledoux, S. F. (2002b). An introduction to the philosophy called radical behaviorism. In S. F. Ledoux. *Origins and Components of Behaviorology—Second Edition* (pp. 25–32). Canton, NY: ABCs. This paper also appeared (2004) in *Behaviorology Today, 7* (2), 37–41.

Morrow, J., Gmiender, S., Sachs, L., & Burgess, H. (1973, April). Elimination of cigarette smoking behavior by stimulus satiation, self–control techniques, and group therapy. Paper presented at the annual convention of the Western Psychological Association.

Skinner, B. F. (1953). *Science and Human Behavior.* New York: Macmillan. The Free Press, New York, published a paperback edition in 1965.

Skinner, B. F. (1969). *Contingencies of Reinforcement: A Theoretical Analysis.* New York: Appleton–Century–Crofts. The B. F. Skinner Foundation (www.bfskinner.org) in Cambridge, MA, republished this book in 2013.

Stuart, R. B. & Davis, B. (1972). *Slim Chance in a Fat World: Behavioral Control of Obesity.* Champaign, IL: Research Press.✑

Chapter 18
The Stimulus Equivalence Relations Horizon

Chapter 17 took us through some applied behaviorology research considerations. We assessed the General Level of Reinforcement (GLR) as well as Progressive Neural Emotional Therapy (PNET). However we spent most of the chapter considering the overall process of developing behaviorological interventions or therapies to address behaviors of concern effectively. The control of smoking behavior provided a classic example of this behaviorological intervention–development process.

Basically we begin that process by thoroughly analyzing the contingencies surrounding the problem behaviors to discover the independent variables that are in explicit functional relations with these problems (i.e., the "causes" of the problems). Then, by applying the established principles and concepts of behaviorology, we arrange environmental countercontrols for those variables in a methodologically sound manner, a necessary step not only so that we can verify countercontrol effectiveness but also so that we can further engineer therapeutic generalization to all relevant settings.

Now, in this chapter, we extend behaviorological research considerations to the well–developed experimental area that we variously call *stimulus equivalence* or *equivalence relations* (Sidman, 1994, 2009). Actually some controversy exists over which of these labels we should prefer, because behaviors and postcedent stimuli also occur in these relations. However, in this introduction to the topic, we stay focused on the antecedent stimuli. In time the self–correction routines of our natural science may select one of these names over the other. Both names tact the functional equivalence of stimuli along with specific relational properties of equivalences. While a name has some effect on our behavior, this large research area has, on the horizon, wide–ranging—but as yet still under–developed—application possibilities for increasing the effectiveness of many human endeavors including not only all types and levels of education but also the policies and practices of sustainable living.

Let's begin by revisiting the minimalist description of equivalence research that we gave in Chapter 1. Then we will expand our account into a more complete picture of these phenomena, with some examples.

A Preview by Review of Equivalence Phenomena

Under some circumstances, after explicitly conditioning some thematically connected functional relations between environmental antecedent or postcedent stimuli and responses, the number of related behavior–controlling functional relations that we can successfully detect is greater than the number originally involved in the explicit conditioning. Stating this agentially, subjects seem to "learn" more than we "teach," although explaining these phenomena requires no inner agents. Researchers in this area have come to call these explicitly and implicitly conditioned relations *equivalence relations*, and these relations contain functionally equivalent stimuli.

Equivalences among stimuli can transpire in fairly simple circumstances. For example, in training a new cloakroom attendant, we might first condition his behavior such that when a regular customer, Ms. Minkowner, appears and puts her polyester, pink mink coat among the coats already on the counter, the Ms. Minkowner stimulus reliably evokes his response of picking up her mink coat. Then, we reinforce the trainee's behavior such that, in the presence of the pink mink coat and several different coat–hanging cubicles, this mink coat reliably evokes the trainee's depositing it in a particular cubicle, say, number seven. *With no further training,* we find that Ms. Minkowner's reappearance at the counter reliably evokes the trainee's movement to cubicle number seven from which he retrieves her pink mink coat.

Beyond such simplistic examples (which actually pertain mostly to the relational property that we call *transitivity)* researchers in this area have demonstrated the phenomena occurring in far more complex circumstances. Using, for example, six sets of three stimuli each, explicit conditioning of a particular 15 environment–behavior functional relations could turn out to condition implicitly an additional 75 behavior–evoking functional relations. In this instance, the effort to condition 15 particular relations might produce a total of 90 testable relations, the 15 from direct conditioning and up to 75 more from indirect conditioning (e.g., see Fraley, 2008, Chapter 16).

The implications of equivalence phenomena for a science–based revolution in, say, education can be substantial. More careful arrangements, which research efforts uncover, of curricular components—what we would scientifically call educational conditioning programs—in, for example, history, language, math, and science, can economize by explicitly conditioning only certain evocative functional relations, relevant to the subject matter, in ways that virtually guarantee the implicit conditioning of many other possible and relevant relations evocable by the same broad set of stimuli. Teachers already engage in approximations of these steps, but that occurs under haphazard and unanalyzed contingencies, which unnecessarily reduces teachers' effectiveness. Imagine the progress that could occur if the vast numbers of teachers had some basic behaviorology in their training programs, enabling them to research and improve the equivalent functioning of stimuli in their teaching activities, a possibility that harbors profound implications for extensive, and scientifically informed changes not only in teaching but also in teacher training. Before pursuing such matters, however, let's turn to a more complete description of equivalence–relation phenomena.

The Three Properties of Equivalence Relations

Equivalence–relation work involves lots of evocation training, often with the response that produces reinforcement (i.e., the "correct" response) dependent on the context that another stimulus, or stimulus class, supplies. We use the term *conditional evocations* for such dependencies. These *conditional evocations* result from the conditioning of function–altering stimuli.

Recall that, using general terms, in the conditioning of function–altering stimuli, reinforcement only follows a response that stimulus A evokes when stimulus A occurs in the presence of stimulus B; we then call stimulus B a function–altering stimulus, because this conditioning makes its presence alter the function of stimulus A, which otherwise does not evoke the reinforcer–producing response. After we complete this conditioning, stimulus A functions as a successful evocative stimulus (i.e., evokes the correct—the reinforcer–producing—response) only when stimulus B is present. This conditioning makes the evocative status of stimulus A dependent on—as in *conditional upon*—the presence of stimulus B. We call the evocative–stimulus relations that this conditioning leaves in place *conditional evocations* (while the users of an older terminology, which we discussed in Chapter 12, called them *conditional discriminations).* For example if the evocative stimulus (i.e., stimulus A, in our more general description) includes two separate fives, then the reinforcer–producing response depends on, or is conditional upon, whether the function–altering stimulus (i.e., stimulus B, in our more general description) consists of a "+" or an *"x"* (i.e., a *"times")* between them. If the "+" function–altering stimulus occurs between the fives (yielding 5 + 5) then the response "10" produces reinforcement, but if the *"x"* function–altering stimulus occurs between the fives (yielding 5 x 5) then the response of "25" produces reinforcement.

In laboratory conditional evocations, the function–altering stimulus occurs as a sample stimulus while the evocative stimulus resides among a group of simultaneously presented "comparison" stimuli. For instance (and note that, in most of the examples in this chapter, we use italics to indicate printed words) if the printed word *cup* is the function–altering (i.e., sample) stimulus, and a graphic (or picture or photo) of a cup is the evocative stimulus that controls, say, a screen–touching response (i.e., a response of touching the location on a computer screen where the graphic of the cup appears) then that graphic of a cup would appear among several alternative, comparison graphics of other items, such as graphics of a dog, a car, a box, a cat, a hat, and a cow, any of which could evoke a touch response. (Following Sidman, 2009, we use these graphics in later examples.) In this case, however, only a touch response on the graphic cup produces reinforcement; that graphic cup evokes that reinforcer–producing response in the presence of the word *cup* or, as this chapter describes, in the presence of other functionally equivalent stimuli after certain further conditioning occurs, conditioning that takes function–altering stimuli and makes them functionally equivalent evocative stimuli.

Conditional evocations can relate to each other as a set in some thematic manner, even if the manner of their relations is or seems arbitrary with only reinforcement in common. For example in that set of graphics we just considered (i.e., a dog, a car, a box, a cat, a hat, and a cow) three items are animals—a fairly clear thematic relationship—while the other three items seem related only by being things, yet all six are related as the graphics of readable words (like *cat)* that can evoke reinforcer–producing responses.

In addition, further conditioning can interconnect sets of these related stimuli such that the stimuli in the sets function equivalently across sets. The stimuli take each other's places in conditional evocations in ways that we describe as certain properties of these relations. We label the three relational properties of such equivalence relations as *reflexivity, symmetry,* and *transitivity,* and conditioning capacitates them. That is, conditioning the appropriate subset of all the possible conditional relations among a given set of stimuli actually conditions many more, often even all, of the remaining relations, although the conditioning produces some relations directly and other relations indirectly. We test for the indirectly conditioned relations to find if they are functional. When a number of functioning conditional evocations show all three of the properties that researchers have found in these relations, then we refer to the set as a metarelation or equivalence relation regardless of whether conditioning has occurred directly or indirectly. Further cooperation between physiologists and behaviorologists will elucidate *how* this indirect conditioning occurs.

To help us get a grip on the direct and indirect conditioning of equivalences and the relational properties that make up equivalence relations, we use some symbolic notation as a sort of shorthand in our discussions. We let the figure "**R**" stand for an equivalence relation. Then, on either side of an **R**, we place representative sample and comparison stimuli that we usually denote with letters like "a" and "b," or "a" and "c," or "b" and "c," and so on. We place the function–altering, sample–stimulus letter right before **R** and we place the evocative, correct comparison stimulus (i.e., the stimulus that, appearing among a number of alternatives, evokes the reinforcer–producing response) right after **R**, as in "a**R**b." For example in the shorthand for our earlier example of a cup graphic (in a comparison set) evoking a reinforcer–producing response in the presence of the sample stimulus word *cup,* we would specify "(printed word) *cup***R**(graphic) cup," or, more symbolically, *cup***R**cup, or a**R**b.

We read those shorthand notations in various ways. While not the only ones, here are four ways to read a**R**b. (a) "Given a, stimulus b evokes the reinforcer–producing response." Or (b) "given function–altering stimulus a, the response that stimulus b evokes produces reinforcement." Or (c) "given the sample stimulus a, the response that comparison stimulus b evokes produces reinforcement." Or (d) "stimulus a is functionally equivalent (**R**) to stimulus b (regardless of whether conditioning has occurred directly or indirectly).

Let's repeat those same shorthand–reading possibilities but this time let's use our cup–stimuli example. (a) "Given the printed word *cup,* the cup graphic evokes the reinforcer–producing response (e.g., of touching the graphic on the screen)." Or (b) "given the function–altering printed word *cup,* the response that the cup graphic evokes produces reinforcement." Or (c) "given the sample printed–word *cup* stimulus, the response that the comparison cup–graphic stimulus evokes produces reinforcement." And finally (d) "the printed word *cup* is functionally equivalent (**R**) to the cup graphic (regardless of whether conditioning has occurred directly or indirectly).

For many people that last alternative, that "**R**" reads as "is functionally equivalent to," provides the most clarity in describing these relations. However, remember that technically the **R** represents the complete set of relations, the equivalence relation or metarelation, while a particular relation, like a**R**b, is only one member of the set. Such notations help us understand the three relational properties of equivalence relations. As we consider each relational property in turn, you will have plenty of opportunity to master reading such shorthand notations in ways like these.

The Reflexive Property

We acknowledge the *reflexive* property when **R** (the equivalence relation) holds for *each single stimulus* in the pattern a**R**a. That is, we see the reflexive property when a sample stimulus is *the same as,* or *equal to,* the comparison stimulus that, from among alternatives, is the stimulus that evokes the reinforcer–producing response. This happens with all of the stimuli in the set of comparison alternatives when each one is also the sample stimulus. Thus the reflexive relational property is in place when a**R**a and b**R**b and c**R**c and so on throughout the set; a is equivalent to a, b is equivalent to b, c is equivalent to c, and so on. If a is the sample stimulus then a, from the comparison–stimulus set, evokes the reinforced response; if b is the sample then b, from the comparison set, evokes the reinforced response; and if c then c, and so on. For example the reflexive relational property holds when **not only** the printed word *cup* is the sample and the printed word *cup,* from among the printed–word comparison stimuli (i.e., from among, say, the printed words *cup, dog, car, box, hat,* and *cow*) evokes the reinforced response, **but also** if this pattern continues for all the other stimuli in the comparison set (i.e., for the printed words *dog, car, box, hat,* and *cow,* all in the pattern of a**R**a, b**R**b, c**R**c, and so on, as in *cup***R***cup, dog***R***dog, box***R***box,* and so on).

However, what if the word *cup* is the sample, and a *graphic* cup—from among the graphic comparison stimuli (say, of *graphics* of a cup, a dog, a car, a box, a hat, and a cow)—evokes the reinforced response (i.e., a pattern of a**R**b)? Rather than the reflexive relational property, this takes us to the symmetric property. Before we go there, remember that the sources in the references provide more details on additional stimulus concerns relevant to these relational properties, details that go beyond our introductory overview here.

The Symmetric Property

We observe the *symmetric* property when **R** (the equivalence relation) holds *in both directions* for two different stimuli in the pattern, if a**R**b, then b**R**a, or if c**R**d, then d**R**c, or if e**R**f, then f**R**e, and so on. For example we can condition a subject such that, when a small pitcher of cream is the sample stimulus, a small bowl of sugar is the stimulus, from among those in a comparison set, that evokes a reinforced response. Then, the symmetric relational property is in effect when the bowl of sugar is the sample stimulus and the pitcher of cream, from among the comparison set, evokes the reinforced response. Said another way, if we condition cream**R**sugar, then our testing should reveal the symmetric relational property of sugar**R**cream (i.e., if cream**R**sugar, then sugar**R**cream). As with the reflexive property, our tests should show the symmetric property holding for all of the functionally related stimulus pairs in a given set (e.g., cream and sugar, steak and potatoes, bread and butter, and salt and pepper).

Consider what happens when you play a "what goes with what" game with a youngster. After a subset of the routine conditioning experiences early in life (which are not tracked but occur nonetheless) you take the child, whom we will call Sam, through the grocery store aisles and hand him various items. The occurrence of the relevant past conditioning will be clear, because after presenting each item, Sam will respond by taking off the shelf the item that goes with the item you presented. When you present sugar, he goes for cream (and vice versa); when you present butter he goes for bread (and vice versa); and so on. (Hopefully, at his age, the conditioning has not yet occurred that induces Sam to take a beer six–pack off the shelf when you hand him peanuts.)

Of course, tests for the presence of the symmetric property cannot include the sample stimulus among the comparison stimuli, as it might evoke a response as an example of the reflexive property (i.e., a**R**a, a is functionally equivalent to a). The possibility of that kind of outcome confounds the reflexivity and symmetry tests. While we predict that both of these relational properties hold, we must test for them separately.

The Transitive Property

We see the *transitive* property when the equivalence relation, **R,** holds for *three different stimuli* in the pattern, if a**R**b and b**R**c, then a**R**c. We (i.e., we, as elements in the contingencies on the subject) directly condition both a**R**b and b**R**c, the first two relations. Testing then reveals the indirect establishment of the third relation, a**R**c. For example let's say we condition the first two relations, *six***R**6, and 6**R**∴. Testing would then reveal the third relation, *six***R**∴. As we said, with italics indicating printed words, we read the stimuli on either side of **R** as the printed word *six,* the numeral 6, and the quantity of six, such as six dots, as here. (Later we will also use spoken words, which we would put in quotes, e.g., "six.") Now, let's use the details; if we condition both "(word)*six*–**R**–(numeral)6" and "(numeral)6–**R**–(quantity)∴" as the first two relations (i.e., *six***R**6 and 6**R**∴) and testing reveals the third relation, *six*R**⁝⁝**, then the transitive property is in place. That is, testing reveals "(word) *six*–**R**–(quantity)**⁝⁝**," which shows the transitive property. The word *six* and the numeral 6 and the quantity **⁝⁝** all become functionally equivalent stimuli.

Is that all that happens? Indeed not. After directly conditioning *six*R6 and 6R**⁝⁝**, additional testing could reveal several more relations established through indirect conditioning. These would include not only *six*R**⁝⁝** (i.e., the transitive relation) but also *six*R*six*, 6R6, and **⁝⁝**R**⁝⁝** (i.e., the reflexive relations) plus 6R*six*, **⁝⁝**R6, and **⁝⁝**R*six* (i.e., the symmetric relations). If testing reveals all of these relations, then directly conditioning only two relations will have indirectly conditioned an additional seven relations, for a total of nine relations.

And what more would happen if we *directly* conditioned *just one more* relation, one using the spoken word, "six," as in "six"**R**6? For the benefit of your own understanding, you should calculate how many additional relations further testing might reveal as indirect–conditioning products.

Now, think about the expanding results of direct and indirect conditioning for numerically related stimuli from one to ten (i.e., up to "ten"/*ten*/10/**⁝⁝⁝**). Further conditioning could then extend and generalize this thematically related set up to 100 and 1,000 and so on. Much of this currently happens coincidentally for most children. With more careful (i.e., scientifically grounded) planning, we would then need minimal conditioning efforts to expand these equivalence relations to cover any quantity of virtually anything (e.g., from atoms to colors to coinage to galaxies). We could even extend further to include the numerical stimuli of another system of verbal behavior (i.e., "language"). The same applies to the vocabulary, reading, and comprehension equivalence relations—written, heard, spoken, seen—that accumulate in the direct and indirect conditioning of one's first verbal–behavior system (i.e., native language) which some further conditioning can then expand to include such equivalence relations in another verbal–behavior system (e.g., an English repertoire expanding to include a Chinese repertoire or a French repertoire). The same likely also holds for the repertoires of environmental concern and sustainable lifestyles in languages and cultures around the world.

In every case remember that while stimuli can function equivalently, an equivalence relation means that all three properties—reflexivity, symmetry, and transitivity—are in place. We will see more of how this works when we consider some examples.

Equivalence Application Area Example

In an introductory tutorial article, Sidman (2009) describes an example that shows some of the extended benefits of equivalence–relation phenomena. The research of this example shows us how we condition reading comprehension without conditioning reading comprehension; that is, after we directly condition some parts of reading upon which comprehension depends, we get reading comprehension indirectly, an example of a widely applicable phenomenon that seems to play a major role in the contingencies that account for many complex human behaviors. This research shows the relevance of equivalence relations to reading; it shows that as a result of conditioning some parts of what we call "reading," we end up having conditioned all the parts of what we call reading. And after this conditioning process repeats by building equivalence relations with other stimulus groups, a little more conditioning then connects the resulting equivalence classes, combining them over several years of informal family reading fun, and formal instruction, in an expanding explosion of stimulus control by printed words over *comprehended* spoken words, the kind of behavior that we call reading behavior. As we will see in the verbal–behavior chapter, the term *textual* describes the verbal relation when written words merely control spoken responses, but the term *reading* describes the result when further conditioning, connecting equivalence relations, adds the stimulus and response classes to which the term *comprehension* applies.

To set the stage for what Sidman's research found, let's start with some related background. On a leisurely drive through the countryside, and with the child looking out the window, some cows in a paddock evoke a parent's vocal response "cow." (Actually, this is a summary of the parent's response; for the sake of this ongoing example, we will keep things simple, at least until we get to that verbal–behavior chapter.) If this "cow" response evokes an appropriate child response, such as pointing to a cow, and later perhaps saying "cow," the parent occasionally provides various, though sometimes subtle, forms of reinforcing attention, thereby conditioning—or further conditioning—the responses. Such scenarios, in nearly every setting, repeat dozens and hundreds and thousands of times or more during the early years of childhood. Indeed a major portion of our verbal behavior (i.e., language) accrues by our verbal community conditioning not only these kinds of responses to spoken–word stimuli but also, in reverse, vocal responses to visual, aural, and other stimuli.

Sidman used such "evocative–stimulus to pointing–response" relations as the starting point in this reading–comprehension part of his research. People say of others that when *printed* (as opposed to spoken) words evoke appropriate responses to their corresponding pictures, then the responders are exhibiting a simple form of reading comprehension. Sidman wanted to find out if such elementary reading comprehension could develop indirectly, without direct conditioning. Working with intellectually handicapped teenage boys, and using the same kinds of printed words and corresponding graphics that we have already described in our examples here (e.g., the words *cup, dog, car, box, cat, hat,* and *cow,* along with *graphics* of a cup, a dog, a car, a box, a cat, a hat, and a cow) he found that, for these boys, printed words did not evoke appropriate responses such as pointing to or touching the corresponding pictures. These boys could not read, so enlisting them as subjects to answer his experimental question could not hurt them if the answer was *no* and could distinctly benefit them if the answer was *yes.*

As we go through this extended example, we will encounter one or another of the three relational properties of stimulus equivalences. Think of their names now, so that you can more easily recognize and name them as they come up.

Sidman's team actually had to cover much ground with the boys before the textual and reading–comprehension part of this research could begin. These efforts ranged from teaching them to sit quietly to teaching them to respond by pointing at specific objects, to teaching them to respond differentially to various standard forms (e.g., circles and squares) to teaching them to respond to letters and words, on up in complexity to matching various graphics *to themselves* and words *to themselves*. (Which relational property is that?)

Sidman's Reading–Comprehension Research Steps

The steps in Sidman's reading–comprehension research study followed the usual conditioning steps for establishing equivalence relations. Again, the question to answer was whether that conditioning would produce the responses indicative of comprehension indirectly, as equivalence–relation products, or would these responses require direct conditioning.

Without going into all the methodological details (which are available in the original referenced sources) the research team began by providing conditioning in which reinforcement occurred when spoken–word samples (which we show here using quotes) spoken repeatedly, evoked the response of touching the graphic representation of the spoken word that was among the set of available graphic comparison stimuli. Ultimately graphics can be of other categories, such as actions and characteristics, as well as of objects, although this example sticks with objects. Next, the team provided conditioning in which reinforcement occurred when the spoken–word samples evoked the response of touching the printed–word version (of the spoken word) that appeared among the set of comparison stimuli, all of which were printed words.

For example, when the spoken word was "cat," and if the comparison–stimulus set contained graphics, then the reinforced response was touching the graphic of a cat. Later, when the comparison–stimulus set contained printed words, then the reinforced response for the spoken "cat" sample stimulus was touching the printed word *cat*. The training set of sample stimuli included 20 spoken words for which all the boys mastered these two types of relations.

In other words the team conditioned the boys' responses to 20 dictated words; each spoken word evoked either touching the graphic of the word, when graphics comprised the comparison–stimulus set (for the first 20 of 40 auditory–visual relations) or touching the printed–word version of the spoken word, when printed words comprised the comparison–stimulus set (for the second 20 of 40 auditory–visual relations). One might be tempted to say that the boys were *reading* some printed words at this point. However, that is not the case. The words have no other appropriate control over their behavior. Only textual responses are occurring; no comprehension is evident. All that is happening is the same that happens when you, without any study of French, correctly match some spoken French words to their printed counterparts. You might even somewhat successfully say these printed French words aloud; but these words would have no other appropriate control over your behavior. No comprehension would be evident; you are not really *reading* French. For example if, without any study of French, hearing the words "retournez–vous" evoked your pointing to the printed words *retournez–vous,* or if these printed words evoked your saying this phrase in what sounds like French, then the action addressed in this phrase would still not occur; no comprehension is evident. (To satisfy your curiosity, turning around would be an indication of comprehension, because this French phrase is a mand for this action.)

Prior to those two research steps, the team had earlier administered sample–comparison tests for reading comprehension but found no comprehension. Presenting printed–word samples evoked no touching of the corresponding graphic objects among a comparison set, which indicated no comprehension was present. Then, turning to those 40 auditory–visual relations, they conditioned spoken–word samples to evoke touching correct comparison *printed words,* and spoken–word samples to evoke touching correct comparison *graphic objects.* Their next step was to repeat the earlier tests for reading comprehension. For these tests, they again presented each of the printed words as a sample to see if it would evoke touching its graphic from among the comparison set. They also presented each of the graphics as a sample to see if it would evoke touching its printed words from among the comparison set. These are two types of visual–visual relations and, with them, the team observed comprehension.

For example, when the graphic stimulus of a cat was the sample, then the correct response was touching the printed–word stimulus *cat* from among the comparison–stimulus set of printed words. Otherwise, when the printed–word stimulus *cat* was the sample, then the correct response was touching the graphic stimulus of a cat from among the comparison–stimulus set of graphics. Think about repeating all these processes with the spoken–word, printed–word, and graphic versions of a cup, a dog, a car, a box, a hat, a cow, and the rest of the 20 stimuli that Sidman used in this research.

Remember, the team had never reinforced those types of visual–visual relations; they had only reinforced the two types of auditory–visual relations of touching the graphic of a spoken word and touching the printed version of a spoken word. Yet the direct preparatory auditory–visual relation conditioning also indirectly conditioned the visual–visual relations that provide evidence of comprehension; that is, the direct preparatory auditory–visual relation conditioning also indirectly conditioned the responses of touching the correct graphic in the presence only of its printed word, and of touching the correct printed word in the presence only of its graphic. As the team directly conditioned the original 40 auditory–visual relations (the 20 spoken words to their graphics and the same 20 spoken words to their printed versions) they were actually *indirectly* also conditioning at least 60 *additional* relations (for a total of 100). These 60 included 20 visual–oral relations (speaking the printed words) in addition to the 40 visual–visual relations (20 printed words to their graphics and 20 graphics to their printed–word versions). Now, especially when the printed words evoked the boys' touching the correct graphics, you can properly describe them as having begun to read with some comprehension. Sidman found that elementary reading comprehension indeed develops indirectly, without direct conditioning (and without needing any traditionally assumed self–agent characteristics or actions). But this is only the beginning.

Some implications of Sidman's Research and Extensions

Having published his early reading equivalence research paper in 1971, Sidman also recognized in his tutorial article (2009) that, in spite of all the experimental data and conclusions that numerous equivalence researchers reported over the last 40 or more years in the peer–reviewed literature, few— including behavior scientists and educators—have been under contingencies that produce giving equivalence–class relations the kind of serious attention that compels action. Few have moved on to the next step, the engineered– application step, even for the most obvious applied arenas ripe for including equivalence–class conditioning among their standard practices. So, in this 2009 article, Sidman explicitly lays out some of the obvious steps. Hoping to evoke further development, I summarize these steps here, which does them only minimal justice when compared to how much more human, and humane, benefits they can provide. Your assistance, dear reader, is needed. You can help by contributing to contingencies that can compel more movement on these steps. That is why we speak of supporting, encouraging, demanding expansion of behaviorology departments and programs at your local colleges and universities. While steps to apply equivalence–relation conditioning take a lot of effort, the range and extent of the predictable outcomes easily compensate. Here we touch on only a few of the equivalence application areas.

While equivalence experimenters often start by using rarely encountered stimuli in their research so as to avoid confusion from preestablished equivalence relations, they later have used more regular and common stimulus materials. Based on the already demonstrated generality of equivalence– conditioning procedures, Sidman suggests extending the work on reading comprehension to the straight–forward task of building verbal–behavior vocabularies. Equivalence conditioning can efficiently expand the number of nouns, verbs, adjectives, adverbs, and so on, in someone's linguistic repertoire. Or, in our standard verbal–behavior terminology, equivalence conditioning can efficiently expand the number of mands, tacts, intraverbals, codics, and duplics in someone's verbal repertoire. In all these cases, "Teach them to match spoken words both to their corresponding printed words and pictures, and without any more instruction, they are able to understand the printed words" (Sidman, 2009, p.13). These responses will likely also generalize to encounters with the related stimuli and stimulus classes in the world beyond both the laboratory and the computer screen.

As a further example, and as previously mentioned, the same procedures apply to numbers in terms of spoken number names, printed number names, printed digits, and quantities. Start by teaching students to match spoken number names (i.e., "one," "two," "three," and so on) both to printed digits (i.e., 1, 2, 3, and so on) and to printed number names (i.e., *one, two, three,* and so on, which we italicize only for reading clarity). Students can then, without further direct conditioning, match the printed digits and the printed number names to themselves and to each other. Then teach (i.e., directly condition) them to match spoken number names to quantities (perhaps starting with numbers of dots, as we previously showed, but later generalizing to numbers of essentially anything); this connects the previous directly and indirectly conditioned equivalence classes (of printed digits and spoken and printed number names) to the equivalence classes of spoken numbers and quantities such that the students can now match both printed digits and printed number names to quantities (first of dots but then of anything) as well as, of course, to spoken numbers, and all of these to each other.

The conditioning produces some of these relations directly, but it produces so many more indirectly, which tickles some folks fancy to the point that they want to say that we get these new relations "for free." But the new relations are not free, in the spontaneous sense, as they inevitably arise naturally, if indirectly, from the conditioning contingencies. Still, they can seem free; indeed the students would never have seen these words, digits, and quantities together before, and you can observe them being thrilled by "getting it," which also thrills the teacher. Teaching can and should be fun also. Remember, if reinforcement fails to follow activities, then the activities extinguish. Let's keep the GLR high for everyone, including teachers.

The same procedures, Sidman continues, are appropriate to teach not only the relations among things like colors, spoken color names, and printed color names, but also the relations among both upper and lower case letters and their spoken and printed names. And these, of course, are foundational to reading and communication. Think about the vast educational ramifications of this.

Note that we tend always to start with the spoken stimuli. This is because the contingencies that build initial vocal language repertoires (i.e., that build what some call native language acquisition) automatically build numerous and extensive equivalence–relation classes. Starting with spoken stimuli takes advantage of these already–conditioned classes to minimize the new conditioning needed to build the greater number of equivalence–relation classes so typical of what we see as the extraordinary—but actually, naturally ordinary—human complex repertoires of behavior.

Then, as we mentioned earlier, add in the verbal behavior of other languages. When conditioning directly leads to matching a new stimulus to some member stimulus of some other class, then the inherent indirect conditioning automatically enables matching that new stimulus to all the members of the class. And any stimulus is likely a member of more than one class, which can vastly expand the indirectly produced automatic matching possibilities. We take appropriate advantage of this phenomenon when we directly condition students to match spoken number names in another language, say French, to the printed digits; they can then automatically match the French spoken and printed number names not only to spoken and printed English number names but also to quantities. We can then repeat this process, perhaps with colors, and later alphabet letters, and then dates and words and so on, directly and indirectly conditioning the verbal–behavior repertoire of the other language.

And the vast force of school teachers, if appropriately conditioned (directly and indirectly) themselves (i.e., with the necessary repertoire in the natural–science fundamental principles, concepts, methods, and practices of behaviorology) can further extend equivalence conditioning in all the other usual subject–matter areas (e.g., spelling, writing, literature, music, math, all the natural sciences, arts, history, civics, sports, and so on). Such equivalence–relation productivity is not a hope; it is a documented and exciting reality. As Sidman puts it, "As a class enlarges, the direct addition of just one new member to the class produces an enormous increase in the number of indirectly established new relations" (2009, p. 14). A little new conditioning produces a lot of new repertoire. However, for equivalence productivity to become an explicitly inherent part of education, society would need to support a vast retooling of teacher training and colleges of education. This would ultimately involve replacing all the non–scientifically informed education curricular components with their appropriate scientifically informed counterparts. This is something that we need now but it will more likely occur down the road a piece, after we succeed in expanding education options in behaviorology itself.

Again, stimuli are seldom members of only one class. Over the last couple of decades, I have often given the example that the word green appears in the classes of novices, colors, and sustainable lifestyle responses. Imagine my pleasure to find that Sidman (2009, p. 15) had also used this example in a nearly identical fashion. Sidman makes a further point, however. Multiple class memberships do not automatically combine those classes; the current settings or circumstances (i.e., the context, such as a speaker's audience or the particular discussion topic) establishes the current stimulus class membership. Furthermore, depending on context, shared class memberships may sometimes lead to classes combining. But in other contexts, shared–membership classes more often might merely intersect while not combining. Sidman reports a delightful example of this phenomenon:

> "If we are discussing *disciplines,* Renoir, Constable, and Pollock go together as artists; Twain, Voltaire, and Byron as writers; and Churchill, Kennedy, and De Gaulle as heads of state. If we are discussing *nationality,* Renoir, Voltaire, and De Gaulle go together as French; Twain, Kennedy, and Pollock as American; and Churchill, Constable, and Byron as British" (emphases added; Sidman, 2009, p. 15; from Bush, Sidman, & de Rose, 1989, p. 31).

To conclude this section, consider that all the efforts and successes that we have discussed, especially those pertaining to what some might mistakenly call mundane academic skills, might even be fun for children and everyone. Skilled video–game programmers would have little difficulty constructing the direct equivalence–relation conditioning programs in active, even thrilling, video–game formats. After all, considering the beneficial emotional effects of the reinforcement components necessarily inherent in effective educational processes and practices, education can be, and should remain, fun. This same approach is even relevant to teaching (i.e., conditioning) the green–behavior skills of sustainable living (see Twyman, 2010, for a recent approximation).

Equivalence Relations, Evolution, and Conclusion

Physiological research gradually continues to elucidate at the cellular and molecular levels *how* respondent and operant conditioning processes contribute to direct and indirect equivalence–relation conditioning. Earlier, natural selection produced the kind of bodies that these processes change in varying degrees, including in relation to equivalence functions. For example, if their genetic endowment happened to include variations that produced neural structures enabling the conditioned mediation of even a small expansion in equivalence relations, leading to extensions of equivalence through operant and respondent processes, then proto–species members could benefit from any likely survival/reproductive advantages that these emergent equivalence–relation extensions confer, thereby passing on these variations in genetically produced neural structures. Over millions of years, the accumulation of such selected variations would result in genetically produced nervous–system structures of increasingly sophisticated potential. As a result, and as an example of our Law of Cumulative Complexity, humans today genetically inherit neural structures that generally mediate a relatively extensive range of equivalence–relation phenomena. These cumulative developments even support the writing, production, distribution, reading, and applying of the contents of books. In turn books, particularly scientifically grounded books, contribute to the further expansion of equivalence relations and the attendant benefits for humanity. (Sadly, equivalence–relation processes also help build superstitions and mysticisms, but at least this helps account for these repertoires.)

With biological selection as the foundation of the physiology through which conditioning processes work, including equivalence relations, untestable mystical constructs—from autonomous behavior–initiating self agents to cognition and the secularized soul called the mind—all remain scientifically unparsimonious, redundant, and potentially harmful, as we have previously discussed. Behaviorological–science accounts for *why* equivalence relations happen, and is bringing them under practical control. Meanwhile a more complete description of the nature of these phenomena awaits the physiological research of neural scientists that will eventually show *how,* in stimulus equivalence, the conditioning of a subset of relations actually changes the nervous system such that the subset conditioning turns out to produce the remaining relations as well. (For more of the many other vital details about equivalence relations, see Sidman, 1994, 2009; also see Fraley, 2008, Chapter 16.) While all behavior–related processes, not just equivalence relations, have substantive implications bearing on quality of life, sustainable practices, and even survival, society has only barely begun to notice the potential benefits of applying these processes and phenomena that basic behaviorological research has discovered in the last 100 years.

Equivalence and Chapter Conclusion

Recall our earlier point that school teachers already engage in approximations of equivalence–relation conditioning, but that their efforts occur under haphazard and unanalyzed contingencies, which needlessly reduces their effectiveness. I suggested that substantial progress in educational best practices could occur if society required both (a) that the vast numbers of teachers study basic behaviorology in their training programs, and (b) that behaviorology inform content throughout education curricula. This would prepare teachers to research and improve the equivalent functioning of stimuli in their subject–matter teaching. With the implied, profound implications this suggestion holds for teacher training, even back in Chapter 1 we accepted that this necessitates increases in programs and departments of behaviorology. These suggestions arose not from some personally pesky preference I have about my discipline but from previous and wider calls for action; I am not the first to suggest such substantive scientific changes in teacher education. In his book, *Equivalence Relations and Behavior,* Sidman expresses deep concern over the failures of school administrations and colleges of education to take seriously the vast and helpful implications from equivalence–relation research findings and conclusions, and on that basis to make appropriate changes (Sidman, 1994, especially the Epilogue, p. 531*ff*). He then makes several suggestions for improving this scenario.

My suggestion, about expanding behaviorology programs and departments to meet the demands of implementing the appropriate changes in teacher education, stems from Sidman's suggestions some of which occurred during the period of shared history (Ledoux, 2002) when natural scientists of behavior, like Sidman, had to work in departments of the non–natural–science discipline of psychology. The continuing incommensurable differences between these two disciplinary groups ultimately led to the separate establishing of behaviorology as the independent natural science of behavior. Since behaviorology is not any kind of psychology, readers of Sidman's 1994 and 2009 suggestions must recast them in this light, which is the point of my "expand departments and programs of behaviorology" suggestion. The applied field of education long ago committed to psychology, when psychology was the only game in town, as though only psychology could inform educational practices. But that has not been the case for some decades now, and any benefits that could be accruing, from behaviorology informing the applied behavior field of education instead, continue to be lost. With their training in psychology rather than behaviorology, educators unsurprisingly ignore equivalence–relation research, and so cannot begin to take its implications seriously. A change from that pattern will happen only when informed citizens like you, dear reader, demand more behaviorology courses, programs, and departments in your local colleges and universities for reasons like these.

Successes with those local demands can also benefit solving more global problems, because these successes will increase the number of behaviorology teachers and researchers, and behaviorologically informed participants— whether citizens or other natural scientists—in solution development and implementation. Let's conclude this chapter with the usual reminder of our ongoing activity of seeing the connection between behaviorology and making and keeping our planet safe for current and future life. Ponder how this behaviorological–science topic that we call stimulus equivalence, or equivalence relations, might interplay with, and thereby assist, the efforts of traditional natural scientists to solve ongoing local and especially global problems. The very complexity of the behavior components of these problems and solutions makes an equivalence analysis of the problems, with equivalence conditioning in the solutions, an appropriate arena in which to extend the horizon of equivalence–relations research.❧

References (with some annotations)

Bush, K. M., Sidman, M., & de Rose, T. (1989). Contextual control of emergent equivalence relations. *Journal of the Experimental Analysis of Behavior, 51,* 29–45.

Fraley, L. E. (2008). *General Behaviorology: The Natural Science of Human Behavior.* Canton, NY: ABCs.

Ledoux, S. F. (2002). An introduction to the origins, status, and mission of behaviorology: An established science with developed applications and a new name. In S. F. Ledoux. *Origins and Components of Behaviorology— Second Edition* (pp. 3–24). Canton, NY: ABCs. This paper also appeared () in *Behaviorology Today, 7* (1), 27–41.

Sidman, M. (1971). Reading and auditory–visual equivalences. *Journal of Speech and Hearing Research, 14,* 5–13.

Sidman, M. (1994). *Equivalence Relations and Behavior: A Research Story.* Boston, MA: Authors Cooperative.

Sidman, M. (2009). Equivalence relations and behavior: An introductory tutorial. *The Analysis of Verbal Behavior, 25,* 5–17.

Twyman, J. S. (2010). TerraKids: An interactive web site where kids learn about saving the environment. *The Behavior Analyst, 33* (2), 193–196.౪

Chapter 19
On Attitudes, Values, Rights, Ethics, Morals, and Beliefs

$\mathcal{L}$ast time, in Chapter 18, we extended our coverage of behaviorological research considerations by examining the well–developed experimental area of *equivalence relations.* Some of the details enable appreciation not only of the functional equivalence of many, even most, stimuli, along with their specific relational properties, but also of some still underdeveloped equivalence–application areas that could drastically increase the effectiveness of a wide range of human endeavors, from all types and levels of education to sustainable living.

Now, in this chapter, we extend our behaviorological concern to a series of interrelated topics including values, rights, ethics, and morals. These, along with attitudes and beliefs, and some topics of later chapters, reflect some ongoing ancient concerns of humans. We humans have for a very long time asked questions and sought answers about these and other concerns, but until recently we have had to manage without reaching really satisfactory answers, the kind of answers that would differ from those of the last several thousand years, the kind of scientific answers that would make a difference in our dealings with ourselves and the world around us. Here we begin to consider such scientific answers to ancient questions. Due to this scientific characteristic, these topics become the first set of topics for which our *Part–II* warning—before Chapter 13, about completing *Part I* before going into *Part II*—is particularly pertinent.

Some Historical Context

Past answers to ancient questions have of course been helpful in various ways, but human historical records—going back, for example, to ancient China and ancient Egypt—show that until quite recently these questions and answers have changed little over at least the last 5,000 years and likely longer, perhaps as far back as the development of verbal behavior, which we currently consider to have occurred at least 50,000 years ago. Across most of that time frame, relatively little else regarding living conditions changed much as well. Also, the extremely gradual changes accruing in cultural contingencies were even, until relatively recently (i.e., about 400 years ago) insufficient to induce much of the behavior patterns that we call science; being thus unavailable science could shed little light on anything, including these ancient questions and answers.

Then, around 1,600 CE, the accumulating cultural–contingency changes began to produce scientific behavior patterns (e.g., behaviors that respected evidence from solely natural events as independent and dependent variables that people can observe or measure). Across the four intervening centuries, developments in science and supportive cultural contingencies mutually accelerated such that, over the last 150 years, science brought generally helpful and quite extensive—relative to the preceding thousands of years—changes to most people around the globe (e.g., changes in communication, food production, construction, sanitation, medicine, and transportation).

However, answers to ancient questions about human nature and human behavior—the general behavioral subject matter—remained outside the purview of the traditional natural sciences (e.g., physics, chemistry, biology) because, perhaps as part of surviving in the still theologically controlled culture around the 1600s, science had left the general behavioral subject matter to the mystically grounded theological and philosophical disciplines. Later, in the 1800s, mystically grounded secular disciplines also laid claims to the general behavioral subject matter of human nature and human behavior that encompasses those ancient questions. Then, in the early 1900s, natural behavior science began to develop and subsequently address these questions (e.g., see Skinner, 1938). For an educational and entertaining account that peeks at the range of these changes in traditional and behavioral natural sciences, and makes comparisons with mystical disciplines, read Joseph Wyatt's 1997 novel, *The Millennium Man.* Recall from Chapter 4, however, that in his novel Wyatt used the term "behavior analysis" as the label for the natural science of behavior to show that this discipline differs incommensurably with the discipline of psychology. However, with the psychology discipline now claiming the "behavior analysis" label, Wyatt's usage confuses current readers. To reduce that confusion, I recommend that readers substitute "behaviorology" when "behavior analysis" appears in Wyatt's book.

No law of nature, however, says that science cannot address ancient human–nature and human–behavior questions. Biology has begun to address parts of them, particularly some parts pertaining to the natural origin of the human body, and behaviorology greatly extends this effort. You can read B. F. Skinner's 1948 novel, *Walden Two,* for an educational and entertaining account of some of the changes that might stem from a behaviorological approach to answering some of the ancient questions confronting humanity. However, again recall from Chapter 4 that while this account is as relevant today as when Skinner wrote it, he often used the word "psychology" to denote *the natural science of behavior* into which he and his colleagues were trying, at the time, to turn the traditional psychology discipline. Since the natural science of behavior was not then—and is not now—any kind of psychology, these usages also confuse current readers. Thus I must again recommend substituting "behaviorology" for these usages as a way to reduce the confusion.

The topics of this chapter involve behavior and ancient questions. We already introduced many basics that a natural science of behavior has discovered in the last 100 years about behavior and the variables of which it is a function, so let's see if we can now begin to get any closer to some satisfactory, difference–making, answers to some of humanity's long–standing questions.

Recognize, though, that behaviorologists are not the only natural scientists concerned with these questions. Various well–known and deserving traditional natural scientists (e.g., Richard Dawkins, Sam Harris) have made efforts to tackle some of these questions (e.g., ethics and morality). However, lacking a minimal background in behaviorology, the kind of background which would provide the natural behavior science that dovetails with their own specializations, they have ended up trying to shoehorn accounts of these phenomena under headings like brain physiology, evolution, and genetics. Each of these clearly plays a role, but that role is more along the lines of *how* the phenomena work. What we address in this chapter instead tracks more of *why* these phenomena happen. Humanity needs this line of inquiry, because it can delineate the accessible independent variables involved in these phenomena.

With some further delineation, those accessible independent variables can evoke more active human participation in the contingencies that compel the development of attitudes, values, rights, ethics, morals, and beliefs in particular directions—some of which may have a distinct bearing on our continued survival—rather than leaving these directions to *coincidence*. Not a new theme, this is the same dichotomy that Skinner and others addressed multiple times in the past in terms of "accident versus design." More recently, given our naturally conditioned discomfort with the non–natural implications of the term "accident," the word "coincidence" fits this context better.

One example of a current topic having a potentially sizable bearing on our continued survival, and showing the interconnection of some of our chapter concepts, concerns the value of sustainable living. Based on the conditioning history that makes conditioned reinforcers out of components of sustainable living, we claim this *value* as a *right* that deserves *ethical* respect and may even be *morally* correct at this point. Such possibilities, of course, threaten the mystical and superstitious assumptions about human nature and human behavior that cultural lore, passed down through generations of cultural conditioning, can induce. This cultural conditioning arose from our 50,000 years—documented for the last 5,000 years—of accumulated verbal conditioning in circumstances that disallowed much thorough reality testing. So now this conditioning includes lots of untestable mystical and superstitious accounts. Perhaps that conditioning still affects your behavior in some ways, possibly even inducing a negative emotional reaction to discussions of these kinds of topics. I can only hope that the counterconditioning that the last 18 chapters have provided will prove to be an adequate intellectual and emotional counterbalance for you.

With that hint about some of the interrelationships of our chapter concepts, let's take, in turn, our introductory look at attitudes, values, rights, ethics, morals, and beliefs. (For far more detail than we can cover here, see Fraley, 2008, Chapter 25. Also see the broad application of the concepts in this chapter to a topic that ultimately concerns everyone, dignified dying, which Fraley presents in his 2012 book, *Dignified Dying—A Behaviorological Thanatology.*)

Attitudes

Past, widespread, scientifically uninformed cultural conditioning compels similarly scientifically uninformed people to respond covertly to (i.e., think about) the verbal stimulus *attitude* as a term that refers to "something" that one could "have," and the having of it then accounts for some behavior pattern. Such standard phrasing, in the presence of your by now more scientifically sophisticated comprehension of contingencies and behavior, likely evoked the thought that this is merely a fictitious account of behavior. "Have" implies *possess,* and an attitude involves nothing to possess; thinking about attitudes in this way inevitably installs some inner agent to do the having, the possessing, the behavior causing. To our credit we long ago scientifically discredited the notion of inner agents and rejected them.

Alternatively, people sometimes say that alleging an attitude merely describes a predisposition to behave in a particular way. This sounds better. However it also begs the question by leaving us wondering: What is a predisposition? If, when we say *predisposition,* we refer to something that yet another putative inner agent has or possesses, then another layer of unscientific distraction only buries us deeper in the mire of mystical accounts. Instead, if by *predisposition,* we are merely making a prediction about the general likelihood of the particular behavior pattern named in the attitude, then we might be on reasonable scientific ground. This holds especially if the outcomes of our past observations of particular behavior–pattern repetitions comprise the stimuli evoking our predicting behavior as part of evoking our verbal responses *attitude* and *predisposition.* As yet another alternative, *predisposition* could refer to nervous–system parts that have the particular structures, from genetics or past conditioning that, when energy traces from the relevant evocative stimuli reach them, readily mediate the particular behavior patterns that we call *attitudes.*

While confirmation of those functional relations progresses with the collaboration of our physiology colleagues, defining an attitude as a neural predisposition to behave in a particular way might satisfy us. But I suspect that conditioning a slightly more elaborate definition serves us better. Let's define the term *attitude* as a verbal–shortcut term for particular behavior patterns that stimuli, thematically–related to the behavior pattern, evoke and consequate, with the theme appearing in the name of the attitude. This requires no mention of a genetic or conditioning history, however valid, that we call a predisposition to behave in a particular way. Think about various attitude labels that circumstances have evoked from you, and you will have a wider range of examples than the standard ones that typically appear as examples in this kind of discussion (e.g., a "positive attitude," or an "aggressive attitude," or a "lazy attitude").

We must of course always assure that, for anyone with whom we are speaking about this or that attitude, the *attitude* term always evokes responses along the lines of essentially that same definition. This is because, while at best the term *attitude* remains a verbal shortcut for a summary of sets of evocative stimuli, and the response patterns that they repeatedly evoke, at worst but more commonly, the term *attitude* remains one of those fictitious explanations of behavior (many of which we discussed in Chapter 4).

Before moving on to values, rights, ethics, and morals, let's consider a measurement problem with attitude research, and some ways to solve it. A phenomenon that researchers in many disciplines have repeatedly observed is the disconnect between the verbal behavior that their survey–based and questionnaire–based studies measure, and the non–verbal behavior that their occasional, subsequent observations measure. Stated agentially, "what people say they will do is often not what people actually do."

In various ways and formats but nearly always agentially, people get asked, "Under such and such conditions or circumstances, what would you do?" This of course is the agential version of the question that we could scientifically phrase, "What is the predicted extent of any evocative effects of such and such conditions or circumstances on 'your' behavior?" In either form the question evokes (a) some neural behaviors of consciousness that we describe as a probability estimate of the evocative effects of the stimuli in the question (i.e., of "such and such conditions or circumstances") and (b) a verbal report of this estimate. Occasionally researchers then observe their subjects with respect to the actual occurrence of the behavior that the survey or questionnaire purports to measure. However, the results of the surveys and questionnaires, which measure verbal behavior, frequently fail to match the results of the observations, which measure non–verbal behavior. Since the results of survey and questionnaire research regularly evoke particular patterns of policy directions or other behavior changes from those who pay for the research, often along with the general public, these discrepancies between verbal behavior and non–verbal behavior can produce negative outcomes.

Perhaps the best way to prevent the negative outcomes that can flow from unrecognized discrepancies, between interviewees' verbal and non–verbal behavior, is always to measure both. An even better solution, some would argue—unless the behavior of concern is merely the *verbal* behavior that appears as survey or questionnaire answers—involves skipping the solo surveys and questionnaires, and spending those efforts (and funds) to observe and measure more carefully the actual behaviors of concern. Now, let's move on to values, rights, ethics, and morals.

Values, Rights, Ethics, and Morals

Many of the concepts that we consider in this chapter relate to each other in a crescendo of complexity, namely the concepts of values, rights, ethics, and morals. For thousands of years now, these topics have evoked questions and discussions among humans. While the answers we consider here are not extensive, they are at least scientifically informed. We start with values because they tie these concepts directly to measurable variables in the prevailing contingencies (see Skinner, 1953, 1971).

Values

Simply put, *values* are reinforcers. For many people though, contingencies cloud this issue. Until contingencies occur that condition this accurate response, the same scientifically uninformed cultural conditioning that causes problems for attitudes also compels people to respond to the term *value* not only as merely a characteristic of various objects, actions, events, people, and so on, but also as a term referring to something that an inner agent possesses. In the usual agential accounts for behavior, the agent (of whatever sort) then directs the body to behavior in ways comporting with the possessed value or value characterizations. We of course exorcize the inner agent as scientifically unworthy, and reject the fictitious accounts for behavior.

So again, simply put, *values* are reinforcers. The things that we value, need, appreciate, hold dear, maintain access to, and so on, function as reinforcing stimuli. For example we often confront a relatively simple stimulus circumstance, such as needing to discover a stimulus that will serve as a reinforcer to improve the behavior of a poor and hungry student. This circumstance evokes our asking this seemingly small question: "What reinforces this student's behavior?" With some observation we may discover an accessible and affordable stimulus type or two that serves this function and so answers this question. The answer also tells us something about what the student values, although as values, reinforcers can become rather complex. Let's say that a broader set of stimuli (which we need not specify here) evokes our asking this seemingly bigger question: "What are this student's values?" With some observations, we may discover and make a sizeable list of his or her values. When we examine the list, though, we find that it contains the names of numerous stimuli that function as reinforcers for her or his behavior, starting with money and food. After all, for what poor and hungry student would money and food not serve as reinforcers? Perhaps you thought that by "poor" I meant that the student was incompetent or lazy, rather than impecunious.

But beyond money and food, a wide range of additional stimuli could appear on this values list, including current and historically based stimuli, all of which could also function as reinforcers for the student's behavior. For example on the list we could find comfortable living quarters, honest friends, fair and capable professors, a quiet place for study, a sophisticated computer, a good sound system, and lots of music to play. We might also find various opportunities for the student on the list, such as opportunities to craft, participate in, or attend entertaining events (e.g., concerts, operas, plays, films) and opportunities to practice fun or practical skills. These could include her musical instrument playing skills that band–related contingencies originally conditioned in middle school, or his target–shooting skills that team–participation contingencies originally conditioned in high school, or her hunting skills that regular field trips, with extended family members, originally conditioned, trips that put food on the table and in the freezer. We might even find on the list a big, strong, excessively safe vehicle that runs reliably although with poor gas mileage; apparently for this student the necessary social conditioning has not yet made a reinforcer out of a smaller carbon footprint. All these things could comprise a portion of the student's values, a portion of her or his reinforcers.

Now look over that list again. The seemingly bigger question (i.e., "What are the student's values?") is essentially the same as the supposedly smaller question (i.e., "What reinforces the student's behavior?"). Both questions concern both the student's reinforcers and the student's values, and these are the same. How many or how few appear on such a list is of little import. The reinforcers are the things that he or she values and, conversely, the values are her or his reinforcers. This applies to everyone. You can see this by making some more lists. Recalling that behaviors, as real events, function as stimuli, make a list of the stimuli that reinforce your behavior, and a list of the stimuli that reinforce the behavior of a friend that you know well, and a list about someone you know poorly, and a list about the members of some thematically related group of people. In every case, you will find that a list of what they value repeats a list of their reinforcers. Values are reinforcers.

As an observation you may also note that the length of those lists gets shorter if you make them in the order that we described. You may be intimately familiar with a long list of your values, and the values of your friends—values that you likely share—while you can spot only a few of the values of persons you know poorly, and possibly you can recognize only the main, and shared, value of a specified group of people. This main value, the main thing that reinforces the behavior of all of the members of the group as a group, likely appears in the group's name. For example, what is the main value—the main thing that reinforces the behavior of all the members of the group as a group— for the group that calls itself the *Death with Dignity Alliance?*

Alternatively some people define a value as the *behavior* that produces a reinforcer. This makes one's values the behaviors that produce one's reinforcers. By our first definition, if a small carbon footprint is among the stimuli that reinforce your behavior, then one of the things you value is a small carbon footprint. By the second definition, one of your values would instead be the behaviors that produce a small carbon footprint. Still, the common phrasing with respect to values implies that they are stimuli, hence my preference for our first definition, which is the one we use here as we move our discussion on to rights and ethics and morals.

Before moving on, however, recognize that we only consider the values that are *unconditioned* reinforcers, which comprise necessary stimuli for individual and species survival (e.g., food, water, sex) as inherently valuable, in the sense of *absolute* values (although even these have at least partial exceptions, as we will see). Other values gain their status as values through the conditioning process, the pairing that conditions the reinforcing function of otherwise non–reinforcing (i.e., initially neutral) stimuli. This process makes these stimuli function as conditioned reinforcers, and thereby also makes them values, but in the sense of conditional or relative values, because without the pairing, they function neither as reinforcers nor as values. This difference will soon show up in the dichotomy between absolute and relative rights, ethics, and morals as well. Now, let's turn to rights.

Rights

While the term *values* refers to reinforcers, the term *rights* refers to access to values, to reinforcers. Given this connection to physical, measurable realities, which includes behaviors and contingencies, we can focus on rights (and, later, ethics and morals) as events amenable to all the scientific consideration that we cover in natural behavior science. We define a right as unhindered access to a value, to a reinforcer. This definition comes from the contingencies, often of deprivation or coercion, that compel particular forms of verbal behavior, forms that we call statements about rights. These rights statements often take the form of claims regarding unhindered access to valued reinforcers (see Vargas, 1975; also see Krapfl & Vargas, 1977). These contingencies of deprivation or coercion also compel non–verbal behavior as specific activities that support the rights and obtain the reinforcers, activities often involved in the exercise of the rights. We generally construe a right as an abstraction while the rights claim or exercise of the right constitutes more explicit behavioral events.

Examining rights a little more deeply, when deprivation accumulates, or when something functions as coercion by getting in the way of our access to a reinforcer, we then claim access to the reinforcer as our right. Sometimes we claim an access right individually, and at other times we claim an access right as a member of a group. Sometimes the right refers to an immediate, personal reinforcer, and at other times the right refers to a long–standing traditional reinforcer. As an example of a right to an immediate, personal and individual reinforcer, after coming home tired at the end of an energy costly work shift, one might claim, as a right, access to a period of some simple peace and quiet in a home that otherwise features the more usual ruckus of kids and blaring stereos or radios or televisions or computers. As an example of a right to a group–shared, long–standing traditional reinforcer, after some government agency makes defending one's person or loved ones illegal (a circumstance which some describe as both deprivation and coercion) the affected group of law–abiding citizens might raise a chorus of claims for restoration of the right of self defense. The affected group in this case really includes everyone, although not everyone participates or even always comprehends the shared long–range interest in the group's endeavor. Even looking just at the twentieth century, many historical examples of tyrannical behavior begin with someone—often not the tyrant— disarming a population, sometimes for what seem like good reasons, with the tyrant coming to power more easily over the disarmed population. However, very few if any historical examples have such tyrants lasting uninterrupted for very long. Tyrannical coercion still induces the full force of countercoercion sooner or later.

Between and beyond those possibilities remain many of the rights that our individual and group contingency history, including our national political contingency history, has conditioned; think about the Bill of Rights in, and Amendments to, the u.s. Constitution. We say a person is exercising a right when stimuli evoke behaviors that produce the reinforcer values to which the right pertains. However, one need not exercise a right to be eligible to make rights claims about the related reinforcers. For example, all law–abiding citizens in the u.s. can make reasonable rights claims about an individual constitutional right to keep and bear arms (i.e., the Second Amendment right); however, only a subset of these citizens may be exercising that right during any given period.

Indeed, not everyone who could make a claim for a right to unimpeded access to particular valued reinforcers needs to exercise the claimed right as the only avenue producing the benefits of that right. For instance, FBI (Federal Bureau of Investigation) statistics for 2011 (which is the most recent year available as I write this) indicate ongoing reductions of violent crimes in general, and of murders in particular, both of which are down about 50 percent over the last 20 years to a more than 40–year low, a period during which the number of u.s. states with right–to–carry–concealed–arms laws increased to over 40. These trends typify states that implement Second Amendment supporting laws. The opposite trend typifies states with governments that enforce laws essentially requiring that citizen adults and children become victims—wounded, raped, maimed, or killed—if the local police are unable to reach the scene before harm occurs during a crime against the citizen. Technically, while some could argue that these data currently convey correlational relationships, others could argue that state legislatures have already organized and carried out repeated and reasonable equivalents of experimental research that makes these statistics convey functional relationships. Either way we can scientifically appreciate the conclusions and implications. All citizens benefit when criminal activity decreases as a result of, or at least in the presence of, laws that recognize the right of responsible armed self defense, as in the right–to–carry–concealed– arms laws of most states. In such states, any potential victim could be a law–abiding citizen legally carrying a concealed self–defense firearm thereby making, through processes like generalization, all such potential victims—

whether carrying or not—less evocative of the illegal behaviors of criminals or would–be criminals.

The group advantages of such benefits often greatly offset the few possible individual disadvantages, say, from firearm accidents, the accidents that, also according to government statistics, have steadily reduced during decades of ongoing and broadly based citizen education programs in firearms safety. (Your own perusal of government reports at your local library, or search of government web sites, will provide you with the latest statistics, which have been along the lines of these examples for decades now.) In addition to governments, other groups implement sanctions against, or establish arrangements to protect, various rights claims. Perhaps the most notable of these groups include religions and corporations (e.g., see Skinner, 1953, Chapters 21–26).

Another set of rights that everyone needs to exercise, and about which everyone could make access claims, pertains to green rights, including rights to clean air and fresh water, to a healthy atmosphere and an intact ozone layer, to pesticide–free food and safe transportation, to enabled recycling and sustainable living, to a population level within the planet's carrying capacity, and so on. Given humanity's long–range requirement for such rights, let's maintain a focus on them even as we move on to consider ethics and morals.

Ethics

While the term *values* refers to reinforcers, and the term *rights* refers to access to values (i.e., to claims of clear access to reinforcers) the term *ethics* refers to the behavior of respecting those rights claims for unfettered access to valued reinforcers. Indeed, we define ethics, and ethical behavior, as behavior respectful of rights claims. Those who respect our rights claims earn the label, "ethical" or, rather, their behavior of respecting our rights claims earns the label, "ethical behavior," and we appreciate the ethics that we say they "show" by respecting our rights claims.

Verbal shortcuts for bodies mediating responses. By now the kind of subtle agential phrasing present in that last sentence likely elicits some reader wincing or annoyance responses, since readers know that no inner agent exists either to display ethics or to order the body to show respect for rights claims. So, what induces that seemingly agential phrasing? Perhaps the complexity of these topics is evoking an authorial extension of verbal–shortcut status to that kind of phrasing due to the economy it provides, as discussed in some early chapters, or perhaps the author is under contingencies to avoid adding dozens of pages at this point, pages that further develop a behaviorological grammar of greater economy but that otherwise go off topic. More likely, the verbal–shortcut agential phrasing, for both the author and the reader, arises from more mundane but scientifically reasonable sources. Consider that, while behavior exists during its occurrence, the body exists before, during, and after the occurrence of the behavior that the body mediates, which it mediates due to the evocative effects of the related stimuli. Thus, due to the simple respondent conditioning that inevitably happens as behavior occurs in the presence of a body, both the behavior and the body, and later perhaps just the body, evoke our energy–economical verbal shortcuts. The success of this conditioning leads to "I" and "we" and other personal pronouns no longer referring to inner agents but merely *to bodies mediating responses,* a usage to which we continue to adhere. The reinforcing value of this accurate scientific phrasing leads to the conditioning of more accurate general verbal behavior, and to claims to rights regarding more accurate verbal behavior; your and my using pronouns this way, with reference to bodies mediating responses, not only helps reduce distracting occurrences of passive voice phrasings but also gives us one type of agent–less active voice as part of conditioning a new, more scientifically friendly grammar. Let's apply this to clarify the basic point about ethics. While stimuli evoke one's (i.e., some body's mediation of) labelling of those who respect (i.e., of bodies that mediate respecting) our rights claims as *ethical,* only the behavior of respecting the rights claims actually earns the label *ethical behavior.* Removing the inner agents enables appropriate use of the economical pronouns.

Ethical communities. Since some further discussions will involve the term *ethical community,* let's define it. We define an ethical community as a group of people who share respect for one or more rights that each one holds in common with the others. While the label only occurs now, we have encountered ethical communities already; recall that broad group of all law–abiding citizens in the u.s. who share respect for the constitutional Second Amendment right of individuals to keep and bear arms. They constitute an ethical community. We can easily recognize other large and small ethical communities, sometimes mutually supportive or overlapping, at other times neutral with respect to each other, and occasionally at odds with each other or in other ways in competition. At one end of the range of such groups you can find, for example, a group that respects a single apartment–complex owner's right to recycle old toilets by using them as outdoor flower pots while another group respects the rights of others (e.g., tourists) to see sights lacking such possibly offensive flower pots. Elsewhere on the general ethical–community continuum, you can find a group respecting the rights of other animals to the preservation of their natural habitats; you can find a group respecting the rights of children to an effective education; you can find a group respecting the rights of medical patients and behaviorally disturbed clients to effective treatments; you can find a group respecting the rights of people to earn and enjoy the fruits of a living wage; you can even find a small group that respects the rights of group members to take whatever they want from others (although the larger group, from whom they take whatever they want, calls them criminals); and, among many more groups respecting various rights, you will find a large group respecting the rights of humanity to a planetary home free of overpopulation and pollution and so on. All of these and so many more constitute ethical communities.

In another impact on ethics from respondent conditioning, the occurrence of coercion, perhaps in the form of punitive enforcement practices, respondently conditions negative emotional reactions, particularly of group members, to the stimuli that accompany responses that disrespect the community's ethics. As a result even slight deviations from the conditioned accepted practices of the ethical community automatically elicit these aversive emotional reactions from which one escapes only by returning to and maintaining the group's ethical practices. For example, after the extensive and sometimes opposed operant and respondent conditioning during life and medical school, a doctor may experience sympathy for a terminally ill patient who requests help in arranging an earlier and more dignified end, rather than waiting for the otherwise guaranteed extremely anguished end. Even before considering such alternatives, the question itself, which contradicts ethical and legal aspects of medical school conditioning, elicits strong negative emotions. Given the doctor's conditioning history, these circumstances evoke medically acceptable steps, such as drugging the patient into a stupor that persists while other processes lead to an earlier or less painful demise. The patient dies with less dignity but the doctor escapes not only the aversive emotional reactions but also the accusations of unethical behavior that could lead to jail time. That others would argue strenuously against such jail time would be of limited consolation to an incarcerated physician. (If you find these themes of interest, Lawrence Fraley addresses them extensively in his 2012 book, *Dignified Dying—A Behaviorological Thanatology.*)

Also, recognize the automatic conditioning of positive emotional reactions, particularly of group members, to the stimuli that accompany responses that respect the community's ethics. Examples would include the emotions experienced as feelings of success and belonging that stimuli elicit when these stimuli indicate other's respect for your values. The same conditioning makes similar stimuli elicit the emotions experienced as feelings of in–group camaraderie, solidarity, and mutual support.

We have now seen some aspects of how natural science addresses ethics (and values and rights, and soon, morals). These include connections of ethics to respondent processes, via the pairing of body and behavior, and the conditioning of positive and negative emotional reactions to stimuli associated with ethical and unethical behavior respectively. These also include some connections between operant conditioning and ethics, via rights, values, added reinforcers, and the subtracted reinforcers that occur as negative–emotion reductions after stimuli evoke escape behavior. These show that, as with all behavior, ethical behavior is a function of the variables operating in past and present operant and respondent contingencies. This is part of why, in Chapter 1, we called behaviorology also a natural science of philosophy, the rubric under which these and other ancient–question topics usually appear, and an area comprised of verbal behavior with which behaviorology deals. No mystical accounts achieve status as relevant explanations of values, rights, or ethics.

The same applies to morals. However, before we move on to that topic, let's consider one more of the many aspects of ethics that a discussion more detailed than our introductory coverage could include. Most of our discussion so far pertains to ethics among people with the relatively equal status of peers. But what about ethics when some of those involved hold power of some sort over the others? Remember, ethics concerns respecting the rights of others, respecting the others' claims to unhindered access to their values, unhindered access to the things they value, their reinforcers. But those holding power could easily be in a position to disrespect the rights of those under them. They can exert disrespect simply by arranging or allowing interference with the rights of others, without the circumstances affecting their own rights, or they can exert disrespect by arranging or allowing interference with the rights of others in ways that enhance their own access to their own reinforcers.

What prevents the occurrence of that kind of power play? Appropriate but competing ethics (e.g., the right to privacy) constrain society's access to the variables that could prevent unethical power plays. Nevertheless news media regularly report about bosses taking advantage of subordinates. But in the resulting general social contingencies reside the variables that induce the overriding, nearly abstract, but insufficiently powerful ethics against such power plays. These variables include virtually everyone's experience of someone who holds power over them having behaved in ways that violate their rights, usually but not always in small ways. This can range from a bigger, older teenage sibling having fun teasing you by hiding your glasses, to a manager applying subtle pressure to get a subordinate to pick up the tab for lunch, to a boss setting sexual favors as the price for promotion.

The contingencies of nearly everyone experiencing those sorts of power–play circumstances induce people to reject the behavior comprising such power plays as generally (and bordering on abstractly) "unethical." Most governments enact laws against the most severe power–play forms; some governments even enact laws against the less severe forms. While those laws add an additional layer of consequences to violations, society's usual ethical training avoids most violations; this conditioning leaves stimuli indicative of a nascent violation eliciting negative emotional reactions. Escape from these reactions hinges on the occurrence, instead, of behavior consistent with society's general ethics. Thus the contingencies induce some resistance to taking advantage of power relations to enhance one's own reinforcers at the expense of others by violating their rights. Most professions also have requirements for ethics courses, or continuing education about ethics, to assure the maintenance of ethical conditioning effects.

However, moving into, and beyond, the area of "society's general ethics" actually moves us along into the topic of morals. When the legal kinds of ethical countercontrols prove inadequate to ensure compliance, society begins to call upon morals and morality to take up the slack. Unfortunately, shifting a behavior from "unethical" to "immoral" not only enables quite an increase in enforcement power, but also enables some dangerous opportunities for unethical coercion and abuse.

Morals

For the sake of valuable pedagogical repetition, the term *values* refers to reinforcers; the term *rights* refers to access to reinforcers that are values; and the term *ethics* refers to the behaviors of respecting rights claims for unfettered access to valued reinforcers. Next in this values–rights–ethics sequence is the term *morals,* a term that refers to ethics that have become abstractions, which may affect their connection with the contingency realities that ground values, rights, and ethics.

Ethical behaviors not only respect others' rights claims but also, as contingency processes generalize their scope, some aspects of them take on the status of *characteristics* of stimuli, especially characteristics that cannot stand alone. Our verbal conditioning then evokes our speaking of this new status as *abstraction.* This phenomenon exceeds the conditioned reach of our "ethics" term, and so evokes a different term. The conditioned term for ethics at an abstract level is *morals.*

Increasingly complex contingencies regularly make functional stimuli out of stimulus characteristics that cannot exist alone. The conditioning that produces abstraction, however, operates long before ethical conditioning reaches abstract levels. For example, early conditioning leaves various behaviors of children under the control of colors. Colors are characteristics that *together with other characteristics* can make a stimulus, but a color itself cannot stand alone. Just ask someone to hand you a red, and only a red, but not a red this or that. A red "this or that" shares the characteristic of "redness" with other characteristics that together make a stimulus that they can hand to you (e.g., a red hat or a red bowl) but the redness cannot exist alone. Even when speaking about this, we respond differentially. When conditions evoke our verbal behavior about this characteristic appearing *along with* other characteristics, we say "red," but when conditions evoke our verbal behavior about this characteristic hypothetically standing *alone,* we say "redness." When a stimulus characteristic cannot exist alone, when it can have no real existence apart from other stimulus characteristics, we then apply the term *abstraction.* You can extend this pattern to a vast list of stimulus characteristics, any of which, when they cannot stand alone, we call abstractions.

Here is an example of ethical conditioning that reaches the point at which ethics become abstract and so evoke speaking of morals. When you were young, your parents might have started with the admonition not to tease your brother; treating your brother poorly constituted unethical behavior. Additional ethical training produced the ethical behavior of treating all family members well, which later extended to people, and pets, in the neighborhood, school, city, and so on. At this point treating any of these poorly constitutes unethical behavior. If this pattern of conditioned ethical extensions grows, it could reach the point where it becomes the more abstract admonitions that treating well every living organism everywhere is moral, and that harming any living organism anywhere is immoral. Such training (i.e., conditioning) converts concrete ethics into abstract morals that tend to override exceptions.

The abstraction that turns such ethics into morals often instantly raises conundrums. For example is your use of antibiotics, to save the life to which you have a right, both ethical (because it respects your rights claim to the valued reinforcer of a cure for an infection threatening your life) and immoral (because it requires the destruction of the micro organisms responsible for the infection)? We will soon see even better reasons for some wariness about excess appreciation of morals.

The jump from ethics to morals also partially derives from the conditioning of stimulus connections between morals and other abstract terms. This conditioning makes the word "moral" evoke responses similar to those that the word "goodness" evokes. This stands in contrast to the responses that the word "badness" evokes, which in turn are similar to the responses that the word "immoral" evokes. All these are abstractions; they cannot exist apart form additional stimulus characteristics. Further extensions carry on to related word dichotomies, such as acceptable and unacceptable, allowable and disallowable, or tolerable and intolerable.

Our conditioning further leads us to respond to the stimuli controlling those morals–related dichotomies as intrinsic qualities. This distinguishes them from ethics, because we respond to behavior as ethical or unethical on the basis of extrinsic criteria regarding specific rights claims. We can measure ethical behavior as actually supporting the claims, and we can measure unethical behavior as actually opposing the claims. In either case, the determination of ethical or unethical depends on specific, external criteria. However, conditioning induces us to respond differentially to moral and immoral behavior on the basis of (i.e., under the control of evocative stimuli regarding) whether the behavior comports with some general, intrinsic goodness or badness characteristics respectively, characteristics that conditioning has made functional but that cannot stand alone. This abstract status of morals, as verbal stimuli, somewhat divorces them from the contingencies that generate them, which can lead to problems, just as rules that no longer reflect the contingencies that the rules describe, because the contingencies changed, can lead to problems.

Another, related result stands out when we compare ethics and morals. *A change in circumstances,* which we can measure, can lead to a change in our assessment of a particular behavior either from ethical to unethical or from unethical to ethical. However, once conditioning processes compel classifying a behavior as moral, we continue to respond to the behavior as inherently good *regardless of changes in circumstances.* Similarly, once contingencies establish classifying a behavior as immoral, we continue to respond to the behavior as inherently bad regardless of changes in circumstances. Morals can extend to the conditioning of large numbers of people, all of whom can then become involved in punishing immoral behavior. However, due to the abstract level, changes that happen in the concrete, ethical contingencies behind some morals often fail to induce respective changes in the related morals. As a result large numbers of people end up punishing behaviors that actually are ethical. For example large numbers of people currently punish behaviors related to humanely, ethically trying to decrease the human population to more sustainable levels.

In addition a moral behavior always occurs along with the body that mediates it, so the inherent goodness gets extended to the body through the usual conditioning pairing process. By the further extension of our culturally conditioned predilection for inner agent accounts, the now inherent goodness of the body gets even further extended to the "person" whom we then consider as inherently good, which evokes even better treatment for her or his moral behavior. The same applies to immoral behavior, extending inherent badness first to the body and then to the "person" whom we then consider as inherently bad, which evokes even worse treatment for his or her immoral behavior. This shows us the increase in enforcement power to which we alluded earlier. When reinforcers become dependent on enforcing the status of a behavior that a powerful group considers unethical, that circumstance evokes the groups' verbal behavior of overextending claims that shift a behavior description from "unethical" to "immoral" for the rest of the culture. Since the immoral behavior remains abstractly bad independently of circumstances, that shift allows, even encourages, more extreme forms of enforcement of the current morality, which opens the door to easy and all–too–often permanent enforcement methods for possibly misconstrued morals violations. We call them possibly misconstrued, because morals are products of behavioral contingency processes and so are not written in stone. They can become harmful when the contingencies change. They can become harmful in ways similar to the ways in which rules become less helpful, even harmful, once the contingencies, which the rules state or describe, have changed, as we discussed near the end of Chapter 12.

These developments should be raising all sorts of red danger flags for you. Pursuing these flags here, however, would take us to levels of detail inappropriate for our introductory analysis. Still, such details help clarify the extent of behaviorology's natural–science analysis of values, rights, ethics, and morals, so I encourage you to pursue them via the references, particularly Chapter 25 of Fraley's 2008 book, *General Behaviorology* (also, see Fraley, 2012).

But before summarizing and concluding this section, let's repeat that presuming some inherent goodness or badness of "persons" also misconstrues "person" as a mystical inner agent (or as a representative of some sort of inner agent). Scientifically we instead construe the "person" as the potential and actual repertoire of behavior that the body is capable of mediating due to both its genetics and its conditioning history, a point to which we will return in a later chapter. Here, however, construing the person as an inner agent shows us a source for some of the common objections to behaviorology, objections often miscast at the level first of ethical concerns, and then at the level of moral concerns, a level that tends to evoke a culture–wide condemnation, a condemnation that backfires to whatever extent humanity's survival requires behaviorology. By definition natural science bars mystical accounts from its explanations, and in behaviorology this means barring inner agents from its explanations of behavior. So anyone whose conditioning has induced accepting inner–agent accounts objects strongly to behaviorology not on some technical or intellectual or scientific grounds but on moral and related emotional grounds. The claim is, "How dare those behaviorologists set up their natural science against the accepted moral reality of not only our mystical, theological maxi–god that moves mountains but also against our mystical, secular mini–gods that move arms and legs!" These mini gods, of course, refer to inner agents of every sort. Such people see the inner agent as good; after all, it started out as the theological soul before its label changed to the more secular psyche or mind or self or person or personality, and so on. So they see the behaviorologists' scientific exorcism of inner agents as automatically and inherently bad, even evil. And the traditional morality of good and evil further conditions people to remain good by fighting evil. As was the case with some other, past scientific perspectives (e.g., Darwin) the available data suggest that society has still insufficiently conditioned resistance to carrying out that admonition about fighting "evil" to unethical, even immoral, extremes; for example witness the attacks on natural science in general and on evolutionary biology and behaviorology in particular. We need some survival enhancing change. Now let's summarize before getting to our last chapter topic, beliefs.

Summary and Conclusion of Reinforcers Through Morals

Let's summarize the interconnected series of reinforcers, values, rights, ethics, and morals. Our reinforcers are or become our values. The stimuli in various contingencies evoke our claims that we have rights regarding unrestricted access to these values, these reinforcers. Then ethics involves the respecting of our rights claims for unobstructed access to our valued reinforcers, and our conditioning evokes our applying the label *ethical* to those who respect our rights claims and further evokes our applying the label *ethical behavior* to their behavior of respecting our rights claims. Then we speak of morals when, through processes like generalization and other conditioning extensions, ethical conditioning reaches the point at which ethics become abstract, that is, come to involve stimulus characteristics that cannot exist alone; this enhances enforcement potential but at the risk of damage to our culture and survival through both insensitivity to reasonable exceptions and determined resistance to needed change.

In conclusion, consider again the difference between unconditioned reinforcers and conditioned reinforcers, for this difference produces the dichotomy not only between absolute and relative morals but also between absolute and relative ethics, rights, and values. We consider unconditioned reinforcers, which function due to genetically produced neural structures, as *inherently* valuable due to their role in species and individual survival. On the other hand, we regard conditioned reinforcers as only *relatively* valuable. We say relative because these stimuli, initially lacking reinforcer functions, gain the status of values only through the conditioning process that bestows a conditioned reinforcing function on them; *without this process occurring, they lack value status.* So, for example, food and sex are inherent values while money and music are only relative values. Does that make food and sex absolute values?

Let's discuss that question. In affecting the status of values, that difference in the origin of various reinforcers flows through the whole sequential framework of these concepts, from values to rights to ethics to morals. In all of these cases, we can only consider those stimuli that are grounded in unconditioned–reinforcer status, and hence have inherent–value status, as having any sort of claim about status as "absolute" values, and hence absolute rights, absolute ethics, and absolute morals. Thus, some might argue that, as unconditioned and so inherent values, food and sex are indeed absolute values (and rights and ethics and maybe morals). Meanwhile other stimuli, being grounded in conditioned–reinforcer status, retain a status of relative, in the sense of arbitrary or conditional; if the conditioning happened, then the status of value—and hence of right and ethic and maybe moral—begins, but if no such conditioning happens, then the stimuli are not values, and so on.

However, are claims to absolute status absolute? Indeed not. While some stimuli that evoke *inherent* also evoke *absolute*, the term *absolute* invites inaccuracy. For example, the unconditioned reinforcer status of sex leads to the claim that procreation (i.e., procreative sex) is an absolute value clearly necessary for species survival. In one sense this is correct, and some people argue that therefore the culture should encourage relationships that support procreative sex as the only legitimate form of relationship. But in another sense the notion that procreative sex is an unquestionable, absolute value, necessary for survival, is dangerously wrong. Beyond debates about issues like free love versus the nuclear family, and heterosexuality versus homosexuality, we can see that procreative sex, rather than being an absolute value, right, ethic, or moral, is an absolute disaster. Consider simply that the current level of procreative sex among humans on this planet is leading to the demise of planetary life, including humanity. As we mentioned in an earlier chapter, our human population is currently already at over 150 percent of the planet's carrying capacity. Such a condition cannot continue for long without portending disastrous effects of truly momentous, even obscene, proportions. Does this not make procreative sex (as opposed to other forms of sex) at least a serious if partial detriment to species survival? Humans will always produce enough babies for species survival. However (and agentially phrased) other, non–procreative forms of sex, between consenting adults of the opposite or same gender, currently contribute to our species survival by not producing babies and so humanely reducing the population growth rate. The resulting survival benefits increase as a function of how quickly new cultural–wide ethics encourage the broadest range of non–exploitative relationships with sexual practices unrelated to pregnancy. Even *"zero* population growth" is now a quite inadequate option.

Humanity has taken too few population–control steps too slowly. Shall we just wait until disasters destroy two or three billion of us? These running–out–of–time circumstances make stimuli that *reinforce alternatives to procreative sex*—thereby producing decreases in population reproduction rates—reasonable values, rights, and ethics to include in any expansion of green interventions for the survival not only of our species at this time but also of the many other species whose survival various web–of–life contingencies tie to whether our actions wreck or save the planet as a place safe for life.

Note that the dangers of *moral* pronouncements, which we already discussed, prevented my including such a pronouncement in those comments on the current survival value—and rights and ethics—of alternatives to procreative sex. Whether or not—or how much, for how long—moral status accrues to those non–procreative sex alternatives remains a question of ongoing cultural conditioning.

The accuracy, inaccuracy, or even reality, of our beliefs, however, may play a role in the success of our green intervention efforts. So let's turn now to the last topic of this chapter, the topic of beliefs.

Beliefs

Like attitudes, with which we opened the chapter, *beliefs* are only indirectly connected to our values–rights–ethics–morals sequence. However, beliefs and attitudes share a connection, and much of that connection involves similarities in the kinds of variables that evoke both terms. After observing an individual's or a group's behavior, we may describe that behavior by saying that they have a serious attitude, for example, toward black magic, or toward free will. Their observation of their behavior, on the other hand, may evoke their saying that they *believe* in black magic, or in free will. At least at this level, the terms attitude and belief are interchangeable; the relevant variables could evoke our saying that they have a belief in free will, or black magic, and such variables could also evoke their saying that they have a serious attitude toward free will, or black magic. While the phrasing of these comments contains some hints regarding attitudes–beliefs interconnections, let's consider beliefs more closely, which might better reveal some of these interconnections.

For starters, while the more common uses of the term belief have circular and agential problems, beliefs, like attitudes, reside in contingencies, not in any inner agent. When contingencies put in place controlling relations that leave particular stimuli strongly, thoroughly, consistently affecting someone, in both intellectual and emotional ways, then we may describe this result as belief.

In addition to beliefs suffering from circularity problems (which we will soon exemplify) and inner–agent problems (e.g., "Who" "does" the believing?) we can also think of beliefs as probabilities. A belief describes a high probability that some stimuli, across one or more contingencies, will consistently evoke particular and predictable patterns of behavior. For example, in some cultures, and for certain variable periods of holiday time, the normal conditioning of young children induces high probability behaviors that we call a belief in Santa Claus. While this normal conditioning is unpretentious and unrecorded, you have nevertheless witnessed it more than once (e.g., as a child and as an adult; as the relevant though culturally bound saying goes, in the four stages of Santa, first you believe in Santa, then you disbelieve in Santa, then you are Santa, and finally you look like Santa). Predictably, in the holiday season, certain stimuli control some standard behavior patterns. The stimuli include—depending on geographical location—the onset of winter and falling snow along with many other variables, any of which might have the effect on a particular child of evoking some seasonally characteristic behavior patterns. More of these similarly functioning stimuli include holiday–theme store displays—which often go up months before the snow flies—and street decorations and Christmas trees and so on. The typical behavior patterns include increasing numbers of children writing and mailing letters to the North Pole, writing notes to leave for Santa Claus, and so on. To the reinforcing delight of their authors, many of the letters appear in local newspapers or on local web pages.

Of course, as is the case with many beliefs (e.g., black magic, free will) reinforcers of many types follow many of the belief–connected activities. Some, even many, of these reinforcers follow responses *coincidentally,* which imparts the status of superstition to the general behavior patterns that we call the belief (in Santa Claus or whatever).

Thinking of beliefs as probabilities, however, tends to distract us from another characteristic of beliefs, which is that they involve circular reasoning. While we implied that attitudes share this characteristic, let's explore it explicitly with respect to beliefs. For both observers and believers, the only evidence for beliefs (and attitudes) comes from the behaviors that the beliefs (and attitudes) putatively explain. These can be verbal behaviors, as belief statements, or non–verbal behaviors as the particular behavior patterns that have a theme that the belief names. Once contingencies evoke the belief statement or the related behavior patterns, both believers and observers infer the beliefs from these verbal and non–verbal behaviors, and then use these inferred beliefs as the causes, the explanations, of these behaviors. You can see such circularity in our Santa Claus example:

"Why did Jane and John write notes to Santa?"
"Because they believe in Santa."
"What makes you say that they believe in Santa?"
"Their writing of notes to Santa makes me say that they believe in Santa."
"But, why did they write those notes to Santa?"
"Because they believe…" and so on.

That circularity shows that beliefs, like attitudes, also come under the category of fictitious explanations of behavior. The belief in Santa is supposed to cause the writing–to–Santa responses. People more easily fall for that falsehood when our language conditioning induces us to regard beliefs as things that people possess (as in "have" or "hold"). If you possess it, then it can cause you to do things. But just what, or who, is doing the possessing or the doing? Why, your favorite untestable mystical inner agent of course. We treat the inner agents that possess beliefs the same way we treat the inner agents that possess attitudes, which is the same way we have treated all these mystical inner agents; we long ago scientifically discredited the whole notion of inner agents and rejected them.

Beliefs and attitudes are further connected in that with beliefs, as with attitudes, we often observe a discrepancy between what we observe people saying that they believe and their behaviors that we observe happening. And as with attitudes, this occurs because different contingencies control these different behaviors. Given these discrepancies between verbal behavior and non–verbal behavior, perhaps the safest approach regards beliefs strictly as, at best, relatively unreliable verbal shortcuts describing, in summary ways, the behaviors to which the beliefs pertain.

Some Ancient Questions Conclusion

In concluding this chapter, let's go beyond the problems of attitudes and beliefs, although we might all benefit from more conditioning that renders a respectful attitude toward and belief in (i.e., that renders behaviors describable as respect and support for) the science behind both acknowledging global warming and developing and implementing sustainable green solutions that can help us survive. Beyond that, let's focus, as part of our usual ongoing activity, on how developing the behaviorology discipline enhances our human potential to deal effectively with the values, rights, ethics and morals of clean air, fresh water, a healthy atmosphere, an intact ozone layer, pesticide–free food, safe transportation, enabled recycling, sustainable living, a population level within the planet's carrying capacity, and so on. For example, let's analyze and adjust our routine cultural contingencies so that they better condition the effectiveness of green values like the reinforcers of lifestyle sustainability at a lower population level. Then let's extend these routine contingencies (a) so that they better condition effective rights claims regarding all components of a healthier planet, and (b) so that they better condition the ethics of respecting these rights claims. However, we should remain aware of the possible dangers of our routine cultural contingencies conditioning too much moral status for those ethics. Such moral status would tend to render the ethics more resistant to change when circumstances change; for instance something better than, but incompatible with, current recycling efforts might develop, but excess moral servitude to present recycling methods might delay or prevent implementation. Thus, overly moral conditioning could contribute to contingencies that work against humanity, in ways similar to the ways in which rules become less helpful once the contingencies, which they state or describe, have changed.

In ways such as those, we can share and apply some of what we have discovered about human nature and human behavior. And these applications show that we may finally be arriving at some scientifically, as well as emotionally, satisfactory and difference–making answers to some of humanity's long–standing questions.❧

References (with some annotations)

Fraley, L. E. (2008). *General Behaviorology: The Natural Science of Human Behavior.* Canton, NY: ABCs.

Fraley, L. E. (2012). *Dignified Dying—A Behaviorological Thanatology.* Canton, NY: ABCs.

Krapfl, J. E. & Vargas, E. A. (Eds.). (1977). *Behaviorism and Ethics.* Kalamazoo, MI: Behaviordelia.

Skinner, B. F. (1938). *The Behavior of Organisms.* New York: Appleton–Century–Crofts. Seventh printing, 1966, with special preface: Englewood Cliffs, NJ: Prentice–Hall. The B. F. Skinner Foundation (www.bfskinner.org) in Cambridge, MA, republished this book in 1991.

Skinner, B. F. (1948). *Walden Two.* New York: Macmillan. In 1976 Macmillan issued a new paperback edition with Skinner's introductory essay, *"Walden Two* Revisited." See the caution in this chapter about a fix for some confusing phrasing in this book.

Skinner, B. F. (1953). *Science and Human Behavior.* New York: Macmillan. The Free Press, New York, published a paperback edition in 1965.

Skinner, B. F. (1971). *Beyond Freedom and Dignity.* New York: Knopf.

Vargas, E. A. (1975). Rights: A behavioristic analysis. *Behaviorism, 3* (2), 120–128.

Wyatt, W. J. (1997). *The Millennium Man.* Hurricane, WV: Third Millennium Press. See the caution in this chapter about a fix for some confusing phrasing in this book.☙

Chapter 20
Language Is Verbal Behavior

Chapter 19 began our coverage of some natural–science answers to some of humanity's ancient questions. We not only addressed attitudes and beliefs but we also introduced the sequence of values, rights, ethics, and morals, the first set of topics for which our *Part II* warning—before Chapter 13, about completing *Part I* before going into *Part II*—was particularly pertinent.

In this chapter we continue to address an aspect of humanity's ancient questions and current scientific answers. Here we cover some basics of the medium in which these questions and answers occur. This medium involves the phenomena that people generally describe as "language." However, we set aside the ancient, cultural traditions that have language coming from a mind—or any other mystical inner agent—as its supposed psychic powers magically take over after any real stimuli and direct the body to produce real responses of its mere choosing. Instead, we take language and analyze it scientifically, under the label *verbal behavior*, because contingencies induce both the ancient questions and their answers to manifest in the form of this kind of behavior.

We introduce verbal behavior, which Vargas (2013) calls *lingual* behavior, in four steps. After providing some general considerations about verbal behavior, such as its definition and some characteristics of its analysis, we analyze elementary verbal operant relationships. Then we cover some applications of verbal–behavior analysis, particularly to teaching additional languages. We finish by listing some of the more advanced verbal–behavior analysis components that the available, more thorough treatments of the topic provide (i.e., Fraley, 2008, Chapter 26; Peterson, 1978; Skinner, 1957).

General Verbal Behavior Considerations

Both terms, language and verbal behavior, describe a topic that many see as among the most, if not *the* most, complex, omnipresent and vital phenomena of our world. Words not only literally surround us, in a myriad of spoken and written and gestured and signed and symbolized forms, but functional word recombinations, which contingencies continuously vary in form, comprise large components of every human's everyday behavior. We go to great lengths to assure this verbal extent and variation (e.g., through education) partly because words are the vehicles that contingencies drive into rules, as we discussed in a previous chapter. The further behaviors that rules evoke build accumulations of knowledge and practices that last beyond individual lifetimes, accumulations that we describe as cultures, accumulations that drastically reduce the need for conditioning to build each individual's repertoire through direct contingency contact. The further behaviors that rules evoke include words combining into larger units, from phrases and sentences to fiction and non–fiction books, and from rhymes and couplets to ballads and epic poetry, with all of these units occurring in relation to nearly every imaginable subject matter, and with many of these units occurring in both the usual overt forms as well as the even more intimately familiar covert varieties that we call thinking, and all under the ordinary, natural control of generally non–coercive contingencies.

Let's begin our introductory coverage of this vast verbal–behavior topic with some of its recent history, and then with some ancient history regarding the relation of evolution and physiology to verbal behavior. These provide a foundation for defining not only verbal behavior but also the verbal community. Then we will describe some characteristics of our verbal–behavior analysis before we begin analyzing elementary verbal forms and applications.

Recent history

The recent history of verbal–behavior analysis sets the stage for us. The term *language,* arising out of traditional, mystical cultural lore, has always been, and continues to be, fraught with agential implications. Out of concern to stress only the real, measurable, naturalistic aspects of linguistic phenomena, including the operation of contingencies in the generation of language repertoires and the production of language products, Skinner (1957) adopted the term *verbal behavior* rather than struggle with the term *language.*

Skinner began considering complex human linguistic behavior early in his career. Some circumstances in 1934 induced his particular interest in this topic. His efforts over the next 20–plus years, including various lectures and courses, culminated in his 1957 book, *Verbal Behavior.* Late in this book, he described the circumstances that induced his interest in verbal behavior:

> In 1934, while dining at the Harvard Society of Fellows, I found myself seated next to Professor Alfred North Whitehead. We dropped into a discussion of behaviorism… and I began to set forth the principal arguments… with enthusiasm. Professor Whitehead was equally in earnest—not in defending his own position, but in trying to understand what I was saying and (I suppose) to discover how I could possibly bring myself to say it. Eventually we took the following stand. He agreed that science might be successful in accounting for human behavior provided one made an exception of *verbal* behavior. Here, he insisted, something else must be at work. He brought the discussion to a close with a friendly challenge: "Let me see you," he said, "account for my behavior as I sit here saying, 'No black scorpion is falling upon this table.'" The next morning I drew up the outline of the present study (pp. 456–457).

The answer to Whitehead's challenge, which runs several pages, provides interesting and fun material that I recommend to you (after completing this chapter, of course).

The appearance of Skinner's *Verbal Behavior,* which many consider his most important work, began to focus attention on how far a natural science of human behavior could reach, particularly in discounting agential explanations of behavior such as those common in traditional linguistics, a bastion of agential habitation wherein the superstitious side of the culture takes language as directly communicating the expressions of the mind, psyche, or self. Following this kind of lead, in 1959 Noam Chomsky published a paper highly critical of Skinner's book, and purporting to be a review of it. By seeming to discredit Skinner's book, especially in the eyes of those who had not, and now would not, read it, Chomsky's paper misled many into further support for fundamentally mystical accounts of language while ignoring Skinner's account, an account that is strictly a natural–science account of linguistic phenomena. Even though Skinner's book accounts for verbal behavior in a largely interpretive manner, those who read Skinner's book held that Chomsky had missed the point; he had argued effectively against things that Skinner had not said, and against viewpoints that Skinner had not held. In 1970, a widely respected paper by Ken MacCorquodale reviewed Chomsky's paper and largely set the record straight. Data supporting the natural–science account that Skinner had offered began to appear in the 1970s and continues to pile up (e.g., see the pages of the currently published journal, *The Analysis of Verbal Behavior,* the first volume of which appeared in 1982). Subsequently others clarified and expanded Skinner's original analysis, and we incorporate some of their work herein (e.g., see Eshleman & Vargas, 1988; Fraley, 2008; Michael, 1982, 1988; Peterson, 1978; Vargas, E., 1988, 1991, 1998, 2013; Vargas, J., 1990).

Before moving on, however, we should note that, after setting aside the inherent inner agents in traditional linguistic analyses, this field has much to offer (e.g., see E. Vargas, 2013) particularly its structural–analysis that sometimes dovetails with the functional analysis that characterizes our coverage of verbal–behavior analysis. We should also recognize a simple convention regarding verbal behavior. Exposure to verbal–behavior material quickly leads to the occasional use of the abbreviation, "vb," in place of "verbal behavior," and occasionally we too will invoke this convention.

VB Evolution and Physiology

Moving from recent history to ancient history, let's add to the comments in Chapter 1 regarding the origins of language and the origins of anatomically modern humans. Until more data say otherwise, researchers generally accept that our species gradually appeared as natural selection produced certain evolutionary changes in the anatomy and physiology of some proto–species members, changes that accumulated to the point where our remote ancestors had neural structures and vocal–musculature structures functioning together as a system that operant contingencies increasingly affected. These contingencies accrued expanding verbal effects during the lifetimes of numerous socially interacting individuals. Then, across the multiple generations of groups of individuals, the contingencies established more and more complex varieties of verbal behavior, the beneficial effects of which maintained the group contingencies, which we now call cultural practices, that drove further verbal behavior developments. Skinner devoted an article to this topic in 1986 that I highly recommend.

While that description gives you a broad view, here are some specifics. The vocal neuro–musculature increasingly came under control of operant contingencies due to the resulting contributions to survival. These contingencies altered various vocal sounds during the individual's lifetime such that particular stimuli came to evoke particular sounds. Postcedent processes would shape and maintain the consistency of these sounds across groups of individuals, each of whom would then say the "right" thing (i.e., the thing others would reinforce) under the appropriate conditions (e.g., the appropriate evocative stimuli). As contingencies expanded an individual's verbal repertoire, the *meaning* of a particular response always stemmed from the particular controlling variables; meaning never resided in agential characteristics or components. And the contingencies produced different sounds that "meant" the same thing across different groups that were separated from, and thus not in contact with, one another. This led to the development of different languages and language groups with a variety of language similarities. The contingencies of physical reality that produce verbal behavior remain similar across all the different verbal communities around the globe, and these similar contingencies result in common linguistic characteristics in different forms worldwide. One example pertains to the presence of verb tenses in various languages that result from similar contingencies regarding past, present, and future events. You will find that a thorough study of the contingencies related to any verbal behavior results in eliminating any relevance for fictional explanations or inner agents in accounting for verbal behavior (see Fraley, 2008, Chapter 26).

All those recent and ancient historical considerations give us some overall perspective regarding verbal behavior. With this in place, let's turn to verbal–behavior analysis definitions and characteristics.

VB Definitions

Let's define two related terms. Of course the first of these terms, verbal behavior, occurs often in this chapter. The other term, verbal community, occurs less often even though it continuously keeps us focused on the inherent social and interactive character of verbal behavior.

Verbal behavior. We basically define behavior as *verbal behavior* when the consequences, particularly the reinforcers, of the behavior occur through the mediation of another organism. In other words, verbal behavior is behavior that produces reinforcers that occur through another organism's behavior. Conversely, non–verbal behavior is behavior that produces reinforcers that occur through the direct effects of the behavior on the environment. For example, a wasp flying into the room through a window serves as an establishing operation that increases the reinforcing value of an open door through which the further response of exiting can occur. While alone in the room, if the wasp stimulus evokes a grasp–the–door–knob–and–turn response, the resulting open–door reinforcer occurs through the *direct* effects of the grasp–and–turn response, so this response is not verbal behavior. However, with another human body present and closer to the door, if the wasp stimulus evokes frantic pointing–at–the–wasp and pointing–at–the–door responses, and this gesturing evokes the other body's door–opening response, then for the gesturing body, the open–door reinforcer occurs *indirectly*, through the response of the other body; the reinforcer that gesturing produces occurs through the mediation of the other body, making the gesturing meet our basic definition of verbal behavior. The wasp could also evoke the vocal response "Open the door!" This response could also produce an open door through another body's physical door–opening response and so it too would fulfill the basic definition of verbal behavior.

That verbal–behavior definition takes verbal behavior beyond the *vocal* behavior that, for many thousands of years, held sway not only over earlier gestural or sign languages but also before contingencies extended verbal behavior into written and other forms. Indeed, this basic definition takes verbal behavior well beyond the very notion of linguistics; under this basic definition, if the reinforcers that a behavior produces occur through the responses of another organism, then we call the behavior *verbal behavior*. For example, when the pet dog scratches the back door and the dog's owner opens the door thereby mediating the reinforcing consequence of the dog's door–scratching behavior, that door scratching technically meets the basic definition of verbal behavior.

Nevertheless, we generally construe behavior like that door–scratching as on the periphery of the kind of behavior that more commonly evokes our response of "That is verbal behavior." Along these lines, we usually add some qualifiers to our definition: *Verbal behavior is behavior that (as a real stimulus event) evokes another organism's responses that mediate—as in provide—the reinforcers for the first organism's behavior, after verbal–community contingencies have conditioned such mediating behavior, and where typically both organisms are verbal community members.* That conditioning of mediating behavior is part of what makes the other person a member of the verbal community, and Skinner incorporated this qualification into a later definition of verbal behavior as "… behavior that is reinforced through the mediation of other people, but only when the other people are behaving in ways that have been shaped by a verbal environment or language" (1986, p. 121).

Verbal community. We define the *verbal community* simply as the group of people whose mutual mediating reinforcements condition the verbal and mediating behaviors of the group as a result of the benefits that accrue to the group from generating and maintaining these verbal behaviors. Many of these benefits become obvious when you consider that this definition broadly refers to all those people conditioned in the set of verbal responses and practices that constitute a particular language (e.g., English); the majority of objects and events in all of our surroundings result from responses that verbal behaviors make possible, including, and perhaps emphasizing, verbal behaviors that result in applications of the written verbal reports of scientific research outcomes. In a more limited sense, the verbal community for any individual consists of only those with whom the individual shares verbal and mediating behaviors.

Nine VB–Analysis Characteristics

Here we consider many characteristics that relate to verbal behavior. The first five apply more to verbal behavior in general while the rest remain particularly pertinent to our subsequent discussion of elementary verbal operant relations.

(*1*) *Function rather than structure.* The long standing tradition of approaching speaking and writing as language has produced some worthwhile results. From this approach linguists know much about the *structures* of many languages, such as their vocabulary, grammar, and syntax. The facts that they have discovered help us compare languages, trace language families, teach about languages, and even improve our teaching about how to write well in a particular language. This structural approach to language, however, usually begins with the traditional assumption that language, especially as language behavior, is the communicative outflow from immaterial, including mentalistic, agents residing inside a body, the body from which the speaking or writing appears merely to emanate magically. Yet linguistic analyses of grammar, syntax, and so on, lack any need for this assumption, an assumption which in any case starts out and remains mystical and thus scientifically inappropriate.

But what if speaking or writing "appears merely to emanate magically" from a body only because no one has scientifically analyzed *why* the speaking and writing occur? What if we analyze *why* scientifically? What if we analyze *why* in terms of real, measurable independent variables? Such *why* questions approach language in terms of *function* rather than structure, and Skinner's verbal–behavior analysis began scientifically to explore the extent to which language behavior might be a function of the same kinds of independent variables of which other behaviors are a function.

In continuing Skinner's analysis, we analyze verbal behavior as a function of mostly evocative and consequential stimuli. This not only gives us details about the variables of which verbal behaviors are a function, but it also tells us about the function of verbal behaviors in affecting other events. By revealing these functions of verbal behavior, behaviorological analysis shows many more of the ways through which people affect the world around them; this makes particularly inescapable the relevance of verbal behavior, and the natural–science behind its analysis, to solving global problems.

(*2*) *No new fundamental principles.* Throughout its historical development, verbal–behavior analysis has succeeded while working only with the same set of independent variables, the same set of concepts and principles and processes, that control non–verbal behavior. Verbal–behavior analysis has not needed any new fundamental independent variables that only apply to verbal behavior. Of course research continues to expand our understanding of behavior, both verbal and non–verbal, and sooner or later the data may require additional fundamental concepts or principles or processes. At this time, however, all the independent variables that make up our concept of contingencies of reinforcement are the only variables needed in accounting for any behavior of humans and other animals, including verbal behavior. These concepts and principles and processes include all those that we have covered in past chapters, such as conditioning, generalization, establishing operations, function–altering stimuli, evocation, reinforcement, schedules, punishment, extinction, shaping, chaining, fading, and equivalence relations.

(3) VB sense modes. When people first hear about verbal behavior, many think of making sounds with the vocal musculature. But verbal behavior occurs in other sense modes as well. While we *hear* vocal verbal behavior, we *see* written verbal behavior, and we *feel* the written verbal behavior of Braille, and we *see* the verbal behavior of gesturing and signing. Any behavior in any of these modes that is reinforced through another person's behavior is verbal behavior. However, sometimes a bodily sound, such as a sneeze, evokes another person's behavior but without any reinforcing effect. Another's subsequent polite comment after a respondent sneeze has little to no effect in making you sneeze more often. Thus such behavior would qualify as non–verbal behavior.

That is one source of confusion about which you should be wary. Another source of confusion comes from mixing stimulus and response words with respect to various sense modes. For example, the terms *auditory* and *visual* describe stimuli while the terms *vocal* and *writing* describe responses.

(4) "Speaker" and "listener." The terms *speaker* and *listener* refer to different behavior repertoires. The vocal verbal community conditions both speaker and listener repertoires in its members, while the signing verbal community conditions both signer and viewer repertoires in its members.

In verbal–behavior analysis, we tend to de–emphasize the "listener," along with the viewer of signs and gestures, because the stimuli from a speaker affect a listener's behavior in ways essentially similar to the ways that any non–verbal stimulation affects a listener or viewer. However, the very notion of verbal behavior focuses attention on the behavior of speakers, along with the behavior of gesturers and signers. Basically, the speaker's behavior is verbal while the listener's behavior often is not verbal; listener behavior responds to another's verbal behavior but otherwise need not itself be verbal. For convenience we often simply speak of the "speaker." Most points that we make, however, usually apply equally well to writers, signers, and gesturers. None of these imply inner agents; these terms *only refer to the body that mediates* (i.e., to the physiology the working of which mediates) *the response* due to current contingencies.

Indeed, you may have noticed that the term, *mediate,* can cause a little confusion, because we use it two ways. Long ago we described bodies (i.e., nervous systems) as mediating behaviors. In this chapter we also describe listeners as mediating the reinforcers of a speaker's behavior, which is what makes the latter verbal behavior. For the key to unlock any confusion, consider that while *bodies mediate behavior, listeners mediate reinforcers* through the behaviors that "their" bodies mediate.

Furthermore, speakers and listeners interact, with the contingencies on each party sometimes producing speaker behavior and at other times producing listener behavior. For example, a conversation might proceed this way: Circumstances evoke a question from party A (a speaker) which evokes an answer from party B (a listener, then a speaker); this answer in turn evokes a comment from party A (now a listener, then a speaker) which evokes a question from party B (a listener, then a speaker) which evokes an answer from party A (a listener, then a speaker) and so forth.

Such interactions show the speaker and listener repertoires coexisting as functionally different and analytically separate repertoires. Still, they become virtually inseparable when stimuli evoke the neural behavior of verbal thinking, which we can describe as covert *simultaneous* speaking–listening.

On a terminological note, Julie and Ernie Vargas (Skinner's daughter and her spouse) in their verbal behavior courses at West Virginia University in the 1990s, replaced "speaker" with "verbalizer," which better addresses all the non–vocal forms of verbal behavior. They also replaced "listener" with "mediator," which better addresses the functional role of mediating reinforcement delivery following the verbalizer's behavior. Given the traditions in which the "speaker" and "listener" terms arose, these newer terms may carry fewer agential implications. While "speaker" and "listener" remain the common terms at present, verbalizer and mediator may become the common terms in the future.

(5) Audience controls. In terms of overt responding, while particular variables control the form of a given verbal response, a general contextual variable controls whether or not any verbal behavior occurs at all. We call this general variable the audience. Without an audience, without a listener, present, no overt verbal response manifests. For example, if you are the last to leave the classroom but cannot open the door due to carrying a large stack of books, no verbal behavior occurs that requests someone to open the door; instead, the current circumstances evoke some other door–opening solution. However, the presence of someone else, the presence of an audience, leads to a verbal door–opening solution. In this sense, the presence of an audience serves as a function–altering stimulus. An audience alters the function of the other relevant stimuli from neutral to evocative of a *verbal* response, while the form of the verbal response depends on the particular controlling stimuli. With an audience present, a closed door that you cannot get open evokes a "Please open the door" request. With an audience present, for example a parent at the zoo, a tiger emerging from a den evokes a child's "Tiger!" statement.

Of course, the automatic establishment of a listener repertoire during the conditioning of a speaker repertoire leaves people, at the covert level of verbal thinking responses, as their own reinforcing listener. Thus, we consider that an audience essentially remains ever present. The general lack of mediation of substantive reinforcers, which requires a separate audience/listener, appears to balance the lower level of energy expenditure associated with covert responses.

The control that an audience exerts also affects other aspects of verbal responding. For instance, the audience controls which parts of a repertoire a particular stimulus evokes. For example, if a physician is the audience, then the discoloration on a patient's leg evokes a "contusion" response from the nurse, but if the patient is the audience, then the discoloration evokes a "bruise" response from the nurse.

We also find that *places* exert an audience–control kind of stimulus control. For example, along with other variables, the local classroom produces a different part of a student's verbal repertoire than the local theater produces or the local pub produces; students talk about different things in those places. They also talk in different ways in those places because, along with other variables, places exert control over other response characteristics. Generally classrooms produce medium sound levels of verbal behavior while theaters produce whispered levels and pubs produce boisterous levels.

Those first five VB–analysis characteristics apply more to verbal behavior in general. The remaining four pertain more to our subsequent discussion of elementary verbal operant relations.

(6) *Responses and response products.* Our upcoming analysis of elementary verbal operants calls for a level of detail that requires a distinction between a response and what we call the *response product*. As we have pointed out before, the occurrence of behavior constitutes a real event that can function as a stimulus; in many cases the response also produces direct and immediate changes in the environment. The term *response product* applies both to such changes and to the stimulus status of responses. Response products are the stimuli that the responses produce.

For example any response may leave one or more response products. Innervated changes in the vocal musculature result in the changes in air–wave patterns that function as auditory stimuli, especially for others; these auditory stimuli are the response products of speaking. Longhand writing involves arm movements that not only produce visual stimuli, especially for others—we see someone writing—but the arm movements also produce marks on paper; both the visible arm–movement stimuli and the visible marks on the paper are response products of the writing responses.

This distinction between responses and response products helps us categorize some formally controlled verbal operants. Formally...? What? "Formal control" refers to one of two general kinds of verbal operant controls. The other is "thematic control," and next we will consider both of them.

(7) *Thematic and formal controls.* When we begin to analyze and categorize verbal operants, we first sort them as either under a *thematic* kind of control or under a *formal* kind of control. While we take "thematic" directly from *themes*, we use "formal" in its *structural* connotation. Thus we call one kind of control *thematic,* because we can generally trace some sort of theme connecting the verbal response with either its controlling variable or its consequence. Meanwhile we call the other kind of control *formal,* because we can trace some fairly explicit structural relations or similarities between the controlling stimulus and the verbal response or response product.

Note, however, that themes can seem arbitrary or ambiguous. So we use the presence or absence of a basic structural feature, the feature that we call point–to–point correspondence (which we discuss next) as the dividing line between thematic and formal controls. We consider verbal operants that lack point–to–point correspondence as under thematic control and verbal operants that have point–to–point correspondence as under formal control.

Consistency across some dimension of stimulus and response establishes a theme, which we name for that to which it pertains. Here are some examples of verbal operants under *thematic* controls. You say "water"—and receive some— after playing soccer for half an hour, which is a theme of getting what you said. You say "dog" when you see a dog, which is a theme of the same thing seen and said. And you say "lights" when you hear someone else say "phone, gas, and…," which is a theme of utilities.

For comparison here are some examples of verbal operants showing *formal* controls (i.e., structural controls, with at least point–to–point correspondence between the stimulus and the response, which we discuss next). You say "bus" when you see the written or printed word *bus.* You say "apple" when you hear the word "apple." And you make the ASL (American Sign Language) sign for cat when you see another signer make this sign. (After the next several pages, you will be able to categorize each of these verbal operants.)

(8) Point–to–point correspondence. When we examine verbal stimuli and verbal responses, we find that they contain parts that by themselves can control other responses in a manner that we describe as *point–to–point correspondence.* That is, each part of the stimulus controls a corresponding part of the response. Here is an example of this controlling correspondence. The three parts of the written–word stimulus *dog* (which is the response product of someone's writing behavior) can control other responses including your saying "dog," or its letters "dee" and "oh" and "gee," or its phoneme sounds. Also, when someone says "dog," that auditory stimulus (which is the response product of someone's vocal musculature movements) can control other responses including your writing *dog* (i.e., *d* and *o* and *g*). In all these cases, you see or hear the parts of the stimulus (written *dog* or spoken "dog") corresponding to, and controlling, the parts of your responses (spoken "dog" or written *dog*). Again, we call this correspondence *point–to–point correspondence,* and it describes some details of the independent and dependent variables of verbal operants, details that help us better organize and categorize verbal operants, and thereby deal more effectively with them.

(9) Formal similarity. Note that none of those stimuli and responses that have point–to–point correspondence are in the same sense mode; they are

either spoken then written or written then spoken. When stimuli and response products that have point–to–point correspondence are *also in the same sense mode* (e.g., they are both spoken, or both written, or both signed) they share something more. They share a kind of physical, structural similarity on a part by part basis that we call *formal similarity.* The parts of the stimulus not only correspond point–to–point with the parts of the response, but each part of the stimulus physically, structurally (i.e., formally) resembles the corresponding part of the response product. Here is an example of this similarity that could also use dogs, as in our point–to–point correspondence example, but instead uses cats. The written–word stimulus *cat* controls your writing–response parts that produce the written word *cat* again. This control occurs through the three parts of the written–word stimulus *cat* (i.e., the written letters *c, a,* and *t*) not only having point–to–point correspondence with, but also having a structural, formal similarity to, the three parts of your writing response *c–a–t.* Similarly the three parts of the auditory stimulus "cat," (i.e., the sounds from the "c," "a," and "t") control your saying "cat," not only through point–to–point correspondence with, but also formal similarity to, the three parts of your vocal response "c–a–t." The same analysis applies to the arm–and–hand movements in the ASL sign for cat when it then controls the same arm–and–hand movements of another signer signing cat. When a stimulus and response product not only show point–to–point correspondence but also are in the same sense mode, they have *formal similarity,* which describes some further details of the independent and dependent variables of verbal operants, details that also help us better organize and categorize verbal operants and thereby deal more effectively with them.

Let's now turn to those verbal operant relations. As we go through them, we will see several of our verbal–behavior analysis characteristics in practice, including the functions of responses and response products, thematic and formal controls, point–to–point correspondence, and formal similarity.

Elementary Verbal Operant Relations

With those nine verbal–behavior characteristics in hand, particularly the last four, let's turn to some major categories of verbal behavior. Here we cover the elementary relations that we call *mands, tacts, intraverbals, codics,* and *duplics,* along with various subtypes in the *codic* and *duplic* categories.

For each verbal–behavior category, we provide the name and main features. Each category, however, includes far more than merely a behavior. Each category comes from the combination of (a) a characteristic kind of process or evocative stimulus, (b) certain kinds of verbal responses, (c) a characteristic kind of reinforcing stimulus, and (d) often additional characteristics as well. Furthermore these features form functional relations with each other, generally operant functional relations although respondent functional relations are also involved. Due to the combination of features that each verbal category includes, we refer to each category as a relation, *a verbal operant relation.* And here we discuss the most basic of such relations, the *elementary verbal operant relations.* Where appropriate our descriptions will refer to one or another of the verbal–behavior characteristics that you mastered from our earlier discussion.

The order in which we cover the elementary verbal operant relations is somewhat arbitrary. We could present the relations in the order in which verbal–community contingencies tend to condition them (i.e., during the agentially labeled phenomena comprising "language acquisition"). This conditioning order, however, can require consideration of extensive and interesting concerns that remain beyond, or peripheral to, our present discussion; we reserve such details for others to provide. From among several other approaches to the order of coverage, we take a strictly pedagogical approach that only gradually increases the number of relevant verbal–relation characteristics because, after all, this is likely your first contact with the natural–science account of verbal behavior. Thus, we will follow a flowchart technique in which a series of four "yes/no" questions leads us through the pertinent analytical characteristics of each verbal–operant–relation category. Our flowchart, which only covers those verbal relations that we describe, appears across the two halves of Chart 20–1, which we label Chart 20–1a and Chart 20–1b.

Recall that non–verbal behavior and verbal behavior differ solely in terms, respectively, of the directness and indirectness of the consequence occurrence. Also recall that all the previous chapters covered non–verbal behavior and many variables of which it is a function. Now, needing only those same variables, we will cover the categories of elementary verbal operant relations. Again, we call them *mands, tacts, intraverbals,* various types of *codics,* and various types of *duplics.* For each verbal–behavior–relation classification, while we use these same names for the responses, we also cover three related factors. We begin with the type of controlling variable (e.g., an establishing operation, a non–verbal evocative stimulus, or a verbal evocative stimulus) and, if relevant, its sense mode (e.g., vocal or visual). Then we consider the possible format of the verbal response (e.g., vocal, written, signed). Lastly we cover the type of mediated consequence (e.g., a response–specific type, or a general type). Of course, we will also review some examples, many of which complete the specifics of examples that we used earlier in the chapter.

Mands

In Chart 20–1a, the first question that we ask concerns whether or not an evocative stimulus, either verbal or non–verbal, controls the response. If the answer is "No" (i.e., if neither a verbal, nor a non–verbal, evocative stimulus controls the response) then we call the response a *mand.*

We use the term *mand* (which derives from com*mand* or de*mand*) for verbal relations that processes like *establishing operations* (e.g., deprivation or aversive stimulation) bring about in the sense of control. Mands can occur in vocal, written, or signed formats. And the mediated consequences for mands involve the *specific type* of stimulus that the mand specifies. Given these characteristics, including that mands lack point–to–point correspondence, we list mands as verbal operants under thematic control.

Here is an example. A closed classroom door that you cannot open due to all the books you are carrying (which is an establishing operation) produces an "open door" mand response (which may take a form that includes additional types of verbal behavior that may go beyond our VB introduction, such as "Please open the door"); the door opening, through the mediation of another person, provides the reinforcer for this mand response, a reinforcer that the mand itself specifies, a reinforcer that shapes and maintains mand behavior.

That example contains all three of the requisite characteristics for a mand: an establishing operation, a verbal response that can be vocal, written, or signed, and a mediated reinforcer that the mand specifies. All mand examples contain these three characteristics. You can trace the three characteristics in this example: Receiving a message as you leave the theater that your ride cannot pick you up is an establishing operation that controls the mand "Taxi!" In most big–city entertainment districts, this mand results in a taxi stopping near you (which is the mand–specified reinforcer).

Again, in Chart 20–1a, the first question that we ask concerns whether or not any evocative stimulus, verbal or non–verbal, controls the response. If the answer is "No," then the response is a mand. If the answer is "Yes," then we must ask another question, which leads us to tacts.

Tacts

In Chart 20–1a, the second question that we ask concerns whether or not the evocative stimulus is a verbal stimulus. If the answer is "No," then we call the response a *tact*.

We use the term *tact* for verbal relations that *non–verbal stimuli* control in the sense of evoke. Tacts can occur in vocal, written, or signed formats. And the mediated consequences for tacts involve general and non–specific reinforcers. Given these characteristics, including that tacts lack point–to–point correspondence, we also list tacts as verbal operants under thematic control. Let's take some earlier examples and be more precise with them.

At the zoo, when the tiger compound seems empty, the appearance of a tiger, which is a non–verbal stimulus, emerging from a hidden den entrance, evokes a "Tiger!" tact response; a comment, such as "Good spotting," that another person mediates, then provides the kind of general and non–specific reinforcer that shapes and maintains tact behavior.

That example contains all three of the requisite characteristics for a tact: a non–verbal evocative stimulus, a verbal response that can be vocal, written, or signed, and a general, mediated, and non–specific reinforcer. All tact examples contain these three characteristics. You can trace the three characteristics in these two examples: For one example, a cookie that is left on the plate, after

Chart 20–1a

A Guide to Elementary Verbal Operant Relationships
(These are under thematic control)

With three of our four questions, this chart covers the three
verbal relations that are under thematic control.

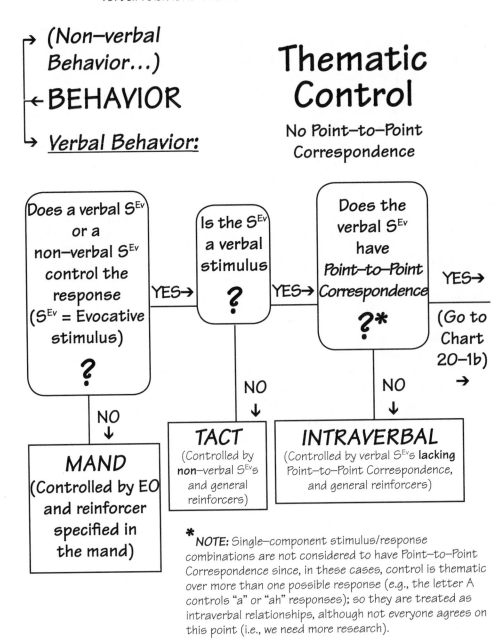

→ (Non–verbal
   Behavior...)

←BEHAVIOR

↳ *Verbal Behavior:*

Thematic Control

No Point–to–Point
Correspondence

Does a verbal S^{Ev}
or a
non–verbal S^{Ev}
control the
response
(S^{Ev} = Evocative
stimulus)

?

YES→

Is the S^{Ev}
a verbal
stimulus

?

YES→

Does the
verbal S^{Ev}
have
Point–to–Point
Correspondence

?*

YES→

(Go to
Chart
20–1b)
→

NO
↓

NO
↓

NO
↓

MAND
(Controlled by EO
and reinforcer
specified in
the mand)

TACT
(Controlled by
non–verbal S^{Ev}s
and general
reinforcers)

INTRAVERBAL
(Controlled by verbal S^{Ev}s **lacking**
Point–to–Point Correspondence,
and general reinforcers)

***NOTE:** Single–component stimulus/response
combinations are not considered to have Point–to–Point
Correspondence since, in these cases, control is thematic
over more than one possible response (e.g., the letter A
controls "a" or "ah" responses); so they are treated as
intraverbal relationships, although not everyone agrees on
this point (i.e., we need more research).

Chart 20–1b

A Guide to Elementary Verbal Operant Relationships (These are under formal control)

With our fourth and last question, this chart covers the two verbal relations, and their five main subtypes, that are under formal control.

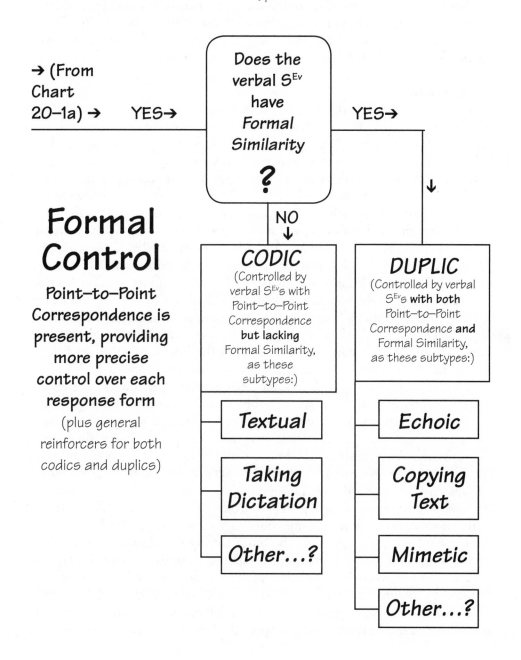

→ (From Chart 20–1a) →   YES→

Does the verbal S^{Ev} have *Formal Similarity* **?**

YES→

NO ↓

↓

Formal Control

Point–to–Point Correspondence is present, providing more precise control over each response form

(plus general reinforcers for both codics and duplics)

CODIC
(Controlled by verbal S^{Ev}s with Point–to–Point Correspondence **but lacking** Formal Similarity, as these subtypes:)

Textual

Taking Dictation

Other...?

DUPLIC
(Controlled by verbal S^{Ev}s **with both** Point–to–Point Correspondence **and** Formal Similarity, as these subtypes:)

Echoic

Copying Text

Mimetic

Other...?

a full meal with cookies for dessert, evokes a "last cookie" tact that produces some general, mediated, and non–specific reinforcer such as the attention from the implied compliment in "Ah, now those were good cookies." For another example the discoloration on a patient's leg evokes either a "contusion" tact in the presence of a doctor, or a "bruise" tact in the presence of a patient, and in both cases a general, mediated, and non–specific reinforcer, such as "I see," completes the verbal episode, or at least this part of it.

Again, in Chart 20–1a, the second question that we ask concerns whether or not the evocative stimulus is a verbal stimulus. If the answer is "No," then the response is a tact. If the answer is "Yes," then we must ask another question, which leads us to intraverbals.

Intraverbals

In Chart 20–1a, the third question that we ask concerns whether or not the evocative verbal stimulus has point–to–point correspondence with the response. If the answer is "No," then we call the response an *intraverbal.* This lack of point–to–point correspondence gives us a fourth characteristic in addition to the previous three of stimulus, response, and consequence types.

We use the term *intraverbal* (which implies a sort of connection *within* sets of verbal responses) for verbal relations that *verbal stimuli* control, in the sense of evoke, but that lack point–to–point correspondence. Intraverbals can occur in vocal, written, or signed formats. And, like tacts, the mediated consequences for intraverbals also involve general and non–specific reinforcers. Given these characteristics we also list intraverbals as verbal operants under *thematic* control.

Here is an example. Hearing someone say, "Red, white, and…," which is a verbal stimulus, evokes a "blue" intraverbal response; a subsequent comment, such as "That's right," or "The people in some countries would say 'green'," that another person mediates, then provides the kind of general and non–specific reinforcer that shapes and maintains intraverbal behavior.

That example contains all four of the requisite characteristics for intraverbals: a verbal evocative stimulus, a lack of point–to–point correspondence, a verbal response that can be vocal, written, or signed, and a general, mediated, and non–specific reinforcer. All intraverbal examples contain these four characteristics. You can trace the four characteristics in these two examples: For one example, after some experiences, seeing the words "Lake Tahoe," perhaps on a brochure in a travel agent's window, evokes a "deep blue water" intraverbal that produces some general, mediated, and non–specific reinforcer such as the attention from a companion suggesting that you both take a trip to the lake. For another example the nurse tacting the discoloration on your leg as a "bruise" evokes your intraverbal response "contusion," (which may also be tacting the leg discoloration) that produces some general, mediated, and non–specific reinforcer such as the nurse, with raised eyebrow, replying "yes."

Notice that the four requisite characteristics that constitute intraverbals lead to some broad implications. You can apply the four–characteristics criterion to verify these implication statements: All definitions have intraverbal status. All translations between languages have intraverbal status. And those sense–mode changes (i.e., from spoken to written or written to spoken) that meet the other criteria have intraverbal status. Spoken or written to signing, and signing to spoken or written, already have intraverbal status, because these changes are translations between languages.

Again, in Chart 20–1a, the third question that we ask concerns whether or not the evocative verbal stimulus has point–to–point correspondence with the response. If the answer is "No," then the response is an intraverbal. If the answer is "Yes," then we must ask our last question, which moves us from thematically controlled verbal operants to formally controlled verbal operants, beginning with codics (in Chart 20–1b).

Codics

In Chart 20–1b, the fourth and last question that we ask concerns whether or not the evocative verbal stimulus, which has point–to–point correspondence with the response, *also* has formality similarity with the response product. If the answer is "No," then we call the response a *codic* of usually the *textual* or *taking dictation* subtypes. For codic relations the addition of point–to–point correspondence maintains a fourth characteristic in our analysis while the lack of formal similarity gives us a fifth characteristic.

The term *codic* derives from the words "code" or "encode," and clarifies the similarity between its subtypes while also differentiating them from the subtypes that come under *duplics.* Let's consider each of the two codic subtypes, and also the related complexities of (a) textuals versus reading, (b) intraverbal reading, and (c) intraverbal writing–what–one–hears.

Textuals. We use the term *textual* for the codic verbal relations in which *written verbal stimuli* control, in the sense of evoke, *vocal responses,* and these share point–to–point correspondence but lack formal similarity. The mediated consequences for textuals also involve general and non–specific reinforcers that shape and maintain textual behavior. Given these characteristics, we list textuals as verbal operants under formal control.

Here is an example. When a parent or teacher is conditioning more vocal responses to marks on a page, the marks *See Jane run* evoke a youngster's textual responses, "See Jane run." A comment, that the parent or teacher mediates, such as "You said each word correctly," then provides the kind of general and non–specific reinforcer that shapes and maintains textual behavior. Note that the "marks" on the page can be not only the visual stimuli of printing or handwriting but also the tactile stimuli of raised Braille dots.

That example contains all five of the requisite characteristics for a textual relation: a written verbal evocative stimulus, a vocal response, point–to–point correspondence between them, but no formal similarity between stimulus and response product, and a general, mediated, and non–specific reinforcer. All textual examples contain these five characteristics.

Textuals versus reading. Textual behavior differs from reading. Before turning to the codic subtype of taking dictation, let's consider some of the differences between textuals and reading. As we saw with Sidman's research in our equivalence–relations chapter, reading is more than textual behavior. The term *textual* describes the verbal relation when written words merely control spoken responses, but the term *reading* describes the result when further conditioning, connecting equivalence relations, adds the stimulus and response classes to which the term *comprehension* applies. After conditioning builds equivalence relations among pertinent stimulus classes beyond textual relations, further conditioning connects the equivalence classes, typically combining them over several years of informal family reading fun, and formal instruction, in an expanding explosion of stimulus control by printed words over spoken words. New readers then increasingly *comprehend* (a neural behavior of consciousness) these overt or covert spoken words, the kind of textual–plus–comprehension behavior that we call reading behavior.

Intraverbal reading and more. Reading behavior involves textual relations in essentially all alphabetic languages including, to some extent, Chinese pin–yin. Textual relations, however, are not involved in reading Chinese characters.

Indeed textual codic relations, as well as the taking–dictation codic relations that we discuss next, are not really possible with languages with non–alphabetic writing systems such as those that use pictures, ideograms, or characters, because point–to–point correspondence, between the controlling stimulus and the response that it evokes, is lacking. Consequently (and checking with our Chart 20–1) reading, and writing what one hears, with non–alphabetic writing systems, generally come under intraverbal relations (with implications for cultural and verbal–community conditioning and education that go beyond our purview here). Fraley (2008) compares alphabetic reading and character reading this way:

> When reading English, various combinations of 26 preconditioned letters and 40 or so preconditioned phonemes evoke word–sounds to which the reader is, in many cases, already conditioned for comprehension at that level. In character languages, the behaviorally defined integrals get no smaller than a word, so to arrive at an opportunity for a word–comprehension response a reader cannot first probe for the word by behaving its constituent elements to a completed sequence. In a character language, the behaving of the word as an integral simply has to be evoked by its character... (p. 982).

Taking Dictation. We use the term *taking dictation* for the codic verbal relations in which *auditory verbal stimuli* control, in the sense of evoke, *written responses,* and these share point–to–point correspondence but lack formal similarity. The mediated consequences for taking dictation also involve general and non–specific reinforcers. Given these characteristics, we list taking dictation as verbal operants under formal control.

Here is an example. When you and a friend try a new recipe, you first check it for any ingredients that your cupboard lacks. As you report these missing ingredients to your friend, the verbal stimuli that your reporting produces evoke your friend's taking–dictation written responses onto a shopping list. A comment, such as "The list seems complete," that you mediate, then provides the kind of general and non–specific reinforcer that shapes and maintains taking–dictation behavior. Note again that the taking–dictation written responses can produce not only the visual stimuli of printing or handwriting but also the tactile stimuli of raised Braille dots.

That example contains all five of the requisite characteristics for taking dictation: an auditory verbal evocative stimulus, a written verbal response, point–to–point correspondence between them, but no formal similarity between stimulus and response product, and a general, mediated, and non–specific reinforcer. All examples of taking dictation contain these characteristics.

Other codics? At present other forms of the codic relation remain rare. Regarding a written form of ASL that William Stokoe developed, Peterson (1978) says, "Rather than the letter corresponding to speech sounds, the 'letter' corresponds to significant features of the signs, and therefore, has point–to–point correspondence that allows for relationships analogous to textual behavior, copying a text [one of the duplic relations that we cover next], and taking dictation. Although these relationships are possible, they have not become common place in the deaf community" (p. 83).

Again, in our Chart 20–1b, the fourth and last question concerns whether or not the evocative verbal stimulus, which has point–to–point correspondence with the response, *also* has formal similarity with the response product. If the answer is "No," then the response is a codic, usually of one of the subtypes, textual or taking dictation. If the answer is "Yes," then we need no more questions; if the answer is "Yes," then the response is a duplic.

Duplics

With a "Yes" answer to the last question in Chart 20–1b, we consider relations featuring both point–to–point correspondence and formality similarity. We then call the response a *duplic* of usually the *ehcoic, copying text,* or *mimetic* subtypes. In these relations the evocative verbal stimulus not only has point–to–point correspondence with the response but also has formal similarity with the response product. For duplic relations the addition of formal similarity maintains the fifth characteristic of our analysis.

The term *duplic* derives from the word "duplicate," which specifies the kind of response–mode similarity that the duplic–relation subtypes each share between their evocative stimuli and their responses and response products. Let's consider each of the three duplic subtypes in turn.

Echoics. We use the term *echoic* for the duplic verbal relations in which *auditory verbal stimuli* control, in the sense of evoke, *vocal verbal responses,* and these share both point–to–point correspondence and formal similarity.

The mediated consequences for echoics also involve general and non–specific reinforcers. Given these characteristics we list echoics as verbal operants under formal control.

Here is an example. A young child, who has just received a cookie, says "cookie" after hearing a parent say "Cookie?" to a sibling. A comment, such as "Yes, this is a cookie," that the parent mediates, then provides the kind of general and non–specific reinforcer that shapes and maintains echoic behavior.

That example contains all five of the requisite characteristics for echoics: an auditory verbal evocative stimulus, a vocal verbal response, point–to–point correspondence between the stimulus and the response, formal similarity between the stimulus and the response product, and a general, mediated, and non–specific reinforcer. All echoic examples contain these characteristics.

Copying Text. We use the term *copying text* for the duplic verbal relations in which *written verbal stimuli* control, in the sense of evoke, *written verbal responses,* and these share both point–to–point correspondence and formal similarity. The mediated consequences for copying text also involve general and non–specific reinforcers. Given these characteristics, we list copying text as verbal operants under formal control.

Here is an example. The written words, *Behavior components of solving global problems,* which a professor has put on the board, evoke a student's writing the words *Behavior components of solving global problems.* A comment, such as "nice note taking," that the professor mediates, then provides the kind of general and non–specific reinforcer that shapes and maintains copying–text behavior.

That example contains all five of the requisite characteristics for copying text: a written verbal evocative stimulus, a written verbal response, point–to–point correspondence between the stimulus and the response, formal similarity between the stimulus and the response product, and a general, mediated, and non–specific reinforcer. All copying–text examples contain these characteristics.

Common copying–text relations involve printing or writing stimuli evoking writing responses. Again, however, copying–text relations can also involve the stimuli that come from the raised dots of Braille writing; when these evoke Braille–writing responses, we still tact the relation as copying text. For the sake of some continuity, also recognize that when written Chinese pin–yin or Chinese characters evoke writing the same thing in pin–yin or characters, respectively, the relations are copying–text relations, while characters evoking pin–yin responses, or vise versa, involve intraverbal relations.

Mimetics. We use the term *mimetic* for the duplic verbal relations in which *signed verbal stimuli* control, in the sense of evoke, *signed verbal responses,* and these share both point–to–point correspondence and formal similarity. The mediated consequences for mimetics also involve general and non–specific reinforcers. Given these characteristics, we list mimetics as verbal operants under formal control.

Here is an example. In a course on ASL, a professor making the sign for cat evokes a student's making the sign for cat. A comment, such as "that's correct," that the professor mediates in English, contingent upon the student's correct ASL sign, then provides the kind of general, mediated, and non–specific reinforcer that shapes and maintains mimetic behavior.

That example contains all five of the requisite characteristics for mimetics: a signed verbal evocative stimulus, a signed verbal response, point–to–point correspondence between the stimulus and the response, formal similarity between the stimulus and the response product, and a general, mediated, and non–specific reinforcer. All mimetic examples contain these characteristics.

Other duplics? Other duplic relations are available. For instance when your finger spelling of a word evokes someone else's response of finger spelling the same word, we have a relation that we would tact as duplic, but we lack a more specific tact for this kind of duplic subtype. By the way, Jack Michael introduced the terms codic, duplic, and mimetic (see Michael, 1982).

While helpful in many respects, Chart 20–1 leaves out many other interesting points. For instance our analysis regularly leaves us tacting the same word form as one verbal relation at one point and another verbal relation at another point. Since this can cause confusion, let's look more closely.

Same Forms but Different Repertoires

Any and all of those verbal relations occur in normal interactions between and among people. Your everyday conversational experience provides you with nearly unlimited examples. Often the verbal responses take different word forms, but sometimes they take the same word form. Must "same form" responses constitute the same verbal relation? For example, must "coffee" in different contexts always be the same verbal relation? With all the material we just went through, you probably made a pretty distinct "No!" response; the same word form need not constitute the same verbal relation, and this helps us see all the verbal relations as involving different repertoires, which the verbal community must condition separately. For example the verbal community must condition coffee as a mand in one stimulus context, as a tact in a different stimulus context, and so on. Contrast this with the ordinary linguistic view that classifies coffee as a vocabulary word in the structural category of nouns that the verbal community supposedly need only condition once, thereby making it available in all contexts regardless of the function of this word.

Which of those alternatives holds in reality is an experimental question, and so far the data support the separate–repertoires alternative. For example say you walk into the hospital room of a friend who recently suffered a head injury. And you are carrying a cup of steaming dark liquid with a distinct and strong aroma. These stimuli evoke the verbal response, "coffee" from your friend. Assuming that your cup contains coffee, is this "coffee" response a tact of the type of liquid or a mand for some of the liquid? We cannot tell from the topography of the verbal response. While the word "coffee" is the same word form in both cases, the verbal relations, in this case tacts and mands, stand as separate repertoires. If your friend usually drinks only tea, then most likely his "coffee" response tacts the type of liquid. However, if your friend drinks only coffee and lots of it, but last drank some yesterday, then most likely his "coffee" response mands some of your liquid.

The presence or absence of different verbal relations with the same word form can sometimes provide particularly important information. Consider your brain–injured friend again. Let's say he manded coffee and you provided several cups thereby leaving him temporarily coffee satiated. Yet upon returning several minutes later with your own cup of coffee, you not only find him showing no interest in your coffee but you also discover that, when he sees and sniffs your coffee, he cannot tell you what is in the cup; he cannot tact the coffee. The presence of the mand repertoire along with the absence of the tact repertoire may provide important clues to your friend's physicians not only about the kind or location or severity of brain injury he received, but also perhaps about appropriate treatments.

Elementary Relations Summary and Conclusion

While the patterns of characteristics that we covered regarding all those verbal operants remain salient, let's summarize them in reverse order and only mention the single most critical characteristic that distinguishes each verbal–operant type from the others. Verbal stimuli with *both* point–to–point correspondence with the response, *and* formal similarity with the response product, evoke duplics (e.g., echoics, copying text, mimetics). Verbal stimuli with only point–to–point correspondence (but no formal similarity) evoke codics (e.g., textuals, taking dictation). Verbal stimuli without point–to–point correspondence (which implies no formal similarity either) evoke intraverbals. Non–verbal stimuli evoke tacts. And establishing operations produce mands. For practice, consider Morse code. How does it relate to each of these elementary verbal operants?

All those elementary verbal operant relations that we discussed, and various more complex ones, constitute verbal behavior, because they have their reinforcing effect on the environment *indirectly;* they produce reinforcement through someone else's behavior. Our discussion also included the general kinds of evocative and consequential stimuli that participate in a verbal community's conditioning of these fundamental components of language.

With those basics in hand, we can start considering applications of our verbal–behavior analysis to the task of building verbal repertoires. The analysis contains practical implications for improving language conditioning in at least three areas. (a) We can improve the conditioning of first–language repertoires for newborn verbal–community members over the first five years of life (e.g., see Hart & Risley, 1995, 1999). (b) We can also improve the conditioning of first–language repertoires for developmentally delayed and autistic verbal–community members, often by starting with the conditioning of a sign–language repertoire (e.g., see Maurice, 1993; also, see Maurice, Green, & Luce, 1996, and peruse the relevant research articles in *The Analysis of Verbal Behavior* journal). And (c) we can improve the conditioning of subsequent, non–native language repertoires, particularly in formal educational settings, which is the topic to which we now turn our attention.

VB Analysis and Teaching Languages

By focusing on improving access to the variables responsible for conditioning verbal behavior, language teachers can improve the conditioning of non–native language repertoires (i.e., "foreign" languages). Here we provide a few possibilities along these lines. While teaching language courses in China in 1979 (at Xi'an Jiaotong University) and again in the early 1990s (at Xi'an Foreign Languages Universtiy) I began developing these possibilities for presentation at some verbal–behavior workshops that I first presented in China in the early 2000s. These possible practices would benefit from more methodologically sound research. So I encourage anyone involved in practices like these to implement them with the kind of sound methodology that supports reporting the outcomes in the behaviorology and language–education literatures.

Let's start with the point of intraverbal vocal or written translations. What should the translator's objective be? Verbal–behavior analysis suggests that the object of translation is more than merely matching word for word or concept for concept, because word forms function in multiple, not single, repertoires. Translation is also more than merely achieving the same effect on the listener or reader, such as getting them to laugh at jokes. Given the difficulty in translating jokes, is telling a different joke in the other language, a joke that also evokes a laugh, really translation? While such questions may remain, verbal–behavior analysis suggests that the object of translation is, as much as possible, to get the listener or reader to behave not only in the same way *but also for the same reasons* (i.e., under control of the same variables) that the listener or reader in the original language behaves.

Imitate Native–Language Conditioning Practices

Verbal community members provide years of round–the–clock conditioning efforts with new members, using mostly coincidental and uncoordinated practices. Nevertheless they achieve substantial success in conditioning basic but fluent native–language speaking repertoires before children go to school. At school most of the conditioning occurs regarding reading, refining speaking fluency, and studying *about* language (i.e., about grammar and syntax and so forth). Why invest all this effort? Verbal communities condition new verbal members not merely for the new members' sake, although new members clearly benefit from the increasing, verbal–behavior induced, effectiveness they experience in dealing with the world around them. The community conditions new members for the sake of the verbal community itself, as it can then interact with new members on a more sophisticated level than non–verbal behavior alone allows. This kind of macrocontingency outcome contributed to the rise of language in the first place, and now it maintains the cultural practice of native–language conditioning, and supports the cultural practice of education, including non–native language conditioning.

More importantly for non–native language conditioning, notice the native–language conditioning sequence: first condition at least somewhat fluent speaking repertoires and then, while refining speaking fluency, work on reading and studying *about* the language. By using scientifically informed and systematically managed methods that follow the same sequence, language teachers can probably more successfully, effectively, and efficiently generate and maintain fluent and extensive non–native language repertoires.

Also, native–language conditioning follows a fairly typical order in producing all the basic, separate and fluent verbal repertoire components. While the reasons for this continue as research topics, the order begins with echoics and then moves on through mands, tacts, intraverbals, and the audience relation. Non–native language conditioning can benefit from following the same native–language repertoire–component conditioning order. While the best conditioning usually occurs outside the classroom, especially for mands, tacts, and intraverbals, the conditioning of these relations continues along with more typical classroom conditioning of the various codic and duplic relations. All these then continue and overlap with more complex verbal relations—ones for which the details go beyond this introductory chapter—including several kinds of mand and tact extensions, and autoclitics, which are verbal behavior about verbal behavior.

As those repertoires accrue, they not only overlap but each one also becomes available, under appropriate stimulus conditions, for combination with, and recombination in, the more common and ever more complex and more extensive verbal response forms (e.g., sentences, paragraphs, conversations, speeches, stories, plays, books). As a result fluent verbal repertoires can seem like magic even in the hands, or rather, mouth, of an inept wordsmith, not to mention the magic–like effects from a well–conditioned one. Nevertheless these repertoires, and all the variables of which they are a function, remain entirely natural.

New Practices for Conditioning Non–Native languages

The conditioning of each verbal relation never really stops; it continues—and involves equivalence processes—during the conditioning of each and all of the subsequent relations. That is, the conditioning program adds new forms to previously conditioned relations while also conditioning new relations.

In general, non–native language conditioning starts simply and proceeds gradually. It starts with single words and shapes more complex responses. For all vocabulary words, begin each as an echoic and then, as appropriate, recondition a word as a mand, then as a tact, then as an intraverbal, and so on. Repeat the cycle with more elaborate, more complete, more common forms.

For example, while the word "book" begins as an echoic, after perhaps a year or two of conditioning a wide range of words and relations, an accurate verbal response, in the presence of a particular book, could be, "This green book discusses behaviorology in a readable style, but I have not yet finished it."

Echoic methods. Non–native language conditioning easily begins in the classroom where students receive instruction about the conditioning methods they will experience because, as fluent *native*–language speakers, this instruction enhances their rule–governed participation in the language–conditioning process. During the interval of instruction–about–instruction, all the usual classroom methods that build echoic repertoires (e.g., imitation and repetition) can begin. These then continue, and overlap, while the work begins on the next repertoire regarding mands.

Mand methods. To condition mands—and tacts and intraverbals as well—make use of already available situations and settings in which manding normally occurs, such as games, shopping, and both formal and informal meals. If you lack sufficient access to such circumstances, then organize your own real ones; prompt students to pretend them in real space or, as a last resort, in imagination. Also, arrange for students to deliver the actual reinforcer characteristic of each mand, or have them explicitly pretend to deliver these. The more closely these instructional settings mimic the real thing, including involving real establishing operations (e.g., a few hours of food deprivation for meal settings, such as the usual time between meals) the more successful the mand conditioning will be, more quickly building more fluency.

Tact methods. To condition tacts, use methods similar to mand conditioning methods. Hold up real objects, or point to characteristics of objects. You may need to model the correct tact to evoke an echoic response from students first. While you might start by manding each tact from students (e.g., "What is this?") this conditioning can easily generalize, leaving them tacting whatever you indicate by pointing your finger. Also point to actions or relations or events so that these more subtle phenomena can evoke appropriate tacts from students, tacts that you can then reinforce. Point at all of these inside the classroom, then start pointing at all of these outside the classroom. For example look out the window and point to the tiny, blue bird flying down from a big, lush, dark–green pine tree to a small lilac bush with a few spring–green leaves budding out (which just happened outside my window; you point to whatever is outside your window). Then ask the students (i.e., mand information) about what just happened, which evokes their tacting as many things as possible. As their repertoires expand, make a contest of which small student group can accurately string the most tacts together (and, at an even more advanced stage, students can write out their tacts and compare them). In this kind of situation, require students to provide several tacts from the at least 21 in the example: four things (bird, tree, bush, leaves) and ten characteristics (tiny, blue, big, lush, dark–green, pine, small, lilac, few, spring–green) and two events (flying, budding) and five relations (down, from, to, out, with).

Reality is just everywhere full of things and events and characteristics and relations and more, any and all of which students can tact. And if you want to teach some tacts that are not available in your accessible environment, then use the product of some recording device (e.g., a drawing, photo, sound, or video).

Intrverbal methods and codic and duplic methods. To condition intraverbals, use methods similar to mand and tact conditioning methods. The range of available evocative stimuli may even be wider than for mands and tacts because, as we mentioned earlier, intraverbals include all definitions and translations. Generally conditioning builds the textual and taking–dictation codics, along with the copying–text duplic, by the instructional methods for reading and writing.

Additional Verbal–Community Methods

Here are more possible methods especially adaptable to foreign–language programs in large educational settings. The setting size is relevant, because these methods work best with verbal communities, for each language, that are larger than a single classroom full of students.

General verbal–community value. A university could have a languages department with many hundreds of majors across the many languages that the department teaches, including Chinese, English, French, German, Italian, Japanese, Korean, Portuguese, Russian, Spanish, and so on. Each language likely has a reasonably large verbal community. This might mean a lot of work to start up new scientifically sound methods and to coordinate them with the ordinary methods. Ultimately it means arranging for a linguistic culture that lasts beyond the tenure of any particular faculty member and beyond the graduation of any particular student. So, why would anyone bother?

Basically we should bother because methods involving verbal communities beyond the classroom can more rapidly build more fluent foreign–language repertoires than just in–class verbal communities. These outside–class efforts, however, receive little attention in educational circles that focus on the in–class efforts of individual faculty. The methods in this chapter help change that emphasis, with the long–term gains thoroughly offsetting the setup difficulties.

The normal value of mutual respectful correction. Verbal–community members gain most of their language fluency through the contingencies in corrective interactions with other community members, *all* of whom not only sometimes make mistakes but also, at other times, can help a fellow member with a correction. Corrections cannot come only through the teacher. Teachers can make mistakes also, so sometimes a student can—and must—help with a correction. For the greater good of the whole verbal community, teachers must reinforce such correcting behavior on the part of students, especially by shaping diplomatic correction forms. As part of this process, teachers may make mistakes as mands to probe for proper (i.e., non–coercive) corrections.

Corrections from students are also appropriate because teachers cannot catch all errors as they cannot always be present. More importantly, more community members benefit from more corrections when any member present participates in correcting errors when they occur.

However, the style of correcting makes a big difference. When a child makes a verbal–behavior error, the correction occurs in a very gentle, non–coercive fashion, often involving little more than a repetition of the correct form; this represents the best correcting style. For example, if the child says, "Billy goed home," then the parent simply says, "Yes dear, Billy went home." The parent's correct form often evokes the child's echoic repetition of the correction. This kind of non–coercive correction is the only kind that will work in foreign–language education in the long run, because any kind of coercive correction interferes with speaker, and possibly even listener, behavior. Consider how the results of coercion (i.e., escape, avoidance, and countercoercion) play out in negative ways in an academic, foreign–language verbal community. How many students never return for the next course in a new–language sequence under coercive conditions? Importantly, note that the non–coercive correction style usually requires some separate conditioning attention to generate and maintain it. Start by conditioning students (and teachers) that, after making mistakes, receiving correction is not only acceptable but also desirable, because corrections benefit everyone present and provide a major avenue to building fluency. A really capable correction–style repertoire can even make correction fun, especially for the listener. These kinds of interaction, with non–coercive corrections, can occur in any of the verbal–community venues to which we now turn our attention.

Verbal–community game rooms. At the typical three to five hours per week, the classroom provides too few opportunities for foreign–language practice. One way to increase these opportunities involves organizing a verbal–community game room. You can arrange a permanent venue operating on a set schedule, or you can schedule a good–sized room for two or three sessions per week, each of several hours duration. Include many tables that each seat four to six people, and keep a variety of board games on hand; options include versions of Clue, mahjong, chess, and Monopoly. Encourage your students to come in and have fun—you can even require some amount of attendance as a course assignment, although that will likely not remain a necessity for long. Also keep track of the intervals over which each student—and teacher, as teachers can have fun too—attends. The room would feature only two "house rules." *When in the game room, verbal–community members—both students and teachers—may only speak the foreign language that they are studying or teaching, and they must non–coercively help each other engage in verbal interactions correctly.* As the room's popularity increases, you may need to arrange the schedule so that only those studying a particular language are there at the same time, perhaps by scheduling different languages for different days. You can double the benefits, perhaps across more languages, by setting up another such room.

Verbal–community dining tables. Another venue for foreign–language practice involves the dining hall. Designate some tables as verbal–community tables. Place a sign on each table indicating which foreign language those using the table will speak; this also encourages the *community* part of verbal community (e.g., one sign could read, *"This is a Chinese–speaking verbal–community table")*. As students gradually experience the benefits of such shared verbal–community activities, they will increasingly participate without anyone needing to assign the activity. Indeed, their non–assigned attendance constitutes one of the two chief barometers of how well your verbal–community contingencies are operating, with the other barometer being the fluency progress you track for them in comparison with the traditional, solely classroom–based methods.

If you can, persuade some native speakers to participate by taking one of their meals each day or each week at an appropriate table; a free meal for these native speakers makes recruitment far easier. At these tables the native speakers would, along with everyone else, operate under the same house rules as the game rooms, that *when at the table, verbal–community members—both students and teachers—may only speak the foreign language that they are studying or teaching, and they must non–coercively help each other engage in verbal interactions correctly*. As more and more students become interested in taking their meals at these tables—and they will if the interactions remain non–coercive—you can convert more and more tables into verbal–community tables.

Verbal–community rooms, floors, wings, dorms. You can also designate some rooms in a dormitory as verbal–community rooms. These could be part–time or full–time common rooms for a particular language community, or a different language community each day, or any number of other arrangements. You could start with just a couple of hours each day and gradually shape longer and longer periods according to student usage, similar to the game room scheduling. These rooms would require users to follow the same two basic verbal–community house rules that cover the game rooms and dining tables.

Or (And?) you could arrange regular dorm rooms in which, for a half term or a full semester, the students living in the room stick as best they can to those same two basic verbal–community house rules. As more students come under functional control of the benefits of such rooms, you can convert more dorm rooms, then a whole dorm floor, a whole dorm wing, or a whole dorm.

Actually that notion is not new. Along with many others, one of my relatives experienced this kind of arrangement when she went to college. Although she had enjoyed studying some French in high school, she had no plans to go on in French. But then, when college started, she stayed in the college's "French House" where residents only spoke French in the common areas (e.g., kitchen, dining room, lounge). In the other parts of the house (e.g., bedrooms) the preferred but not required language was French. As part of the experience, of course, the students would politely (i.e., non–coercively) correct each other to help improve each other's fluency. This fluency–building experience of daily speaking only French in a supportive verbal community, however, led her first to major in French, then to earn her Ph.D. in French, and then to work as a university French professor for over 30 years. I must suspect that teachers would like to see this the kind of fluency occur with their students.

Language–Teaching Application Conclusion

In many more ways than we have mentioned here, non–native language students should experience a fundamental reality of effective language study, which is that they benefit most when they speak their non–native language *whenever* they are in the presence of other members of their language community. Immersion programs, in a country where everyone speaks the language that students are studying, are indeed fun and valuable for students, but such trips are usually short–duration affairs, and the students are often not ready to get the most out of them (e.g., verbal–behavior analysis may not as yet be informing their training programs). Yet so many opportunities exist closer to home that provide even better experiences, as we have described.

Let me reiterate that if you implement any of these practices in a methodologically sound manner, you can make more meaningful contributions by reporting the outcomes at your professional conferences or through a peer–reviewed article, perhaps in the *Journal of Behaviorology* or in *The Analysis of Verbal Behavior* journal. (Contact the former at www.behaviorology.org).

Advanced VB Topics and Chapter Conclusion

This introductory chapter cannot substitute for more thorough coverage of this vast and complex and ever so distinctly—but not exclusively—human area of behavior. We still have many verbal–behavior analysis topics to cover. Their advanced status, however, requires that they receive a full treatment in more advanced resources. Here we touch on a couple of these topics, some only a little, while barely mentioning the names of others.

Verbal Behavior and Covert Events

The conditioning of verbal behavior proceeds smoothly as long as the controlling processes or stimuli (e.g., the establishing operations and evocative stimuli) remain *overt* (i.e., they affect different people in the same way, such as everyone present seeing a cat pounce, or everyone present hearing a dog bark). Many controlling relations, however, include *covert* (i.e., private or internal) events that only affect the behavior of the body in whom they occur. Remember, the skin is not a boundary to the laws of the universe. How can the verbal community condition appropriate verbal responses under the control of such covert events (i.e., events inside the skin)? This question contains far too many implications for us to address thoroughly, such as implications relevant to the vast area of covert behaviors that occur as elicited emotions and evoked thoughts (e.g., see Fraley, 2008, pp. 1068–1085). Instead we will touch on five of the methods that we currently recognize for conditioning verbal responses to covert events; none of these, however, contains the same level of precision available when conditioning verbal responses to overt events. We call these five methods *collateral stimuli, collateral responses, common properties, response reduction,* and *parallel stimuli* (see Peterson, 1978, pp. 109–119, for some elaboration; also see Skinner, 1953, Chapter 17).

Collateral stimuli. The term, *collateral stimuli*—or "public accompaniments" as some authors prefer—refers to the verbal community's conditioning of verbal responses to covert stimuli *coming under the control of overt stimuli* that occur along with the covert stimuli that evoke a verbal response. For instance when we observe an object directly affecting someone (e.g., a baseball bangs into his or her elbow) or when we see tissue damage (e.g., a scraped knee) we typically condition (i.e., reinforce) a response like "It hurts," which the victim's neural stimulation, from the elbow bang or the knee scrape, evokes.

Collateral responses. The term, *collateral responses*—or "inferences from overt behavior" as some authors prefer—refers to the verbal community's conditioning of verbal responses to covert stimuli *coming under the control of overt responses* that occur along with the covert stimuli that evoke a verbal response. For instance, while we may not have observed the ball strike the elbow, which would have been a collateral stimulus, if the person has called time out and is holding the elbow and moaning, which are collateral responses, we would still typically reinforce (i.e., condition) a response like "It hurts," which the victim's neural stimulation from the elbow bang evokes. Similarly we would typically reinforce a response like "toothache" when someone is holding the jaw and moaning; presumably some tissue damage (i.e., some collateral stimuli, which may only be visible to a properly instrumented dentist) is contributing to evoking the "toothache" response.

Common properties. The term, *common properties*—or "coinciding properties" as some authors prefer—refers to the verbal community's conditioning of verbal responses to covert stimuli *coming under the control of characteristics or properties of the covert stimuli that first control verbal responses to overt stimuli* that the verbal community already reinforces. For example, first an overt stimulus, with a property or characteristic of oscillating in intensity (e.g., a computer sleep–mode light oscillating between light and dim) evokes a public response of "throbbing," which the verbal community conditions by reinforcing it. Later when a covert stimulus, with this oscillating–in–intensity characteristic, evokes the verbal response of "throbbing" (e.g., as in a "throbbing headache") the verbal community also reinforces it on the basis of the common properties. We variously tact other overt stimulus characteristics that covert stimuli can share with terms like sharp, dull, pointed, subtle, intense, and so on. When covert stimuli share one or another of these characteristics, the verbal community reinforces the verbal response that the covert stimulus evokes on the basis of the shared characteristics; we would typically reinforce a response like a "sharp pain" or an "intense feeling."

Reduced responses. The term, *reduced responses*—or "response reduction" as some authors prefer—refers to the verbal community's conditioning of verbal responses to covert stimuli *coming under the control of responses that are reduced in intensity.* This more complex basis for the verbal community's conditioning of verbal responses to covert stimuli stems from each of us functioning as our own reinforcing listener as the rest of the community conditions our speaker repertoire. At the same time, as a public of one, each of us remains somewhat privy to the occurrence of the covert (e.g., purely neural) stimuli and responses that happen inside our skin. And also at the same time, our conditioning includes a gradual reduction of verbal–response intensity until various stimuli evoke our behaving verbally only neurally (a type of thinking). If we take these together, then when another person's covert stimuli seem to be evoking a verbal response, our reinforcement delivery depends in part on our own covert response of observing the extent to which our own covert stimuli might evoke a similar response from us.

Parallel stimuli. The term, *parallel stimuli*—or "covert proprioceptive stimuli" as some authors prefer—refers to a particular version of collateral stimuli wherein the verbal community conditions a member's *non–verbal* behavior through the ever present parallel pairing of overt stimuli and proprioceptive stimuli. That is, when we reinforce the overt *non–verbal* responses that overt—including verbal—stimuli evoke, we are also automatically reinforcing covert responses to proprioceptive stimuli (i.e., covert stimuli that the position and movement of body parts produce, in space and with respect to each other; we will cover the description and classification of covert stimuli and their neural receptors in a later chapter).

Take, for example, the occasion in which a sighted verbal–community member conditions—via overt verbal evocative and reinforcing stimuli—a blind community member's safely moving around the furniture in a room. The blind member moves successfully from place to place when the sighted member reinforces the blind member's responses to parallel *proprioceptive* stimuli. The sighted member conditions the blind member's covert responses to proprioceptive stimuli at the same time that the sighted member reinforces overt, non–verbal movement responses compliant with the parallel *overt stimuli* of verbal instructions regarding overt movements around other overt stimuli (i.e., the furniture items) that control the sighted member's verbal responses. As this occurs the proprioceptive stimuli gain control over the blind member's movements. Contingencies involving the aversive consequences of running into objects in one's path still hold for the blind member's movements. For sighted members, the control of moving behavior resides mainly with similar contingencies where, if vision–stimulating, path–blocking objects fail to evoke evasive maneuvers, then the same aversive consequences occur.

VB and covert events conclusion. The wide range of covert–event types include physiological changes (e.g., perhaps as establishing operations occur) and physical changes in body parts (e.g., the rise in skin temperature while sun tanning) and *stimulus–controlled* felt emotions, urges, images, thoughts, and even tendencies to behave more or behave less. As a result of conditioning verbal responses to covert stimuli, verbal–community members benefit; they mand observation of, and reporting the occurrence of, these stimuli, and so each member finds out what others are thinking and feeling and so forth, all of which helps all of them—and us—better deal with the world.

Other Advanced VB Topics

Currently Skinner's 1957 book, *Verbal Behavior,* remains the most comprehensive resource for covering the widest range of advanced topics (although you should also consult Fraley, 2008, Chapter 26). Read through Skinner's book—this chapter should have made that easier—to discover details about fragmentary responses, autoclitic grammar and syntax, editing, literature, poetry, composition, and many other advanced topics. Here are others.

Humor. Humor is the label that we use when verbal responses that are under *multiple controls* provoke laughter from listeners. Consider these two examples from unknown sources: Does the name Pavlov ring a bell? And how do you make a baby buggy? (Answer: tickle its feet.) Can you surmise the multiple controls in each case? Multiple controls comprise a big part of the difficulty translating jokes. The same multiple controls may not function, or even exist, in both languages.

Non–human verbal behavior. Humans are not the only species capable of verbal behavior, including humor. Studies with a wide range of primate subjects over the last 50 years have accumulated much reliable data (e.g., consider the recorded response history of Koko, who is a gorilla; see www.koko.org). Unfortunately many humans react negatively to this data, and deny its validity, usually based on either the presumption that only humans can "communicate," or that other animals are "only responding to conditioning." Of course humans also "only respond to conditioning," as all our previous chapters attest. And the assumption that only humans communicate (i.e., behave verbally) which implies that non–humans cannot, results from age–old inner–agent accounts and coincidental conditioning that mystically places humans outside of nature. That kind of conditioning contributes to an earlier demise of humanity (i.e., at least earlier than otherwise likely) by slowing the spread of the natural sciences, including behaviorology, that can prevent such a demise.

Verbal behavior extensions. All the characteristics of all stimuli, even unrelated stimuli that we call novel, which are present when reinforcers follow verbal responses, themselves become somewhat conditioned as evocative and reinforcing stimuli. As a result much verbal behavior occurs in the presence of novel stimuli that share varying numbers of characteristics with stimuli previously present when reinforcement followed verbal responses. Thus novel stimuli participate in further, and more complicated, verbal relations. We use the term *extensions* to describe such relations. These extensions can occur with any of the elementary verbal operant relations (e.g., mand extensions and tact extensions). And we speak of three kinds of extensions, *generic* extensions, *metaphoric* extensions, and *metonymic* extensions, as convenient descriptions of three extension levels. While generic extensions involve novel stimuli that retain all the characteristics of a stimulus that previously controlled responding, metaphoric extensions involve novel stimuli that retain only some of the characteristics of a previously controlling stimulus, and metonymic extensions involve novel stimuli that retain none of the characteristics of previously controlling stimuli. Peterson (1978, pp. 87–108) provides both defining and irrelevant features for each extension type, along with several examples.

Autoclitics. We use the term, *autoclitic,* for yet another kind of verbal relation. Autoclitics involve verbal behavior *about* verbal behavior. We sometimes use the term "secondary verbal behavior" for autoclitics, with all the elementary verbal relations making up primary verbal behavior. Secondary verbal behavior involves ongoing verbal behavior affecting other verbal behavior. Some aspect of a verbal relation can itself control a verbal response. When we analyze all the verbal components making up sentences, we find that many components serve autoclitic functions, including punctuation. Autoclitic relations reduce to zero any residual role for inner agents in accounting for verbal behavior. Further details appear in our preferred sources (e.g., Peterson, 1978, pp. 161–180; Fraley, 2008, pp. 1025–1068; Skinner, 1957, pp. 311–369).

Automatic reinforcement. In a process that we call automatic reinforcement, *sounding similar to the sounds around you* reinforces the response topographies that produced those similar sounds. We see *automatic reinforcement* as responsible for accents and dialects as well as contributing to children speaking the same native language that their parents speak to them.

Multiples, multiples. Perhaps the most complex aspect of verbal–behavior relations involves the multiple functions of both stimuli and responses. While sometimes a single stimulus controls a single response, more often a single stimulus controls multiple responses, or multiple stimuli control single responses, or multiple stimuli control multiple responses. Any of our preferred sources will provide you with plenty of examples; please experience the value of exploring these further.

Chapter Conclusion

By revealing the functions of verbal behavior, this analysis reveals many more of the ways through which people bring about effects on their world. As we finish this chapter, pause and ponder, as part of our usual ongoing activity, what behaviorology can contribute to helping traditional natural sciences solve global problems by taking into account the verbal–behavior analysis introduced in this chapter. For instance one aspect of the importance of verbal behavior to solving global problems resides in the necessary role of verbal behavior both in the usually non–coercive contingencies of education and persuasion, and in the usually coercive contingencies of lawsuits and law enforcement. While we can already acknowledge these roles, other roles require further research.✦

References (with some annotations)

Chomsky, N. (1959). Review of B. F. Skinner's *Verbal Behavior. Language, 35,* 26–58.

Eshleman, J. W. & Vargas, E. A. (1988). Promoting the behaviorological analysis of verbal behavior. *The Analysis of Verbal Behavior, 6,* 23–32.

Fraley, L. E. (2008). *General Behaviorology: The Natural Science of Human Behavior.* Canton, NY: ABCs. Chapter 26 (pp. 949–1094) details major components of verbal behavior, including extensive coverage of autoclitics as well as the private verbal behavior of thinking.

Hart, B. & Risley, T. R. (1995). *Meaningful Differences in the Everyday Experience of Young American Children.* Baltimore, MD: Paul H. Brookes.

Hart, B. & Risley, T. R. (1999). *The Social World of Children Learning to Talk.* Baltimore, MD: Paul H. Brookes.

MacCorquodale, K. (1970). On Chomsky's review of Skinner's *Verbal Behavior. Journal of the Experimental Analysis of Behavior, 13,* 83–99.

Maurice, C. (1993). *Let Me Hear Your Voice—A Family's Triumph over Autism.* New York: Ballantine Books.

Maurice, C., Green, G., & Luce, S. (Eds.). (1996). *Behavioral Intervention for Young Children with Autism.* Austin, TX: Pro–Ed.

Michael, J. L. (1982). Skinner's verbal operants: Some new categories. *VB News* (subsequently *The Analysis of Verbal Behavior) 1,* 1.

Michael, J. L. (1988). Establishing operations and the mand. *The Analysis of Verbal Behavior, 6,* 3–9.

Peterson, N. (1978). *An Introduction to Verbal Behavior.* Grand Rapids, MI: Behavior Associates.

Skinner, B. F. (1957). *Verbal Behavior.* New York: Appleton–Century–Crofts. The B. F. Skinner Foundation (www.bfskinner.org) in Cambridge, MA, republished this book in 1992.

Skinner, B. F. (1986). The evolution of verbal behavior. *Journal of the Experimental Analysis of Behavior, 45,* 115–122.

Vargas, E. A. (1988). Verbally–governed and event–governed behavior. *The Analysis of Verbal Behavior, 6,* 11–22.

Vargas, E. A. (1991). Verbal behavior: A four–term contingency relation. In W. Ishaq (Ed.). *Human Behavior in Today's World* (pp. 99–108). New York: Praeger.

Vargas, E. A. (1998). Verbal behavior: Implications of its mediational and relational characteristics. *The Analysis of Verbal Behavior, 15,* 149–151.

Vargas, E. A. (2013). The importance of form in Skinner's analysis of verbal behavior and a further step. *The Analysis of Verbal Behavior, 29,* 167–183. This paper helps us appreciate the value of traditional linguists' work.

Vargas, J. S. (1990). Cognitive analysis of language and verbal behavior: Two separate fields. In L. J. Hayes & P. N. Chase (Eds.). *Dialogues on Verbal Behavior* (pp. 197–201). Reno, NV: Context Press.☙

Chapter 21
Accounting for Consciousness

*I*n Chapter 20 we continued providing current, scientifically consistent, answers for some of humanity's ancient questions by introducing the medium of verbal behavior in which such questions and answers occur. This was the second topic for which our *Part II* warning—about completing *Part I* before going on to *Part II*—was particularly pertinent.

We introduced verbal behavior in four steps. After providing some general considerations about verbal behavior, such as its definition and some characteristics of its analysis, we analyzed elementary verbal operant relationships, and we covered some applications of verbal–behavior analysis, particularly to teaching additional languages. We finished by listing some of the more advanced verbal–behavior analysis components that the available, more thorough treatments of the topic provide. Now we will see the relevance of verbal behavior to consciousness.

Our *Part II* warning also pertains to the present chapter on consciousness. We touched briefly on some basic points about consciousness in Chapter 1, and here we revisit some of those points while elaborating some further introductory details, such as the interface between physiology and behaviorology. This interface enhances our understanding of consciousness, and research in both of these natural sciences continually expands and further clarifies their accounts. Physiology addresses *how* consciousness happens while behaviorology addresses *why* consciousness happens. Recall our Chapter–1 example of physiology addressing *how* striated muscle contractions occur (e.g., as functions of neural processes that constitute the mediation of overt behavior) while behaviorology addresses *why* a body mediates behavior, that is, *why* those innervated muscle contractions occur (e.g., as functions either of eliciting stimuli or of evocative and consequential stimuli). Behaviorology accounts for the functional relations between independent variables, such as a table blocking the path through a dining room, and the dependent variables of body–mediated *behavior,* such as the muscle contractions that this obstacle evokes thereby taking the body around the table. Such entirely natural events, at both levels of analysis, extend throughout the range of overt and covert behaviors during every human body's whole lifespan, so we will see this kind of pattern when accounting, even interpretively, for the neural behaviors of consciousness.

Also, behaviorology accounts for consciousness without conjuring or invoking any kind of mystical inner agent. We set aside the ancient, cultural traditions that have consciousness coming from, or being a characteristic of, a mind, or any other mystical inner agent. Consciousness is not a vehicle through which an agent's supposed psychic powers magically consider the effects of real stimuli and then initiatively direct the body to produce real responses of the agent's knowing and self–observant choosing. Instead, we introduce the term *consciousness* as our tact for the natural chaining of cascades of neural impulses that comprise the neural behavior–behavior relations that play functional supplemental roles in the contingencies controlling more accessible overt behaviors. A common chain of consciousness behaviors includes speedy but identifiable parts that we tact as raw sensation, awareness, recognition, comprehension, observation, and covert—then overt—reporting, with the complexity of such chains dependent on the presence of verbal repertoires from prior overt conditioning.

Given the substantial significance that so many people attach to consciousness, the details with which we perhaps should cover consciousness range far wider than we can tackle in this introductory coverage. In the hope that increased knowledge will reinforce your contact with more material on this topic, I encourage you to check out the 180–page "book" on consciousness that Professor Fraley wrote; he realized, however, that to follow it, a reader required some extensive and consistent contact with behaviorology first (for which the book you are reading may suffice) so he only published his consciousness material as a chapter in his 2008, three–course text, *General Behaviorology— The Natural Science of Human Behavior* (i.e., Chapter 27, pp. 1097–1276).

Pre–Scientific Views of Consciousness

Before considering several aspects of consciousness that involve reviewing realities "inside the skin," we first review a little about the pre–scientific views of consciousness. While these arose in the past, many continue in the present.

Past Pre–Scientific Views of Consciousness

The traditional views usually regard consciousness as a characteristic of an untestable inner agent, which at various times people described with general terms such as soul or psyche or mind or self, or with more specific terms such as "the mind's eye." Whatever the mystical agent, people often considered consciousness as part of the causal armaments with which the agent directed a body's behavior, and with thought and language either indicating consciousness or comprising some of the main means through which consciousness operated.

Before the contingencies of life produced the kinds of scientific verbal–behavior supplements available today, which now enable a scientific accounting for consciousness, the contingencies extant long ago produced those mystical notions about consciousness. These notions have persisted, because their occurrence also included the conditioning of a prevailing fallacious presumption that they were the best account, as well as the only possible account, of human nature. Gradually the contingencies of traditional cultural lore, affecting ever larger numbers of individuals, built organized forces of theological mysticism that survived across generations for many thousands of years. This survival occurred through means such as the political powers of various religious states as each backed its own version of mysticism through a history of coercive contingencies, including war, arrest, torture, or death, whenever contingencies of cultural conditioning or persuasion failed to produce behaviors consistent with their particular mysticism version. Nevertheless, as the increasingly complex contingencies of life gradually gave rise to natural science, mystical perspectives increasingly became describable as superstitious, which brings us to some *continuing* pre–scientific views of consciousness.

Continuing Pre–Scientific Views of Consciousness

More recently those organized forces of theological mysticism and superstition have expanded to include forces of secular superstition and mysticism. Such forces arise through disciplines like psychology, which makes claims to science status merely because it employs some scientific methods even though at the same time it adamantly disregards, and even actively opposes, the assumptions of naturalism that both fundamentally inform natural science and give scientific methods most of their value. This lack of connection with natural science is a major part of both why behaviorology separated from psychology when contingencies showed the doomed–to–fail status of attempts to change psychology, and why this natural science of behavior is not, and never really was, any kind of psychology. The contingencies on all these forces of superstition, theological and secular, induce much opposition to natural science in spite of the thorough reliance of these forces on the vast array of science–based products that support living quality lives while rendering that opposition. For example science opposers take modern transportation to anti–science meetings held in beautifully designed and elegantly engineered buildings with expertly appointed facilities, all of which are products of science that the science opposers would find aversive to live without. And the superstitions that these forces support include continuing the traditional views of consciousness as somehow indicative or characteristic of inner agents.

In turn natural science today inherently opposes superstition. Indeed, the histories of psychology and behaviorology (see Fraley & Ledoux, 2002) show that shortly after the traditional views began expanding from theological to secular superstitions, some general cultural contingencies supporting natural science began inducing development of the natural science of behavior. As a result of such developments, including the rise of the philosophy of science that we call radical behaviorism, Thompson (2008) provided an article summarizing some solvable difficulties in moving "self awareness" into natural science.

Consciousness in other animals. An extension of the ongoing, pre–scientific views regarding consciousness involves denying any possibility of consciousness in other animals, simply because they are not human and only humans have consciousness, or so the reasoning goes. Yet we find consciousness in humans because humans have the body parts (e.g., brain parts) that are capable of mediating consciousness behaviors. So the first question to ask regarding consciousness in other animals concerns whether or not they also have the requisite body parts. If their physiology contains these parts, then scientifically we should proceed by presuming that they are capable of consciousness responses. Thereafter, however, the questions and answers go well beyond our introductory purview (e.g., see Fraley, 2008, pp. 1245–1249). In any case elaborating the natural science analysis of consciousness helps counter all the mystical, superstitious trends that misconstrue consciousness, and this service necessarily leads us inside the skin.

Scientific Views of Consciousness—Inside the Skin

Behaviorology subsumes a multitude of independent and dependent, stimulus and response, variables. While some occur outside the skin as overt events, and others occur inside the skin as covert events, *all* are natural events.

The covert events make up parts of functional behavior chains. These chains begin with stimulus energy changes outside the skin that, by affecting sensory neurons, continue inside the skin where they affect nervous system structure through a series of events that lead back outside the skin through the mediation of behaviors that occur via innervated muscle contractions. There, outside the skin, these muscle contractions affect further events the occurrence of which induces other stimuli, other energy changes at receptor cells (e.g., reinforcing stimuli functioning also as evocative stimuli) that provide more links in the chains of functional behavior events.

In general behaviorologists focus on the events constituting the overt parts of a functional behavior chain while deferring to physiology colleagues regarding the internal events. However, with topics like consciousness, we must focus more on the functional chains of behavioral events occurring inside the skin, for our discussion pertains to dealing with these covert events, but in terms of the laws of behavior rather than the laws of physiology.

That discussion of events outside and inside the skin should help us recall a related though basic classification of behavior in terms of the manner of behavior mediation. The more familiar behavior type, which we call *neuro–muscular* behavior (i.e., innervated muscle behavior) occurs through the mediation of evoked or elicited neural impulse chains that ultimately include motor–neuron impulses inducing muscle contractions. The other type, the type on which this chapter focuses, we simply call *neural* behavior as it occurs through the mediation of evoked or elicited neurons, or bundles or cascades of neurons, firing without inducing muscle contractions; indeed the neurons firing *is* the mediation and we cannot conceptually separate the firing and the mediation from each other. Each of these behavior types, neuro–muscular and neural, overt and covert, can occur as respondent or operant behaviors.

With our focus turned to inside the skin, we begin our account of consciousness with an over–simplified definition of consciousness as neural behavior. The behaviors of consciousness manifest as *pure neural processes* generally lacking muscle contraction components. Our discussions will expand this definition.

Reduced Accessibility

Before proceeding, however, recall our Chapter–1 discussion about our reduced access to real but covert events. This reduced access makes research more difficult but not impossible. The skin remains a scientifically unimportant boundary; the laws of the universe operate the same way on either side of the skin. Besides, covert status only reduces—not eliminates—access to these events. Overt events, occurring outside the skin, can affect more than one human organism; for *any* affected humans, many of these events evoke observing and reporting responses. On the other hand, the covert events affect only the one human organism in which they occur; but each such human functions as a *public of one,* so many of these events evoke the observing and reporting, responses of the one human in which they occur. Many have access to overt events, and the publics of one have access to the covert events inside them.

Those reporting responses, which stimuli evoke, involve verbal behavior. When discussing such reporting in our verbal–behavior chapter, we promised to review our most common sense–mode receptors and the types of stimuli that affect them. Here is that review, as it may help, among other concerns, to elucidate the sources of the overt and covert stimulation that evoke observing and reporting responses, for either the public or the public of one.

The usual five sense modes. We are already familiar with some stimulus– and–receptor pairs as our usual five sense modes. We traditionally cast sense modes in agential response terms (e.g., see, hear, taste, smell, touch). While the response status accurately portrays these modes, the agential casting—although occasionally satisfactory as vernacular verbal shortcuts—remains scientifically unacceptable. So here we cast sense modes as stimulus–and–receptor pairs from which we can move into non–agential response terms.

Here are the traditional five sense modes, with stimulus–energy type and related receptors: (a) Light (or visual) stimuli affect photoreceptors, producing vision responses (or, more agentially, visualizing, imaging, and seeing responses). (b) Sound (or auditory) stimuli affect phonoreceptors, producing aural responses (or, more agentially, hearing responses). (c) Chemical stimuli on the tongue affect chemoreceptors on the tongue, producing gustatory responses (or, more agentially, tasting responses). (d) Chemical stimuli in the nasal passage affect chemoreceptors in the nasal passage, producing olfactory responses (or, more agentially, smelling responses). And (e) surface pressure (or tactile) stimuli affect mechanoreceptors at the body surface, producing contact responses (or, more agentially, touch, or touching responses).

Five more sense modes. Other stimulus–and–receptor pairs are less familiar than the traditional five sense modes. While our list here remains incomplete, these other types of receptors, and the type of stimulus energy that respectively affects them, include these five, the first of which is an in–body counterpart to the surface sense of touch: (a) Deep pressure stimuli affect mechanoreceptors in the body. (b) Stimuli from movements in muscles, tendons, and joints affect kinesthetic receptors, and relate to coordination. (c) Stimuli from movements of the body in space affect vestibular receptors, and relate to balance. (d) Heat and cold (i.e., less heat) stimuli affect thermoreceptors. And (e) aversive stimuli, both at the body surface and in the body, affect free nerve endings, receptors also at the body surface and in the body, the firing of which leads to pain responses (or, more agentially, "hurt" responses as in "I'm hurt").

Note that our discussion uses general, common terms to highlight valuable physiological facts. You can find the related details and technical terms in a good physiology textbook (i.e., a text that not only covers laws and facts from the natural science of physiology, but that also rejects and excludes superstitious material from either theological cultural lore or fundamentally mystical secular disciplines that can taint content with agentialism).

Sense–mode covert behaviors. In all the sense modes, before a stimulus energy trace can affect a receptor, the energy trace must reach the necessary physiological threshold. When the energy reaches this threshold, the receptor transduces the energy to neural impulses. Sometimes the energy trace, through these impulses, ultimately produces an overt response with few if any covert chain components; this can happen through a process that we call *direct stimulus control,* which is the process that, for example, can produce some relatively safe driving while covert daydreaming responses are occurring. At other times a stimulus affecting a receptor induces nerve impulses that, upon reaching the central nervous system, produce the first behavior, a covert behavior, in a chain that includes covert behaviors then overt behaviors. We call these first covert behaviors *raw–sensation behaviors,* which involve a crude kind of awareness behavior, and we can conceptualize these raw–sensation behaviors as respondent behaviors that the stimulus elicits by affecting a receptor.

In a later chapter, we discuss the implications of those raw–sensation behaviors for the nature of reality. Here, however, our interest concerns what happens next, what happens when, and after, a stimulus affects a receptor. The answer pertains to a phenomenon that we call behavior–behavior relations (see Hayes & Brownstein, 1986).

Behavior–Behavior Relations

What happens when and after a stimulus affects a receptor? Upon reaching threshold the receptor transduces the energy into nerve impulses. A transduced energy transfer to neural structures of any sense mode can function as an eliciting or evocative stimulus—for simplicity, here we emphasize evocative stimuli—for either neuro–muscular responses or neural responses or both, depending on the history of such energy changes (e.g., past conditioning) and the state of the nervous system at the time. As real events, any such evoked responses, neuro–muscular or neural, can function evocatively for other neuro–

muscular responses or neural responses, and so on. This happens regardless of the kind of behavior–mediating neural structures involved. As a real energy change, any response can evoke other responses either when *a directly genetically produced* neural structure mediates it or when a neural structure *that various continuously operating conditioning processes have changed* mediates it. If the necessary gene–produced structure or the necessary conditioning (i.e., neural restructuring) has occurred, then once some stimulation evokes a response, that response—as a real event—can evoke a further response, which can evoke yet another response, and so on, chaining according to the current set of operating functional relations. In this way a sequence, or chain, of behaviors occurs. We call these relations *behavior–behavior* relations, because behaviors, as real events, are functioning as stimuli for further real events. This connects directly with consciousness, but we need several more background points before the connections easily become fully apparent.

Physiologists continue to provide evidence and descriptions of the kinds of structural, nervous–system changes that stimuli produce, and the function of these structures and structural changes in mediating behaviors. Explanations for behavior, however, reside not in the structures themselves, but in the functional effects of stimuli in changing structures, and in the mediating function of the neural structural changes between the energy traces at receptor cells and the responses that the energy traces evoke or elicit.

For example, on the behaviorological level, we acknowledge the necessity of practice that produces reinforcers if practice is to improve or refine a behavioral skill. At the physiological level, the reinforcers that follow practice change the neural structure in ways that support skill improvements; Thompson (2008) describes the outcome of a series of physiological studies as demonstrating "that dendritic spine growth and synaptogenesis occur in [the] motor cortex as a consequence of reinforced practice, and that when reinforcement ceases, the number of such newly formed synapses regresses" (p. 142). The synaptic growth not only stems from the occurrence of reinforcement but also provides mediation of a performance that we describe as more skillful. The explanation for why the more skillful performance occurs, however, resides not with the improved structure but with both the stimuli that evoke the performance and the reinforcers that the performance produces, reinforcers that improve the structure. No inner agent drinks in the reinforcing attention from practice and, wanting more, decides to direct the body in the production of a more skillful performance. The how and why of that skillful performance inhere in the physiological and behaviorological accounts at their respective levels of analysis. Even our still–developing accounts of these relations require no mystical events, no inner agents.

Note that our discussion emphasized covert behaviors as operant, evoked behaviors. However, covert behaviors, and the behavior–behavior relations that construct covert behavior chains, can occur as either respondent or operant behaviors. Let's look at each of these in turn.

Covert Respondent Relations

Some covert responses occur respondently. When a stimulus, an overt or covert stimulus (which might be a previous covert response) elicits a covert response that a genetically determined neural structure mediates, we describe both the stimulus and the response as *unconditioned*. On the other hand, some overt or covert stimuli produce no effect on the genetically determined neural structure, so we call these stimuli *neutral*. The occurrence of some pairing (i.e., conditioning) of a neutral stimulus with a stimulus that already elicits the covert response, however, alters neural structure such that when this previously neutral stimulus occurs, the now changed structure mediates the covert response. We again say that the stimulus elicits the response although, in this case, we describe both the stimulus and the response as *conditioned*.

For example let's say that some unspecified overt or covert stimulus induced a covert leg–muscle cramp just when the blaring horn and screeching brakes from a rapidly decelerating truck—in front of which you had just stepped—elicited the covert chemical dump of a strong emotional reaction, a reaction far in excess of the usual reaction that a bothersome leg cramp elicits. Later you are reading quietly in a library when another unspecified stimulus induces another covert leg cramp. We would not then be surprised if this new covert leg–cramp response now served the conditioned stimulus function of also eliciting another covert chemical dump as a conditioned response, a dump also exceeding the usual leg–cramp reaction.

How does that example work? That conditioned respondent involves the pairing (i.e., the overlapping or successive occurrence) of an eliciting stimulus (i.e., the blaring horn and screech of brakes) and a neutral stimulus (i.e., the leg–muscle cramp that lacks an eliciting function with respect to the kind of strong emotional reaction that a blaring horn and screech of brakes can elicit). The energy traces from the pairing of these stimuli transfers energy to nervous–system structures, energy that alters the structure such that the previously neutral stimulus comes to function as an eliciting stimulus of the strong emotional response (because the body has changed, not the stimulus). When this happens we describe the previously neutral stimulus as a conditioned stimulus and the response that it elicits as a conditioned response.

Respondent seeing. When we consider the covert neural behaviors of consciousness, we find that some, perhaps many, start as covert respondent behaviors. While this applies to any sense mode, the vision mode seems the easiest to conceptualize. The unconditioned stimulus energies at our photoreceptors elicit the initial "unconscious" seeing responses as raw sensations. This is unconditioned respondent neural vision behavior. How contingencies improve these, from ambiguous blotches of shifting shape and color to focused and detailed tactable objects and events, goes beyond our purview (again, see Fraley, 2008, Chapter 27). When other stimuli occur together with the stimuli that elicit unconditioned seeing responses, in the pairing process, then those other stimuli come to elicit the vision responses as conditioned seeing.

For example take a bird watcher who regularly sees a member of an ordinary, uninteresting species fluttering among the leaves of a particular kind of tree. Seeing this bird shows little evidence of functioning as a reinforcer, while the pairing of this bird type and this tree type has been extensive. We should not then be surprised if this bird watcher reports seeing such a bird in a tree of this kind when a video camera running at the time shows no such presence. The conditioned tree stimulus elicited the bird–seeing response. Seeing that bird in its absence constitutes covert conditioned respondent neural vision behavior.

Covert Operant Relations

Many covert responses (i.e., neural behaviors) occur operantly. We have covert operant behavior when an overt or covert antecedent stimulus (which might be a previous covert response) evokes a covert response. Such responses often evoke further covert responses as consequences. At other times, such responses chain to further responses that include overt responses the mediation of which affects—as in "operates on"—the environment in ways that produce stimulus energy changes as a consequence of the response; these changes are stimuli that transfer energy back to the nervous system thereby producing (i.e., through the operant–conditioning process) neural structural changes that further establish the functioning of similar future evocative antecedent stimuli. With all the stimuli in our internal and external environments, and the continuous overt and covert interactions among them, both respondent and operant conditioning processes continuously affect each of us on a moment by moment basis from before we are born until we die. This not only builds the extensive, different and still natural history that each of us accumulates as we live, as well as the extensive overt behavior repertoire, but it also builds the extensive covert behavior repertoire, parts of which we call consciousness.

Operant seeing. When Skinner (1963) described consciousness as *"seeing that we are seeing,"* which we call *"conscious* seeing," he was alluding to the operant contingencies of the verbal community that condition both our conscious seeing and our reporting of what we see; the thing seen elicits our initial "unconscious" seeing responses which in turn evoke the seeing/reporting conscious responses. Skinner could have used any sense mode but, again, the vision mode seems the easiest to conceptualize at present.

Once stimuli have elicited raw–sensation vision responses, these responses can chain to vision–awareness responses (i.e., Skinner's "seeing that we are seeing" responses). All of our operant laws of behavior apply to these responses, and they produce many kinds of covert operant behavior. However, let's clear up a potential confusion: The neural behavior does not produce the vision; instead it *is* the vision. Extend this with respect to consciousness; just as the functioning of neural structure does not mediate behavior but rather *is* the mediation of behavior—we cannot separate the two—so also neural behavior does not produce consciousness; instead some natural functioning of neural structure *is* consciousness behavior; again we cannot separate the two.

Before adding more details, here is one example of covert operant neural vision behavior. As you wait at the airport to pick up a friend who has been away for a year abroad, some people in the crowd of strangers may share some facial features with your friend and, as a result, several of those strangers may evoke your seeing your friend and calling out her name. As a public of one, you can see that you *saw your friend* and called her name, but others can only hear you call her name in the presence of strangers. You can analyze *seeing her* as the stimulus that evoked your calling her name, but others analyze the similar facial features as evoking your calling her name. In either case such seeing and calling produces reinforcement only the last time when one of the strangers turns out actually to be your friend. Such seeing exemplifies operant seeing.

Again, you also *saw* that you were seeing your friend. You could report that you *knew* you saw your friend each time you called out her name, and the consequences of your calling response evoked your knowing after each response that it was incorrect, except for the last calling response. We can take such *seeing that you were seeing* your friend as an example of a neural response of consciousness; you were conscious of seeing your friend. We can also say that you were aware, or self aware, of seeing your friend (although no *self* saw…).

"Self awareness"

While we can define the term *self awareness* scientifically, the very use of "self" in the term carries inherent agential implications. The implied problems get compounded when people use self awareness interchangeably with consciousness. To reduce confusion, we generally avoid the self–awareness term. However, some uses of it seem legitimate so let's explore them a little.

The term *self awareness* has different usages, some of them problematic. In one case when stimuli in a mirrored space evoke appropriate responses, such as young children or pigeons responding appropriately to their mirror images (which we discussed in Chapter 1) people say that the child is self aware; the question of the pigeon being self aware only emphasizes some problems with this term. In another case when stimuli evoke a person's overt behavior that produces overt stimuli, and these stimulus products then evoke, from the person, overt verbal–behavior reports about the stimulus products, people say the person is self aware. For example after stimuli evoke musical instrument playing, and the quality of the produced sounds evokes the verbal report, "I played that piece really well that time," people say that the player is self aware. In such cases "self aware" may simply, and poorly, substitute for "a public of one." Acknowledging the player's public–of–one status, we may also accept that the quality sounds first evoked the neural behavior of observing, as in hearing, the sounds, which then evoked the overt verbal report.

In Chapter 1 we also mentioned a constraint on the relevance of past or future events on any behavior, including self–awareness behavior. While past or future stimulus events seem to affect our behavior, they actually *cannot* directly evoke or consequate responses, because both responses and stimuli occur only in the present, which makes all behavior *new* behavior (with responses grouping into response classes for experimental analysis). Regarding the future—be wary of teleology—neither stimuli nor responses have yet happened, although presently operating contingency rules can seem to bring future events into play. And regarding the past, neither stimuli nor responses exist in any kind of storage (e.g., file cabinets or flash drives of the mind). Some people prefer the metaphor of neural structure being the storage facility, but this still too easily misleads us. *Current physiology*—neural structure—mediates every behavior, overt or covert, and every behavior occurs under the functional control of *current* overt or covert stimuli regardless of the complexity, multiplicity, or interactivity of these stimuli or responses. Even memories are not stored responses. They are *new* responses that *current* stimuli evoke and that *current* neural structures mediate, *neural structures that have their current structure because conditioning processes changed them both at, and since, the time of the original instance.* For this reason they often fail as accurate re–behavings of past events, a point pregnant with implication for eye–witness testimony.

Knowing. Where the terms self awareness and consciousness really begin to overlap, and where we begin really to prefer to speak of consciousness, concerns what we mean by the term *knowing*. The movement of objects (e.g., rain on a window) or the behavior of another organism (e.g., some bees hovering around some flowers) can evoke your body mediating some covert behavior such as seeing the rain splattering or the bees hovering. With a history of appropriate conditioning (e.g., conditioning that has made stimulus indications of strong weather, or the dangers of insect stings, effective evocative stimuli) this covert seeing behavior then evokes your body's mediating a covert observing response of the rain or the bees; and this covert observing response then evokes a covert verbal report of the covert observation behavior, such as, "An umbrella would be handy" or "Keep away from those flowers." If audience variables are present, this covert verbal report may merge into an overt verbal report of the same or similar content. Sometimes the audience variables will be potent enough that the covert observing response directly evokes the overt verbal report. In any case, we would say both that you *know* what evoked the overt verbal report (i.e., the rain or the bees) and that the covert verbal report constituted a kind of consciousness responding that we call thinking.

That same kind of thing happens with a body's "own" behavior. When a stimulus evokes the body mediation of some overt behavior, and this behavior then evokes the body mediation of some covert behavior, such as seeing or hearing or feeling the overt behavior, and this covert behavior then evokes the body mediation of a verbal report (possibly covert then overt) of the covert behavior, we can describe this sequence in two main ways. Traditionally we would agentially say that *you* are *self* aware, that *you* know or are aware of what *you* were *doing*. But no inner *self* exists to be aware, no inner *you* exists to *do* the behavior or to know about it. Instead we can say scientifically that the body (i.e., "you," as a verbal shortcut for the mediating body) *knows* or is conscious of the behavior that stimuli make the body mediate, not in the sense of giving the body inner–agent status, but merely in the sense that a stimulus evoked a sequence of completely natural (i.e., non–mystical) behavior–behavior functional relations, a sequence that included purely neural consciousness responses that we call *knowing*.

Brain, Chaining, and Consciousness

Covert verbal behavior constitutes a common kind of purely neural consciousness behavior. We often use the term *thinking* to describe this kind of neural behavior, although thinking can happen in other sense modes as well. While overt verbal behavior, along with other overt behavior, involves the body parts we call muscles, covert behavior involves what body part?

The role of the brain. When we say that consciousness manifests as purely neural behavior, lacking muscle–contraction components, we allude first to the brain as a behavior–capable organ. However, the brain is far more complex than a muscle, and the brains of some species are far more complex than the brains of other species. So the behaviors of which brains are capable cover a vast range. Even so, these behaviors share one characteristic; they are all covert, which necessarily means less accessible than overt behaviors.

While physiological equipment and procedures provide a kind of access to what is happening in the brain when behaviors, both overt and covert, occur, we must remain respectful of disciplinary expertise. I am a behaviorologist, not a physiologist. So I defer to my physiology colleagues for more accurate elaboration of points pertaining to the brain per se. And they defer to behaviorologists for elaboration of the functional effects of stimuli on the operation of brain parts, particularly as this operation pertains to behavior, covert and overt, and its mediation because, with respect to behavior, that comprises a major, perhaps *the* major, brain function.

The brain, as the major component of a nervous system, mediates— not originates or initiates—behavior that occurs as a function of other real variables. This puts the brain, as a physiological organ, at the interface of exchanges between environmental energies and overt behavior. Furthermore, as a behavior–capable body part, the brain mediates covert, neural behaviors—inseparable from the mediation—including those we consider as consciousness behaviors such as thinking. In this particular disciplinary interface, behaviorology accounts for specific functional relations between real, independent variables and real, dependent variables on both sides of the skin, including in the brain, while brain physiology accounts for the structural changes that are occurring as those behaviorological–level independent and dependent variables interact. Both disciplines, of course, also concern so much more, in their separate subject–matter domains, than behavior and mediation.

Neural behavior chaining. Contingencies in the external and internal environments of peoples' daily existence feature the occurrence of energy exchanges that already affect people's behavior, or that condition people's behavior, such that some behaviors, including neural behaviors, function as real, independent variables evoking further behaviors, including more neural behaviors, as real, dependent variables, which evoke further behaviors, including more neural behaviors, and so on. This process can build strings of behavior, and these behavior strings include some of the types of consciousness behaviors that we have already mentioned, such as thinking—verbally or non-verbally and in any sense mode—and feeling, and observing, and knowing, and seeing that we are seeing, and hearing that we are hearing, and so forth. These behavior strings provide typical examples of the natural phenomena of responses linking together in a chain of responses (i.e., responses, as real events, evoking other responses) usually including neural responses, all in the present, all new, and none requiring that the thing seen or heard, or felt (and so on) be the current source of evocative stimulation.

Those neural chains interweave with reinforcement feedback loops featuring functional relations outside the skin between stimuli and neuro-muscular behaviors. The neuro-muscular behaviors change the environment, producing the reinforcement feedback into the nervous system such that neural structures change. These changes leave the neural structures, which mediate the covert and overt responses, more susceptible to the energy traces from the covert and overt evocative stimuli, which a behavior could provide. And some of these reinforcing effects may affect responses supportive of survival.

Behaviors and Sequences of Consciousness

Having dealt with many pieces of our consciousness puzzle separately, the time has come to consider them together. Our account of consciousness in terms of neural–behavior chains usually involves several types of consciousness behaviors. In the sequence in which they usually occur, after respondent raw sensation and basic operant awareness, the common terms for two types of consciousness behaviors that usually occur first are recognition and comprehension. Other operant types of consciousness behaviors typically occur after these two, but their order depends on the prevailing contingencies. The common terms for some of these other types are knowing, observation, explaining, and thinking, which also covers covert verbal reporting. Any of these—sensation, awareness, recognition, comprehension, knowing, observation, explaining, thinking, and so on—can each evoke covert or overt non–verbal behavior, and most can evoke further covert and overt verbal behavior. After raw sensation, the sequence of awareness, recognition, and comprehension often chains further to verbal, and still natural, knowing, observing, thinking, or explaining. These usually lead to overt behaviors including reporting, which supplements other controls and thereby induces more effective responding to the world around us.

For example let's focus on the common sequence of sensation, awareness, recognition, and comprehension. Continuing with the convenience of our visual sense mode, consider a living adult human body with a behavior repertoire from the usual life–long and thus extensive conditioning history (i.e., an adult *person*). If an object passes by, a foot or two from the person's face, the raw sensations—occurring when energy traces bounce off the object and strike the retina—evoke *awareness* responses that manifest as shifting spots of shape and color. These awareness responses then evoke (i.e., as real events these responses function as stimuli and evoke) the *recognition* responses of basically knowing—from past conditioning—that the properties (i.e., the shifting spots of shape and color of the awareness responses) comprise a butterfly; the recognition responses may even include the covert or overt verbal behavior of the simple tact "butterfly." These recognition responses then evoke *comprehension* responses that could involve any or all of the wider covert aspects of this body's behavior repertoire—if previously conditioned—with respect to butterflies. Thus the recognition responses could evoke comprehension responses such as (covertly, and not necessarily verbally) "That is a Monarch butterfly, a really big and beautiful one, the first one in the garden this season, and seeing one this early makes for wonderful feelings about the potential success of pollination this season." Other comprehension responses could include further poetic and scientific implications of the presence of Monarch butterflies in that location at that time of year, and so forth. Much of this we might further tact as parts of thinking and explaining. Also, other covert or overt stimuli could evoke overt verbal forms of these responses at any point during their covert occurrence.

We can conceptualize the neural awareness–recognition–comprehension behavior chain as seeing (or hearing, or other sense mode) some object or event (i.e., awareness), knowing what it is (i.e., recognition), and understanding how it relates to other events (i.e., comprehension), all in the sense of the covert differential responding to those aspects of it. *We summarize such covert responding with the term consciousness.*

A related concern, however, pertains to the ongoing conditioning involved in these sequences, which continuously increases the speed at which each chain component—awareness, recognition, and comprehension—evokes the next. The time duration of these components was never extensive, given the speed of nerve impulses, even masses or cascades of them. Hence the conditioning quickly succeeds in shrinking the duration of each component to the point that sometimes they all *appear* to occur simultaneously, even instantaneously; people come to seem to instantly behave the awareness, recognition, and comprehension not separately, sequentially, but all at once. This raises the need for us to repeat that we all know better by now not to see ghosts—inner agents—in these events, or in our phrasing of them, for all these events are still, and inevitably, occurring quite naturally. No other reasonable option exists.

We should also take note of some other characteristics of consciousness responding. These concern consciousness appearing as a single–channel covert process often separate from a concurrent occurrence of other overt behaviors happening under direct stimulus control. While we have encountered each of these terms (i.e., single–channel process, and direct stimulus control) before, here we relate them to consciousness in the context of our favorite daydreaming–while–driving example, along with repeating that this combination is dangerous and drivers should resist it.

To begin, recall that consciousness responses provide a supplementary focusing of the evocative effects of stimulus control. Without the consciousness responses, some stimuli fail to evoke behavior, or evoke only part of a response pattern, possibly the wrong part in the sense of a part that fails to produce reinforcement. On the other hand, the unchanging or familiar stimuli of a regular or consistent environment seldom require the supplements that consciousness responses provide; in this case the controlling stimuli may not even evoke consciousness responses, and instead control responding directly. Directly operating stimuli control vast areas of human behavior, although people tend to consider these areas less important than the equally vast human behavior areas that benefit from consciousness responding, a point which you may have already surmised from the content of the examples that we used in earlier parts of this book.

Consciousness responses, however, involve easily evoked neural responses that other stimulus–response chains replace just as easily. To keep them going, they must quickly, even continuously, produce reinforcers. In the most common scenarios, each next evoked neural response in a chain serves not only the function of evoking another response but also serves the function of reinforcing the previous response. We called this *dual function of a stimulus* when we saw it among overt behaviors as we described the intervention technique of backward chaining. Here, however, the covert consciousness responses, as real events, serve these stimulus functions. Nevertheless, the relatively constant stimulus barrage that we are under from the internal and external environments can easily induce fleeting changes in stimulus control from one neural response chain to another to yet another, and so on, unless a particular chain is producing some compensating reinforcing effect that maintains it.

A contributing factor to the brief duration of many consciousness response chains is the apparent single–channel nature of these chains. Only one consciousness chain happens at a time; the contingencies, or our physiology, or both, seem unable to manage even two, let alone several, consciousness chains at the same time. On those occasions when two seem to be occurring simultaneously, closer inspection reveals the chains to be rapidly alternating and so still occurring only one at a time.

For example consider driving on the 100–mile–long trip from Sacramento to Lake Tahoe on u.s. Highway 50, a rather scenic two–lane highway that, over the Sierra Nevada mountains, occasionally grips the edge of cliffs. As the car approaches some curves, those curves, as stimuli, evoke steering responses for which staying safely on the roadway provides appropriate reinforcement. This consequence of staying on the road is contingent upon the necessary steering responses, which in turn are contingent upon the evocative stimuli that changes in road direction—the curves—supply. If the curves are the sharp ones along the steep cliffs near the pass that leads down to the lake, then adequate steering responses may require the supplementation that consciousness response chains provide; your eyes and all your consciousness resources remain glued to the road with the intensity of a pilot on final approach to landing. The children arguing in the back seat, and even the beautiful, high sierra scenery, fail to evoke other consciousness chains, neither ones alternating with each other nor ones alternating with the chain focused on the curves. The single consciousness channel remains completely preoccupied for the duration of those particular cliff–hugging curves. On the other hand, once you reach the valley floor, and the curves become simple, gentle direction changes, the scenery and the children's activity begin to control not only some of your overt behavior chains but also your single–channel covert–behavior chains, either in slow or rapid alteration, but not simultaneously.

Actually the available data (e.g., see Fraley, 2008, Part III) are not yet entirely adequate for full confidence in the current single–channel conclusion. So far this is the conclusion to which the available data lead. However we may discover, from accumulating additional data, some circumstances under which more than one simultaneous channel is possible for consciousness response chains. Or we may discover particular processes or procedures the potency of which allows successfully conditioning the occurrence of simultaneous, rather than alternating, consciousness response chains (i.e., dual, or multiple, channels). At the time of this writing, however, consciousness response chains occurring only as a single–channel capability remains the most parsimonious option. Perhaps a general limitation is one "channel" *for each behavior–capable body part, including brain parts;* this still allows a wide range of simultaneous behaviors that involve various, separate body parts and brain parts.

Returning to our driving example, and the gentle curves on the valley floor, another consideration is that *you need not be aware* of those curves for them to exert their evocative stimulus function on your driving responses. Perhaps the children are napping and your single–channel neural consciousness responses are engaged, say, daydreaming about a week of fishing in the beautiful lake waters, and the great trophy fish you are sure to catch. The daydreaming consists of a sequence of thematically related, often mostly visual responses, each one evoking the next while each one also reinforces the previous one. Even while the neural daydreaming response chain completely occupies your consciousness response–chain channel, preventing awareness responses with respect to the curves, the curves still function as effective evocative stimuli for your overt steering and other driving responses (and continuing to stay on the road still functions as the effective reinforcer). Again, our term for the curves successfully evoking your steering responses is *direct stimulus control.*

Another example of single–channel consciousness chains, with direct stimulus control over concurrent overt responses, involves some types of needlework. This activity can occur while the television has control of your single–channel consciousness response chains; the stimulus aspects of the needlework exert direct stimulus control over your needlework–related responses. On the other hand, if you were working crossword puzzles while listening to the news, then each of these would be controlling its own consciousness response chain, but alternating with each other rather than actually occurring simultaneously, as *both* chains involve the same body parts.

Returning again to our driving example, at some point a change in energy–trace types refocuses the consciousness channel. As you stop at a red light, you wonder (another neural behavior of consciousness) how you got there, or where the last mile or three went. You neither crashed nor heard sirens wailing and get pulled over, so your driving responses must have been adequate. However, the driving responses occurred under the direct stimulus control that the curves exerted; those overt driving responses occurred right along with—that is, at the same time as—your covert daydreaming responses. Neuro–muscular (i.e. overt) behaviors, and neural (i.e., covert) behaviors seem to occupy different neural channels, involving different behaving body parts. While neural behaviors like consciousness seem limited to a single channel, neuro–muscular behaviors seem to have many channels available (i.e., many different overt responses, involving many behaving body parts, often occur simultaneously). In any case these daydreaming responses prevented your being aware of the steering responses because the needed channel, so to speak, was already occupied with these daydreaming responses. You may have experienced something like this many times. It is not magic (although it seems magical); it is just the behaviorological laws of nature at work, laws which operate without our needing to be aware of them or the variables involved.

Since we have talked about the necessity and value of reinforced practice, you can experience the benefit of practice by analyzing how parts of our previous examples line up with various parts of the general consciousness behavior chain of raw sensation, awareness, recognition, and comprehension, possibly mixed in with knowing, observing, thinking, explaining, covert reporting, and overt behaviors. For example analyze the music–practice chain we covered earlier in this chapter under "self awareness." Also analyze the rain chain, or the bee chain, both of which we covered earlier under "knowing." The more thoroughly you comprehend the scientific account of consciousness, the more comfortable you will be with the interpretive accuracy of your analyses.

Consciousness Conclusion

Conditioning involves energy transfers between the environment (internal and external) and the body in ways that, as our physiology colleagues can show, trigger cascades of neural impulses that variously induce several processes. Some impulses induce—and constitute—chains of consciousness responses, while other impulses induce the greater energy expenditures involved in bodily movements, and still other impulses induce the altered neural structures that constitute a different body that mediates (not initiates) both covert and overt behavior differently on future occasions (with those greater energy expenditures coming from nutritionally derived bodily reserves). When happening, all these intertwined environmental/neural/behavioral processes move along at such a rapid pace that they may seem undetermined, particularly when we try to encompass behavior in general across a time frame beyond a few moments, because events can quickly outpace our measurement technology. This, however, is a problem not with nature but with our residual ignorance (see Fraley, 1994) as these processes are *all lawful, all entirely natural.*

As all those events unfold, including chained consciousness behaviors, no capricious inner agent makes them happen; any behaviors that happen had to happen (and if a behavior failed to happen, that was because it could not have happened) under the present functional independent variables. As we pointed out in earlier chapters, including self agents in accounts for behavior is not only redundant and misleading but also dangerous and irresponsible because the resulting reduced effectiveness in problem solving can cause harm.

On the other hand, as natural scientists we respect the natural functional history even of extremely complex and multiply–controlled response chains such as those involved in consciousness responding. These cascading chains of sequential relations, though seemingly obscure, are not mystical. They simply involve two kinds of behaviors, covert and overt, each serving evocative, and reinforcing, stimulus functions for ongoing responses in chains of either type. And, when contemplating interventions, behaviorologists quickly track back along the links in any causal chain to search in the *accessible* environment for the functional public antecedents and consequences of the covert events. By tracing the functional relations back to events in a more accessible part of the environment, they locate potentially changeable independent variables. This affords control over the subsequent internal and otherwise inaccessible parts of the sequence as well as control over the external parts.

All this complexity in behavioral and physiological events can seem amazing, wondrous, awesome, even overwhelming, but it is all natural, and exemplifies our *Law of Cumulative Complexity:* The natural physical/chemical interactions of matter and energy sometimes result in more complex structures and functions that endure and naturally interact further, resulting in an accumulating complexity. Our account of consciousness as covert, neural–behavior chains—variously with raw sensation, awareness, recognition, comprehension, knowing, observing, thinking, explaining, and reporting components—is *cumulatively complex* and *entirely natural.*

Intelligence, Consciousness, and Survival

Consciousness clearly can and has benefited us humans, particularly as a primary process supportive of the verbal behavior through which we accumulate both individual and cultural archives of knowledge that extend our intellectuality. However, the age–old misconstrual of consciousness, as involving mystical inner agents, detracts from that intellectuality by helping superstition to go unchallenged. The age–old misconstrual of the nature of human nature, of life, of living, of behavior, and of consciousness behavior, might just represent the greatest analytical error of humanity. The implications of this error compound across so many aspects of human culture and survival.

Coincidental contingencies have led a surprising array of disciplinary domains into superstitious, analytical dead–ends. During times when humanity's numbers comprised little threat to environmental stability, the superstitious assumptions continued relatively benignly, and possibly with occasional short–term benefits. However, humanity's population has climbed well beyond the Earth's carrying capacity, and the resulting environmental stresses currently, and predictably on into the future, present us with increasing numbers of survival tests at an increasing rate. Have we passed them? Can we continue to pass them? Do humanity and the rest of life have a future?

Current environmental data suggest clearly that humanity can no longer afford its traditional reliance on, and modern compromises with, superstition and mysticism. Culturally, our intellectual history provides the basis, the natural sciences, for understanding, solving, and preventing our problems. These sciences include behaviorology, the science that not only accounts for consciousness but also shows that by itself consciousness cannot provide a complete solution; it is one factor among many, albeit a potentially helpful one.

Behaviorology itself can help solve global problems. But how helpful can its existence be when superstitions are so thoroughly entrenched in the culture that a person can get advanced degrees in various superstitious disciplinary domains at most major universities, and yet rarely encounter behaviorology at these same institutions? Meanwhile even many natural scientists are not as yet under contingencies to assure that their own students contact the natural science of behavior, rather than the currently more prevalent mystical disciplines of behavior, by assuring the availability of sufficient behaviorology courses and the necessary professor–producing programs. Will humanity be a victim or a beneficiary of the resolution of such questions? Will we resolve our conundrum between science and mysticism in a timely enough fashion? Fraley (2008, p. 1098) puts the question this way. "Will we witness it [the resolution] safely from the secure perspective of an intellectual alternative that portends cultural survival or from the comfortably seductive perspective of an intractable mysticism that leaves us blissfully imperiled?"

With humanity, evolution has produced a species with extensive capacity for both consciousness and intellectuality. Are these enough to prevent the destruction to which coincidental contingencies are leading by producing collective behavioral mistakes? Without invoking inner agents, having written this book makes me think that I must think that they are enough, but I can certainly be wrong. They will be enough only if more humans, especially more natural scientists, participate in expanding educational options for far greater numbers of humans in the sciences, especially, at this juncture, in behaviorology, which helps us understand ourselves and our shared place in the web of existence that we call life, and thus helps the realities of global problems evoke more effective solution behaviors among more people.

Chapter Conclusion

As we finish this chapter, pause and ponder, as part of our usual ongoing activity, what else behaviorology can contribute to helping traditional natural sciences solve global problems when we consider the account of consciousness introduced in this chapter. All these considerations about consciousness stem from the philosophy of science that we call radical behaviorism (which we described in Chapter 1) that continues to inform the natural science of behavior, behaviorology. This natural science experimentally studies and interprets human nature and human behavior, and provides the derivative engineering technologies for effectively addressing accessible independent variables in ways that bring about improvements in behavior. The contingencies of global problems are inducing us to implement these technologies not only at home and work, in education and diplomacy, and in interpersonal relationships, but also in the global problem–solving applied–behavior fields of recycling, sustainable lifestyles, the management of dangerous asteroids, resource and biodiversity protection, dealing with overpopulation, and so on.☘

References (with some annotations)

Fraley, L. E. (1994). Uncertainty about determinism: A critical review of challenges to the determinism of modern science. *Behavior and Philosophy, 22* (2), 71–83.

Fraley, L. E. (2008). *General Behaviorology: The Natural Science of Human Behavior.* Canton, NY: ABCs. See Chapter 27 for more on consciousness.

Fraley, L. E. & Ledoux, S. F. (2002). Origins, status, and mission of behaviorology. In S. F. Ledoux. *Origins and Components of Behaviorology— Second Edition* (pp. 33–169). Canton, NY: ABCs.

Hayes, S. C. & Brownstein, A. J. (1986). Mentalism, behavior–behavior relations, and a behavior analytic view of the purposes of science. *The Behavior Analyst, 9,* 175–190.

Skinner, B. F. (1963). Behaviorism at fifty. *Science, 140,* 951–958.

Thompson, T. (2008). Self awareness: Behavior analysis and neuroscience. *The Behavior Analyst, 31* (2), 137–144.☙

Chapter 22
Cultural Concerns of Life, Personhood, and Death

Chapter 21 extended our coverage of current scientifically consistent answers, for some of humanity's ancient questions, into the area of consciousness. Basically, we described the term *consciousness* as our tact for the natural chaining of cascades of nerve impulses that comprise the neural behavior–behavior relations (i.e., behaviors, as real events, evoking further behaviors) that play functional, supplemental roles in the contingencies on more accessible overt behaviors. A common covert chain of consciousness behaviors includes raw sensation, awareness, recognition, comprehension, observation, and reporting (and then overt reporting) with the complexity of such chains dependent on the presence of verbal repertoires from prior conditioning. This topic comprised the third of our topics for which our *Part II* warning—about completing *Part I* before going into *Part II*—was particularly pertinent.

That *Part II* warning pertains as well to the topics of this chapter, ancient topics that invite and demand scientifically consistent answers, the topics of life, personhood, and death, with the warning applying particularly to personhood. Our coverage introduces scientific accounts for each of these topics in turn, even though these topics invariably overlap each other extensively. (To explore these extensive overlaps, especially the implications for our culture of having invested so thoroughly, and for so long, in superstitious alternatives to natural science, see Chapter 28 of Fraley, 2008.) Let's look at these topics, beginning with the living body and what we mean by *living*, what we mean by *life*.

Life

Life inevitably intertwines with personhood, but agentialism can taint each of these, and death as well. As usual we begin by setting aside the ancient, cultural traditions that imbue these terms with agentialism. Early and long–enduring, pre–scientific contingencies compelled ultimately unhelpful descriptions of each of these topics: *Life* referred to animation from the breath of a god. *Personhood* referred to a mystical behavior–directing, and so responsible, inner entity, religious or secular, that inhabited the animated (i.e., alive) body. And *death* referred to the departure of the entity from the body, leaving it no longer alive. Today, instead, *life* scientifically tacts a range of levels of chemical complexity that feature processes involving energy exchanges among simple and complex chemical units that change these units and their surroundings, processes such as evolution and conditioning, even consciousness and culture.

Similarly, *personhood* scientifically tacts the whole repertoire of behavior that a physiological body, by its nervous system—all products of natural life processes—mediates when environmental energy changes produce behavior. And *death* scientifically tacts the results when natural processes reduce complex chemical units to simpler levels of chemical structures and functions, levels that no longer evoke the tact "it's alive" but instead evoke the tact "it's dead."

Even in current vernacular usage, the term *life* involves the status of living that an untestable inner agent animates; supposedly the agent animates a lump of matter making it alive. Of course we reject this notion as it not only confounds pre–scientific and scientific accounts of the involved events but it also confounds our interrelated topics of the alive body and its mediated behavior, life, and personhood. To enable a scientific discussion that clarifies the interrelationships, we treat the topics separately and without the agentialism. For starters, while we define personhood behaviorally, we define life, the alive body, biologically.

A Scientific Meaning of the Term "Life"

In general scientific terms, *living* refers to the body–mediated behavioral activity that results from environmental energies affecting a lump of matter that shares enough characteristics for us to tact it as alive, as an alive body, as a life form or, more pertinently, as a form of life. Yet what are these characteristics? What is this *life* to which we refer when we say "a form of life?"

Incorporating and extending the understanding of life and its characteristics as our biology colleagues see them, for us life refers to forms that reside on part of a continuum of chemical complexities that extends from non–life forms to life forms, with an area between these where realities continue to evoke debate about tacting some forms as living, as life. This chemical continuum derives from the circumstances specified in our *Law of Cumulative Complexity* (i.e., the natural physical/chemical interactions of matter and energy sometimes result in more complex structures and functions that endure and naturally interact further, resulting in an accumulating complexity):

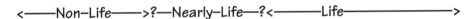

On that continuum, as lumps of matter accumulate complexity, non–life forms grade gradually into life forms. At some point forms of chemical combinations still show too few chemical characteristics to evoke the tact "life" from even the most thoroughly conditioned biologist. After this point come some forms of chemical combinations that show increasingly complex chemical characteristics that evoke tacts such as "not quite non–life" to "almost life" to "proto life" to "maybe life" and to "life, except..." The constant shifting on how to define "life" characterizes the stimuli in this "gray" part of the continuum, stimuli that thus evoke these kinds of ambiguous responses. Biologists who study the origins of life find a range of increasingly complex chemical structures and functions (e.g., lipid barriers, RNA, DNA, prokaryotic cells, eukaryotic cells) the accumulation of which leads to the definition for an unambiguous notion of "life." At this point some lumps of matter finally evoke the tact "it's alive," or "it's life," while some just slightly simpler forms next to it evoke ambiguous tacts rather than the "it's not alive" or "it's not life" tacts.

Indeed, no thin line exists on one side of which we find non–life and on the other side of which we find life. While some forms of chemical combinations show complex chemical characteristics that unambiguously evoke the tact "life," no qualitative difference, with forms that fail to evoke this tact, has occurred; only a continuous gradation in chemical complexity has occurred, all entirely naturally. Thus the term *life,* and the chemical realities that evoke this term, merely help control our responding to complex phenomena in ways that evoke other responses that deal successfully with the world around us, ways that we call scientific, ways that may relate supportively to our survival.

Some Implications of the Term "Life"

This entirely natural conception of life, and by extension personhood and death, if popularized throughout the culture in a timely manner, may prove to be one of the most culturally valuable developments of science, for it can countercontrol that most pernicious cumulative error in human history, the error that we tact as theological and secular agentialism, and the anti–science superstitions that these engender. This error threatens to prevent adequate and timely human solution responses to the stimuli of global problems. Thus the entirely natural conception of life and personhood, and so forth, a conception which behaviorological analysis presently provides, supports the sustainable survival of life as we know it.

The continuum from non–life to life has, of course, another end, the end where physiological structures, and the functions they support, begin a disintegration on the level of an individual organism. This disintegration occurs relatively rapidly compared to the organism's lifespan, and the tact for such disintegration is *death.* We will return to the topic of death, with respect to some practical matters for individual humans. Meanwhile this other end of the continuum from non–life to life also applies at the species level. At this level biological *extinction* becomes the appropriate tact.

Our emphasis on behaviorologically answering ancient questions includes the status of all human scientific endeavors as behaviors under the control of natural laws. Here that emphasis carries to considering life, personhood, and death mostly in relation to individual human organisms in the same manner in which, throughout this book, we have considered the laws of environment–behavior functional relations (e.g., antecedent and postcedent stimuli and their effects on non–verbal and verbal behavior including values, rights, ethics, morals, and consciousness). Having touched on life, and before considering death—which we will cover in terms of dignified dying—let's elaborate on life in terms of personhood.

Personhood

Again, setting aside the cultural traditions that misconstrue the term *personhood* as representing some mystical inner agent or the behavior–initiating activities of one, our scientific construal of personhood refers to the full potential and actual *repertoire of behavior* that a body is capable of mediating due to the structures, neural and non–neural, resulting from its genetics and conditioning history. These constitute the *person,* and thus the person only continues to exist while the body remains capable of reacting to the environmental energy changes that evoke the body's mediating of these behaviors.

Some Personhood Considerations

Those behaviors include not only the general run of neuro–muscular non–verbal and even verbal behaviors that our laws of behavior well explain in accessible and applicable detail, but also the more complex neural behavior which those laws account for interpretively under labels such as thinking, consciousness, and "sentience." Neural behaviors like these, rather than representing actions of untestable inner agents, instead refer to the most complex manifestations of personhood, even if only observable by the public of one. Indeed such neural behaviors, along with all behaviors, rather than implying a person, *are* the person.

Sentience. However, while we covered thinking and consciousness already, especially in our last chapter, the term *sentience* deserves some separate attention. People reserve this term nearly exclusively for fellow humans, using sentience to indicate something supposedly special about people. In this case, though, what people see as special about people remains outside science. Sentience implies a supposed and unmeasurable *qualitative* difference between humans and other species, an implication rather in contradiction with the finding, ever since Darwin, of only *quantitative* differences among species, including the human species. Sentience says that something about humans resides above or outside of nature, a status generally characteristic of putative inner agents. While the kinds of stimuli that evoke verbal–response terms like consciousness and thinking can also evoke the term sentience, such stimuli seldom actually evoke this term. Instead, the stimuli that evoke the term sentience are generally the same stimuli that evoke such terms as mind, psyche, or self. While some might see "sentience" as another term involving neural behavior, the term sentience retains agential implications that we may better handle by simply setting it aside as a non–scientific category.

Goodness and badness. Furthermore the scientific construal of personhood separates behavior repertoires from the notion of inherent goodness or badness of either "persons" or bodies. Bodies only mediate behavior that environmental variables evoke, and that behavior is the person. Characteristics of goodness and badness stem from the kind of contingency relations that we described as values, rights, ethics, and morals. Many cultures often consider an as yet unconditioned body, such as a newborn infant, as essentially innocent of such characteristics, and possibly not yet even "human," not until some observable response establishes personhood. Some cultures may even specify various indicators that they accept as establishing personhood. For example, in the Navajo culture, many acknowledge a baby's first laugh as representing the first expression or indication of personhood (e.g., Simpson, 2003, p. 54).

Some Personhood Implications

Our traditional, agent–laden cultural conditioning generally compels us to respond as though a body's personhood, as a characteristic of an inner agent, retains a continuous and steady–state status. Science, however, having compelled setting inner agents aside, also compels recognition responses that we actually see personhood occurring occasionally, with ups and downs. With personhood inseparable from the neural mediation of environmentally evoked behavior, and *with behavior occurring occasionally*—more at some times and less at other times—the variable status of personhood fails to surprise us. This variable status derives from a range of sources. While environmental contingencies remain responsible for much of the occasional status of behavior occurrences, and hence for some part of the variable status of personhood, another part of this variable status happens in accordance with various factors that change the body. Good nutrition, bad nutrition, disease occurrence, disease recovery, more stress, less stress, and many other factors (e.g., more education) drive changes in a body's capacity to mediate behavior and, as a result, we see personhood manifesting not only intermittently but also with changes in magnitude.

With personhood thus occurring as behavioral process, we see those factors bringing about repeated waxings and wanings of the person, and we can track these on various time scales throughout the time a body evokes the tact "living." As the bodily effects of conditioning begin to accumulate early in life, building a more and more complex neural and neuro–muscular behavior repertoire, we see the person generally waxing, coming into existence, not as any sort of inner agent but as a sophisticated and interrelated collection of behaviors. At the point when environmental events fail completely to evoke any behavior, particularly neural behavior, due to bodily degradation leaving the body incapable of ever again mediating such behavior, the person wanes completely, going out of existence permanently, a phenomenon that we call person death, which differs from body death. Sometimes a body seems temporarily to lose the physiological capacity to mediate evoked neural behavior, but should the body eventually regain that capacity and so mediate some of that neural behavior, the person may then manifest again. All this helps us to comprehend the nature of a person, the nature of personhood.

Many additional culture–related implications stem from the contradiction between the mystical and natural–science views of personhood. Fraley considers his own 2008 treatment of the topic—to which he devoted over 80 pages—as still incomplete (personal communication). Some of these implications include the abstract person, the relative worth of persons, and the cultural costs of superstitiously misconstruing personhood. These cultural costs include the financial costs of developing and maintaining undergraduate and graduate degree programs in mystically grounded subject matters, both theological and secular, in so many colleges and universities. Other implications involve the costs of the behaviorological engineering that the culture needs to reduce some of the costs of personhood misconstrual. These engineering costs include establishing undergraduate and graduate degree programs in behaviorology subject matters in colleges and universities, programs that can produce demonstrable products and outcomes which reimburse society some of the costs of the programs. And we still must mention the implications of our natural–science account of personhood as it applies to helping not only to foster humanity's survival along with the rest of life on this planet, but also to foster our own further evolutionary development—biologically, emotionally, intellectually and, perhaps most important, culturally, if we are to continue to survive, particularly beyond our current global crises.

Does Life Have any Meaning?

Before moving on from considering life and personhood to considering death, let's extend that last implication of our personhood analysis, about fostering our development and supporting our survival beyond our current global crises, to a common ancient question: *Does life have any meaning?* Many answers exist for such a question, and they may all evoke a response of at least seeming partially right, although only some will evoke a response of also seeming scientifically reasonable. Certainly, some people feel that life has meaning, and others feel it lacks meaning. Indeed the same person may feel one way at times and the other way at other times. Having addressed feelings as part of the natural science of behavior, we can recognize not only the validity of all of these feelings but also the circumstances that produce them.

Traditional (i.e., pre–scientific) cultural contingencies, however, in competition with scientific contingencies, sometimes evoke a superstitious reaction saying that a purely natural–science analysis removes any meaning from life. Natural–science analysis, though, only reasonably removes the already invalid *unnatural or supernatural meaning* to life, which remains a purely natural phenomenon. What that superstitious reaction leaves unstated, though, proves most meaningful, because by leaving no mystical meaning to life, the purely natural–science analysis leaves only the meaning to life that remains ever present. This meaning puts our inherent nature as *a part of nature* under contingencies that induce interactions with the rest of nature that affect at least the viability of our continued status as a *living* part of nature, as opposed to the status, in nature, of fossils of yet another extinct species. Will we become another extinct species (and likely take lots of other species into extinction with us) or will the complexities of behavior, with which nature endows us, enable us to survive? Only time will tell although, again, we are running out of time. Will the contingencies, of which we are a part, evoke enough listening and other appropriate activities supportive of solutions to our problems in the required timely manner? Participating in these contingencies in ways contributing to these solutions seems at least to me as more than adequately meaningful, a potential meaningfulness that ever remains present for everyone to a degree, again, that the contingencies determine. In this sense one basic meaning of life inheres in the contingencies on helping our species remain unfossilized, remain a non–extinct, viable species.

Your successfully plowing through this book to this point implies that such meaning already affects you as well as me. On the other hand, if the cultural contingencies in effect for your living fail to induce such meaning for you, then so much the worse for the culture, so much the worse for the species, so much the worse for the planet. If the lack of this meaning of life extends too far among the population, then extinction may follow too soon. And that leads us now to consider death.

Death

With that grip on some scientifically valid meaning of personhood and life, you can begin to appreciate the greater care we may need to take as a culture regarding quality of life in comparison to quantity of lives. This is a theme to which science fiction writers have long alerted us. Until rather recent times (e.g., 100 years ago to ten years ago) the contingencies on species survival induced and rationalized reproduction. This favored large quantities of living human bodies trumping the quality of these human lives. However, now that we have exceeded the planet's carrying capacity with overpopulation numbers that threaten our continued existence, the contingencies on species survival actually favor fewer humans. If the contingencies on the culture induce letting its natural scientists and other citizens begin to help everyone humanely resolve the overpopulation problem, then we might possibly avoid the kind of decimation of population numbers that otherwise remains the likely outcome of current circumstances (e.g., through extreme weather and any accompanying disasters, pestilence, war, and other outcomes). Producing more behaviorologists to work on such problems would be of particular help to the culture, because the overpopulation problem is, among all our global problems, the most pernicious and perhaps the most thoroughly imbued with behavior components, as we previously discussed (e.g., at the end of Chapter 19). If the quantity of humans reduces carefully and gradually rather than insensitively and drastically, the necessary shift in contingencies can also begin to favor efforts at improving the quality of life.

That shift in contingencies, to favor concerns about the many aspects of life quality, may have already begun with the increasing concern, in many communities around the globe, about end–of–life issues for individuals. As a result of natural–science developments leading to improvements in medicine, an ever increasing number of human bodies are living long enough to face death through one or another terminal illness. Let's review some of the implications from natural behavior science regarding not only death but also the processes that ever more people are calling dignified dying. For details beyond our introduction here, you can consult the book *Dignified Dying—A Behaviorological Thanatology* (Fraley, 2012).

Dignified Dying

Widespread cultural contingencies have traditionally emphasized the suffering of those left behind when someone dies, rather than the suffering of those who are dying. Of course, throughout the historical and pre–historical tenure of our species—up until a few hundred years ago—many deaths occurred violently and most deaths occurred relatively swiftly; few lived long enough to develop the kinds of conditions that produce the slow, protracted dying of a terminal illness. The recent change in this pattern, wherein more of us are living long enough to die a protracted death from a terminal illness, has altered the contingencies, giving rise to the movement for dignified dying, a movement that emphasizes reducing the suffering of the dying individuals at least as much as, if not more than, the suffering of their survivors whose life continues and for whom many helpful cultural resources already exist. Scientifically our point is to provide some additional intellectual foundations to add to the usual emotional foundations for better treating—that is, treating with more dignity—those who must undergo a protracted dying process.

A natural–science thanatology can clarify some of the ethical and other issues in support of dignified dying. As a start to carrying these concerns toward practical and effective levels, we will introduce five areas of consideration. (a) We consider three basically different kinds of death concerning the body, the person, and social relations. (b) We consider the extra strain on both the dying and the survivors from the gradual degradation in social contingencies that accompanies the gradual degradation of the dying person and the dying body. (c) We consider the confounding of the person and the body. (d) We consider some improvements that behaviorology can provide to reduce the aversiveness of these burdens on both the dying and the survivors, including a new cultural practice that Fraley (2012) introduced, and which he calls a "Foreniscon." And (e) we consider some questions regarding medical ethics.

Three kinds of death. Bodily mediation of the behavior repertoire that constitutes personhood leads us to differentiate between the first two kinds of death, body death and person death. The usual deterioration in social contingencies during the slow dying of person and body leads to the third kind of death, social death. Indeed, all three kinds of death can occur at different times across a extended time span.

The term *body death* tacts the irreversible cessation of all body–sustaining physiological activity, the kind of activity that defines what we call living. At the cellular level, this death can take from minutes to hours to complete. With substantial medical maintenance, the early parts of this process can stretch from days to years, because modern medical life–support equipment and procedures can keep enough respondent and physiological processes operating to prevent complete cellular death. After complete cellular death, the body continues various gradual processes of disintegration that we call decomposition. Prior to body death, a period of physiological functioning and basic respondent reactions may proceed *only* due to medical life supports. During such a period, operant neural behavioral function may not only be completely missing but may also no longer be possible; this takes us to person death.

The term *person death* tacts the irreversible termination of the physiological functions that mediate operant neuro–muscular and neural behavior. This of course includes the complex neural behaviors of consciousness. At this point neither internal nor external stimuli can ever again evoke the behaviors that they previously evoked, the behaviors that constituted the person. The person is gone. Meanwhile other physiological functions may continue at a level that leaves the body alive. Person death occurs when the person stops happening because the body parts that mediate the person degrade to the point where they are permanently unable ever to mediate the person again. Often body death and person death coincide. This happens, for instance, under extreme trauma such as upon contacting the ground after falling from a height, or after suffering an irreversible heart function failure. However, person death from the degradation of the body parts that mediate the person can occur slowly, particularly during protracted dying from terminal illness, and lead to complete person death before, and sometimes long before, body death occurs. Changes that occur in some of the contingencies operating during such protracted person death take us to social death.

When person death precedes body death, the term *social death* tacts the interpersonal results that sometimes occur due to the gradual loss of contingency components between the dying person and his or her survivors prior to person death. The interpersonal relations, between the terminally ill person and her or his loved ones and friends or associates, gradually fail as the degradation of body and person brings about a reciprocal degradation of the social contingencies responsible for those interpersonal relations, a degradation that all parties find confusing and distressful. Let's consider further how social death happens.

Social contingency disruptions. Every person (i.e., the evoked behavior that every body mediates) participates in interconnected webs of mutually controlling, and generally reinforcing, contingencies with a lesser or greater number of other persons. These social contingencies comprise the bulk of interpersonal relations. At times in these contingencies, one's behavior evokes the behavior of others that reinforces one's behavior. At other times in these contingencies, another's behavior evokes one's behavior that reinforces the other's behavior. That is, at some times your behavior produces reinforcers from others while at other times your behavior provides the reinforcers for another's behavior. Interpersonal relations not only involve such social contingencies but also require physiologies healthy enough to mediate the related behaviors.

When someone becomes terminally ill, the resulting progressive physiological deterioration reduces the body's capacity to mediate the necessary social behaviors. The physiology of the dying person increasingly becomes incapable of mediating behavior that provides the relevant reinforcing consequences that it provided to others in the past. The dying person's physiology also becomes increasingly incapable of mediating behavior that evokes the reinforcing behavior of others. As each of these occurs, they together progressively eat away at previously well established social–contingency relations. Business and professional associates, friends, and loved ones become increasingly disconcerted as their previously occurring behaviors, which reinforced their dying loved one, fail to occur (i.e., because the evocative stimuli can no longer occur, given the reduced capacity of the dying person's physiology to behaviorally produce those stimuli). They also become disconcerted as reinforcers for their current behaviors also fail to occur (i.e., because the dying person's physiology is also unable to mediate these reinforcers). Sometimes the coercive quality of these contingency changes on some of these others produces their withdrawal from further social contact with the dying individual, and not always at an appropriate or helpful time. We describe the end results of this process, the resulting social isolation even in the presence of some surviving friends and loved ones, as the social death of the person.

The social–death process, the gradual degradation in social contingencies that accompany the gradual degradation of the dying person and body, exacerbates the strain on both the dying and the survivors. While nothing can completely eliminate the stress of a friend or loved one's—or your own—protracted dying, some steps might reduce some unnecessary components of this stress. The behaviorological analysis of the contingencies surrounding protracted dying prompt new practices related not only to ethics and person/body death, but also to easing the bereavement of survivors and especially to managing the social death experience. While we will leave a thorough discussion of such practices to other sources (e.g., Fraley, 2012) let's touch on them only after considering the confounding of person and body that natural processes automatically condition.

Confounding the person and the body. A complicating factor in all the relations surrounding protracted dying involves the early occurring, general and long–term confounding of the person and the body, a confounding in which either can seem to evoke behavior from other persons, due to their ever–present pairing. While the body can serve evocative stimulus functions for the behavior of others, these functions remain separate from the status, living or dead, of the body (e.g., both a living and a dead body can evoke the response "That body stretches over two meters from head to toe").

However, no body ever initiates the behavior that it exhibits; a body only mediates behavior, with all the behavior that a body mediates constituting the total person. At various times various parts of this behavior, this person, evoke behavior—relevant to the living status of the person—from other persons (e.g., the person–behavior of calling out "Anyone home?" at an open door evokes another person's response of saying "Come on in"). Thus, only the *person* that a body mediates (i.e., only the behavior, which occurs one or another response at a time) evokes the person–relevant behavior of other persons.

The behavior *is* the person; the person is the collective set of behaviors that respondent and operant processes have conditioned a particular body to mediate under the range of external and internal environmental controls, and these behaviors evoke the behavior of other persons. Inevitably, however, this means that if the person occurs (i.e., if behavior occurs) then the body is also occurring, for we cannot separate the behavioral person, a process, from the physiological, mediating body. Meanwhile, the behavior that constitutes the person continually evokes behaviors from other persons. This continual occurring–together status of person and body, *this repeated pairing of the person and the body,* can leave the *body* also evoking responses from other persons, responses that otherwise only the person could evoke, as if that *body* is the person it mediates, which it is not (i.e., the body only constitutes the necessary physiological, mediating interface between environmental energies and behavioral responses, particularly person–related behaviors). However, a problem arises in that the permanent pairing, during life, of person and body leaves *even a dead body* evoking responses from other persons as if that dead body were the person it had mediated before death, for example, "Wake up. Speak to me. Tell me how to help you." Since such responses produce no reinforcers from the dead body, they undergo fairly rapid extinction.

Furthermore, a body that remains biologically alive (e.g., via medical life support) but person dead (i.e., it permanently lacks the operant, and much of the respondent, capacity for behavioral interaction with the environment) can also still evoke the previously person–relevant behaviors of others, due to the constant pairing during life of the person and the body. Whether in this state or also biologically dead, a body provides a most unreinforcing stand–in for the previously mediated person, and provides a not entirely necessary increment in the general aversiveness of the situation for survivors. But, reasserting our interest in reducing aversiveness not only for survivors but even more for the terminally ill, let's now consider a couple new and scientifically grounded cultural practices that help us better handle end–of–life issues.

Aversiveness reductions including the Foreniscon. Current cultural practices concerning death and dying often originate in traditional folklore, much of which comes from a range of superstitions. Since these practices arose when few efforts could stop pre–death suffering, these traditional practices also typically accept the horrors that nature visits on the dying, including the terminally ill, and their survivors. In addition these superstitious practices even push participants, both the dying and survivors, to endure every measure of pain and suffering that death, including protracted dying, has to offer, even to the extent of making a putative virtue of such endurance. We must begin to replace these practices with newer, more humane practices that afford substantive relief from these horrors. Some modern practices already move in this direction (e.g., hospice care, and the sensitive counseling that even some scientifically uninformed professionals provide, which nevertheless coincides with scientific realities). Other practices to reduce the aversiveness of these burdens on both the dying and the survivors emerge as implications of our behaviorological analysis.

A good place to start involves increasing, through the inclusion of behaviorology in standard educational programs, everyone's comprehension of the natural processes that are operating during person death and, especially, social death, as well as body death. This comprehension can reduce the confounding of person and body, and mitigate some of the ensuing confusion and distress, especially during periods of slow dying.

Furthermore, such comprehension can also show us the previously unrecognized scientific rational for the general, intuitive practice of saying good–byes before the factors leading to social death deprive the dying person of all capacity for such interactions. Fraley (2012) introduced one new cultural practice directed precisely at enhancing, formalizing, and culturally institutionalizing this general, early good–byes practice. He calls it the *Foreniscon* (short for "**For**mal **En**ding of **I**ntimate **S**ocial **Con**tacts") and it revolves around improving the social–death experience by programming parts of it for both the dying and their survivors while following the pattern of ceremonial recognition of life's major events (e.g., birth, marriage). The Foreniscon involves one or more arranged, ceremonial occasions (somewhat like Wakes that you attend) at which you say your good–byes to various and increasingly intimate groups *while you still can say goodbye* and, importantly, after which you do not see those in each group again.

An endeavor like a series of Foreniscons takes substantial planning, an effort likely beyond the capacity of a terminally ill person. Yet a person in this position likely also needs help with several other end–of–life considerations, probably including business, legal, and counseling as well as medical and other concerns. A coordinated, team approach would seem to be in order to assist in these matters with the kind of ethical and concerned professionalism that promotes the fair and dignified conclusion of these affairs.

Along with the business, legal, medical, and other experts on such a team, at a technical level a behaviorologist team member can provide not only the scientific thread that ties the team together but also the necessary and appropriate professional therapeutic tools to help the dying person—and the survivors—comprehend all that is happening and deal, emotionally and intellectually, with the different parts effectively. The behaviorologist should also apply her or his knowledge and skills to help reduce the problems that any survivors face during the bereavement process, particularly after the complete body death of their intimate, which can even reduce the length and negative effects of the bereavement–process period.

Questions of medical ethics. While all those possibilities receive further development in some of the resources listed in the references, we should mention one other important area for which behaviorology is relevant. All the protracted–dying events that happen under current cultural practices generally occur under medical supervision. Some of these practices, which I cannot recommend (Can you?) can include treating terminally ill patients in ways that compel them, often against their preference statements, to experience all the suffering and horror of body degradation and person death and even social death. Sometimes this even extends to shipping such patients off to often privately owned, for–profit "hospitals" specializing in procedures specifically designed mainly to keep patients bodily "alive" for as long as possible, regardless of their experience, while milking every possible dollar from the "health" care system. Such treatment also compels survivors to experience additional kinds of suffering and horror during the social death and person death, and even *beyond* the body death, of their terminally ill intimates.

Those circumstances appropriately evoke reviews of all applicable ethics, including medical ethics. For instance traditional medical ethics put many capable doctors in desperate conflict with the contingencies driving their own preferences for reasonable, compassionate, and dignified patient care. Sometimes ethically acceptable medical action, or inaction, under strained medical ethics that need updating, alleviates the *very end* of the protracted dying experience for the terminally ill, but this leaves what some would call undignified patient abuse necessarily occurring during the periods of social death and person death. Improvements in these medical ethics would not only generally improve end–of–life experiences, for the terminally ill and their survivors, but such improvements also bear substantial implications for resolving questions about medically assisted death. Some countries (e.g., Belgium, Luxemburg, and the Netherlands) along with several states in the U.S.A., have already made notable progress in resolving questions of medically assisted death. While you can apply the values, rights, ethics, and morals parts of an earlier chapter to these concerns, the two *Fraley* resources listed in the references contain more thorough material worthy of your attention.

Chapter Conclusion

Our treatment of this chapter's topics can evoke grave concerns about certain historical trends in our culture. In general terms these concerns arise from observing some long–standing contingencies driving our cultural investment so thoroughly, and for so long, in superstitious alternatives to natural science in essentially all areas.

Those observations leave us concerned about whether or not more recent scientific contingencies can reverse the current trends that threaten our survival. Far beyond this chapter's concern with life, personhood, and death, such contingencies extend to our present scientific efforts to solve all aspects of global problems, including behavioral aspects. While we cannot know the final outcome beforehand, we can make predictions from certain past patterns of contingencies that included compelling *continued trying* (e.g., the contingencies that compelled the extensive behavioral efforts that gave rise to natural science in the first place). We can predict that these recent patterns of contingencies compelling further extensive scientific, including behaviorological, efforts to solve global problems, will succeed, because these contingencies also include components compelling *continued trying*. While only "giving up" guarantees failure and extinction, ask "yourself" what efforts your contingencies are promoting regarding supporting the natural sciences, including behaviorology, in solving global problems.❧

References (with some annotations)

Fraley, L. E. (2008). *General Behaviorology: The Natural Science of Human Behavior.* Canton, NY: ABCs. See Chapter 28 in general, and particularly pages 1293 through 1301 regarding questions about death.

Fraley, L. E. (2012). *Dignified Dying—A Behaviorological Thanatology.* Canton, NY: ABCs. Chapter 3 is particularly relevant to medical ethics.

Simpson, G. K. (2003). *Navajo Ceremonial Baskets.* Summertown, TN: Native Voices.☙

Chapter 23
The Unexpected Nature of Reality and Robotics

*I*n Chapter 22, we addressed some scientifically consistent answers to ancient human questions about life, personhood, and death. We used *life* to tact a range of levels of chemical complexity that feature processes involving energy exchanges, among simple and complex chemical units, that change these units and their surroundings, processes such as evolution, conditioning, consciousness, and culture. We next used *personhood* to tact the whole repertoire of behavior that conditioning has made a physiological body, through its nervous system, mediate when environmental energy changes produce behavior. Then we used *death* to tact the results when natural processes reduce complex chemical units to simpler levels of chemical structures and functions, levels that no longer evoke the tact "it's alive" but instead evoke the tact "it's dead" or, at the species level, the tact "it's extinct." Beyond such body death, we also considered person death, and social death, in the context of scientifically supporting death with dignity for the terminally ill.

All those areas not only comprise natural products of natural life processes but they also come under the topics for which our *Part II* warning—about completing *Part I* before going on to *Part II*—was particularly pertinent. That *Part II* warning pertains as well to the related topics of this chapter—reality and robotics—one ancient and the other more recent. The reality topic in particular involves answers to ancient human questions. Our coverage introduces scientific accounts for each of these topics in turn. (To explore these accounts more extensively, see Chapters 29 and 30 of Fraley, 2008.)

Reality

Ahhh, reality. "Not only is reality stranger than we imagine, but it also is stranger than we can imagine." With words quite like these, several science commentators over the last century have characterized the results of closely examining the nature of reality. Our commentary on reality may show them as more *literally* correct than they ever imagined. Yet a century from now, science commentators will likely report both how this chapter fits right in with their characterization, and how it too falls short of the realities regarding reality. Ahhh, the value of self–correcting science.

Knowing Versus Existence

Even at this early point, however, a caution may be in order. This chapter's introductory–level conclusions about reality, to which our behaviorological analysis leads, *only* concern what we can *know* about reality and how we can know it; our conclusions remain largely irrelevant to *existence* questions, such as questions concerning whether or not reality "exists" or what "really" exists in, or makes up, reality. But if a reader's past conditioning imbues such existence questions with value, then our conclusions can come across as answers— which they are not—to those questions, a result that can produce confusion, disappointment, and other negative emotional reactions. Fraley's (2008) treatment of reality goes far enough to involve these questions but, again, here we stay close to the level where conclusions about reality only concern what we can know about reality and how we can know it.

About reality. Prior to stating our behaviorological conclusion about reality, we should recognize that it develops as a direct extension from the principles, concepts, methods, practices, research, and philosophy of science that make up the behaviorology discipline. This reality conclusion develops from taking all of the laws of behavior, in everything that we have covered in this book, and seeing where the combination leads. Even if our past conditioning makes the present conclusion elicit emotional discomfort, we can still only ultimately benefit from improving our knowing of reality.

To say it baldly, our current behaviorological conclusion about reality states that the firings of sensory neurons evoke neural behaviors that *are* (i.e., that establish) our reality. In other words, due to the firings of sensory neurons, we individually *neurally behave* reality. That is, "Reality *establishes.*" Reality establishes, due to our neurally behaving it due to the firings of sensory neurons. This sounds confusing only because English seems to lack a better term than "establishes" for these events, and some typical substitutes (e.g., "presents," or "occurs," or "happens") provide little help. Reality establishes due to the firings of sensory neurons evoking our neurally behaving reality. For example seeing, hearing, tasting, and so forth, are all neural behaviors that begin occurring due to the firing of sensory neurons, and these neural behaviors establish what a body "sees," "hears," "tastes," and so on. That is, with the firing of sensory neurons, what one "sees," "hears," "tastes," and so on *establishes.* Furthermore, since we have no sensors going beyond the sensory–neuron parts of our nervous system, this evoked neural behaving of reality is our only source of knowledge about reality. This means that this evoked neural behaving of reality is *also* the only source of knowledge about not only the reality of our sensory neurons, but also the reality of any stimuli or stimulus energy traces that produce the firing of sensory neurons that thereby evoke behaviors, as well as the reality of any behavior—neural or neuro–muscular—that the firing of sensory neurons subsequently induces. For instance, as a result of our still only contacting, *through our sensory neurons,* all the equipment that we consider as enhancing our sensors (e.g., from microscopes to telescopes and spectroscopes, as part of the visual sense mode in science) we still only know about the reality of this equipment, and about whatever the equipment produces, through the firings of sensory neurons. And if this paragraph fails to leave you at least a little confused, then you can turn to the last chapter, as the remainder of this chapter mostly attempts to disentangle some of that confusion (and add a little about robots and robotics).

Reviewing "knowing." A little review, however, about *knowing* can help. We discussed knowing more in the chapter on consciousness than elsewhere. This word relates to epistemology. More specifically, it tacts parts of the cascade of purely neural responses of consciousness that overlap in rolling, sequential chains that external or internal stimuli evoke when energy traces bring about raw sensation, awareness, recognition, comprehension, thinking, observing, reporting, and so on.

Actually our concern with the behaviorology of reality began back in Chapter 1 where we touched on consciousness and knowing. There we pointed out that behaviorological analysis leads to the same conclusion about reality that Hawking and Mlodinow reached, through the logic of naturalism in physics, in their book *The Grand Design* (2010). With a little updating, let's restate our Chapter 1 conclusion about reality: Our neurally behaving reality is the sole source of knowing about reality, because we can get no closer to reality than the neural behaviors that the firings of sensory neurons evoke.

Three different perspectives. That conclusion starts with its beginning, according to its own, new perspective, which Fraley (2008) calls the *robotic* perspective. Yet, from a more common, current perspective—one which we call the *environmental* perspective—that conclusion seems to start with its end point. To deal with why that conclusion starts with "neurally behaving reality"—a statement which would seem to end a logical sequence rather than begin it (e.g., energy traces induce sensory neuron firings that evoke our neurally behaving reality)—we must differentiate three related perspectives.

The first perspective is the age–old *agential* perspective. This perspective constitutes a *pre–scientific* environmental perspective in which various forces in an environment, including magical forces, exert mere influences on an inner agent while this agent spontaneously initiates directives that tell the body what behaviors it wants done, that tell the body what behaviors to emit. (Due to this residual agential connection, the word, *emit,* rarely appears in this text.) Since the main variables in this perspective, the inner agents, reside untouchably (i.e., neither testable nor even measurable) in a mystical realm, the perspective remains divorced from effective application in practical matters.

Having set aside the agential perspective as pre–scientific, the next perspective is the current, agent–free scientific *environmental* perspective. This perspective informs the other chapters in this book. In this perspective only energy traces from natural, internal or external environmental stimulus events produce the bodily changes that *are* the mediation of behaviors. This perspective occurs throughout the natural sciences, and currently enables the practical, including behaviorological, benefits through which our culture justifies the expenses of scientific and engineering activities. Furthermore, in this perspective, environmental energy traces induce sensory neuron firings that evoke our neurally behaving reality, including the environment. While nothing precludes the existence of the environment beyond our capacity to neurally behave both it and knowledge of its reality, the perspective nevertheless shows that organisms lack the capacity for more direct contact with the environment beyond what our sensory neuron firings furnish.

The third perspective is the newer, scientific perspective that we call the *robotic* perspective. This perspective *expands from* the point of recognizing that we behave reality, because we can get no closer to reality than the behaviors that the firings of sensory neurons evoke. This means that these neuron firings constitute the *start* of behavior, including the start of knowing and any other consciousness behaviors, such that only behavior itself is available to establish reality (i.e., to establish the sights, sounds, tastes, smells, touches, and feels, including movements and dreams, to which we otherwise seem to be reacting). Furthermore, the robotic perspective currently seems somewhat removed from practicalities, although not inherently so as is the case with the agential perspective, yet it also seems to be the most consistently naturalistic perspective, and further research will clarify its potential practical benefits.

Fraley tacts this perspective as the *robotic* perspective, because it describes the limitations on perspective for a completely natural lump of organized matter, a category into which all known life forms classify, including people. I suspect that you can already sense the difficulties that some language problems are going to cause us; our language arose with contingencies inducing the agential perspective, and scientific contingencies have only just begun to induce linguistic "catching–up," so to speak, with the environmental perspective. Yet here we are trying to talk about the thoroughly naturalistic robotic perspective that, from the comfort of the familiar environmental perspective, seems to turn things all around. For this reason, and due to the introductory level of this book, we will try—admittedly without complete success—to stick with the environmental perspective. Perhaps solace can arise from recognizing that any shifting in perspective in this writing stems from contingencies evoking phrasing that seems to cover a point not only in the best way available but also with the least induced confusion.

In spite of all those difficulties, that description (i.e., that the *robotic* perspective describes the limitations on perspective that a completely natural lump of organized matter can have) applies broadly, equally well, to everything. With adjustments for quantitative differences, it applies to me and to you and to all other evolving biological lumps of "living" and behaving matter, as well as to any lumps of matter, such as electro–mechanical or digital lumps (e.g., computers and other robotic devices or robots) that contingencies induce other biological matter lumps (e.g., people) to build such that these robots also behave as well as meet scientific criteria justifying the tacts "living" and "it's alive." For example, a misbehaving computer often evokes such tacts as "it's alive" and "it has a life of its own." From the robotic perspective, these constitute entirely legitimate tacts that examples of one relatively primitive and naturally constructed (through contingencies on human behavior) life form evokes from examples of another, perhaps less primitive though equally naturally constructed (through biological evolution) life form.

Recomputing the Behaving of Reality

Now, this may be a good time to recall that these conclusions about reality only consider what we can know about reality and how we can know it. Here we keep these conclusions, and their connections with knowing behaviors, entirely separate from questions regarding whether or not reality really exists, or what really exists in any reality that exists.

Still, the point of elucidating those three perspectives concerned why our reality conclusion starts with what our scientific environmental perspective leaves us responding to as an end point rather than a starting point. Turning that conclusion around might more easily show the sequence that reaches that conclusion, although turning it around switches from the robotic perspective to the environmental perspective. Consider together some of our behaviorological discoveries that we have otherwise covered across a variety of chapters: (a) In the process that we called *direct stimulus control,* some behaviors, especially neuro–muscular behaviors, occur under direct control of particular stimuli, as in our driving–while–daydreaming example, without supplementation from neural–only and single–channel consciousness–behavior chains. (b) Energy traces from internal or external stimuli evoke neural chains of consciousness behaviors, which cannot happen any other way. And (c) these behaviorological processes ultimately address the accessible independent variables that are responsible for *why* behavior happens, while physiological processes address the accessible independent variables that are responsible for *how* behavior happens (i.e., how the physiology mediates behavior).

When we put those three puzzle pieces together, we get an environmental–perspective view of reality. *This perspective assumes* (which is just another kind of behavior) *reality in the first place* without necessarily characterizing it further. In this perspective we still behave reality, with our reliance on the firing of sensory neurons—as the sole source of knowledge about reality—constraining our behaving reality. However, in this perspective, reality "establishes" further into the event chain. Following a standard environmental–perspective time line, the environmental–perspective chain begins with energy traces, from whatever exists that constitutes reality, inducing the firings of sensory neurons; these firings evoke neural behaviors that establish reality and that chain to other neural and neuro–muscular behaviors.

A difference in perspectives. While the environmental perspective *begins* with the energy traces, the robotic perspective *begins* with neural behaviors establishing reality, including the reality of the sensory–neuron firings, the reality of the energy traces that induced the firings, the reality of the things/ events that produced the energy traces, and even the reality of the sensory neurons. That is, in the robotic perspective, the *sources* of the energy traces, the energy traces, the sensory neurons, and the sensory–neuron firings due to these energy traces, all remain parts of the reality that the neural behavior establishes. However, from the environmental perspective, the neural behavior establishes the reality of these parts only *after* the energy traces induce sensory–neuron firings that evoke the reality–establishing neural behavior. Confusing?

To remove one source of confusion, let's stick with the familiar environmental perspective. Even though our comments often relate to both perspectives, we cannot easily talk about both of them, or from both of them, at the same time. For instance in both perspectives—but stated from the environmental perspective—improvements in our neurally behaving reality result from the ongoing conditioning that affects the neural structures the functioning of which comprises the neural behaviors that establish reality.

Some reality implications. The current evolved state of our physiology provides no other access for inputs (i.e., energy traces) of any kind from further afield than our sensory neurons, as extensive or as limited as they might be. All inputs must come through our sensory neurons. This carries some important implications. For instance, any inputs that induce sensory–neuron firings that fail to evoke the *necessary* neural behaviors of knowing with respect to them (i.e., to the inputs) remain inputs about which we know nothing. They affect other of our neural and neuro–muscular behaviors (e.g., direct stimulus control) but their reality remains *un*established. As another and related implication, inputs that induce sensory–neuron firings that fail to evoke *accurate* neural behaviors of knowing with respect to them (i.e., to the inputs) remain inputs for which our neural knowing behaviors reflects an inaccurate establishing of reality. This leaves their reality *mis*established even while they then affect other of our neural and neuro–muscular behaviors, also inaccurately. We will see examples of both ***unestablished*** reality and ***misestablished*** reality.

Technical reality summary. Energy traces from environmental events affect sensory neurons the firing of which evokes neural behaviors, of which some establish reality. Other neural behaviors of consciousness constitute what we previously described as knowing. However, since the only contact points available to physiological organisms (e.g., the famous carbon units so common in science fiction) are the sensory neurons at the interface of the physiology and whatever else exists, we can only know of what exists by behaving it; we can get no closer to reality than the behaviors that the firings of sensory neurons evoke.

Here is how that would come out if we stated it from the robotic perspective. Humans (and likely—to the extent possible—other animals) neurally behave reality; that is, their neural behavior establishes the reality of whatever elicits or evokes (or consequates) their further neural or neuro–muscular behaviors, including whatever they "know" about reality (e.g., the sights, sounds, tastes, smells, touches, feels, movements, and dreams, as well as the sensory–neuron firings that evoke their neural behavior and the energy traces that induce the sensory–neuron firings, and even the sensory neurons).

Some reality examples. Regardless of which perspective we use, natural sciences have on several occasions clarified that realities differ, and change, according to the repertoire of the body that is behaving the reality. For example, in a dining room either from the first half of the twentieth century or from today, non–physicists behave the reality of an inch–thick tabletop through which they cannot push a coffee cup regardless of how much force they can personally exert in the effort; they behave the table top as "solid." Yet professional physicists of the first half of the twentieth century, making an instrumented examination of a small sample of the same tabletop, behave a complex atomic structure consisting mostly of empty space. And professional physicists of today, making a more advanced, instrumented examination of that same small sample of the same tabletop, behave a reality of even greater complexity, greater precision, and greater obscurity to the non–physicist.

Returning to ***un****established* reality, a lack of behaviors of knowing, regarding particular inputs, leaves the reality of these inputs unestablished. This means that, while these inputs can still affect us, they remain unknown to us.

As an example, once again consider daydreaming and driving. Energy–trace inputs from the road ahead induce sensory–neuron firings that evoke only driving responses, under direct stimulus control, while other stimuli are evoking knowing responses on the single–consciousness channel (i.e., we "know," and can often later describe, the theme, content, and direction of our daydream). As a result, we remain unaware of (i.e., we lack knowledge of) those road stimuli. They remain unestablished, and we cannot report them later even though they controlled our driving responses with at least minimal adequacy.

Returning to ***mis****established* reality, "inaccurate" behaviors of knowing (i.e., behaviors of knowing that other stimuli evoke, rather than the stimuli that induced the sensory–neuron firings) regarding particular inputs, leave the reality of these inputs misestablished. This means that these inputs likely affect subsequent behavior with similar inaccuracy.

As one example, once again consider our seeing–ghosts account (from an earlier chapter). The energy–trace inputs from vague and partial stimulus sources, which in this case were from both the external and the internal (due to conditioning history) environment, induced sensory–neuron firings that evoked neural behaviors of consciousness and knowing, along with neuro–muscular behaviors; all of these behaviors were quite unrelated to (i.e., inaccurate with respect to) at least the external–environment inputs. As a result, the behavior-established reality (i.e., what became known) was not the actual inputs. Reality was misestablished as seeing ghosts.

Here is another example, one like so many mundane—and so likely missed—examples that each of us has experienced. Indeed, these events happened to me and, under contingencies compelling the recording of simpler examples than those previously entered in my notes, I recorded this observed example. However, like all *"I"* personal–pronouns, the *I*s in this example never imply inner agents; they only imply a physiology mediating evoked consciousness behaviors. Here is the written record of this example, which follows the actual time line, and includes reports of my observed covert neural behaviors as well as my overt neuro–muscular behaviors: One of the bicycle paths to my office crosses a small, forested island. Walking that path one day shortly after the college had cleared the detritus left over from the now completely melted snow, I observed that I was seeing, perhaps 50 feet ahead at the end of a curving part of the path, a stone the size of a large watermelon at the edge, and in the shadow, of a bush that was about five feet tall and about eight feet away from the path edge. That is, I neurally behaved the stone, presumably from light–energy traces inducing some sensory–neuron firings in my eyes, which established the reality of the stone, along with, of course, the path and bush and so on. If the path had not taken me close past the stone, I likely would never have behaved it again, and thus left it a stone forever so far as I could *know.* However, the path did take me closer to the stone. As I started to pass the stone, I observed that I was seeing letters on it. That is, I was behaving letters on it, establishing the reality of the letters, and also observing further neural responses such as "What? Letters? On a stone?" Then, as the letters were evoking these neural questions, I also observed that I was behaving the "stone" changing shape—Whoa! Spooky! (in the respondent, elicited–emotion sense)—apparently in a sudden gust of wind the reality of which I was also behaving. Nearly simultaneously, I began behaving (i.e., establishing the reality of) the "stone" as a discarded gray grocery bag with the name of a store printed on it. Was it a grocery bag all along? Careful; tricky question. So far as I *know,* no, it was not, although so far as I or anyone might *presume,* yes, it was. However, presuming involves a further behavioral step away from the events, and it is *not knowing.* It is an additional behavior that comes from less direct but combined lines of suggestive evidence. Having behaved the object as a stone, one can only know it to be a stone; only after behaving it later as a bag can one know it to be a bag, and presume that it had always been a bag.

We all behave "realities" such as the variety in those examples, even those of us from whom such examples tend to evoke snide giggles. Ahhh, reality; not as sure a thing as past conditioning compels us to think, or so present conditioning compels us to think. And so the question arises, what if any supports are available for the behaved confidence reality evokes in its apparent existence? To answer, as with those examples, recall that we are still addressing, not *whether* reality exists, but only how we can *know* about any reality.

Reality supports? The behaviorological characterization of reality as minimally introduced in this chapter—that we neurally behave reality under the limitation of sensory–neuron firings—allows, in that it fails to preclude, the non–directly knowable existence of a reality, an environment, a realm beyond our neural behaving. Inevitably a discussion, of how we can know any reality, evokes (Provokes?) questions of the *existence* of reality, of the environment, of something, some realm, outside us, a realm, of course, that everyone else must inhabit since otherwise they are only inside our own head.

To prevent this chapter's content from inducing some perhaps unnecessary stresses, let's look at some starting points, for further discussion elsewhere, about these questions. And questions they must remain, because the kind of answers that would fully satisfy us require contact points forever unavailable to us. They remain unavailable because, again, the only source of *knowledge* that *anything* exists—including any particular thing as well as any reality, environment, or realm outside us—is our evoked behaving of them due to the firings of sensory neurons; we are stuck inside that constraint. In the same way that we cannot *know* that reality (and so forth) exist, we also cannot *know* that they do not exist; *we lack contact points* with the relevant places—if those places exist—beyond our sensory neurons, the reality of which we must also behave. Can we say nothing else about any reality of reality? Technically, the sensory–neuron contact–point constraint leaves little else to say in terms of what we might otherwise consider the most direct evidence of reality. However, we can consider some possible, suggestive, indirect lines of evidence, even though these too are under the same constraint: to know them we must behave them.

Some seemingly sensible supports for the reality of reality seem to exist, although these more indirect lines of evidence still ultimately merely involve more behavior that also remains subject to the same sensory–neuron firings constraint. Nevertheless, in support of the self–correction routines of natural science, we will list the labels of some of these indirect supports, and follow this list with a variety of practical examples; we will not, however, further analyze the examples to connect them with particular types of indirect support.

Such indirect supports strengthen our confidence that, despite lacking more direct proof, something (e.g., reality, environment, realm) exists outside us. Of course, confidence is but another one of the variety of neural behaviors of consciousness. These indirect supports include the contiguity, consistency, reliability, potency, and even patterns extant in our ultimately behaved data, along with the process and outcomes of direct stimulus control and our very continuation and survival.

Here are several examples supportive of our confidence with respect to the reality of reality. They take the form of a question, "Why…, *if no reality exists* that somehow supplies the energy traces that induce the sensory–neuron firings that evoke neurally behaving the reality of whatever provided those energy traces?" Here are several question like this: Why would someone behave a noisy car (or a screeching cat, or an arguing neighbor, or a crashing burglar, or the sweet sounds of a capable street musician, or an alarm clock) disturbing their sleep, if… (i.e., if *no reality exists* that…)? Why would a letter carrier behave a dog appearing from behind a bush and biting her or his behaved leg, if…? Why would a college graduate behave bird droppings landing on his or her behaved, long–haired head and then dripping onto her or his behaved new suit as he or she behaves walking into a behaved building where behaved job interviews occur, if...? Why would you start down your basement stairs (Yes, I know, "behaved basement stairs.") on a hot August afternoon only to stop short as you behave a four–foot layer of dirty water covering the floor, if...? Why not just behave clean water, and take a swim? No? Why would that not work?

Such examples can go on virtually endlessly. Here is a last set: Why would a body, when driving a car over the crest of a hill on a curving road, behave a flock of turkeys, or a deer, or a family of skunks, or a batch of baby lambs on the road around the bend at the top, which may or may not also require behaving a swerve, or a hard stop, or getting out to pet the lambs, or hitting the deer and wrecking the car and maybe terminating the reality behaving of a passenger, if...?

With respect to reality and knowing it and behaving it and questions of its existence, what are we to make of that very last possibility? When we behave someone dying, the reality behaving of that body ends; she or he behaves no more reality. (Or, saying the same thing without the confusing pronouns, the now dead physiology no longer mediates any reality.) Yet *we* continue to behave reality. Is this another indirect line of evidence that reality exists? We wonder because, while our reality behaving ends at death, we presume—a behavior that our death stops—that our death leaves everyone else still behaving their reality, even in their reality, which presumably then exists beyond us, and so on?

We might never behave satisfactory answers because, again, all those example questions continue, "..., *if no reality exists* that somehow supplies the energy traces that induce the sensory–neuron firings that evoke neurally behaving the reality of whatever provided those energy traces." Yet in each of those, and any other, examples, we can get no closer to reality than neurally behaving it—and the neural consciousness behaviors of knowing—that the firings of sensory neurons evoke. However, altogether, as natural–science self–correction routines operate, such supports may strengthen our behavior of confidence that something really exists as reality, although still subject to the sensory–neuron firings constraint that invariably limits what we can *know* about reality; we still can only neurally behave reality. So let's move on from this topic and take a little look at robotics.

Robots, Robotics, and Chapter Conclusion

So far we talked about reality, including some about the robotic perspective. Here we will now talk a little about robots and robotics. The robotic perspective and robotics are related but not identical topics. And unlike the nature of reality, the nature of robotics is not so "unexpected."

The term *robotics* refers to the study of life forms that we call *robots* that arise naturally through the conditioned behavior of other life forms (e.g., humans) rather than through biological evolution. However, in a vital scientific sense, the term *robot* applies to both of these life–form types, a finding that results from comparisons of these two types of life–form. We also entertain some implications that these comparisons hold for the building of robots.

While we could discuss robotics from any of our available perspectives—agential, environmental, or robotic—we derive our discussion of robotics from our standard environmental perspective. We thereby avoid any terminological confusion that could stem from trying to deal with both robotics and the robotic perspective at the same time. More importantly, trying to address robot construction along with the prediction and control of robot behavior from the agential perspective is as doomed to failure as is trying to predict and control human behavior, or the behavior of any other biological robot (i.e., organism) from the agential perspective, as we have regularly noted.

The term *robot* applies both to life forms arising through biological evolution and to life forms arising through engineered construction. Both life–form types are entirely natural, complex clumps of organized chemical matter. Both operate through entirely natural processes. Both biological evolution and engineered construction also constitute entirely natural processes, although the latter depends on the extent of the conditioning processes that the former made possible (e.g., so far only humans, but not rabbits, construct robots). And both life–form types rate the label *natural organism,* regardless of whether they are biologically evolved or factory engineered. We can even see Mark Twain on this same track when (in his anonymously published last book *What Is Man?)* he supported the view that circumstances entirely determine an individual's life, that humans are biological machines. We simply add the clarification that humans lack inner agents. On the other hand, science fiction writers regularly treat machines as human, and posit evolving machine cultures lacking any further presence of the engineers—biological or machine—that originally built the earliest cultural members. Nothing in nature says that such notions must be wrong; indeed many laws of nature provide the basis upon which to predict that such notions may prove right, although current contingencies (even beyond those supporting organized superstition) may prove inadequate to the task of leaving citizens emotionally comfortable with these possibilities.

So, how is a robot, particularly a non–biological robot, made? Leaving structural considerations to others (simply because our science of function has less to say about them) professionals in the field of robotics, whom we can tact as *designers,* have been at this task for decades, and continue to make progress. However, they also regularly run into conundrums. Without a natural science of behavior available to them, the traditional contingencies under which they labor still wreak some havoc by inducing further doomed–to–failure attempts to replicate the workings of spontaneously initiative inner agents. Others emphasize gradually successful attempts at more and more comprehensive *pre*programming to determine a robot's functioning.

Overall our elucidation of the laws of behavior and how they work with biologically evolved robots provides the kind of input that can make much easier the task of engineering the kind of functioning life form that we would traditionally tact as a robot. When basically familiar with behaviorology, or even merely under contingencies to avoid the problems of agential controls, some robot designers have seen, and others may well see, the value of designing for conditioning rather than preprogramming, particularly designing for the feedback mechanism of operant conditioning. From that point the discussion can considerably broaden. The value of stimulus–control processes (e.g., generalization and evocation) quickly becomes clear.

Fraley (2008, pp. 1547–1563) extends the behaviorology–for–robot–construction discussion in several areas. These include energy sources, reproduction, pre–installed experience, planned diversity, social life, quality of life considerations, and even providing robots with respondent emotions, events that, by design, would affect the intensity of their current responding in the same way that emotions affect the intensity of a human's current responding. As more researchers and robot designers, particularly more thoroughly behaviorologically informed researchers and robot designers, become involved in robotics, developments can only expand further.

Chapter Conclusion

We should be clear that examining reality and robotics may prove at the moment to be not so much a practical matter as a matter of acknowledging these greater extents of behaviorology. Whatever the status of reality, and however we know it or cannot know it, we daily deal successfully with reality, something which our continued existence shows. While our reality conclusions may presently bear only on some aspects of the robotics field, only time will tell whether of not either of these produce any other practical benefits. For instance, as part of our ongoing activity, we may have to recognize that how our improved natural–science understanding of reality and robotics might benefit solutions to global problems remains an unanswered question. Perhaps behaviorology's more accurate and complete, natural–science account of behavior, including scientific behavior, will help produce even more accurate and complete accounts in the subject matters of other natural–science and engineering disciplines. Perhaps contingencies will induce some readers to take the next steps along these lines even while the rest of us implement solution aspects from the other parts of this book, other parts of behaviorology's contribution to helping the natural–science team efforts to solve global problems.⚘

References (with some annotations)

Fraley, L. E. (2008). *General Behaviorology: The Natural Science of Human Behavior.* Canton, NY: ABCs. See Chapter 29 for more on reality, and see Chapter 30 for more on robotics.

Hawking, S. & Mlodinow, L. (2010). *The Grand Design.* New York: Bantam. Especially see Chapter 3.⁊

Chapter 24
Evolutions and Epilogue

$\mathcal{B}$efore it finished with a brief tour of robotics, Chapter 23 explored the result of observing where the accumulating laws of behavior would take us with respect to reality, or at least with respect to what we can know of reality and how we can know it. That chapter was the last of several chapters that began to address some natural–science answers to some of humanity's long–standing questions with some interpretive applications of behaviorology; earlier topics in these chapters included verbal behavior, the values–rights–ethics–morals series, consciousness, personhood, life, and death. We addressed all these topics in light of the concepts, principles, methods, and practices of behaviorology from even earlier chapters. After a topical overview (in Chapter 1) of the first 100 years of behaviorological–science history, the topics of those earlier chapters included the value of a scientific philosophy of science, and the ubiquity of not only the emotional and intellectual behavior related to respondent and operant conditioning but also reinforcement schedules, the stimulus controls of evocation and generalization, direct stimulus control, direct–acting contingencies and rule–governed behavior, equivalence relations, a range of intervention practices grounded in these lawful relations, and the research methods behind these discoveries and developments.

Now, in this last chapter of this book, and out of respect for our Law of Cumulative Complexity, we begin by taking a look at some of the larger context that affected those discoveries and developments, the larger context in which they happened. This context covers three different and overlapping disciplinary levels of evolution that result from three different but related kinds of selection, all involving consequences of one sort or another. Then we finish the chapter, and the book, with an Epilogue that again ties the discoveries and developments of behaviorology to the "Great Work" of addressing solutions for many current problems that otherwise seem like the doom of our times.

Evolutions

Our three kinds of evolution involve three kinds of selection, each of which is actually a kind of *selection by consequences*. For the first kind of selection, we use the more than 150–year–old standardized name, *natural selection*. Lacking standardized names for the remaining two kinds of selection, here we use *behavioral selection* and *group–practices selection*. While all are *natural*, natural selection produces biological evolution, behavioral selection produces repertoire evolution, and group–practices selection produces cultural evolution.

The broad extent of those interrelated topics makes thorough coverage of them unsuitable for an introductory–level book. Hence we only cover them briefly here, leaving details to other resources (e.g., the references).

Let's consider each evolution in turn, beginning with a quote that provides an overview of our first two evolution types. Originally a part of B. F. Skinner's Herrick Lecture at Denison University in 1960, this point was one that Skinner included at the end of Chapter 9 in his book, *Contingencies of Reinforcement:* "All human behavior, including the behavior of the machines which man builds to behave in his place, is ultimately to be accounted for in terms of the phylogenic contingencies of survival which have produced man as a species and the ontogenic contingencies of reinforcement which have produced him as an individual" (1969, p. 297).

Biological Evolution, Natural Selection

While our other two kinds of evolution and selection arose, as topics of study, rather recently, biological evolution and natural selection command a certain familiarity simply from having been studied already for over 150 years. Natural selection, at the species level, involves selection by consequences, through contingencies of survival, leading to *biological evolution.* This kind of evolution describes species changes as basically a function of gene selection through reproductive success. The result is the accumulation of changes that we see in species and speciation, a result that we call biological evolution. Any good biology textbook provides additional details.

Repertoire Evolution, Behavioral Selection

Behavioral selection, at the level of individuals, involves selection by consequences, through the extensive variety of "contingencies of reinforcement," leading to *repertoire evolution.* This kind of evolution describes the general changes in the behavior of individuals that occur through control by contingencies in the broad sense that covers all contingent environment–behavior relations, including reinforcement and the way it affects the effectiveness of evocative stimuli, along with schedules, punishment, function–altering stimuli, and so forth. The result is the gradual changes that accumulate as the individual's personhood (i.e., the full, evocable repertoire of behavior) a result that we call repertoire evolution. Our earlier chapters provide some of the basic details, particularly on the processes of operant conditioning.

Natural selection not only selects genes that produce physiological structures that mediate specific behaviors in a particular environment, but it also selects genes that produce structures that environmental feedback loops can affect. These loops, such as those involving the operant conditioning processes of behavioral selection, can alter these physiological structures such that they mediate new behaviors to new features in a changing environment. Hence behavioral selection can enhance natural selection, although only some behaviorally selected behaviors actually enhance survival, and even then only sometimes. For example, the social repertoires that operant conditioning processes produce enhance survival prospects when they involve imitative behaviors of moving with the crowd when some external threat induces crowd movement; however, such repertoires reduce survival prospects when they induce superstitious repertoires, like cutting down all the trees in a deity–appeasement activity such as may have happened on Easter Island. (For additional details, see Skinner, 1987, particularly Chapter 4 on "Selection by consequences," and Chapter 5 on "The evolution of behavior.")

Cultural Evolution, Group–Practices Selection

Group–practices selection, at the level of groups of individual organisms, involves selection by consequences operating as effects on the behaviors of most or even all of the individual group members—with both the effects, and the group–behavior products, outlasting the lifetime tenure of the individual group members—through the extensive variety of social "contingencies of reinforcement." This leads to *cultural evolution.* This kind of evolution describes the general changes in the behavior of individuals that combine as cultural practices, producing group effects, through control by the social aspects of that same broad range of contingencies in the same broad sense of involving all contingent environment–behavior relations.

Those broad, contingent relations produce shared group–level successes such as lower energy and other costs, increased efficiency or effectiveness, and group survival that bring about group–practice changes. The shared social contingencies emphasize modeling and imitation, but actually involve no new behavioral processes. The result is gradual change in cultures and the practices that characterize them, a result that we call cultural evolution.

Let's examine cultural evolution a little more closely. For problems at the individual level, solutions reinforce the behaviors that produced them. Then the behaviors that solve problems serve as models for the behaviors of others that share the problems. That is, the occurrences of reinforcing solutions for some individuals function as evocative stimuli for the imitative solution behaviors of other individuals. When contingencies on the group make these solution practices become widespread across a group, they begin to affect the group *as a group,* often becoming socially institutionalized (e.g., *formal* education of group members). Rather than depending mainly on the reinforcing consequences on individual group members, group benefits, including group survival, come to depend on the effects on the group, the group effects, that the group practices produce. These changes in group effects select changes in group practices, and this group–practices selection then drives cultural evolution. (See Skinner, 1987, for more examples and details. Also see Skinner, 1978, and 1989.)

Evolution and Selection by Consequences Summary

Considering *selection by consequences* across genetic, behavioral, and cultural levels helps us to appreciate its status as a fundamental and universal process.

At the behavioral level, past chapters along with the work of innumerable other natural, behavior–science researchers have shown this process at work in the conditioned behavior of other phyla as well as our own, including the conditioned behavior of languages. At the cultural level, the process is at work in the conditioned behavior that accumulates as cultural practices, the practices that define cultures. At the biological level, we see the process all around us in speciation and extinction. And we understand the selection process better as we build interdisciplinary connections with our physiology and other natural–science colleagues, which brings us to a look at the closest disciplinary overlaps among behaviorology and its neighbors. Before looking at these overlaps, however, here is how B. F. Skinner once summarized the position of behavior in evolution and selection processes:

> In summary, then, human behavior is the joint product of (1) the contingencies of survival responsible for the natural selection of the species and (2) the contingencies of reinforcement responsible for the repertoires acquired by its members, including (3) the special contingencies maintained by an evolved social environment. (Ultimately, of course, it is all a matter of natural selection, since operant conditioning is an evolved process, of which cultural practices are special applications) (1987, p. 55).

Beneficial Disciplinary Overlaps

Among the natural sciences, behaviorology is one of the foundation life sciences (along with biology) rather than one of the foundation physical sciences (such as physics or chemistry). The life sciences stretch across a continuum of analysis levels, from molecules to cultures. We find the sub–cellular and cellular levels of the organism at one end of this continuum. In the middle we find the level of individual organisms. And on the other end we find the level of groups or populations of organisms.

"Culturology." While names already exist for the sub–individual level, and individual level, of the life–science continuum, biology and behaviorology respectively, no name has covered the behavior–oriented natural–science group or population level. Sociology might have worked, but attempts to turn it into a natural science (see Fraley & Ledoux, 2002) remain unsuccessful. One area of another contender, anthropology, contains a natural–science philosophy of science, namely the cultural materialism of Marvin Harris (Harris, 1979; also see Vargas, 1985); however, no separate disciplinary name for a natural–science anthropology has arisen. So, since 1986 (Fraley & Ledoux, 2002, p. 147) we have been using the term *culturology* as the label to fill this gap. This label provides a conveniently short replacement for "anthropology informed by cultural materialism" although, in due time, natural–science anthropologists will likely provide a better name for their discipline. Stay tuned.

Each of those three life–science disciplines studies functional relations at its own level of analysis. Biology studies the functional relations both in the history of species and in the physical and chemical processes of individuals from the sub–cellular parts to the whole organism. Behaviorology studies the functional relations between environments (both internal and external) and the behavior (both overt and covert) of individual organisms during their lifetimes. And culturology studies the functional relations in the behavior of social and cultural groups, particularly group–produced effects that can outlast the lifetimes of the individuals that make up the group.

The overlaps. However, each of those disciplines also overlaps somewhat with the others. Biologists and behaviorologists share interests in the physiological mechanisms through which the body mediates behavior, particularly purely neural behavior. Behaviorologists and culturologists, meanwhile, share interests in the operation of the laws of behavior because, while the same laws apply at both levels, outcomes can differ due to the complexity increment that comes from dealing with groups of interacting individuals rather than with single individuals. Furthermore, some applied fields (i.e., an area where one applies a foundation science discipline) of interest to behaviorologists, such as solving global problems, reside as well, if not more so, in the province of culturologists. Figure 24–1 illustrates the positions of these three disciplines along a life–science continuum.

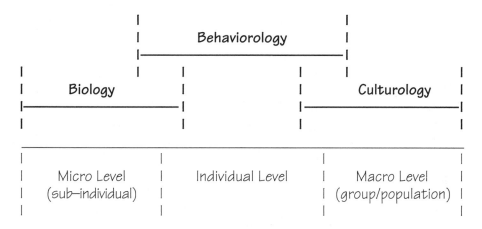

Figure 24–1. Disciplinary coverage for the three main levels of analysis in the life sciences.

The study of ecosystems, species evolution, and the behavior of animals in groups by some biologists points to a disciplinary overlap also between biology and culturology. So you might redraw Figure 24–1 as three intersecting circles. Try it. Each circle would represent one of these disciplinary domains, while the areas where the circles overlapped could then represent the shared–interest area of the intersecting disciplines. (See Chapter 6 of Fraley & Ledoux, 2002, for additional details, including yet another diagrammatic option for indicating the overlap between biology and culturology.) These disciplinary overlaps provide further areas for applications. Let's take a peek at an important one.

Cultural—and Green—Engineering

One of our continuing themes concerns developing the behaviorology–culturology overlap and applying it to the cultural–practice engineering that supports solving global problems, an area that we call green behavior—or behaviorological—engineering, an area particularly relevant to many pressing issues including the humane reduction of population levels and the building of sustainable lifestyles. Perhaps the culture could currently derive the most benefits by first expanding behaviorology into this educational area, from which graduates could then extend it into the needed practical areas. Consider a degree in *Green Behavior and Engineering* (or *Behaviorology and Green Engineering)* that includes basic coverage of the full roundtable of foundation natural science and engineering disciplines (e.g., physics, chemistry, biology, behaviorology) so that graduates can contribute to any and every area of solutions for global problems. We are sitting on the brink of a breakthrough to substantive successes in slowing global warming and solving global problems by building a more complete science and engineering team—a team with members from *all* the natural sciences—to address these concerns. How long will we sit on this brink?

We have perhaps been sitting on that brink ever since Rachel Carson's book, *Silent Spring,* appeared over 50 years ago. William Souder's 2012 biography, *On a Farther Shore: The Life and Legacy of Rachel Carson* (from Crown Publishers) takes us back to that time, refocusing our attention on *Silent Spring* as the origin of the movement to save and preserve our environment. (You can find Julianne Lutz Warren's review of Souder's biography on pages 146–147 of the March–April 2013 issue of *American Scientist* under the title "Crafting a narrative of care.") This biography also reiterates the legacy of controversy surrounding the movement from its very beginnings, controversy rooted in the notion that humans are supposedly the masters of nature, a notion that stems from the traditional but erroneous cultural view that we are somehow—usually agentially—above, or outside of, nature. This view continues to pressure us even as we finally face current realities with the kind of humility that must be a part of long–term solutions.

Sitting on that brink, we have spent less time devising and implementing long–term solutions while spending more time arguing about short–term interests, an activity akin to fiddling while Rome burns. Can we now move beyond this brink, and work seriously—with all the relevant natural sciences including behaviorology—on the long–term solutions instead?

Education may comprise a more helpful arena, at least one with less controversy. Educational campaigns, about steps that ordinary citizens can take to help solve global problems, will be vital components in the efforts to clean up and protect our planetary home. Currently these campaigns stress some of the crucial *behaviors* that contribute to solutions for these problems. For example the folks at the Environmental Defense Fund (EDF) broadly disseminate a range of materials. Among these are two lists of such behaviors, one containing ten "precycling" tips and the other presenting ten steps that fight global warming (with both lists still likely available at www.edf.org). Other organizations, such as the Union of Concerned Scientists (UCS) and the Natural Resources Defense Council (NRDC) provide similar and related materials.

To emphasize the importance of recognizing that *behavior* components comprise a major portion of the problems and their solution that demand our attention, components for which behaviorology provides the relevant natural science, here is a quick version of the ten EDF steps that fight global warming:

- Recycle used materials…
- Wash clothes in colder water…
- Install low–flow shower heads…
- Run the dishwasher when full, and without heat to dry…
- Replace standard light bulbs with CFL or LED bulbs…
- Plug window and door air leaks…
- Replace appliances with energy efficient models…
- Walk, bike, carpool, or use public transport when possible…
- Adjust the thermostat seasonally…
- Share these simple steps with others, and increase awareness…

Each of those ten steps involves an explicit *behavior* (e.g., recycle, wash, install, and so on). Providing lists of behaviors, like that list, addresses the middle term in our fundamental three–term contingency regarding basic environment–behavior functional relations. Additional steps concern also addressing, orchestrating, and engineering not only the first term of the needed contingencies (i.e. the stimulus changes that *evoke* these behaviors successfully) but also the third term (i.e., the stimulus changes that *reinforce* these behaviors successfully) with the culture increasingly supplying the full contingencies that generate and maintain these behaviors as increasingly standard, even institutionalized, cultural practices. With such practices we can begin to reverse the retardation of human intellectuality that the age–old cultural dalliance in superstition breeds, particularly with respect to the theological and secular purveying of agential superstitions and the activity these breed against the productive cultural practice that we call natural science.

EPILOGUE

We have now come just about full circle. The topics of the last several chapters convince us of our readiness to apply, as a culture, the behaviorological technologies that we can derive from the principles and concepts of the natural science of behavior (covered in earlier chapters) to the widest range of humanity's concerns. How can we make this happen? Basically, the next step involves you and me and other readers supporting, perhaps even agitating for (Dare I say "campaigning for"?) establishing more university behaviorology programs and departments; we all require these to meet growing needs—see the WANTED poster at the start of Part II, and show it to those positioned to help—so that soon an increasing proportion of our world population operates with an expanded, behaviorologically informed repertoire as a component of their general science education. This would lead not only to more traditional natural science in the education of behaviorologists, but also to more behaviorology—and less superstition—in the education of the general population and traditional natural scientists.

Why is that important? Recall the value of the interdisciplinary approach to the engineering of solutions for global and local problems that we covered near the end of Chapter 1. This approach is not only valuable but necessary, because humanity is running out of time to solve these problems before their implications overwhelm us, forcing us to experience their worst effects. To design and implement such solutions effectively requires *all* the relevant natural sciences to coordinate their efforts. And these sciences include not only the traditional ones focused on energy, matter, and life forms (e.g., the sciences surrounding physics, chemistry, and biology) but also the science focused on life functions (i.e., behaviorology). Given the acknowledged, substantial behavior components of global problems and their solutions, humanity needs all of these natural sciences working together, with reasonable familiarity with each other, if the solutions are to occur in a timely fashion.

That familiarity enhances the interdisciplinary field that we call green behavior engineering (which, to stress the catch–up relevance of behaviorology, is short for behaviorological green engineering). The culture, or at least cultural survival, puts this interdisciplinary cooperation at the forefront of efforts to solve global problems. Examples of concerns on which traditional natural scientists and behaviorologists work together, or even on which just *behaviorologically informed* traditional natural scientists work, include humanely reducing overpopulation (the necessary foundation for solving *many* global problems), establishing sustainable lifestyles, keeping the air and water clean, and preserving habitats and resources and species diversity, to name but a few.

Beyond helping solve local and global problems, recall the other contributions behaviorology makes to the capabilities of traditional natural scientists. Once they become basically familiar with behaviorology, they become more able to remain naturalistic in dealing with subject matters at the edge of, and beyond, their particular specializations; thus they can avoid slipping into the compromising use of common culturally conditioned, superstitious agential accounts. Furthermore, knowing some behaviorology helps them add supportive details to accounts within their specializations (for example, how various natural sciences each help account for phenomena that we tend to see as limited to humans, such as language, and ethics). Also, behaviorology provides the students of natural scientists with a natural–science alternative to the non–natural disciplines that most of these students study when covering subject matters related to behavior, including scientific behavior. (For additional related considerations, I encourage you to take a quick look again at the last section of Chapter 1.) In addition, go beyond the coverage of this book. Its real value might be in its setting the stage for your repertoire in the behaviorology discipline, and its applications, to expand beyond these essentially beginner's considerations. Delve deeper, studying and applying the full range of conceptual and practical details that grace the pages of behaviorological science journals and disciplinary textbooks (for example, see Skinner, 1971; also see Fraley, 2008; and check out the wide range of resources in the bibliography at the end of this book).

Even if people—especially those whom we call scientists, because they have a thoroughly conditioned repertoire in one or another traditional natural science—have only a minimal conditioned repertoire in behaviorology, they would still be more likely to produce, or consider as viable, solutions to global problems that at least intuitively, or better, through design, took behaviorological realities into account. As a result the behavior–related components of the solutions could develop as reasonable behaviorological interventions with an increased likelihood of success, thereby supporting the physical, chemical, and biological solution components.

The alternative of continuing to stumble along with relatively little success from attempting solution that stem from *only* traditional natural sciences, or from superstitious cultural lore, could be reduced or entirely avoided. This would be a major benefit even if it only meant that solutions would not involve the usual knee–jerk suggestions such as "Make pollution (or whatever) illegal and punishable" (since nature will punish plenty for failures anyway). Instead, solutions would involve designing interventions that included new or enhanced reinforcing contingencies, for behaviors consistent with overall, improving environmental health. These contingencies should involve the long–term best interests of most people, and would help generate and stabilize the behaviors required to maintain environmental health. Indeed, as many more people gain more extensive behaviorological repertoires, and begin to apply them to global problem solutions, so much more is possible and will, I think (and we all hope) get done; TIBI and professional behaviorologists already direct their energies this way, and they stand ready to assist you as well (contact them through their professional–organization website, www.behaviorology.org).

And so, dear reader, we approach the end of the book and our journey together, a journey that introduced behaviorology to you out of concern for our planetary home, a journey about this natural science of WHY human behavior happens, a natural science to help solve global problems in a timely manner. Virtually every chapter included wonderful realities about human behavior, and ended with an *ongoing–activity* admonition to consider the pertinence of behaviorology, as part of the roundtable of natural sciences, to solving the behavioral components of global as well as local problems.

Of course, that may have left you wondering why I never spelled out exactly how to apply behaviorology to solve all those problems. Why does this book only introduce the principles, concepts, methods, and practices of this natural science? The reasons are several; here are two of them. (a) The topic of how, thoroughly, to apply behaviorology, to cover its share of the efforts to solve global problems, requires at least one more book, and likely several more books, with a book like this book as a prerequisite. And (b) since a proper treatment of this topic extends well beyond my own expertise, and likely beyond the expertise of any single professional, it would best come from a team of authors, a team from all the basic natural sciences—including many new, fully trained behaviorologists—working to solve these problems. Possibly you will be a member of that team.

In any case our current global problems, with their behavior components, loom in our collective face. If contingencies compel enough of us to participate in the production and implementation of solutions, then we can together prevent humanity, and life on this planet, from running out of time.❦

References (with some annotations)

Carson, R. (1962) *Silent Spring.* Boston, MA: Houghton Mifflin.

Fraley, L. E. (2008). *General Behaviorology: The Natural Science of Human Behavior.* Canton, NY: ABCs.

Fraley, L. E. & Ledoux, S. F. (2002). Origins, status, and mission of behaviorology. In S. F. Ledoux. *Origins and Components of Behaviorology—Second Edition* (pp. 33–169). Canton, NY: ABCs.

Harris, M. (1979). *Cultural Materialism: The Struggle for a Science of Culture.* New York: Random House.

Skinner, B. F. (1969). *Contingencies of Reinforcement.* New York: Appleton–Century–Crofts. Chapter 7, "The phylogeny and ontogeny of behavior," first appeared, without the extensive notes that accompany it in this book, in 1966 in *Science, 153,* 1205–1213. The B. F. Skinner Foundation (www.bfskinner.org) in Cambridge, MA, republished this book in 2013.

Skinner, B. F. (1971). *Beyond Freedom and Dignity.* New York: Knopf.

Skinner, B. F. (1978). *Reflections on Behaviorism and Society.* Englewood Cliffs, NJ: Prentice–Hall.

Skinner, B. F. (1987). *Upon Further Reflection.* Englewood Cliffs, NJ: Prentice–Hall.

Skinner, B. F. (1989). *Recent Issues in the Analysis of Behavior.* Columbus, OH: Merrill. Especially see Chapter 5, which is entitled "Genes and behavior."

Vargas, E. A. (1985). Cultural contingencies: A review of Marvin Harris's *Cannibals and Kings. Journal of the Experimental Analysis of Behavior, 43,* 419–428.❧

Appendix

*N*ote: Fred Skinner provided this short article at the request of Stephen F. Ledoux and Carl D. Cheney. They were working on their 1987 book, *Grandpa Fred's Baby Tender, or Why and How We Built Our Aircribs* (Canton, NY: ABCs) the point of which was to increase the dissemination of accurate information on Aircribs. To support this point, Dr. Skinner contributed his article. The Aircrib design that their book followed was a design that Dr. Skinner had made and modeled but not actually constructed. Ledoux and Cheney placed Dr. Skinner's article as the Foreword in their book. While the book is at present out of print, we reprint the Foreword here with permission also to increase the dissemination of accurate information on Aircribs, both by further elucidating this development in the history of the natural science of behavior, and by bringing to the attention of new generations of readers this enduring behaviorological contribution to childcare.♣

The First Baby Tender

B. F. Skinner

I designed what we called the baby tender as a laborsaving device. We wanted to have a second child, but my wife said she rather hated the chores of the first year or two. I suggested that we simplify the care of a baby. All that was needed during the early months was a clean, comfortable, warm, and safe place for the baby, and that was the point of the baby tender. I started to build it about the time we started the baby, and in spite of war–time shortages finished it just before our daughter Deborah was born.

As soon as she came home from the hospital, we put her in the baby tender. We discovered immediately that the labor we saved was far less important than the advantages for her. She slept on a tightly stretched canvas covered with a sheet (later replaced with a single plastic cloth that felt rather like linen). There were no nightclothes, sheets, or blankets, and she wore only a diaper. There was no danger that she would smother, as there occasionally is in a standard crib. She breathed clean air, which we humidified and maintained at just the right temperature. She was free of colds for many years, and I am inclined to think that it was due primarily to the warm humid air she breathed as a child. In the winter in a northern climate a house is about 30 degrees below body temperature, and the air the baby breathes is chilled further by evaporation from moist surfaces in the air passages. It is possible that the superficial layers of the bronchi and lungs grow as much as 40 degrees below body temperature, and that could make a great difference. The species originated in the tropics, where warm, moist air was standard, and there may not have been enough time for further evolutionary changes.

The space was quiet, and Deborah was free to move about and take comfortable positions at any time of day or night. She soon began to exercise much more vigorously than would have been possible in a standard crib, and she grew very strong. Our pediatrician commented on her unusual strength. Her skin stayed dry, and she never had any diaper rash. She never objected to being put into the baby tender and almost never cried.

Her rapid physical development was matched by behavioral gains. She was free to explore all parts of the space and there was a large window through which she could watch life around her. At one point she seemed to pass through a phase in which she used her feet prehensilely. Another couple who made and used a baby tender sent us a photograph of their baby holding its bottle with its feet while it drank. I made toys which Deborah used very early. By pulling a ring that hung from the ceiling, she produced a whistle. By twisting a T–bar that hung from the ceiling, she made small banners spin. Later, by pulling a ring she operated a music box, tone by tone.

She was not socially isolated. She was taken out for feeding and play, of course, and we could allow the neighborhood children to talk and gesture to her through the window without passing on their viruses. The labor we saved not only made it easy for us to treat her affectionately but encouraged us to spend more time with her. She spent a lot of time outside the baby tender, especially as she grew older. Eventually she slept in it only at night and for naps.

During her second and third years, when we could predict her bowel movements, she slept without clothing. Urine passed through the plastic cloth (which could be quickly washed and dried) into a tray to be thrown out the next morning. She learned to postpone urination, in part, I think, because of the consequences. Urination in a diaper is immediately followed by a pleasurable warmth; it is only after several minutes that a damp diaper grows cold and uncomfortable. Without a diaper urination immediately moistens the skin and chills it. Deborah began to go for long periods of time without urinating, and by the time she first slept in a bed she had learned to keep herself dry and never wet her bed. All the supposed psychological problems connected with toilet training were avoided.

I have seen many young people who spent part of their first years in similar spaces, and most of them were rather tall and strong. It would be extraordinary if those first years of rapid growth could have made that kind of difference, but it is certainly something worth exploring further.

The response to my article in *The Ladies Home Journal,* written when Deborah was nine–months–old, drew hundreds of letters asking where a "baby tender" could be purchased or how one could be made. I sent out hundreds of crude instructions. There were only a very few critical letters. I have never found anyone who, upon seeing a baby in an Aircrib, did not immediately think it was a wonderful idea. But misunderstandings began to spring up and were widely circulated. The *Journal* had given my article the title "Baby In A Box" and some of the misunderstanding came from a confusion with the equipment used in operant research. Misunderstandings are still common. Here is a sample from an article published by a reputable psychologist: "In the late '40s, Professor Skinner invented the 'air–crib,' a Skinner box for babies. It was a large, soundproof, germproof, air–conditioned box for giving children mechanical care for the first two years of life." Every statement in the passage is wrong. I designed and built the box in 1944. It is not an experimental apparatus. It is not soundproof; Deborah was shielded from loud noises, but we could hear her at all times. It is not germproof, although it was a kind of shield against sudden large doses of infection. "Air–conditioned" suggests cooling, but the air was only warmed. It is no more mechanical than a standard crib, and there was nothing mechanical about the care we gave our child. Deborah may have spent a bit more time in the Aircrib than she would have spent in a standard crib, because she was freer and more comfortable there, but in her second year she merely slept in it, at night and for naps. (Perhaps I should add that rumors that she committed suicide or became psychotic are equally wrong. Now 43 [in 1987] she is a happily married, talented artist and writer.)

It is possible to build a better world for a baby and the baby tender was a step in that direction.

B. F. Skinner
January, 1987.

Endnote [included in the 1987 book]: The Skinners' first daughter, Julie, is also happily married, is engaging in a successful career as a Behaviorologist, and has used an Aircrib with her own children.

Endnote (for this Appendix): In a 1988 interview, Skinners' first daughter, Dr. Julie Vargas, cleared up several persistent misunderstandings about the aircrib. The other interviewees were Drs. Lawrence Fraley, Stephen Ledoux, and Ernie Vargas, Julie's husband. You can find the interview, which is entitled, "August 1988 Public Radio interview of the organizers of the first behaviorology convention," in (2013) *Journal of Behaviorology, 16* (1), 15–20; you can also order a CD recording of it through the TIBI website (www.behaviorology.org).

References

Skinner, B. F. (1945, October). Baby in a box. *The Ladies Home Journal.* Also in Skinner, B. F. (1999). *Cumulative Record—Definitive Edition* (pp. 613–620). Cambridge, MA: B. F. Skinner Foundation.

Reader's Notes

Glossary of Some Terms & Phrases

$\mathcal{T}$he text resulting from an author's evoked writing responses controls various behaviors of readers. Beyond some general controls including grammar, syntax, and common vocabulary, scientific technical terms and phrases function for readers as precise thematic behavior–controlling stimuli. After appropriate conditioning of the author and the reader (the latter often through an author's writing–behavior products) the appearance of particular disciplinary terms or phrases affects the reader's nervous–system structure in ways that evoke a *particular* response or combination of responses or chain of responses. Those responses are formal and informal varieties of the definitions and applications of the terms and phrases. Regarding any related set of scientific disciplinary terms, when such definitional responses of a group of people show a certain commonality on an evidential basis, that group defines a scientific disciplinary verbal community that shares a technical vocabulary. The accumulation of a list of some parts of a technical vocabulary (i.e., many of its terms and phrases) and their related definitions results in a glossary.

This is a glossary of some of the most common terms and phrases that the scientific verbal community of behaviorologists uses at this time. These definitions may change as research developments accrue, and they may also differ somewhat from those used at different points in various chapters of this book. Such differences happen because, across the chapter coverage, topics expand through enhanced repetition and so accumulate descriptive parts into increasingly comprehensive definitions. This sometimes even results in an over–elaboration of the stated or implied definitions in later chapters. In any case, and as an indication of the value of your expanding behaviorological repertoire, observe that both the whole *as well as the parts* of each of these alphabetically–listed definitions evoke appropriate and helpful responses.✦

ABOLISHING OPERATION — An antecedent process (a) that momentarily *lowers* (abolishes) the effectiveness of a reinforcer, and (b) that momentarily *lowers* (abolishes) the effectiveness of any stimulus that in the past evoked the behavior that the reinforcer follows. (*Satiation* is an example. See ESTABLISHING OPERATION.)

ANTECEDENT — An event that occurs *before* some other event. (See POSTCEDENT.)

AVERSER — (See PUNISHER.)

AVOIDANCE — (See ESCAPE.)

BACKWARD CHAINING — The procedure of taking advantage of the DUAL FUNCTION OF A STIMULUS to build chains of responses by linking them with stimuli, that is, by strengthening the function of the evocative stimulus for a response and then using that stimulus, which has also become a conditioned reinforcer, to strengthen another response before it, and repeating this process *from end to beginning* until the chain is complete (with each stimulus functioning both as an evocative stimulus for the next response and as a conditioned reinforcing stimulus for each added "previous" response) and also by always proceeding to the end of the chain regardless of where in the chain a practice run begins.

BEHAVIOR — (See RESPONSE.)

BEHAVIOR PASSIVITY — The nature of all behavior as inevitable reactions that happen as a function of other also real variables; if a behavior happens, then it had to happen, and if a behavior does not happen, then it could not have happened.

BEHAVIORISM — (See RADICAL BEHAVIORISM.)

BEHAVIOROLOGY — Minimally, the natural science of environment–behavior functional relations and the beneficial interventions derived therefrom, both for understanding and improving both normal and problematic behaviors.

BEHAVIOROLOGICAL ENGINEERING PROCEDURE / INTERVENTION — (See PROCEDURE.)

BEHAVIORAL PROCESS — (See PROCESS.)

CHAINING — (See BACKWARD CHAINING.)

CLASS (See RESPONSE CLASS or STIMULUS CLASS) — A group the members of which share something in common.

COINCIDENTAL REINFORCER / REINFORCEMENT — Reinforcers / reinforcement that occurs through a functional chain of natural events that is otherwise unrelated to the behavior that the occurrence of these reinforcers affects; conversely, the effects of the behavior on the environment did not include the production of these reinforcers. (See SUPERSTITIOUS BEHAVIOR.)

CONDITIONED — The pairing (i.e., the occurring at about the same time) of stimulus events such that the neutral (i.e., non–functional) stimulus events start functioning like the already functional stimulus events (e.g., an operant S^r or a respondent cs). (Operant synonym: "secondary.") (Also see UNCONDITIONED.)

CONTINGENCIES OF REINFORCEMENT — The generic term that refers to *all* behavioral contingencies at once, including the many types of contingencies that do not involve reinforcement.

CUMULATIVE COMPLEXITY — (See LAW... OF...)

DIFFERENTIAL REINFORCEMENT — The term for response class members producing different consequences such that one member of a response class earns reinforcement while other class members go without reinforcement leading to their extinction. (See SHAPING.)

DISCRIMINATION — An older and less precise term for EVOCATION, due to the potential agential implications of "discrimination."

DISCRIMINATIVE STIMULUS — An older and less precise term for EVOCATIVE STIMULUS, due to the potential agential implications of "discriminative."

DUAL FUNCTION OF A STIMULUS — A stimulus serving two functions as a result of being regularly present when a REINFORCER follows a response; such a stimulus comes to function both as a conditioned reinforcing stimulus (for the previous behavior) and as an EVOCATIVE STIMULUS (for the next behavior).

ENVIRONMENT — The natural domain (on both sides of the skin) that the existence of theoretically measurable independent variables defines in behavior–controlling functional relations.

ENVIRONMENTS, EXTERNAL and INTERNAL — The environments that occur on each side of the scientifically unimportant boundary which the skin of the behaving organism specifies. (Also see ENVIRONMENT.)

ENVIRONMENTAL EXCLUSIONS — The concept of the behavior–controlling environment excludes all *non*–natural events.

ESCAPE (and AVOIDANCE) — Escape Behavior is behavior that results in the termination of unconditioned or conditioned punishers (i.e., unconditioned or conditioned aversive stimuli). Traditionally we differentiate between escape and avoidance on the basis of whether the unconditioned aversive stimulus is present, as in affecting the nervous system, or not; that is, we separate these two on the basis of whether the occurring–then–terminated stimulus is unconditioned (escape) or the occurring–then–terminated stimulus is conditioned (avoidance). Again, when the termination of *conditioned* punishers simply occurs, or accompanies a delay or prevention of the occurrence of unconditioned punishers, we can call it, and we traditionally have called it, "avoidance," even though it is a form of escape, because the conditioned punisher is occurring and the occurrence of the response terminates (i.e., escapes) this occurrence of the conditioned punisher.

ESTABLISHING OPERATION — A process (a) that momentarily *raises* (establishes) the effectiveness of a reinforcer, and (b) that momentarily *raises* (establishes) the effectiveness of any stimulus that in the past evoked the behavior that the reinforcer followed. (*Deprivation* is an example. See ABOLISHING OPERATION.)

EVOCATION — The term that either indicates (a) the process of a particular stimulus evoking a particular response while other present stimuli have less or no such effect, at least on that particular response, or indicates (b) the process by which a particular stimulus *comes more readily to evoke* a particular response, because that stimulus has been reliably present when reinforcers follow that response; in either case, the change occurs in the physiology that mediates the behavior rather than in the stimulus itself, which remains unchanged.

EVOCATIVE STIMULUS — A stimulus that evokes a response, because it is a stimulus in the presence of which a reinforcer follows a response (which tends to make that stimulus also a conditioned reinforcer). (See DUAL FUNCTION OF A STIMULUS.)

EXTINCTION [operant] — A behavior change process in which the previously operating reinforcers cease to accompany or follow members of a response class. As a result a decrease occurs in the rate or relative frequency of that behavior across subsequent occasions. (See FORGETTING; also see PRECLUSION.)

EXTINCTION, RESPONDENT — A behavior change process in which the unconditioned stimulus, that previously occurred at least occasionally with the conditioned stimulus, ceases to occur, with the result that the occurring conditioned stimulus gradually ceases to elicit the response and returns to the status of a neutral stimulus with respect to the response.

FADING — The gradual change of or in stimuli that results in the transfer of stimulus control from one stimulus or dimension of a stimulus to another stimulus or dimension of a stimulus.

FORGETTING — A behavior change process in which the evocative stimuli have not occurred and so, as a result of the accumulation of normal degradation in physiological structures, the evocative stimuli *cannot* evoke the response, which therefore cannot occur. (See PRECLUSION; also see EXTINCTION.)

FUNCTION–ALTERING STIMULUS — A stimulus whose presence alters the function of other present stimuli.

FUNCTIONAL RELATION — The relation in which a dependent variable changes in some orderly, systematic way when orderly, systematic changes in an independent variable occur.

GENERALIZATION (STIMULUS and RESPONSE) — A STIMULUS CONTROL process with two parts; in stimulus generalization *different stimuli* evoke the same response, while in response generalization the same stimulus evokes *different responses.*

GENERALIZED (conditioned) REINFORCERS — Conditioned reinforcing stimulus events that function regardless of satiation or deprivation due to pairing with (i.e., occurring at the same time as) *several* other reinforcers. (See CONDITIONED REINFORCERS.)

INTERMITTENT REINFORCEMENT — Term for reinforcers following only some responses. (See SCHEDULES OF REINFORCEMENT.)

INTERNAL and EXTERNAL ENVIRONMENTS — (See ENVIRONMENTS, EXTERNAL and INTERNAL.)

INTERVENTION — (See PROCEDURE.)

LAW (or theory) OF CUMULATIVE COMPLEXITY — *The natural physical/chemical interactions of matter and energy sometimes result in more complex structures and functions that endure and naturally interact further, resulting in an accumulating complexity.* (An origin of life is an outcome of the *Law of Cumulative Complexity.* On this planet other examples of this law include the vast range of life forms available for study, and the interrelations of physiology and behaviorology; all these are cumulatively complex; all are entirely natural.)

NATURAL EVENT — An event that is definable in terms of time, distance, mass, temperature, charge, and/or a few other properties taken into account by theoretical physicists. Measurable physical properties define a natural event, which occurs only as the culmination of a sequential history of similarly definable events; natural events cannot occur spontaneously (i.e., magically, mystically, or superstitiously).

NATURALISM — The name for the general philosophy of science that informs all natural sciences, from the traditional natural sciences (e.g., physics, chemistry, biology) to behaviorology, by providing systematic verbal supplementary controlling stimuli, for science–related behaviors, in the form of assumptions that the results from experimental research and successful engineering applications have conditioned over the last several centuries (all in opposition to the convenient, anytime inventing of mystically or superstitiously grounded assumptions). (Also, see RADICAL BEHAVIORISM.)

OPERANT — A relation in which a response operates on the environment in a manner that provides energy feedback into the nervous system changing it in ways that we see as *increased or decreased* rates of that kind of response on future occasions *controlled by increased or decreased evocative effects* of the stimuli that evoked the response in the first place.

PNET — (See PROGRESSIVE NEURAL EMOTIONAL THERAPY.)

POSTCEDENT — An event that occurs *after* some other event. (See ANTECEDENT.)

POSTCEDENT ENVIRONMENT — The environment as it exists beginning immediately after a response.

PRECLUSION — A behavior change *procedure* (i.e., intervention) in which interveners prevent the evocative stimuli from occurring and so, as a result of the accumulation of normal degradation in physiological structures, the evocative stimuli *cannot* evoke the response, which therefore cannot occur. (See FORGETTING; also see EXTINCTION.)

PREMACK PRINCIPLE — A high probability behavior can reinforce a low probability behavior. (AKA "Grandma's Principle.")

PROCEDURE (Behaviorological Engineering Procedure/Intervention) — The procedural *prescription* of a behavioral process by which a functional change in environmental events (e.g., an added reinforcement procedure) produces a change in responding; the functional change occurs in an organism's internal or external environment, in the presence of facilitation by the evoked responding of one or more other organisms, which could include the subject, as these are all natural parts of the environment. (See PROCESS.)

PROCESS (Behavioral) — Any change in environmental events that produces a change in responding. The change in behavior is a function of the change in the environmental events (e.g., an added reinforcement procedure). Behavioral processes occur in an organism's internal or external environment, in the absence *or* presence of facilitation by the evoked responding of one or more other organisms, which could include the subject, as these are all natural parts of the environment. (See PROCEDURE.)

PROGRESSIVE NEURAL EMOTIONAL THERAPY (PNET) — A refined, step–by–step procedure generating and maintaining neural and biochemical (emotional) homeostasis that behaviorologists standardized after declaring independence from agential disciplines in 1987. In the past people inadequately described the range of diverse earlier versions of this type of procedure as kinds of "relaxation training." We now describe this refined and standardized procedure as successive muscular–emotional re–conditioning.

PUNISHER — A relative change in the environment (i.e., a *stimulus*) that provides an energy change at receptor cells during or immediately after a response (with reducing effects as the time between these events increases) that results in a *decrease* in the RATE or RELATIVE FREQUENCY of the behavior across subsequent occasions. Can be UNCONDITIONED (i.e., operative due to genetically produced neural structure) or CONDITIONED (i.e., operative due to pairing with—occurring at the same time as—other punishers; the pairing changes the neural structures, not the stimuli). (See OPERANT.)

PUNISHER: ADDED PUNISHER — The addition (e.g., presentation) of a stimulus in the environment during or immediately after a response that results in a decrease in the rate or relative frequency of the behavior across subsequent occasions.

PUNISHER: SUBTRACTED PUNISHER — The subtraction (e.g., withdrawal or termination) of a stimulus in the environment during or immediately following a response that results in a decrease in the rate or relative frequency of the behavior across subsequent occasions.

PUNISHMENT — A behavior change process in which a punisher occurs during or immediately after a response and results in a decrease in the rate or relative frequency of that behavior across subsequent occasions. Can be UNCONDITIONED or CONDITIONED…

PUNISHMENT: ADDED PUNISHMENT — A behavior change process involving an added punisher occurring in the environment during or immediately following a response that results in a decrease in the rate or relative frequency of the behavior across subsequent occasions.

PUNISHMENT: SUBTRACTED PUNISHMENT — A behavior change process involving a subtracted punisher occurring in the environment during or immediately following a response that results in a decrease in the rate or relative frequency of the behavior across subsequent occasions.

RADICAL BEHAVIORISM — The philosophy of science that B. F. Skinner introduced that extends naturalism, the general philosophy of science in the traditional natural sciences (e.g., physics, chemistry, biology) to inform the study of behavior, including human nature and human behavior, under the more recently emerged (in the early twentieth century) natural science of behavior that we now call behaviorology.

REFLEX — A stimulus AND the response that it elicits.

REINFORCER — A relative change in the environment (i.e., a *stimulus*) that provides an energy change at receptor cells during or immediately after a response (with reducing effects as the time between these events increases) that results in an *increase* in the RATE or RELATIVE FREQUENCY of the behavior across subsequent occasions. Can be UNCONDITIONED (i.e., operative due to genetically produced neural structure) or CONDITIONED (i.e., operative due to pairing with—occurring at the same time as—other reinforcers; the pairing changes the neural structures, not the stimuli). (See OPERANT.)

REINFORCER: ADDED REINFORCER — The addition (e.g., presentation) of a stimulus in the environment during or immediately after a response that results in an increase in the rate or relative frequency of the behavior across subsequent occasions.

REINFORCER: SUBTRACTED REINFORCER — The subtraction (e.g., withdrawal or termination) of a stimulus in the environment during or immediately following a response that results in an increase in the rate or relative frequency of the behavior across subsequent occasions.

REINFORCEMENT — A behavior change process in which a reinforcer occurs during or immediately after a response and results in an increase in either the rate or relative frequency of that behavior across subsequent occasions. Can be UNCONDITIONED or CONDITIONED...

REINFORCEMENT: ADDED REINFORCEMENT — A behavior change process involving an added reinforcer occurring in the environment during or immediately following a response that results in an increase in the rate or relative frequency of the behavior across subsequent occasions.

REINFORCEMENT: SUBTRACTED REINFORCEMENT — A behavior change process involving a subtracted reinforcer occurring in the environment during or immediately following a response that results in a increase in the rate or relative frequency of the behavior across subsequent occasions.

RESISTANCE TO EXTINCTION — An effect of intermittent reinforcement schedules, in which the more a schedule resembles extinction the more responding on that schedule resists extinguishing after reinforcement stops; that is, as the number of unreinforced responses on the schedules increases, the number of responses during extinction, after reinforcement stops, also increases.

RESPONDENT — A relation involving a response that an unconditioned stimulus or a conditioned stimulus elicits.

RESPONSE — *EITHER* any covert or overt *innervated* muscular movements of an organism resulting from energy transfers within the organism that other energy changes—initially from beyond the affected body parts—evoke or elicit, *OR* patterns of neural activity resulting from energy transfers within the organism that other energy changes—initially from beyond the affected body parts—evoke or elicit; in all cases *responses are instances of behavior.*

RESPONSE CLASS — A group or variety of responses that have in common either the same eliciting stimulus, for respondent behavior, or the same effect on the environment (e.g., each class member produces the same environmental change) for operant behavior.

RESPONSE GENERALIZATION — (See GENERALIZATION.)

(RESPONSE) **RATE** — The quotient when a count of responses is divided by a count of time units during which the responses occurred:

$$\frac{\text{Count (dividend)}}{\text{Time (divisor)}} \ = \ \text{Rate (quotient)}$$

(RESPONSE) RELATIVE FREQUENCY — The quotient derived from the ratio of fulfilled opportunities to respond to total opportunities to respond. The result may be expressed as a percentage (by multiplying the quotient by 100).

$$\frac{\text{Fulfilled opportunities (dividend)}}{\text{Total opportunities (divisor)}} \ = \ \text{Relative Frequency (quotient)}$$

SCHEDULES OF REINFORCEMENT — Aside from the CRF (Continuous Reinforcement) schedule in which a reinforcer follows every response, "Schedules of Reinforcement" refers to the possible patterns of intermittent reinforcers (i.e., when reinforcers follow only some responses). In the four most common patterns, either reinforcers can occur based on the *ratio* of responses to reinforcers (i.e., on how many responses occur with respect to each reinforcer that occurs) or reinforcers can occur based on the *interval* of the amount of time that elapses since the last reinforcer occurred, during which time no reinforcers are available, before a reinforcer follows the next occurring response. Both ratio schedules and interval schedules can be fixed or variable. On *fixed ratio* (FR) schedules a reinforcer follows each fixed number of responses, while on *variable ratio* (VR) schedules a reinforcer follows a number of responses, with that number varying around some average. On *fixed interval* (FI) schedules a reinforcer follows the first response that occurs after each fixed–duration interval elapses, while on *variable interval* (VI) schedules a reinforcer follows the first response that occurs after each interval elapses, with the duration of each interval varying around some average. Numerous other types of reinforcement schedules are possible (e.g., mixed, multiple, chained, tandem, and concurrent) including time schedules (fixed or variable, FT or VT) which involve only intervals of time elapsing before each reinforcer occurs, regardless of responding; any reinforcers that follow responses on time schedules are coincidental, and we call any increase in responding superstitious.

SHAPING (& DIFFERENTIAL REINFORCEMENT) — Shaping is the gradual change in behavior that occurs when the criteria for the successive approximations of behavior that earn reinforcement gradually shift along some dimension of the behavior thereby producing a more refined behavior or a very different behavior or both. When shaping occurs as part of an intervention, we can say that it involves repeatedly applying the differential reinforcement procedure for each change in the reinforcement criteria. DIFFERENTIAL REINFORCEMENT involves response class members producing different consequences such that one member of a response class earns reinforcement while other class members go without reinforcement leading to their extinction.

STIMULUS — An internal or external environmental event that affects responding (e.g., elicits or evokes or consequates a response).

STIMULUS CLASS — A group of different stimuli that share some characteristic in common (e.g., a pile of different pictures in which each picture somehow shows people, and a pile of different pictures in which no picture shows people, would comprise *two* stimulus classes).

STIMULUS CONTROL — The interaction of an environmental event with the nervous system that compels the mediation of a behavior. We describe the stimulus as exerting functional control by evoking the response (see EVOCATIVE STIMULUS), or by having some other effect on another stimulus (see FUNCTION–ALTERING STIMULUS). We include respondent elicitation under stimulus control, and even as a kind of evocation. Also encompasses stimulus and response GENERALIZATION.

STIMULUS GENERALIZATION — (See GENERALIZATION.)

SUPERSTITIOUS BEHAVIOR (& COINCIDENTAL REINFORCERS) — *Superstitious behavior* is the term we use for the behavior when the occurrence of one or more coincidental reinforcers conditions (i.e., generates or maintains) a behavior. *Coincidental reinforcers* are reinforcers that occur through a functional chain of natural events that is otherwise unrelated to the behavior that the occurrence of these reinforcers affects; conversely, the effects of the behavior on the environment did not include the production of these reinforcers.

TIME OUT — A procedure (i.e., an intervention) in which the interveners are under contingencies compelling arrangements for the subject to be in a setting that disallows contact with any reinforcers (or with as few reinforcers as possible) for a short time (on the order of 30 seconds to one minute).

UNCONDITIONED — Stimulus events that already function due to nervous–system structure that genes—rather than conditioning processes—produce (e.g., an operant S^R or a respondent ucs). (Operant synonym: "primary.") (Also see CONDITIONED.)☙

A Basic & Occasionally Annotated Bibliography

$\mathcal{T}$his bibliography contains a range of behaviorological–science references to help interested readers expand their repertoire. Beyond that, you can find a more extensive bibliography in the 30–chapter, three–course textbook, *General Behavoirology—The Natural Science of Human Behavior,* which is itself a good next book. (It is listed here under Fraley, 2008; autographed copies are available from its author whom you can contact through www.behaviorology.org which is the web site of *The International Behaviorology Institute.*)

Note that *Behaviorology* and *Behaviorological Commentaries* are fully peer–reviewed journals. *Behaviorology Today* (ISSN 1536–6669) on the other hand, peer reviewed most articles only minimally for its first 14 volumes, while fully peer reviewing only the occasional article that it then explicitly so labeled. However, beginning with Volume 15, Number 1 (Spring 2012) *Behaviorology Today* fully peer reviewed *all* articles. Also, beginning with Volume 16, Number 1 (Spring 2013) the name became *Journal of Behaviorology* (ISSN 2331–0774).✣

Barrett, B. H. (1991). The right to effective education. *The Behavior Analyst, 14,* 79–82.

Baum, W. M. (1995). Radical behaviorism and the concept of agency. *Behaviorology, 3* (1), 93–106.

Bjork, D. W. (1993). *B. F. Skinner: A Life.* New York: Basic Books.

Bjork, D. W. (1993). Toward a biography of B. F. Skinner: Rationale and interpretation. *Behaviorology, 1* (1), 7–11.

Cautela, J. R. (1994). General level of reinforcement II: Further elaborations. *Behaviorology, 2* (1), 1–16.

Cautela, J. R. & Ishaq, W. (Eds.). (1996). *Contemporary Issues in Behavior Therapy: Improving the Human Condition.* New York: Plenum.

Cheney, C. D. (1991). The source and control of behavior. In W. Ishaq (Ed.). *Human Behavior in Today's World* (pp. 73–86). New York: Praeger.

Comunidad Los Horcones. (1986). Behaviorology: An integrative denomination. *The Behavior Analyst, 9,* 227–228.

Daniels, A. C. (1989). *Performance Management (Third Edition, revised).* Tucker, GA: Performance Management Publications. This book applies behaviorological science in business and industry.

Epstein, R. (1996). *Cognition, Creativity, and Behavior.* Westport, CT: Praeger.

Eshleman, J. W. (1993). Science history: Review of *The Timetables of Science. Behaviorology, 1* (1), 61–67.

Eshleman, J. W. (2002). If telling were teaching. *Behaviorology Today, 5* (1), 30–32.

Eshleman, J. W. & Vargas, E. A. (1988). Promoting the behaviorological analysis of verbal behavior. *The Analysis of Verbal Behavior, 6,* 23–32.

Feeney, D. R. (2002). Creative life–style management through on–line and real–time application of the behaviorological education practices of precision teaching. In S. F. Ledoux. *Origins and Components of Behaviorology—Second Edition* (pp. 259–295). Canton, NY: ABCs.

Ferreira, J. B. (2012). Progressive neural emotional therapy (PNET): A behaviorological analysis. *Behaviorology Today, 15* (2), 3–9.

Ferster, C. B. & Skinner, B. F. (1957). *Schedules of Reinforcement.* Englewood Cliffs, NJ: Prentice–Hall. The B. F. Skinner Foundation (www.bfskinner.org) in Cambridge, MA, republished this book in 1997.

Fraley, L. E. (1980). The role of measures in the contingencies on teacher behavior. In L. E. Fraley (Contributing Editor). *Behavioral Analysis of Issues in Higher Education* (pp. 9–45). Reedsville, WV: Society for the Behavioral Analysis of Culture. This paper describes a range of measures including "gain scores" and "achieved percent of possible gain" to help faculty improve the teaching effectiveness of their course designs.

Fraley, L. E. (Contributing Editor). (1980). *Behavioral Analysis of Issues in Higher Education.* Reedsville, WV: Society for the Behavioral Analysis of Culture.

Fraley, L. E. (1983). The behavioral analysis of Mens Rea (doctrine of culpable mental states). *Behaviorists for Social Action Journal, 4* (1), 2–7. (Also, see Fraley, 2013.)

Fraley, L. E. (1984). Belief, its inconsistency, and the implications for the teaching faculty. *The Behavior Analyst, 7,* 17–28.

Fraley, L. E. (1987). The cultural mission of behaviorology. *The Behavior Analyst, 10,* 123–126. (Also, see Fraley, 2012a, 2013.)

Fraley, L. E. (1988). Covert mini–courts within judicial and law enforcement operations. *Behavior Analysis and Social Action, 6* (2), 2–14. (Also, see Fraley, 2013.)

Fraley, L. E. (1988). Introductory comments: Behaviorology and cultural materialism. *The Behavior Analyst, 11,* 159–160.

Fraley, L. E. (1991). The behaviorology movement. *Behaviorological Commentaries, Serial No. 1,* 3–13.

Fraley, L. E. (1992). Behavior analysis and behaviorology. *Behaviorological Commentaries, Serial No. 2,* 22.

Fraley, L. E. (1992). The religious psychology student in a behaviorology course. *Behaviorological Commentaries, Serial No. 2,* 18–21. This paper also appeared (2009) in *Behaviorology Today, 12* (2), 11–13.

Fraley, L. E. (1994). Behaviorological corrections: A new concept of prison from a natural science discipline. *Behavior and Social Issues, 4* (1 & 2), 3–33. (Also, see Fraley, 2013.)

Fraley, L. E. (1994). Uncertainty about determinism: A critical review of challenges to the determinism of modern science. *Behavior and Philosophy, 22* (2), 71–83.

Fraley, L.E. (1998a). A behaviorological thanatology: Foundations and implications. *The Behavior Analyst, 21* (1), 13–26. This is Part 1 of a "trilogy" of articles; see Fraley 1998a, 2006, 1998b, 1998c, 2001. (Also, see Fraley, 2012a.)

Fraley, L.E. (1998b). New ethics and practices for death and dying from an analysis of the sociocultural metacontingencies. *Behavior and Social Issues, 8* (1), 9–31. This is Part 3a of a "trilogy" of articles; see Fraley 1998a, 2006, 1998b, 1998c, 2001. (Also, see Fraley, 2012a.)

Fraley, L.E. (1998c). Pursuing and interpreting the implications of a natural philosophy and science with the values associated with other epistemologies. *Behavior and Social Issues, 8* (1), 33–39. This is Part 3b of a "trilogy" of articles; see Fraley 1998a, 2006, 1998b, 1998c, 2001. (Also, see Fraley, 2012a.)

Fraley, L.E. (2001). Behaviorological principles for the analysis of bereavement. *European Journal of Behavior Analysis, 2* (II), 143–153. This is Part 4 of a "trilogy" of articles; see Fraley 1998a, 2006, 1998b, 1998c, 2001. (Also, see Fraley, 2012a.)

Fraley, L. E. (2002). Defining the behaviorology movement: Critical distinctions from 1990. *Behaviorology Today, 5* (1), 54–59.

Fraley, L. E. (2002). The discipline of behaviorology and the postulate of determinism. *Behaviorology Today, 5* (1), 45–49.

Fraley, L. E. (2003). The strategic misdefining of the natural sciences within universities. *Behaviorology Today, 6* (1), 15–38.

Fraley, L.E. (2006). The ethics of medical practices during protracted dying: A natural science perspective. *Behaviorology Today, 9* (1), 3–17. This is Part 2 of a "trilogy" of articles; see Fraley 1998a, 2006, 1998b, 1998c, 2001. (Also, see Fraley, 2012a.)

Fraley, L. E. (2008). *General Behavoirology—The Natural Science of Human Behavior.* Canton, NY: ABCs. This is the 1,600–page, 30–chapter, three–course textbook that for the first time systematically and comprehensively presents most of the major facets of the separate, independent, natural science discipline of behaviorology.

Fraley, L. E. (2012a). *Dignified Dying—A Behaviorological Thanatology.* Canton, NY: ABCs.

Fraley, L. E. (2012b). The evolution of a discipline and our next step. *Behaviorology Today, 15* (1), 23–28.

Fraley, L. E. (2013). *Behaviorological Rehabilitation and the Criminal Justice System.* Canton, NY: ABCs.

Fraley, L. E. & Ledoux, S. F. (2002). Origins, status, and mission of behaviorology. In S. F. Ledoux. *Origins and Components of Behaviorology—Second Edition* (pp. 33–169). Canton, NY: ABCs. This multi–chapter paper also appeared across 2006–2008 in these five parts in *Behaviorology Today:* Chapters 1 & 2: *9* (2), 13–32. Chapter 3: *10* (1), 15–25. Chapter 4: *10* (2), 9–33. Chapter 5: *11* (1), 3–30. Chapters 6 & 7: *11* (2), 3–17.

Fraley, L. E. & Vargas, E. A. (Eds.). (1976). *Behavior Research and Technology in Higher Education.* Reedsville, wv: Society for the Behavioral Analysis of Culture.

Fraley, L. E. & Vargas, E. A. (1986). Separate disciplines: The study of behavior and the study of the psyche. *The Behavior Analyst, 9,* 47–59.

Glenn, S. S. & Madden, G. J. (1995). Units of interaction, evolution, and replication: Organic and behavioral parallels. *The Behavior Analyst, 18,* 237–251.

Harris, M. (1974). *Cow, Pigs, Wars, and Witches: The Riddles of Culture.* New York: Random House.

Harris, M. (1977). *Cannibals and Kings: The Origins of Culture.* New York: Random House.

Harris, M. (1979). *Cultural Materialism: The Struggle for a Science of Culture.* New York: Random House.

Hayes, S. C. & Brownstein, A. J. (1986). Mentalism, behavior–behavior relations, and a behavior–analytic view of the purposes of science. *The Behavior Analyst, 9,* 175–190.

Holland, J. G. (1960). Teaching machines: An application from the laboratory. *Journal of the Experimental Analysis of Behavior, 3,* 275–287.

Holland, J. G. (1967). A quantitative measure for programmed instruction. *American Educational Research Journal, 4,* 87–101.

Holland, J. G. & Skinner, B. F. (1961). *The Analysis of Behavior.* New York: McGraw–Hill. This book is the original comprehensively programmed text; the authors successfully applied the laws of behavior that it teaches to its design and use.

Holton, G. (2000). B. F. Skinner, P. W. Bridgman, and the "lost years." *Behaviorology, 5* (1), 1–14.

Ishaq, W. (Ed.). (1991). *Human Behavior in Today's World.* New York: Praeger.

Johnson, P. R. (2012). A behaviorological approach to management of neuroleptic–induced tardive dyskinesia: Progressive neural emotional therapy (pnet). *Behaviorology Today, 15* (2), 11–25.

Johnston, J. M. & Pennypacker, H. S. (1980). *Strategies and Tactics for Human Behavioral Research.* Hillsdale, nj: Erlbaum.

Keller, F. S. (1993). Education by torchlight. *Behaviorology, 1* (2), 1–8.

Krapfl, J. E. & Vargas, E. A. (Eds.). (1977). *Behaviorism and Ethics.* Kalamazoo, mi: Behaviordelia.

Latham, G. I. (1994). *The Power of Positive Parenting.* Logan, ut: P & T ink.

Latham, G. I. (1998). *Keys to Classroom Management.* Logan, ut: P & T ink.

Latham, G. I. (1999). *Parenting with Love.* Salt Lake City, ut: Bookcraft.

Latham, G. I. (2002). *Behind the Schoolhouse Door: Managing Chaos with Science, Skills, and Strategies.* Logan, ut: P & T ink. This book includes two earlier pieces: *Eight Skills Every Teacher Should Have* and *Management, Not Discipline: A Wakeup Call for Educators.*

Latham, G. I. (2002). China through the eyes of a behaviorologist. In S. F. Ledoux. *Origins and Components of Behaviorology—Second Edition* (pp. 297–302). Canton, NY: ABCs. This paper also appeared (2002) in *Behaviorology Today, 5* (1), 17–20.

Ledoux, S. F. (1985). Designing a new *Walden Two*–inspired community. *Communities—Journal of Cooperation, No. 66,* Spring (April), 28–32, 84.

Ledoux, S. F. (2002). A parable of past scribes and present possibilities. *Behaviorology Today, 5* (1), 60–64. This is a parable on the 20–year, billion–dollar American education research effort called *Project Follow Through,* the outcomes of which the American education establishment tends to ignore, to the detriment of students, teachers, schools, and communities across the country and even around the world.

Ledoux, S. F. (2002). An introduction to the origins, status, and mission of behaviorology: An established science with developed applications and a new name. In S. F. Ledoux. *Origins and Components of Behaviorology—Second Edition* (pp. 3–24). Canton, NY: ABCs. This paper also appeared (2004) in *Behaviorology Today, 7* (1), 27–41.

Ledoux, S. F. (2002). An introduction to the philosophy called radical behaviorism. In S. F. Ledoux. *Origins and Components of Behaviorology—Second Edition* (pp. 25–32). Canton, NY: ABCs. This paper also appeared (2004) in *Behaviorology Today, 7* (2), 37–41.

Ledoux, S. F. (2002). Behaviorology curricula in higher education. In S. F. Ledoux. *Origins and Components of Behaviorology—Second Edition* (pp. 173–186). Canton, NY: ABCs. This paper also appeared (2009) in *Behaviorology Today, 12* (1), 16–25.

Ledoux, S. F. (2002). Behaviorology in China: A status report. In S. F. Ledoux. *Origins and Components of Behaviorology—Second Edition* (pp. 187–198). Canton, NY: ABCs. A Chinese translation appeared (2002) in *Behaviorology Today, 5* (1), 37–44. This paper also appeared in English (2009) in *Behaviorology Today, 12* (2), 3–10.

Ledoux, S. F. (2002). Carl Sagan is right again: A review of *The Millennium Man. Behaviorology Today, 5* (2), 23–25.

Ledoux, S. F. (2002). Defining natural sciences. *Behaviorology Today, 5* (1), 34–36.

Ledoux, S. F. (2002). Increasing tact control and student comprehension through such new postcedent terms as added and subtracted reinforcers and punishers. In S. F. Ledoux. *Origins and Components of Behaviorology—Second Edition* (pp. 199–204). Canton, NY: ABCs. This paper also appeared (2010) in *Behaviorology Today, 13* (1), 3–6.

Ledoux, S. F. (2002). Multiple selectors in the control of simultaneously emittable responses. In S. F. Ledoux. *Origins and Components of Behaviorology—Second Edition* (pp. 205–241). Canton, NY: ABCs. This paper also appeared (2010) in *Behaviorology Today, 13* (2), 3–27. (This paper includes material about a procedure that further develops behaviorological single–subject designs; Chapter 6 includes parts of this paper.)

Ledoux, S. F. (2002). Successful smoking control as an example of a comprehensive behaviorological therapy. In S. F. Ledoux. *Origins and Components of Behaviorology—Second Edition* (pp. 243–258). Canton, NY: ABCs. This paper also appeared (2011) in *Behaviorology Today, 14* (1), 3–13.

Ledoux, S. F. (2002). *Origins and Components of Behaviorology—Second Edition.* Canton, NY: ABCs.

Ledoux, S. F. (2004). The future and behaviorology. *Behaviorology Today, 7* (2), 4–8. This paper also appeared (2005) in Jón Grétar Sigurjónsson, Jara Kristina Thomasdóttir, and Páll Jakob Líndal (Editors). *Hvar er hún nú? Arfleifð atferlisstefnunnar á 21. öld* (pp. 98–116). Reykjavik, Iceland: Háskólaútgáfan.

Ledoux, S. F. (2004–2007). [Syllabi for 13 online TIBI behaviorology courses spread across six issues:] *Behaviorology Today, 7* (2) to *10* (1).

Ledoux, S. F. (2012). Behaviorism at 100. *American Scientist, 100* (1), 60–65. This article extends B. F. Skinner's 1963 article "Behaviorism at fifty." The Editor introduced this article with excerpts, on pages 54–59, which he listed as an "*American Scientist* Centennial Classic 1957" that came from Skinner's 1957 *American Scientist* article "The experimental analysis of behavior." (Skinner's complete 1957 paper was also available online.)

Ledoux, S. F. (2012). Behaviorism at 100 unabridged. *Behaviorology Today, 15* (1), 3–22. With *Behaviorology Today* becoming fully peer–reviewed with this issue, this fully peer–reviewed version of the paper that originally appeared in *American Scientist* included the material set aside at the last moment to make more room for the Skinner article excerpts that accompanied the original article. Chapter 1 of this book includes and extends this paper. (*American Scientist* posted this paper online along with the original version.)

Ledoux, S. F. (2013). Human multiple operant research equipment. *Journal of Behaviorology, 16* (2), 3–9. This paper describes newer equipment that behaviorological experimenters could use in continuing the research program described in Chapter 6.

Ledoux, S. F. (2014). *Running Out of Time—Introducing Behaviorology to Help Solve Global Problem.* Ottawa, CANADA: BehaveTech Publishing.

Ledoux, S. F. & Cheney, C. D. (1987). *Grandpa Fred's Baby Tender or why and how we built our aircribs.* Canton, NY: ABCs. (See the Appendix for this book's Foreword, "The first baby tender," by B. F. Skinner.)

Lee, V. L. (2000). Using scientific visualization software with human operant data. *Behaviorology, 5* (1), 93–109.

Lloyd, K. E. (1985). Behavioral anthropology: A review of Marvin Harris' *Cultural Materialism. Journal of the Experimental Analysis of Behavior, 43,* 279–287.

Logue, A. W. (1988). A behaviorist's biologist: Review of Philip J. Pauly's *Controlling Life: Jacques Loeb and the Engineering Ideal in Biology. The Behavior Analyst, 11,* 205–207.

Malott, R. W. (1988). Rule–governed behavior and behavioral anthropology. *The Behavior Analyst, 11,* 181–203.

Michael, J. L. (1982). Distinguishing between discriminative and motivational functions of stimuli. *Journal of the Experimental Analysis of Behavior, 37,* 149–155.

Moore, J. (1981). On mentalism, methodological behaviorism, and radical behaviorism. *Behaviorism, 9,* 55–57.

Moore, J. (1984). On privacy, causes, and contingencies. *The Behavior Analyst, 7,* 3–16.

O'Heare, J. (2010). *Changing Problem Behavior.* Ottawa, CANADA: BehaveTech Publishing. While this book, a new edition of which is in preparation, openly addresses the behavior of a wide range of species, Dr. O'Heare has half a dozen other books that more specifically deal with dogs.

Peterson, N. (1978). *An Introduction to Verbal Behavior.* Grand Rapids, MI: Behavior Associates.

Potter, B. & Hixson, M. (1996). Science and our culture: Review of Carl Sagan's *The Demon–Haunted World: Science as a Candle in the Dark. Behaviorology, 4* (1), 68–80.

Schlinger, H. & Blakely, E. (1987). Function–altering effects of contingency–specifying stimuli. *The Behavior Analyst, 10,* 41–45.

Sidman, M. (1960). *Tactics of Scientific Research.* New York: Basic Books. Authors Cooperative, in Boston MA, republished this book in 1988.

Sidman, M. (1994). *Equivalence Relations and Behavior: A Research Story.* Boston, MA: Authors Cooperative.

Sidman, M. (2001). *Coercion and its Fallout—Revised Edition.* Boston, MA: Authors Cooperative.

Sidman, M. (2003). Reinforcement in diplomacy: More effective than coercion. *Behaviorology Today, 6* (2), 30–35.

Skinner, B. F. (1938). *The Behavior of Organisms.* New York: Appleton–Century–Crofts. Seventh printing, 1966, with special preface: Englewood Cliffs, NJ: Prentice–Hall. The B. F. Skinner Foundation (www.bfskinner.org) in Cambridge, MA, republished this book in 1991.

Skinner, B. F. (1948). *Walden Two.* New York: Macmillan. (In 1976 Macmillan issued a new paperback edition with Skinner's introductory essay, *"Walden Two* Revisited.") This novel provides a fictional description of a culture the design of which is based on behaviorological science. While this story is as relevant today as when it was written, the author often used the word "psychology" in this novel to denote the natural science of behavior into which he was trying, at the time, to turn the traditional field of psychology. Since these usages can confuse readers today, they should substitute "behaviorology" for these usages.

Skinner, B. F. (1953). *Science and Human Behavior.* New York: Macmillan. The Free Press, New York, published a paperback edition in 1965.

Skinner, B. F. (1957). The experimental analysis of behavior. *American Scientist, 45* (4), 343–371.

Skinner, B. F. (1957). *Verbal Behavior.* New York: Appleton–Century–Crofts. The B. F. Skinner Foundation (www.bfskinner.org) in Cambridge, MA, republished this book in 1992.

Skinner, B. F. (1963). Behaviorism at fifty. *Science, 140,* 951–958.

Skinner, B. F. (1966). The phylogeny and ontogeny of behavior. *Science, 153,* 1205–1213.

Skinner, B. F. (1968). *The Technology of Teaching.* New York: Appleton–Century–Crofts. The B. F. Skinner Foundation (www.bfskinner.org) in Cambridge, MA, republished this book in 2003.

Skinner, B. F. (1969). *Contingencies of Reinforcement: A Theoretical Analysis.* New York: Appleton–Century–Crofts. The B. F. Skinner Foundation (www.bfskinner.org) in Cambridge, MA, republished this book in 2013.

Skinner, B. F. (1971). *Beyond Freedom and Dignity.* New York: Knopf.

Skinner, B. F. (1972). *Cumulative Record: A Selection of Papers (Third Edition).* New York: Appleton–Century–Crofts. The B. F. Skinner Foundation (www.bfskinner.org) in Cambridge, MA, republished this book as "... *Definitive Edition,"* in 1999. (Several of the papers in this collection contain material concerning Skinner's invention of various pieces of equipment. The paper, "A case history in scientific method"—on pages 108–131 in the *Definitive Edition*—is particularly relevant to the invention of the **Cumulative Recorder.)**

Skinner, B. F. (1974). *About Behaviorism.* New York: Knopf.

Skinner, B. F. (1976). *Particulars of My Life.* New York: Knopf. (This is the first of three volumes in Skinner's autobiography.)

Skinner, B. F. (1977). Why I am not a cognitive psychologist. *Behaviorism, 5* (2), 1–10.

Skinner, B. F. (1978). *Reflections on Behaviorism and Society.* Englewood Cliffs, NJ: Prentice–Hall.

Skinner, B. F. (1979). *The Shaping of a Behaviorist.* New York: Knopf. (This is the second of three volumes in Skinner's autobiography, and includes some interesting material about his equipment inventions.)

Skinner, B. F. (1983). *A Matter of Consequences.* New York: Knopf. (This is the third of three volumes in Skinner's autobiography.)

Skinner, B. F. (1983). Can the experimental analysis of behavior rescue psychology? *The Behavior Analyst, 6,* 9–17.

Skinner, B. F. (1984). The shame of American education. *The American Psychologist, 39,* 947–954.

Skinner, B. F. (1987). *Upon Further Reflection.* Englewood Cliffs, NJ: Prentice–Hall.

Skinner, B. F. (1987). The first baby tender. In S. F. Ledoux & C. D. Cheney. *Grandpa Fred's Baby Tender or why and how we built our aircribs* (pp. iii–v). Canton, NY: ABCs. This paper also appeared (2004) in *Behaviorology Today, 7* (1), 3–4. (See the Appendix; also see Ledoux & Cheney, 1987.)

Skinner, B. F. (1989). *Recent Issues in the Analysis of Behavior.* Columbus, OH: Merrill.

Skinner, B. F. (1993). A world of our own. *Behaviorology, 1* (1), 3–5. This is the published version of Skinner's 1989 "declaration of independence" address to the Association for Behavior Analysis (see Ulman, 1993, p. 54).

Skinner, B. F. & Vaughan, M. E. (1983). *Enjoy Old Age.* New York: W.W. Norton.

Thompson, L. (2010). Climate change: The evidence and our options. *The Behavior Analyst, 33* (2), 153–170.

Ulman, J. D. (1993). The Ulman–Skinner letters. *Behaviorology, 1* (1), 47–54.

Ulman, J. (1998). Toward a more complete science of human behavior: Behaviorology plus institutional economics. *Behavior and Social Issues, 8*, 195–217. This paper includes consideration of macrocontingencies.

Vargas, E. A. (1975). Rights: A behavioristic analysis. *Behaviorism, 3* (2), 120–128.

Vargas, E. A. (1982). Hume's "ought" and "is" statement: A radical behaviorist's perspective. *Behaviorism, 10* (1), 1–23.

Vargas, E. A. (1985). Cultural contingencies: A review of Marvin Harris's *Cannibals and Kings. Journal of the Experimental Analysis of Behavior, 43*, 419–428.

Vargas, E. A. (1987). "Separate disciplines" is another name for survival. *The Behavior Analyst, 10*, 119–121.

Vargas, E. A. (1991). Behaviorology: Its paradigm. In W. Ishaq (Ed.). *Human Behavior in Today's World* (pp. 139–147). New York: Praeger.

Vargas, E. A. (1991). Verbal behavior: A four–term contingency relation. In W. Ishaq (Ed.). *Human Behavior in Today's World* (pp. 99–108). New York: Praeger.

Vargas, E. A. (1996). A university for the twenty–first century. In J. R. Cautela & W. Ishaq (Eds.). *Contemporary Issues in Behavior Therapy: Improving the Human Condition* (pp. 159–188). New York: Plenum.

Vargas, E. A. (2013). The importance of form in Skinner's analysis of verbal behavior and a further step. *The Analysis of Verbal Behavior, 29*, 167–183. This paper helps us appreciate the value of traditional linguists' work.

Vargas, E. A. & Fraley, L. E. (1976). Progress and structure: Reorganizing the university for instructional technology. *Instructional Science, 5*, 303–324.

Vargas, J. S. (1972). *Writing Worthwhile Behavioral Objectives.* New York: Harper & Row.

Vargas, J. S. (1990). Cognitive analysis of language and verbal behavior: Two separate fields. In L. J. Hayes & P. N. Chase (Eds.). *Dialogues on Verbal Behavior* (pp. 197–201). Reno, NV: Context Press.

Watkins, C. L. (1997). *Project Follow Through: A Case Study of Contingencies Influencing Instructional Practices of the Educational Establishment.* Cambridge, MA: Cambridge Center for Behavioral Studies.

West, R. P. & Hamerlynck, L. A. (Eds.). (1992). *Designs for Excellence in Education: The Legacy of B. F. Skinner (Limited Edition).* Longmont, CO: Sopris West.

Wood, S. W. (1976). Responsibilities of the college teacher: A behavioral perspective. In L. E. Fraley & E. A. Vargas (Eds.). *Behavior Research and Technology in Higher Education* (pp. 11–20). Reedsville, wv: Society for the Behavioral Analysis of Culture.

Wyatt, W. J. (1997). *The Millennium Man.* Hurricane, wv: Third Millennium Press. In this novel the author often used the term "behavior analysis" as the name for the natural science of behavior to show that it is different from, and not any kind of, psychology. Today, however, with the psychology discipline claiming "behavior analysis" as part of itself, such phrasing could confuse readers. To avoid confusion readers today should substitute "behaviorology" when "behavior analysis" appears in this book.

Wyatt, W. J., Hawkins, R. P., & Davis, P. (1986). Behaviorism: Are reports of its death exaggerated? *The Behavior Analyst, 9,* 101–105.

Youth Policy Institute. (1988, July/August). *Youth Policy, 10* (7). This special issue is devoted to data–based reports on the successes of the educational applications of behaviorological science.ℰℑ

Index

(Also access topics via the *Detailed Table of Contents*)

*T*his Index features a *selected* set of topical entries, not a comprehensive set. In addition the page numbers only represent content of some substantive nature rather than every appearance of the entry in the book. On the other hand, for many topics, the content entries include multiple access points.♣

G

H

I

College 2012 portrait Photo courtesy of SUNY–Canton

About the Author

$\mathcal{T}$he author, Dr. Stephen Ledoux (pronounced "la–dew") began his professional activities in the early 1970s after earning his B.A. and M.A. degrees, in 1972 and 1973 respectively, at *California State University, Sacramento*. At various times he has taught courses in behaviorology, education, English, and psychology, all (except English) at both the undergraduate and graduate levels, with behaviorology taught at the high school level as well. After several years of university teaching, he returned to full–time study, and earned his Ph.D. from *Western Michigan University,* in 1982, in The Experimental Analysis of Behavior. (He says that, in part, the Ph.D. seemed necessary, because his California M.A. degree was signed by "Ronald Reagan" and "James Bond"...)

Prof. Ledoux has held positions both at home and abroad. For four years, 1975–1979, he taught in Australia, starting at the *University of Queensland* in Brisbane, and then at the *Gippsland Institute of Advanced Education* near Melbourne. He also taught in China, both in 1979 (at *Xi'an Jiaotong University* where he was their first foreign English teacher) and in 1990–1991 (at *Xi'an Foreign Languages University*) as part of a faculty exchange. He has been teaching since 1982 at the Canton campus of the *State University of New York.*

Over the last several decades, Dr. Ledoux has prepared numerous publications and presentations, and been involved in many other ways in professional work. With colleague Dr. Carl Cheney, of Utah State University, Logan, he worked extensively in the early 1980s with the concept and function of the *Aircrib* (A.K.A. the "baby tender") an invention of the late Prof. B. F. Skinner. For several decades he has been very involved with the movement formally establishing the independent discipline of behaviorology, including a three–year term (1988–1991) as the first elected president of *The International Behaviorology Association.* In 1997 he published two books, *Origins and Components of Behaviorology* (which went into its second edition in 2002) and, with his spouse, Dr. Nelly Case (a professor of the Crane School of Music at SUNY–Potsdam) as the primary author, a book about the year they taught in China, with their then five–year–old son, entitled *The Panda and Monkey King Christmas—A Family's Year in China.*

In 1998 Prof. Ledoux was elected Chair of the Board of Directors of *The International Behaviorology Institute* (TIBI) which is a non–profit educational corporation that he helped establish to support behaviorologists by providing training in behaviorology. In 2001 he declined to continue as Chair so that he would have the time to fulfill appointments as the Editor both of the TIBI journal *Behaviorology Today,* which he edited for ten years, and of the TIBI web site (www.behaviorology.org) which he still edits. After also completing a half dozen study–question books for various textbooks, in 2005 he accepted the task of editing Lawrence Fraley's 1,600–page, three–course text, *General Behaviorology: The Natural Science of Human Behavior,* which was completed and published in late 2008. This was the first extensive and systematic text of the independent behaviorology discipline, and an inspiration for the present book. Dr. Ledoux also edited two other books by Prof. Fraley, *Dignified Dying—A Behaviorological Thanatology* (2012) and *Behaviorological Rehabilitation and the Criminal Justice System* (2013) before undertaking the present book. (You can find full references for these books in the Bibliography.)

In 2012 Prof. Ledoux updated Fred Skinner's 1963 article, Behaviorism at Fifty, by publishing his article, Behaviorism at 100, in *American Scientist.*

Two months later *Behaviorology Today* published the peer–reviewed version of this article using the title, Behaviorism at 100 Unabridged. The present book, *Running Out of Time—Introducing Behaviorology to Help Solve Global Problems*, was the next step.

Beyond teaching and writing, Dr. Ledoux's professional interests include verbal behavior (especially as applied to language teaching), pedagogical effectiveness, and the experimental analysis of simultaneously evoked and simultaneously selected human operants. His hobbies involve star gazing, meteorites, and arts from China, Japan, and Southwest Native Americans. He lives in Canton, NY, with his family.℀

Photo courtesy of SUNY–Canton

The author, Stephen F. Ledoux, Ph.D., in January 2012 upon receiving the volume 100, number 1, issue of *American Scientist* containing his "Behaviorism at 100" article, which became the core of Chapter 1.

Reader's Notes